Catalyst-free Organic Synthesis

Green Chemistry Series

Editor-in-Chief:
James H. Clark, *Department of Chemistry, University of York, UK*

Series Editors:
George A. Kraus, *Iowa State University, USA*
Andrzej Stankiewicz, *Delft University of Technology, The Netherlands*
Peter Siedl, *Federal University of Rio de Janeiro, Brazil*

Titles in the series:
1: The Future of Glycerol: New Uses of a Versatile Raw Material
2: Alternative Solvents for Green Chemistry
3: Eco-Friendly Synthesis of Fine Chemicals
4: Sustainable Solutions for Modern Economies
5: Chemical Reactions and Processes under Flow Conditions
6: Radical Reactions in Aqueous Media
7: Aqueous Microwave Chemistry
8: The Future of Glycerol: 2nd Edition
9: Transportation Biofuels: Novel Pathways for the Production of Ethanol, Biogas and Biodiesel
10: Alternatives to Conventional Food Processing
11: Green Trends in Insect Control
12: A Handbook of Applied Biopolymer Technology: Synthesis, Degradation and Applications
13: Challenges in Green Analytical Chemistry
14: Advanced Oil Crop Biorefineries
15: Enantioselective Homogeneous Supported Catalysis
16: Natural Polymers Volume 1: Composites
17: Natural Polymers Volume 2: Nanocomposites
18: Integrated Forest Biorefineries
19: Sustainable Preparation of Metal Nanoparticles: Methods and Applications
20: Alternative Solvents for Green Chemistry: 2nd Edition
21: Natural Product Extraction: Principles and Applications
22: Element Recovery and Sustainability
23: Green Materials for Sustainable Water Remediation and Treatment
24: The Economic Utilisation of Food Co-Products
25: Biomass for Sustainable Applications: Pollution Remediation and Energy
26: From C–H to C–C Bonds: Cross-Dehydrogenative-Coupling
27: Renewable Resources for Biorefineries
28: Transition Metal Catalysis in Aerobic Alcohol Oxidation
29: Green Materials from Plant Oils
30: Polyhydroxyalkanoates (PHAs) Based Blends, Composites and Nanocomposites

31: Ball Milling Towards Green Synthesis: Applications, Projects, Challenges
32: Porous Carbon Materials from Sustainable Precursors
33: Heterogeneous Catalysis for Today's Challenges: Synthesis, Characterization and Applications
34: Chemical Biotechnology and Bioengineering
35: Microwave-Assisted Polymerization
36: Ionic Liquids in the Biorefinery Concept: Challenges and Perspectives
37: Starch-based Blends, Composites and Nanocomposites
38: Sustainable Catalysis: With Non-endangered Metals, Part 1
39: Sustainable Catalysis: With Non-endangered Metals, Part 2
40: Sustainable Catalysis: Without Metals or Other Endangered Elements, Part 1
41: Sustainable Catalysis: Without Metals or Other Endangered Elements, Part 2
42: Green Photo-active Nanomaterials
43: Commercializing Biobased Products: Opportunities, Challenges, Benefits, and Risks
44: Biomass Sugars for Non-Fuel Applications
45: White Biotechnology for Sustainable Chemistry
46: Green and Sustainable Medicinal Chemistry: Methods, Tools and Strategies for the 21st Century Pharmaceutical Industry
47: Alternative Energy Sources for Green Chemistry
48: High Pressure Technologies in Biomass Conversion
49: Sustainable Solvents: Perspectives from Research, Business and International Policy
50: Fast Pyrolysis of Biomass: Advances in Science and Technology
51: Catalyst-free Organic Synthesis

How to obtain future titles on publication:
A standing order plan is available for this series. A standing order will bring delivery of each new volume immediately on publication.

For further information please contact:
Book Sales Department, Royal Society of Chemistry, Thomas Graham House, Science Park, Milton Road, Cambridge, CB4 0WF, UK
Telephone: +44 (0)1223 420066, Fax: +44 (0)1223 420247
Email: booksales@rsc.org
Visit our website at www.rsc.org/books

Catalyst-free Organic Synthesis

By

Goutam Brahmachari
Visva-Bharati University, India
Email: brahmg2001@yahoo.co.in; goutam.brahmachari@visva-bharati.ac.in

ROYAL SOCIETY OF CHEMISTRY

THE QUEEN'S AWARDS FOR ENTERPRISE: INTERNATIONAL TRADE 2013

Green Chemistry Series No. 51

Print ISBN: 978-1-78262-412-7
PDF eISBN: 978-1-78801-278-2
EPUB eISBN: 978-1-78801-318-5
ISSN: 1757-7039

A catalogue record for this book is available from the British Library

The Royal Society of Chemistry is a charity, registered in England and Wales, Number 207890, and a company incorporated in England by Royal Charter (Registered No. RC000524), registered office: Burlington House, Piccadilly, London W1J 0BA, UK, Telephone: +44 (0) 207 4378 6556.

For further information see our web site at www.rsc.org

Printed in the United Kingdom by CPI Group (UK) Ltd, Croydon, CR0 4YY, UK

Preface

Since its first proposition in the 1990s, the topic of green and sustainable chemistry has grown considerably over the past 25 years and has become more popular among researchers working in all branches of chemical science. Dedicated research endeavors have yielded innumerable green products and processes so far, and over the past quarter century, well-documented advances in the fields of green chemistry and green engineering have framed a solid platform that motivates and empowers today's researchers to carry forward the flag with much enthusiasm and potential! The development of processes for the sustainable production of chemicals and materials, innovations in green chemistry and engineering, at the molecular, process, and systems levels, are being reported at an ever-increasing rate. Hence, the next 25 years of green and sustainable chemistry are expected to become even more powerful and promising with the potential of finding many applications in the industrial sectors.

During the recent past, various greener alternatives to the traditional chemical syntheses and transformations have been described across a diverse field of chemistry, achieving sustainability through newer concepts such as step- and atom-economy and E-factor. Greener features for designing an alternative protocol for useful organic molecules and/or materials include the use of bio-renewable resources, benign reaction media (use of greener solvents or no solvent), recyclable magnetic heterogeneous catalysts and minimization of byproducts or waste generation with the efficient isolation of target products. Furthermore, consideration has been given to the energy-efficiency in performing a chemical reaction, by judicious screening of reaction conditions capable of carrying out the transformation at ambient conditions, or with the use of microwaves (MWs), ultrasound (US), visible lights and mechanical mixings (ball-milling) as green tools for alternative heating and activation of reactants. Besides the on-going

Green Chemistry Series No. 51
Catalyst-free Organic Synthesis
By Goutam Brahmachari

Published by the Royal Society of Chemistry, www.rsc.org

progress with all of these views, a somewhat challenging concept has also been emerging in recent times concerning whether a certain transformation could be attained efficiently without the aid of any catalyst or additive by a careful selection of reaction conditions. If so, then it would be very interesting with many levels of benefits. Designing catalyst-free synthetic processes is a step forward toward safe, cost-effective, waste-free, simple, and sustainable environment. The concept of the present book originated from this perception with an aim to summarize the key advances developed in recent times in this field of topical interest.

The book *Catalyst-Free Organic Synthesis* is the first attempt to offer recent cutting-edge advances in developing organic synthetic protocols achieved without the aid of any catalyst or additive, and it features an in-depth and thorough coverage of a huge number of such organic synthetic protocols. This book is unique in its authoritative, thorough, and comprehensive inclusion of a wide variety of more than 130 comprehensively screened catalyst-free organic synthetic methods for the generation of carbon–carbon and carbon–heteroatom bonds, which result in a wide spectrum of chemical compounds — aliphatic, aromatic, alicyclic and heterocycles. Clearly structured for easy access to the information, each selected reaction is discussed in a very compact manner through point-wise discussion covering all possible aspects. This book consists of six chapters. Chapter 1 is an introductory chapter that focuses on the essence of catalyst-free organic synthesis and offers an overview of the topics covered in the book as well as guiding the readers on how to use it. The five technical chapters are classified based on their varying reaction conditions. Chapter 2 presents catalyst-free organic transformations occurring under room-temperature conditions, while Chapter 3 covers those catalyst-free organic reactions accomplished with conventional heating. Catalyst-free organic transformations performed by means of the applications of microwave irradiation, ultrasound irradiation and ball milling are summarized in Chapter 4, 5 and 6, respectively.

Green chemistry is changing the practice of chemistry within industry and academia. What is also an equally important and essential task for educators of chemistry is to infuse the green chemistry advances into the existing chemistry curriculum so that these new discoveries can be disseminated to the future generation of molecular innovators who will eventually tackle the up and coming challenges in the fields of sustainable chemistry and engineering. Proper training and nurturing of the future generations of chemists, by educating them about these new concepts, are crucial for future innovations in research and industrial applications that will be needed to meet society's growing demand for sustainable products and processes. Although notable progress has been made toward infusing green chemistry concepts, strategies and tools into the modern chemistry curriculum, there is still some way to go.

The book successfully integrates research advances in designing catalyst-free reaction procedures for useful organic transformations which satisfy several green chemistry principles and their feasibility into industrial applications and process developments. A wide spectrum of such important synthetic methodologies involving carbon–carbon and carbon–heteroatom bond forming reactions dealt with in this book would surely make the work of much interest to the scientists deeply engaged in organic synthesis and related fields. This timely volume also serves the purpose as an outstanding source of information with regard to the industrial applications. It will serve not only as a valuable resource for researchers in their own fields, but will also motivate young scientists and advanced chemistry students in the dynamic field of organic synthesis and practice of green chemistry.

I would also like to express my deep sense of appreciation to all of the editorial and publishing staff associated with the Royal Society of Chemistry for their keen interest in producing the work, as well as their all-round help to ensure that the highest standards of publication have been maintained in bringing about this book. My effort will be successful only when it is found helpful to the readers at large. Every step has been taken to make the manuscript error-free; in spite of that, some errors may have crept in and any such error is, of course, my own. Constructive comments regarding the approach of this book from the readers will be highly appreciated.

Finally, I should thank my wife and my son for their understanding and for allowing me enough time throughout the entire period of writing; without their support, this work would not have been successful.

Goutam Brahmachari
Visva-Bharati University
India

To my most beloved sister *KAMALA* who lives with me at all times!

Contents

Green Chemistry Series No. 51
Catalyst-free Organic Synthesis
By Goutam Brahmachari

Published by the Royal Society of Chemistry, www.rsc.org

CHAPTER 1

Catalyst-free Organic Synthesis: An Introduction

1.1 Introduction

The terminology '*organic synthesis*' is used in a broad sense to refer to constructions of organic molecules by chemical means that follow certain distinct synthetic protocols designed for those purposes. These organic synthetic processes have been at the core of the chemical industry for hundreds of years in the production of numerous organic compounds of varying skeletons finding immense applications such as fine chemicals, medicinal and pharmaceutical agents, dyes and pigments, polymeric substances, food additives, petrochemicals, agrochemicals, and many more.[1–26] Over the past two centuries, synthetic organic chemistry has seen a tremendous all-round development, and the credit obviously goes to the synthetic chemists at large!

However, with the advent of the '*Green Chemistry Concept*',[27] the central theme of an organic synthetic process has now encountered a '*complete rethinking*' or a '*new look*' that not only considers the desired product(s) in optimum yield but also gives pertinent emphasis to the greenness and sustainability of the process! Hence, the overall outcome of an organic synthesis, *i.e.* the productivity, cost, safety, wastes, hazards, energy, and all other green chemistry parameters along with environmental-concerns, is largely dependent on the generality and effectiveness of its synthetic protocol.[28–45] And encouraged with this motivation, the field of organic synthesis has already gained notable developments in recent years in designing equally efficient processes blended with the flavors of eco-friendliness and sustainability that avoid the extensive use of toxic and hazardous reagents and solvents, harsh reaction conditions, and expensive and sophisticated catalysts.[46–60] '*Green Chemistry*', thus, offers a broad

Green Chemistry Series No. 51
Catalyst-free Organic Synthesis
By Goutam Brahmachari

Published by the Royal Society of Chemistry, www.rsc.org

platform encompassing a series of considerations in the design of efficient, eco-friendly and sustainable processes with parameters such as product yields, atom efficiency, E-factor, energy consumption, cost-effectiveness, number of reaction steps, availability of starting materials and their consumption (extent of use of bio-renewable resources), man-power (automation), and reactor usage (*e.g.* flow *versus* batch reactions).[61–67] Innovative green chemical techniques, if they can be applied as alternatives to traditional synthetic processes to generate old and new chemicals of industrial importance, would be highly beneficial to mankind.

Implementation of green chemistry in designing alternative synthetic protocols for value-added products or compounds is really a great challenge to today's organic chemists and the young researchers! The fundamental challenge for developing a sustainable chemical enterprise will be finding creative ways to minimize human exposure to, and the environmental impact of, harmful chemicals while enhancing scientific progress. The task not only involves the replacement of such kinds of reagents and/or catalysts, but also makes use of eco-friendly alternative ways in an innovative fashion that can promote the desired chemistries. As a result, microwave (MW) irradiation, ultrasound (US) irradiation, and ball-milling have been gaining extensive uses in organic synthesis over the recent years,[68–85] and these alternative sources of energy are now known as '*green tools*'. A completely different outlook based on careful selection of reaction conditions is the cornerstone of this advanced scenario of modern organic synthesis.[86–97]

Advanced organic synthesis is dedicated now to design improved and novel alternative protocols that aim to avoid the use of catalysts or to replace expensive and toxic catalysts with cheap and eco-friendly homogeneous and/or magnetically recoverable heterogeneous catalysts; to avoid hazardous solvents or to use water and other benign solvents such aqueous ethanol, glycerol or polyethylene glycol (PEG), ionic-liquids (ILs), supercritical carbon dioxide, and deep-eutectic mixtures (DEM); to minimize side-reactions and wastes leading to target products in high yields (to attain high atom efficiency and lower E-factor); to set reaction conditions at ambient temperature and pressure to minimize energy consumption or to apply microwaves, ultrasound and ball-milling as other alternatives green tools.[98–131] A significant amount of advancement toward such green chemistries has already been made, and concerted efforts are ongoing among synthetic chemists to attain more – this is, indeed, a new-born branch of chemical sciences that is growing rapidly. The present book is designed with the aim to offer recent cutting-edge advances in developing organic synthetic protocols under catalyst-free conditions.

1.2 Catalyst-free Organic Synthesis – A Step Forward

When we think of a chemical reaction, it is very much synonymous of thinking about a catalyst as well! The role of catalysts, both homogeneous and heterogeneous, in organic synthesis is obvious, and thus they find huge applications and uses. Catalysts usually promote faster chemical reactions and, for some reactions, the desired selectivities (regioselectivity or chemoselectivity) can be obtained using specific selective sites of them. Conventional catalysts/additives are usually associated with much costs, toxicity, and non-reusability, thereby generating wastes. From the green chemistry perspectives, considerable efforts have been made to improve overall suitability of catalytic substances from suitable modifications and/or innovation of new kinds of catalysts with multiple benefits. However, the most fruitful way-out would be to go for designing an organic reaction protocol without the aid of a catalyst, if feasible! With this unique and challenging view, the last decade has seen the dedicated attempts of chemists becoming successful in this venture with notable advancement.[132–150] Catalyst-free synthetic processes have many-folds of benefits so as to get rid of toxicity and wastes associated with using these catalysts. Hence, designing of catalyst-free synthetic processes is a step forward toward safe, cost-effective, waste-free, simple, and sustainable environment!

To design a chemical process that would be at the same time facile, efficient and high-yielding without the use of any catalyst/additive is really challenging! And for this purpose, one must carefully select reaction conditions and starting materials. It is often observed that appropriately selected starting materials can undergo self-catalysis in suitable solvents (preferably in aqueous or aqueous ethanolic medium) in many situations and/or the solvents can also impart catalytic benefits to certain reaction processes from their unique inherent properties. Reactions can also be promoted by simple conventional heating in the presence or absence of solvent(s), and also by the applications of microwave irradiation, ultrasound irradiation and mechanochemical mixings.

1.3 Overview of the Book

More than 130 catalyst-free organic reactions yielding a variety of useful organic molecules have been thoroughly researched and discussed in this book under five distinct chapters – Chapter 2 to Chapter 6 – classified based on their varying reaction conditions. Chapter 2 presents catalyst-free organic transformations occurring under room-temperature conditions, while Chapter 3 discusses those catalyst-free organic reactions accomplished under conventional heating. Catalyst-free organic transformations performed by means of the applications of microwave irradiation, ultrasound irradiation and ball-milling are presented in Chapter 4, 5 and 6, respectively.

1.4 How to Read

As mentioned above, this single volume incorporates more than 130 comprehensively screened organic synthetic protocols with catalyst-free conditions for the generation of carbon–carbon and carbon–heteroatom bonds which result in a wide spectrum of chemical compounds — aliphatic, aromatic, alicyclic and heterocycles. The reactions are classified in five distinct chapters (Chapter 2 to 6) based on reaction conditions (*viz.* at room temperature, conventional heating, microwave irradiation, ultrasound irradiation, and ball-milling). Clearly structured for easy access to the information, each selected reaction is discussed in a very compact manner through point-wise discussion such as: reaction type; reaction conditions; reaction strategy; keywords; general reaction scheme; mechanism; representative examples; experimental procedure; characterization data of representative entries; critical views; literature. Literature references are continuously numbered and presented at the end of each chapter. Reaction scheme, plausible mechanism (if any) and illustrative examples relating to a particular reaction are presented under that reaction and are self-explanatory in nature. Each organic synthesis is supplemented with all its details including the experimental procedure, representative examples and their physical and spectral properties so that one can reproduce the same with ease.

1.5 Concluding Remarks

Ongoing developments in greener and more efficient methodologies for the syntheses of organic compounds of interest are vital for making chemical processes more sustainable. Among various developments in this direction, designs for catalyst-free organic methodologies have drawn the attention of the researchers in their fields in recent times. The book successfully integrates cutting-edge research advances in designing catalyst-free reaction procedures for useful organic transformations with the inclusion of a comprehensive range of examples and chemistries that illustrate the significant strides made in this research area over the past few years. This research area is growing progressively, and many different alternatives to further advancements in the field of greener synthetic processes are to come! Further developments in more innovative, cost-efficient, and sustainable strategies would surely be disclosed in the near future. Under this purview, the present book is an endeavor to boost the ongoing green chemistry research and also to motivate the young minds to this truly dynamic field of chemistry!

References

1. J. I. G. Cadogan and R. K. Mackie, *Chem. Soc. Rev.*, 1974, **3**, 87.
2. L. F. Tietze, *Chem. Rev.*, 1996, **96**, 115.
3. R. M. Abdel-Rahman, *Pharmazie*, 1999, **54**, 791.
4. R. M. Abdel-Rahman, *Phosphorus, Sulfur Silicon Relat. Elem.*, 2000, **166**, 315.
5. A. de Meijere, *Chem. Rev.*, 2000, **100**, 2739.
6. C. Bolm, J. Legros, J. Le Paih and L. Zani, *Chem. Rev.*, 2004, **104**, 6217.
7. G. J. ten Brink, I. W. C. E. Arends and R. A. Sheldon, *Chem. Rev.*, 2004, **104**, 4105.

8. K. C. Nicolaou, D. J. Edmonds and P. G. Bulger, *Angew. Chem., Int. Ed.*, 2006, **45**, 7134.
9. D. Ravelli, D. Dondi, M. Fagnoni and A. Albini, *Chem. Soc. Rev.*, 2009, **38**, 1999.
10. C. Vargas, A. M. Balu, J. M. Campelo, C. Gonzalez-Arellano, R. Luque and A. A. Romero, *Curr. Org. Synth.*, 2010, **7**, 568.
11. T. Bach, D.-C. Jörg and P. Hehn, *Angew. Chem. Int. Ed.*, 2011, **50**, 1000.
12. J. M. Bothwell, S. W. Krabbe and R. S. Mohan, *Chem. Soc. Rev.*, 2011, **40**, 4649.
13. H. M. Davies and Y. Lian, *Acc. Chem. Res.*, 2012, **45**, 923.
14. W. A. Nugent, *Angew. Chem., Int. Ed.*, 2012, **51**, 8936.
15. M. O. Simon and C. J. Li, *Chem. Soc. Rev.*, 2012, **41**, 1415.
16. S. Kotha and P. Khedkar, *Chem. Rev.*, 2012, **112**, 1650.
17. Z. L. Shen, S. Y. Wang, Y. K. Chok, Y. H. Xu and T. P. Loh, *Chem. Rev.*, 2013, **113**, 271.
18. M. B. Gawande, A. K. Rathi, P. S. Branco and R. S. Varma, *Appl. Sci.*, 2013, **3**, 656.
19. L. Pan, X. Bi and Q. Liu, *Chem. Soc. Rev.*, 2013, **42**, 1251.
20. W. Tang and M. L. Becker, *Chem. Soc. Rev.*, 2014, **43**, 7013.
21. I. Bauer and H.-J. Knölker, *Chem. Rev.*, 2015, **115**, 3170.
22. K. L. Skubi, T. R. Blum and T. P. Yoon, *Chem. Rev.*, 2016, **116**, 10035.
23. D. A. Bakhotmah and R. M. Abdel-Rahman, *Mini-Rev. Org. Chem.*, 2016, **13**, 62.
24. M. Büschleb, S. Dorich, S. Hanessian, D. Tao, K. B. Schenthal and L. E. Overman, *Angew. Chem., Int. Ed.*, 2016, **55**, 4156.
25. G. Brahmachari, in *Green Synthetic Approaches in Organophosphorus Chemistry: Recent Developments with Energy-efficient Protocols*, ed. D. W. Allen, D. Loakes and J. Tebby, RSC Specialist Periodical Report – Organophosphorus Chemistry, 2016, vol. 45, pp. 438–491.
26. G. Brahmachari, in *Green Synthetic Approaches in Organophosphorus Chemistry: Recent Developments*, ed. D. W. Allen, D. Loakes and J. Tebby, RSC Specialist Periodical Report – Organophosphorus Chemistry, 2017, vol. 46, pp. 418–431.
27. P. T. Anastas and J. C. Warner, *Green Chemistry: Theory and Practice*, Oxford University Press, New York, 1998.
28. *Green Synthetic Approaches for Biologically Relevant Heterocycles*, ed. G. Brahmachari, Elsevier, Amsterdam, The Netherlands, 2015.
29. R. Sheldon, *Nature*, 1999, **399**, 636.
30. B. M. Trost, *Acc. Chem. Res.*, 2002, **35**, 695.
31. W. Zhang, *Green Chem.*, 2009, **11**, 911.
32. N. R. Candeias, L. C. Branco, P. M. P. Gois, C. A. M. Afonso and A. F. Trindade, *Chem. Rev.*, 2009, **109**, 2703.
33. M. Sankar, N. Dimitratos, P. J. Miedziak, P. P. Wells, C. J. Kiely and G. J. Hutchings, *Chem. Soc. Rev.*, 2012, **41**, 8099.
34. T. J. Collins, *J. Chem. Educ.*, 1995, **72**, 965.
35. P. J. Dunn, S. Galvin and K. Hettenbach, *Green Chem.*, 2004, **6**, 43.

36. S. K. Ritter, *Chem. Eng. News*, 2007, **85**, 38.
37. H. Chen, M. Armand, G. Demailly, F. Dolhem, P. Poizot and J. M. Tarascon, *ChemSusChem*, 2008, **1**, 348.
38. P. Anastas and N. Eghbali, *Chem. Soc. Rev.*, 2010, **39**, 301.
39. S. Everts, *New Sci.*, 2010, **15**, 57.
40. P. Anastas, *Chem. Eng. News*, 2011, **89**, 62.
41. R. A. Sheldon, *Chem. Soc. Rev.*, 2012, **41**, 1434.
42. S. K. Ritter, *Chem. Eng. News*, 2014, **92**, 24.
43. National Research Council, *A Framework to Guide Selection of Chemical Alternatives*, National Academies Press, Washington, DC, 2014.
44. R. Cernansky, *Nature*, 2015, **519**, 379.
45. D. Larcher and J.-M. Tarascon, *Nat. Chem.*, 2015, **7**, 19.
46. B. M. Trost, *Science*, 1991, **254**, 1471.
47. K. Tanaka and F. Toda, *Chem. Rev.*, 2000, **100**, 1025.
48. C. J. Li, *Chem. Rev.*, 2005, **105**, 3095.
49. B. P. Mason, K. E. Price, J. L. Steinbacher, A. R. Bogdan and D. T. McQuade, *Chem. Rev.*, 2007, **107**, 2300.
50. S. A. Chowdhury, J. L. Scott and D. R. MacFarlane, *Pure Appl. Chem.*, 2008, **80**, 1325.
51. M. A. P. Martins, C. P. Frizzo, D. N. Moreira, L. Buriol and P. Machado, *Chem. Rev.*, 2009, **109**, 4140.
52. M. Tobiszewski, A. Mechlińska and J. Namieśnik, *Chem. Soc. Rev.*, 2010, **39**, 2869.
53. P. J. Dunn, *Chem. Soc. Rev.*, 2012, **41**, 1452.
54. P. Galletzot, *Chem. Soc. Rev.*, 2012, **41**, 1538.
55. S. Barata-Vallejo and A. Postigo, *Eur. J. Org. Chem.*, 2012, 1889.
56. M. B. Gawande, P. S. Branco and R. S. Varma, *Chem. Soc. Rev.*, 2013, **42**, 3371.
57. Y. Gu and F. Jérôme, *Chem. Soc. Rev.*, 2013, **42**, 9550.
58. R. García-Álvarez, P. Crochet and V. Cadierno, *Green Chem.*, 2013, **15**, 46.
59. M. B. Gawande, P. S. Branco, I. D. Nogueira, C. A. A. Ghumman, N. Bundaleski, A. Santos, O. M. N. D. Teodoro and R. Luque, *Green Chem.*, 2013, **15**, 682.
60. S. G. Newman and K. F. Jensen, *Green Chem.*, 2013, **15**, 1456.
61. T. Collins, *Science*, 2001, **291**(5501), 48.
62. R. A. Sheldon, *Green Chem.*, 2005, **7**, 267.
63. C. Blum, H. Friege and H.-C. Stolzenberg, *Chem. Watch Glob. Bus. Brief.*, 2015, **79**, 11.
64. R. S. Verma, *ACS Sustainable Chem. Eng.*, 2016, **4**, 5866.
65. J. A. Haack and J. E. Hutchison, *ACS Sustainable Chem. Eng.*, 2016, **4**, 5889.
66. G. J. Ruiz-Mercado, A. Carvalho and H. Cabezas, *ACS Sustainable Chem. Eng.*, 2016, **4**, 6208.
67. C. Blum, D. Bunke, M. Hungsberg, E. Roelofs, A. Joas, R. Joas, M. Blepp and H.-C. Stolzenberg, *Sustainable Chem. Pharm.*, 2017, **5**, 94.

68. G. Rothenberg, A. P. Downie, C. L. Raston and J. L. Scott, *J. Am. Chem. Soc.*, 2001, **123**, 8701.
69. G. Cravotto and P. Cintas, *Chem. Soc. Rev.*, 2006, **35**, 180.
70. S. V. Chankeshwara and A. K. Chakraborti, *Org. Lett.*, 2006, **8**, 3259.
71. P. J. Wallis, K. J. Booth, A. F. Patti and J. L. Scott, *Green Chem.*, 2006, **8**, 333.
72. P. J. Walsh, H. M. Li and C. A. de Parrodi, *Chem. Rev.*, 2007, **107**, 2503.
73. D. Dallinger and C. O. Kappe, *Chem. Rev.*, 2007, **107**, 2563.
74. V. Polshettiwar and R. S. Varma, *Acc. Chem. Res.*, 2008, **41**, 629.
75. V. Polshettiwar and R. S. Varma, *Chem. Soc. Rev.*, 2008, **37**, 1546.
76. I. T. Horváth, *Green Chem.*, 2008, **10**, 1024.
77. S. Alesi, F. Di Maria, M. Melucci, D. J. Macquarrie, R. Luque and G. Barbarella, *Green Chem.*, 2008, **10**, 517.
78. D. A. Fulmer, W. C. Shearouse, S. T. Medonza and J. Mack, *Green Chem.*, 2009, **11**, 1821.
79. T.-Y. Chen, C.-S. Chao, K.-K. T. Mong and Y.-C. Chen, *Chem. Commun.*, 2010, **46**, 8347.
80. A. Stolle, T. Szuppa, S. E. S. Leonhardt and B. Ondruschka, *Chem. Soc. Rev.*, 2011, **40**, 2317.
81. P. Cintas and G. Cravotto, *Nat. Chem.*, 2012, **4**, 77.
82. S. L. James, C. J. Adams, C. Bolm, D. Braga, P. Collier, T. Friscic, F. Grepioni, K. D. M. Harris, G. Hyett, W. Jones, A. Krebs, J. Mack, L. Maini, A. G. Orpen, I. P. Parkin, W. C. Shearouse, J. W. Steed and D. C. Waddell, *Chem. Soc. Rev.*, 2012, **41**, 413.
83. A. M. Balu, D. Dallinger, D. Obermayer, J. M. Campelo, A. A. Romero, D. Carmona, F. Balas, K. Yohida, P. L. Gai, C. Vargas, C. O. Kappe and R. Luque, *Green Chem.*, 2012, **14**, 393.
84. F. Alonso, Y. Moglie, G. Radivoy and M. Yus, *Green Chem.*, 2012, **14**, 2699.
85. A. DeVierno Kreuder, T. House-Knight, J. Whitford, E. Ponnusamy, P. Miller, N. Jesse, R. Rodenborn, S. Sayag, M. Gebel, I. Aped, I. Sharfstein, E. Manaster, I. Ergaz, M. Gebel, A. Harris and L. Nelowet Grice, *ACS Sustainable Chem. Eng.*, 2017, **5**, 2927.
86. W. Leitner, *Science*, 1999, **284**, 1780b.
87. V. V. Namboodiri and R. S. Varma, *Green Chem.*, 2001, **3**, 146.
88. G. Franc and A. K. Kakkar, *Chem. Soc. Rev.*, 2010, **39**, 1536.
89. M. Tobiszewski, A. Mechlinska and J. Namiesnik, *Chem. Soc. Rev.*, 2010, **39**, 2869.
90. A. M. Balu, B. Baruwati, E. Serrano, J. Cot, J. Garcia-Martinez, R. S. Varma and R. Luque, *Green Chem.*, 2011, **13**, 2750.
91. M. Sankar, N. Dimitratos, P. J. Miedziak, P. P. Wells, C. J. Kiely and G. J. Hutchings, *Chem. Soc. Rev.*, 2012, **41**, 8099.
92. M. B. Gawande, V. D. B. Bonifácio, R. Luque, P. S. Branco and R. S. Varma, *Chem. Soc. Rev.*, 2013, **42**, 5522.
93. C. Jimenez-Gonzalez, C. S. Ponder, Q. B. Broxterman and J. B. Manley, *Org. Process Res. Dev.*, 2011, **15**, 912.

94. P. A. Julien, C. Mottillo and T. Friščić, *Green Chem.*, 2017, **19**, 2729.
95. M. J. H. Worthington, R. L. Kucera and J. M. Chalker, *Green Chem.*, 2017, **19**, 2748.
96. A. M. Ferreira, A. Filipa, M. Cláudio, M. Válega, F. M. J. Domingues, A. J. D. Silvestre, R. D. Rogers, J. A. P. Coutinho and M. G. Freire, *Green Chem.*, 2017, **19**, 2768.
97. J. Niemeier, R. V. Engel and M. Rose, *Green Chem.*, 2017, **19**, 2839.
98. P. A. Grieco, *Organic Synthesis in Water*, Blackie Academic and Professional, London, 1998.
99. J.-L. Luche, *Synthetic Organic Sonochemistry*, Plenum Press, New York, 1998.
100. P. Tundo, P. Anastas, D. S. Black, J. Breen, T. Collins, S. Memoli, J. Miyamoto, M. Polyakoff and W. Tumas, *Pure Appl. Chem.*, 2000, **72**, 1207.
101. T. Tsukinoki, S. Nagashima, Y. Mitoma and M. Tashiro, *Green Chem.*, 2000, **2**, 117.
102. F. Bigi, M. L. Conforti, R. Maggi, A. Piccinno and G. Sartori, *Green Chem.*, 2000, **2**, 101.
103. M. Jeselnik, R. S. Varma, S. Polanc and M. Kocevar, *Chem. Commun.*, 2001, 1716.
104. A. R. Katritzky, D. A. Nichols, M. Siskin, R. Murugan and M. Balasubramanian, *Chem. Rev.*, 2001, **101**, 837.
105. K. Manabe, S. Iimura, X. M. Sun and S. Kobayashi, *J. Am. Chem. Soc.*, 2002, **124**, 11971.
106. T. J. Mason and J. P. Lorimer, *Applied Sonochemistry*, Wiley-VCH Verlag GmbH, Weinheim, 2002.
107. V. V. Namboodiri and R. S. Varma, *Org. Lett.*, 2002, **4**, 3161.
108. U. M. Lindstrom, *Chem. Rev.*, 2002, **102**, 2751.
109. R. Breslow, *Acc. Chem. Res.*, 2004, **37**, 471.
110. S. Narayan, J. Muldoon, M. G. Finn, V. V. Fokin, H. C. Kolb and K. B. Sharpless, *Angew. Chem.*, 2005, **117**, 3339.
111. M. Carril, R. SanMartin, I. Tellitu and E. Dominguez, *Org. Lett.*, 2006, **8**, 1467.
112. C. J. Li and L. Chen, *Chem. Soc. Rev.*, 2006, **35**, 68.
113. Y. Hayashi, *Angew. Chem., Int. Ed.*, 2006, **45**, 8103.
114. Z. V. Todres, *Organic Mechanochemistry and its Practical Applications*, Taylor & Francis, Boca Raton, 2006.
115. V. V. Boldyrev, *Russ. Chem. Rev.*, 2006, **75**, 177.
116. R. D. Rogers and G. A. Voth, *Acc. Chem. Res.*, 2007, **40**, 1077.
117. Jung and R. A. Marcus, *J. Am. Chem. Soc.*, 2007, **129**, 5492.
118. J. Zhang, Z. W. Cui, F. Wang, Y. Wang, Z. W. Miao and R. Y. Chen, *Green Chem.*, 2007, **9**, 1341.
119. C. Gonzalez-Arellano, J. M. Campelo, D. J. Macquarrie, J. M. Marinas, A. A. Romero and R. Luque, *ChemSusChem*, 2008, **1**, 746.
120. A. Chanda and V. V. Fokin, *Chem. Rev.*, 2009, **109**, 725.
121. Y. Hirai and Y. Uozumi, *Chem. Commun.*, 2010, **46**, 1103.

122. J. K. Beattie, C. S. P. McErlean and C. B. W. Phippen, *Chem. – Eur. J.*, 2010, **16**, 8972.
123. M. M. Savant, A. M. Pansuriya, C. V. Bhuva, N. Kapuriya, A. S. Patel, V. B. Audichya, P. V. Pipaliya and Y. T. Naliapara, *J. Comb. Chem.*, 2010, **12**, 176.
124. R. N. Butler and A. G. Coyne, *Chem. Rev.*, 2010, **110**, 6302.
125. F. Rajabi, S. Naserian, A. Primo and R. Luque, *Adv. Synth. Catal.*, 2011, **353**, 2060.
126. M. Y. Wu, K. Li, N. Wang, T. He and X. Q. Yu, *Synthesis*, 2011, 1831.
127. M. B. Gawande, A. K. Rathi, P. S. Branco, I. D. Nogueira, A. Velhinho, J. J. Shrikhande, U. U. Indulkar, R. V. Jayaram, C. A. A. Ghumman, N. Bundaleski and O. M. N. D. Teodoro, *Chem. – Eur. J.*, 2012, **18**, 12628.
128. R. B. N. Baig and R. S. Varma, *Chem. Soc. Rev.*, 2012, **41**, 1559.
129. *Handbook on Applications of Ultrasound: Sonochemistry for Sustainability*, ed. D. Chen, S. K. Sharma and A. Madhoo, CRC Press, Boca Raton, FL, 2012.
130. G. Brahmachari, *Room-Temperature Organic Synthesis*, Elsevier, Amsterdam, the Netherlands, 2015.
131. *Mechanochemical Organic Synthesis*, ed. D. Margetic and V. Štrukil, Elsevier, Amsterdam, 2016.
132. S. Kumar, P. Sharma, K. K. Kapoor and M. S. Hundal, *Tetrahedron*, 2008, **64**, 536.
133. G. Choudhary and R. K. Peddinti, *Green Chem.*, 2011, **13**, 276.
134. D.-P. Li, G.-L. Zhang, L.-T. An, J.-P. Zou and W. Zhang, *Green Chem.*, 2011, **13**, 594.
135. J. Liu, M. Lei and L. Hu, *Green Chem.*, 2012, **14**, 2534.
136. M. B. Gawande, V. D. B. Bonifácio, R. Luque, P. S. Branco and R. S. Varma, *ChemSusChem*, 2014, **7**, 24.
137. G. Brahmachari and B. Banerjee, *Curr. Green Chem.*, 2015, **2**, 274.
138. L. Xia, H. Cai and Y. R. Lee, *Tetrahedron*, 2015, **71**, 6894.
139. S. Nagaraju, N. Satyanarayana, B. Paplal, A. K. Vasu, S. Kanvah and D. Kashinath, *RSC Adv.*, 2015, **5**, 81768.
140. B. A. Dar, *J. Ind. Eng. Chem.*, 2016, **36**, 194.
141. N. Molleti and J. Y. Kang, *Org. Biomol. Chem.*, 2016, **14**, 8952.
142. N. Shabalala, S. Maddila and S. B. Jonnalagadda, *New J. Chem.*, 2016, **40**, 5107.
143. M. A. Ibrahim, M. Elagawany and T. S. Ibrahim, *Green Chem. Lett. Rev.*, 2016, **9**, 91.
144. D. J. Faizi, A. Issaian, A. J. Davis and S. A. Blum, *J. Am. Chem. Soc.*, 2016, **138**, 2126.
145. K. N. Tu, J. J. Hirner and S. A. Blum, *Org. Lett.*, 2016, **18**, 480.
146. D. J. Faizi, A. J. Davis, F. B. Meany and S. A. Blum, *Angew. Chem., Int. Ed.*, 2016, **55**, 14286.
147. F.-S. Wu, H.-Y. Zhao, Y.-l. Xu, K. Hu, Y.-M. Pan and X.-l Ma, *J. Org. Chem.*, 2017, **82**, 4289.

148. D. Kong, Q. Wang, Z. Zhu, X. Wang, Z. Shi, Q. Lin and M. Wu, *Tetrahedron Lett.*, 2017, **58**, 2644.
149. L. Wang, L.-X. Shi, L. Liu, Z.-X. Li, T. Xu, W.-J. Hao, G. Li, S.-J. Tu and B. Jiang, *J. Org.* Chem., 2017, **82**, 3605.
150. D.-L. Kong, G.-P. Lu, M.-S. Wu, Z.-F. Shi and Q. Lin, *ACS Sustainable Chem. Eng.*, 2017, **5**, 3465.

CHAPTER 2

Catalyst-free Organic Reactions under Room Temperature Conditions

2.1 Introduction

With the advent of the concept of '*green and sustainable chemistry*',[1] modern organic synthesis encompasses a number of agenda, such as the avoidance of extensive use of toxic and hazardous reagents and solvents, harsh reaction conditions, and expensive and sophisticated catalysts.[2] The past decade has seen a tremendous effort toward savings in energy consumption, use of eco-friendly solvents, proficiency in atom economy, and minimization of wastes from reactions in order to design novel green synthetic protocols for organic compounds of interest.[3–10] Among various energy-efficient processes, the most effective way to save energy is to develop strategies/protocols that are capable of carrying out the transformations at ambient conditions (*i.e.* room temperature and pressure).[11] In addition, room temperature offers a mild reaction condition, essentially required for many temperature-sensitive organic substrates as a key step in multistep sequence reactions. Designing reactions at room temperature and pressure coupled with other green aspects is, thus, a current area of emphasis. The concept of developing reaction strategies at ambient conditions is now an emerging field of research in organic chemistry and is progressing considerably. It becomes more interesting when such an energy consideration is coupled with the catalyst-free strategy in performing syntheses of certain useful organic molecules.[12] With this view, the present chapter offers detailed descriptions of a wide variety of 75 comprehensively screened catalyst-free organic synthetic protocols, which occur at room temperature and pressure, for the generation

Green Chemistry Series No. 51
Catalyst-free Organic Synthesis
By Goutam Brahmachari

Published by the Royal Society of Chemistry, www.rsc.org

of carbon–carbon and carbon–heteroatom bonds resulting a wide spectrum of synthetically and pharmaceutically useful chemical compounds – aliphatic, aromatic, alicyclic and heterocycles.

2.2 Room Temperature Organic Transformations Under Catalyst-free Conditions

An appreciable number of room temperature organic reactions leading to the synthesis of a variety of organic compounds under catalyst-free conditions have been reported in the literature. This section presents such useful organic transformations in an entry-based format, highlighting the key aspects for each of them.

2.2.1 Entry-1: Synthesis of α-Amino Nitriles

Type of reaction: C–C and C–N bond formation
Reaction conditions: Catalyst-free, water, room temperature
Synthetic strategy: One-pot multicomponent reaction
Keywords: Aldehydes, ketones, amines, acetone cyanohydrin, aqueous medium, catalyst-free, room temperature, Strecker reaction, α-amino nitriles, one-pot three-component reaction, chemoselectivity

2.2.1.1 General Reaction Scheme

Synthesis of racemic α-amino nitriles in a chemoselective and convenient manner through a one-pot, three-component Strecker reaction of carbonyl compounds (**1/5**), amines (**2**), and acetone cyanohydrin (**3**) in water at room temperature in the absence of any added catalyst was documented by Galletti and coworkers (Scheme 2.1).[13]

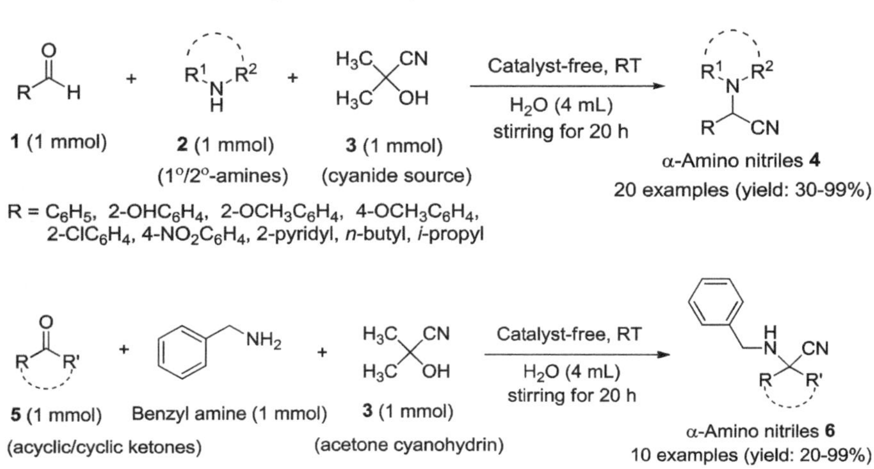

Scheme 2.1 Synthesis of α-amino nitriles (**4/6**) *via* catalyst-free Strecker reaction.

2.2.1.2 Representative Examples

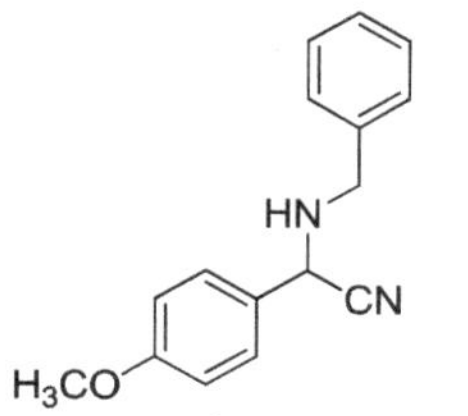

4a

2-(Benzylamino)-2-(4-methoxyphenyl) acetonitrile (**4a**): solid; yield: 68%

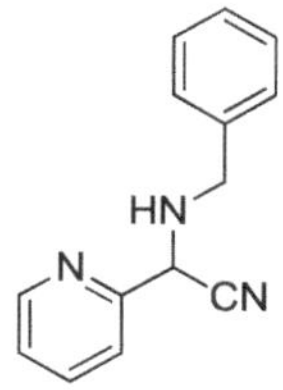

4b

2-(Benzylamino)-2-(pyridin-2-yl)acetonitrile (**4b**): oil; yield: 98%

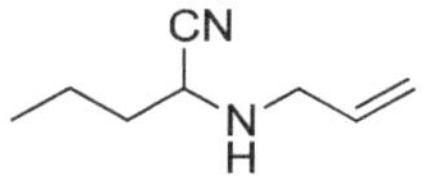

4c

2-(Allylamino)pentanenitrile (**4c**): oil; yield: 99%

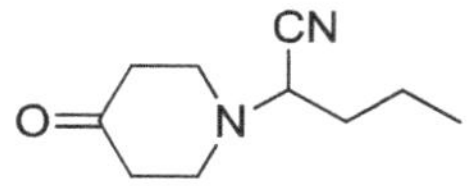

4d

2-(4-Oxopiperidin-1-yl)pentanenitrile (**4d**): oil; yield: 30%

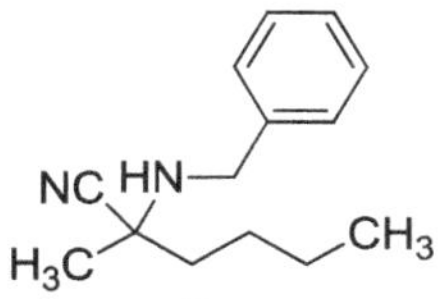

6a

2-(Benzylamino)-2-methylhexanenitrile (**6a**): oil; yield: 47%

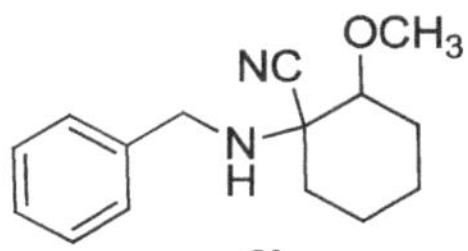

6b

1-(Benzylamino)-2-methoxycyclohexanecarbonitrile (**6b**): Oil; yield: 61% (*dr* = 57:43)

2.2.1.3 Experimental Procedure

A vial equipped with a screw cap was charged with a magnetic stir bar, aldehydes (**1**; 1 mmol) or ketones (**5**; 1 mmol) and amines (**2**; 1 mmol), and stirred well to mix the reactants thoroughly. After 10 minutes, water (4 mL) and acetone cyanohydrin (1 mmol) were added, and the cap was closed. The resulting mixture continued to be stirred for up to 20 h to complete the reaction (monitored by TLC). The reaction mixture was then poured into brine (5 mL) and extracted with ethyl acetate (2×10 mL), dried over sodium sulfate, and concentrated under reduced pressure. The crude product was purified either by flash chromatography or simply by recrystallization to yield pure α-amino nitriles (**4**/**6**). All the products were characterized by means of elemental analyses and detailed spectral studies including IR, ^{1}H NMR, ^{13}C NMR and HRMS.

2.2.1.4 Characterization Data of Two Representative Compounds

2-(Benzylamino)-2-(pyridin-2-yl)acetonitrile (**4b**): oil; yield: 98%; IR (NaCl): 3300, 2228, 1673 cm^{-1}; ^{1}H NMR ($CDCl_3$, 400 MHz): δ 2.75 (1H, br. s, NH), 3.97 (1H, d, J_{AB} = 12.8 Hz, CHHPh), 4.04 (1H, d, J_{AB} = 12.8 Hz, CHHPh), 4.79

(1 H, s, CHCN), 7.25–7.48 (7H, m, ArH), 7.76 (1H, dd, $J = 10.0$, 10 Hz, ArH), 8.60 (1H, d, $J = 8.0$ Hz, ArH); ^{13}C NMR ($CDCl_3$, 100 MHz,): δ 51.4, 54.9, 122.0, 123.9, 127.7, 128.5, 128.6, 128.8, 137.4, 138.0, 149.9, 153.7; LCMS: *m/z* (*tr* = 7.1 min): 224 $[M+1]^+$; $C_{14}H_{13}N_3$ (223.27): Calc. C, 75.31; H, 5.87; N, 18.82; found C, 75.44; H, 5.94; N, 18.98.

2-(Benzylamino)-2-methylhexanenitrile (**6a**): oil; yield: 47%; IR (NaCl): 3318, 2219, 1455 cm^{-1}; ^{1}H NMR ($CDCl_3$, 400 MHz): δ 0.95 (3H, t, $J = 7.2$ Hz, CH_3), 1.39 (2H, quint, $J = 7.2$ Hz, $CH_2CH_2CH_2$), 1.44–1.57 (6H, m, NH, CH_2, CH_3), 1.67–1.82 (2H, m, CH_2), 3.90 (2H, s, CH_2Ph), 7.27–7.40 (5H, m, ArH); ^{13}C NMR ($CDCl_3$, 100 MHz,): δ 13.7, 22.5, 24.6, 25.8, 39.5, 49.0, 55.6, 122.1, 127.2, 128.1, 128.3, 139.1; $C_{14}H_{20}N_2$ (216.32): Calc. C, 77.73; H, 9.32; N, 12.95; found C, 77.82; H, 9.41; N, 12.91.

2.2.1.5 Critical Views

The investigators reported on the development of a convenient and catalyst-free protocol for the synthesis of racemic α-amino nitriles with high chemoselectivity through one-pot, three-component Strecker reaction in water at room temperature. Both aliphatic and aromatic aldehydes, and cyclic ketones, in combination with primary and secondary amines were found to undergo smooth reaction. In some cases, pure α-amino nitriles can be obtained just by direct separation from water. An unusual application of the Strecker reaction to 1,2-diamines to obtain 1,2-diamino nitriles, and to cyclic secondary amines was also reported. This cyanation process is atom-economic and operationally simple. Dialkyl and alkyl aryl ketones practically did not undergo this reaction but cyclic ketones afforded excellent yields, which is in accordance to the difference in the reactivity and internal strain effect (I-strain) of linear *versus* cyclic ketones in nucleophilic addition reactions. The present method is overall relatively superior to the previously reported methods,[14–25] mainly in respect of using (i) mild reaction conditions, (ii) water as reaction medium, and (iii) no catalyst. The mild reaction conditions and the operational simplicity, thus, anticipate making this atom-economic cyanation process attractive to the development of cleaner and environmentally friendlier processes for the synthesis of α-amino nitriles and their analogs of synthetic importance.[26]

2.2.2 Entry-2: Synthesis of Tetraketones

Type of reaction: C–C bond formation
Reaction conditions: Catalyst-free, water, room temperature
Synthetic strategy: One-pot pseudo-multicomponent reaction
Keywords: Aldehydes, 1,3-cyclic diketones, aqueous medium, catalyst-free, room temperature, tandem Knoevenagel condensation and Michael addition, tetraketones, no column chromatography

2.2.2.1 General Reaction Scheme

A catalyst-free synthesis of a wide range of tetraketones (**3**) from the reaction between aldehydes (**1**) and cyclic-1,3-diketones (**2**) through tandem

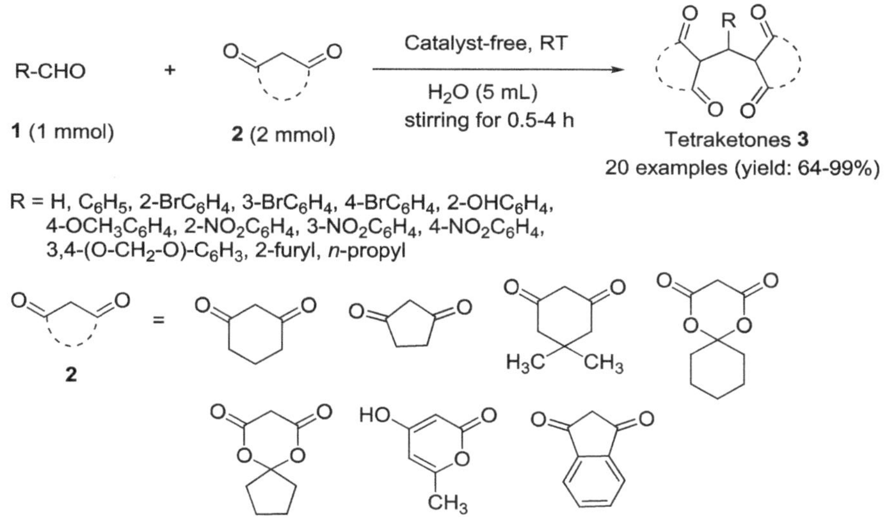

Scheme 2.2 Synthesis of tetraketones **3**.

Knoevenagel condensation and Michael addition in water at room temperature was reported by Yu *et al.* (Scheme 2.2).[27]

2.2.2.2 Plausible Mechanism

The authors suggested a plausible mechanism for this transformation (Scheme 2.3). Besides acting as solvent, water is also thought to facilitate the enolization process of 1,3-diketone (**2**) through making hydrogen bonds with the enolic OH so as to enhance nucleophilic character of the methylene carbon of this C–H activated acid. In addition, water at the same time can increase the electrophilic character of the carbonyl carbon of the reacting aldehyde (**1**) by forming hydrogen bonds with the carbonyl oxygen. A tandem Knoeveangel condensation and Michael addition ultimately afforded the desired tetraketonic derivative (**3**) in good yields.[27]

2.2.2.3 Representative Examples

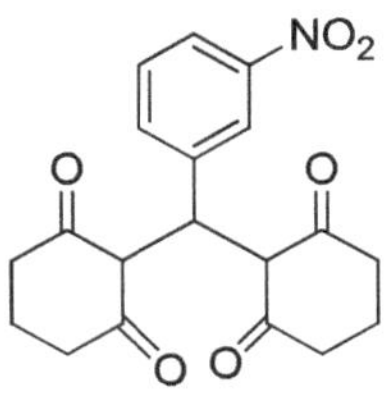

3a

2,2'-((3-Nitrophenyl)methylene)*bis*(cyclohexane-1,3-dione) (**3a**): Sand yellow solid, mp 204–205 °C; reaction time: 0.5 h; yield: 99%

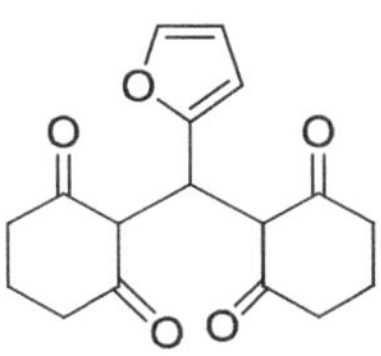

3b

2,2'-(Furan-2-ylmethylene)*bis*(cyclohexane-1,3-dione) (**3b**): Brown solid, mp 142–144 °C; reaction time: 4 h; yield: 98%

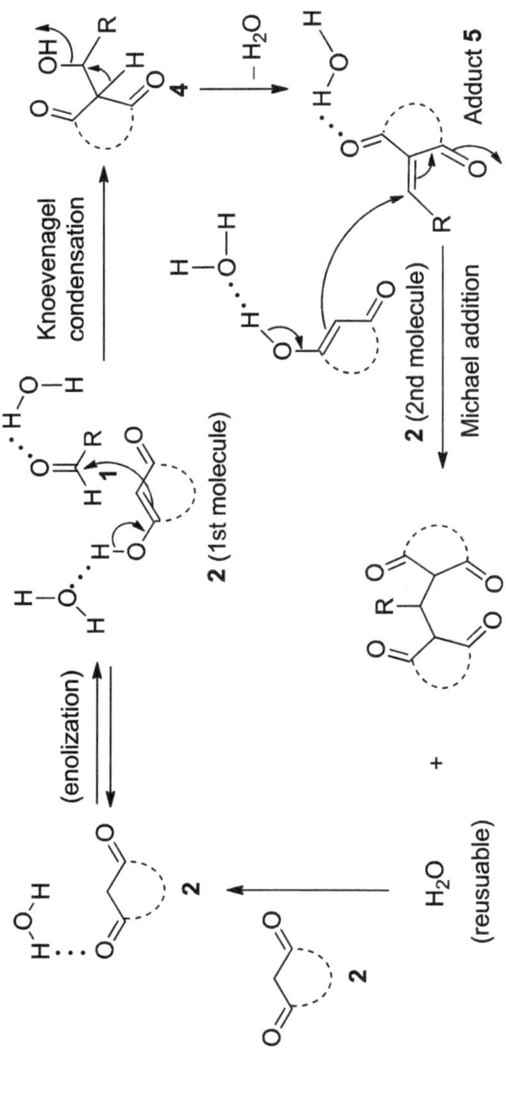

Scheme 2.3 Plausible mechanism for the synthesis of tetraketonic derivative (3).

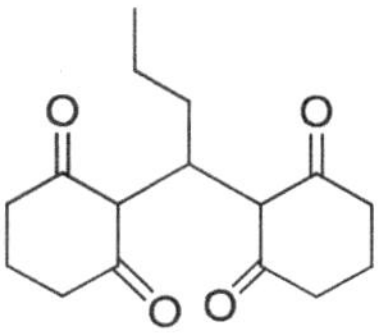

3c

2,2'-(Butane-1,1-diyl)*bis*(cyclohexane-1,3-dione) (**3c**): White solid, mp 97–98 °C; reaction time: 4 h; yield: 64%

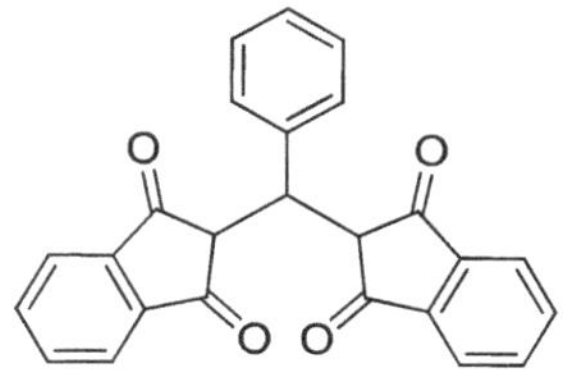

3d

2,2'-(Phenylmethylene)*bis*(1*H*-indene-1,3(2*H*)-dione) (**3d**): Green solid, mp 156–158 °C; reaction time: 4 h; yield: 95%

2.2.2.4 Experimental Procedure

A 20 mL flask was charged with a magnetic stir bar, aldehydes (**1**; 1 mmol), 1,3-cyclic diketone (**2**; 1 mmol) and water (5 mL). The resulting mixture was then stirred at room temperature for the stipulated time (0.5–4 h) to complete the reaction (monitored by TLC). The crude precipitate was filtered, washed by cold water and dried in a vacuum, and finally purified just by recrystallization to give the desired tetraketone **3** in good yields. All the products were characterized based on spectral studies.

2.2.2.5 Characterization Data of Two Representative Compounds

2,2′-(Furan-2-ylmethylene)bis(cyclohexane-1,3-dione) (**3b**): brown solid, mp 142–144 °C; Yield: 98%; ^{1}H NMR ($CDCl_3$, 400 MHz): δ 1.93–2.02 (4H, m, H-5/H-5′), 2.27–2.48 (4H, m, H-4/H-4′), 2.50–2.66 (4H, m, H-6/H-6′), 5.36 (1H, s, H-7), 5.95 (1H, s, H-3), 6.28 (1H, t, $J=4$ Hz, H-4), 7.26 (1H, d, $J=8.0$ Hz, H-5), 12.47 (2H, br s, –OH); ^{13}C NMR ($CDCl_3$, 100 MHz): δ 20.0 (C-5/C-5′), 29.5 (C-7), 32.7 (C-4/C-4′), 33.3 (C-6/C-6′), 106.3 (C-3), 110.1 (C-2/C-2′), 115.3(C-4), 140.9 (C-5), 151.6 (C-2), 190.9 (–COH), 191.1 (CO); EIMS: *m/z* 302.1 (M^+, 36).

2,2′-(Phenylmethylene)bis(1*H*-indene-1,3(2*H*)-dione) (**3d**): green solid, mp 156–158 °C; Yield: 95%; ^{1}H NMR ($CDCl_3$, 400 MHz): δ 3.92 (1H, t, $J=7.2$ Hz, H-6), 4.27 (2H, d, $J=7.6$ Hz, H-2, H-2′), 7.19 (1H, t, $J=7.2$ Hz, H-4), 7.23 (1H, t, $J=7.6$ Hz, H-3/H-5), 7.38 (2H, d, $J=7.2$ Hz, H-2/H-6), 7.79–7.83 (4H, m, Ar–*H*), 7.89–7.94 (4H, m, Ar–*H*); 13C NMR (DMSO-d_6, 100 MHz): δ 41.7(C-6), 55.0(C-2/C-2′), 123.1(Ar–C), 127.2 (C-4), 128.3(C-2/C-6), 129.3(C-3/C-5), 136.3(Ar–C), 142.1(C-4/C-5/C-4′/C-5′), 145.9(C-1), 199.1(CO); EIMS: *m/z* 380.1 (M^+, 26).

2.2.2.6 Critical Views

There are several reports available in the literature for the efficient synthesis of tetraketones, most of which suffer from the use of catalysts, surfactants, and aqueous organic solvent as co-solvent.[28–31] Conversely, this catalyst-free

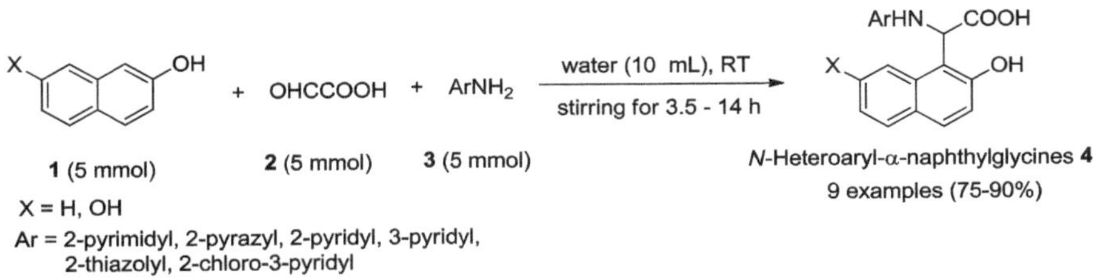

Scheme 2.4 Synthesis of *N*-heteroaryl α-arylglycines (**4**).

protocol developed by Yu *et al.* is not only eco-friendly but operationally simple and highly efficient in pure water just at ambient temperature. In their proposed mechanism, the authors suggested the probable catalytic activity of water apart from acting as solvent in facilitating the reaction.[27] There is no need for column chromatographic purification of the products, thereby eliminating the use of toxic organic solvents. However, dimethyl malonate and acetylacetone were found not to produce the desired tetra-ketones under the reaction conditions of this method.

2.2.3 Entry-3: Synthesis of *N*-Heteroaryl α-Naphthylglycines

Type of reaction: C–C and C–N bond formation
Reaction conditions: Catalyst-free, water, room temperature
Synthetic strategy: One-pot multicomponent reaction
Keywords: Naphthols, glyoxalic acid, heteroaryl amines, catalyst-free, water, room temperature, one-pot multicomponent reaction, no column chromatography, *N*-heteroaryl α-naphthylglycines

2.2.3.1 General Reaction Scheme

Olyaei *et al.*[32] reported on the synthesis of a series of *N*-heteroaryl α-arylglycines (**4**) from the one-pot three-component condensation reaction of β-naphthols (**1**), glyoxalic acid (**2**) and heteroaryl amines (**3**) in water at ambient conditions in the absence of any catalyst with moderate to high yields (Scheme 2.4).

2.2.3.2 Representative Examples

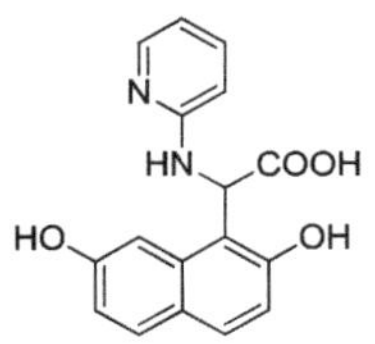

4a

2-(2,7-Dihydroxynaphthalen-1-yl)-2-(pyridin-2-ylamino)acetic acid (**4a**): colorless crystals, mp 179–180 °C, reaction time: 4.5 h, yield: 89%

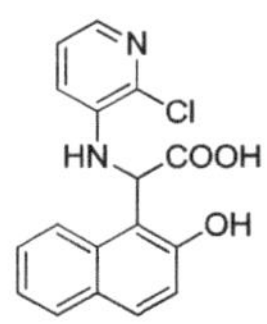

4b

2-((2-Chloropyridin-3-yl)amino)-2-(2-hydroxynaphthalen-1-yl)acetic acid (**4b**): colorless crystals, mp 200–202 °C, reaction time: 8 h, yield: 83%

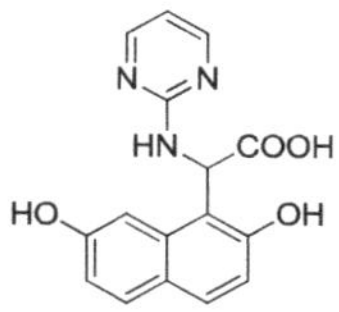

4c

2-(2,7-Dihydroxynaphthalen-1-yl)-2-(pyrimidin-2-ylamino)acetic acid (4c): colorless crystals, mp 196–198 °C, reaction time: 3.5 h, yield: 85%

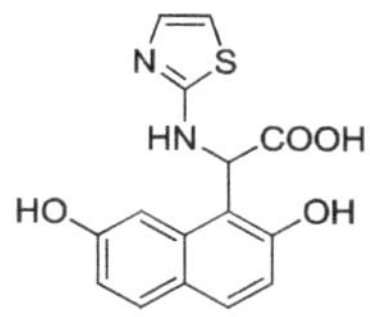

4d

2-(2,7-Dihydroxynaphthalen-1-yl)-2-(thiazol-2-ylamino)acetic acid (4d): colorless crystals, mp 169–170 °C, reaction time: 4 h, yield: 78%

2.2.3.3 Experimental Procedure

To a solution of heteroarylamine (**3**; 5 mmol) and glyoxalic acid (50% aqueous solution) (**2**; 5 mmol) in water (10 mL) was added β-naphthols (**1**; 5 mmol) and the mixture was stirred for stipulated time-frames (3.5–14 h) at ambient temperature. Progress of the reaction was monitored by TLC, and upon completion of the reaction the crude precipitate as obtained was filtered, washed with cold ethanol, dried and purified by recrystallization from DMSO-H_2O to achieve colorless crystals of *N*-heteroaryl α-naphthylglycines (**4**) with moderate to high yields. All the compounds were characterized on the basis of analytical and spectral studies.

2.2.3.4 Characterization Data of Two Representative Compounds

2-((2-Chloropyridin-3-yl)amino)-2-(2-hydroxynaphthalen-1-yl)acetic acid (**4b**): colorless crystals; Yield: 83%; mp 200–202 °C; IR (KBr): 3388, 3246–2569, 1704, 1627, 1583, 1490, 1436, 1380, 1288, 1253, 1147 cm^{-1}; ^1H NMR (DMSO-d_6, 500 MHz): δ 6.07–8.19 (11H, m, methine-H, pyridine-H and naphthalene-H, NH), 10.48 (1H, s, OH), 13.13 (1H, s, COOH); ^{13}C NMR (DMSO-d_6, 125 MHz): δ 52.15, 118.88, 119.52, 122.75, 123.14, 123.55, 124.68, 127.68, 129.28, 129.62, 130.78, 133.30, 136.94, 136.99, 140.28, 154.34, 173.88; MS (EI): *m/z* 328 $(M)^+$, 310, 283, 247, 200, 183, 172, 156, 144, 128, 127, 115, 92; Anal. calc. for $C_{17}H_{13}ClN_2O_3$: C, 62.10; H, 3.95; N, 8.52; found: C, 62.20; H, 4.02; N,8.50.

2-(2,7-Dihydroxynaphthalen-1-yl)-2-(thiazol-2-ylamino)acetic acid (**4d**): colorless crystals; Yield: 78%; mp 169–170 °C; IR (KBr): 3203–2617, 1608, 1517, 1481, 1460, 1375, 1305, 1222, 1164, 1137 cm^{-1}; ^1H NMR (DMSO-d_6, 500 MHz): δ 6.21 (1H, s, methine-H), 6.61–7.96 (8H, m, thiazole-H and naphthalene-H, NH), 9.67 (1H, s, OH), 10.25 (1H, s, OH); (DMSO-d_6, 125 MHz): δ 53.58, 105.84, 107.62, 114.97, 115.72, 115.81, 123.84, 130.25, 130.87, 135.41, 138.81, 154.57, 156.93, 169.30, 173.83; MS (EI): *m/z* 298 $(M-18)^+$, 268, 253, 199, 172, 160, 144, 115, 100; Anal. calc. for $C_{15}H_{12}N_2O_4S$: C, 56.96; H, 3.79; N, 8.86; found: C, 56.89; H, 3.75; N, 8.90.

2.2.3.5 Critical Views

The investigators developed a greener and eco-friendly protocol for the water-mediated synthesis of a series of novel *N*-heteroaryl α-arylglycines containing naphthol ring in the absence of any catalyst at room temperature *via* one-pot three-component reaction. From their experimental observations relating to the mechanistic aspects of the condensation reaction, they also assumed that an iminoacid generated *in situ* from the reaction of amine and glyoxalic acid, reacts with 2-naphthol, affording the desired α-naphthylglycine.[32]

2.2.4 Entry-4: Synthesis of bis(Hydroxyethyl)thioethers

Type of reaction: C–S bond formation
Reaction conditions: Catalyst-free, water, room temperature
Synthetic strategy: Coupling of epoxides with ring opening
Keywords: Epoxides, sodium sulfide, catalyst-free, water, room temperature, coupling of epoxides with ring opening, bis(hydroxyethyl)thioethers

2.2.4.1 General Reaction Scheme

Azizi *et al.*[33] demonstrated a water-mediated catalyst-free highly efficient coupling of epoxides (**1**) with sodium sulfide (**2**) followed by ring opening to produce substituted bis(hydroxyethyl)thioethers (**3**) in excellent yields without the formation of any undesirable by-products (Scheme 2.5).

2.2.4.2 Representative Examples

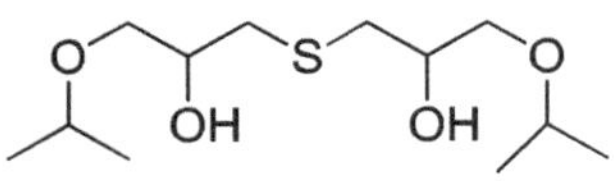

3a

3,3'-Thio*bis*(1-isopropoxypropan-2-ol) (**3a**): Yield: 95%

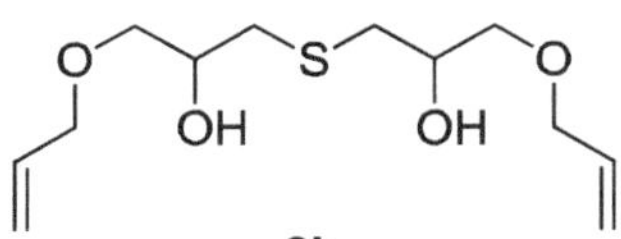

3b

3,3'-Thio*bis*(1-(allyloxy)propan-2-ol) (**3b**): Yield: 88%

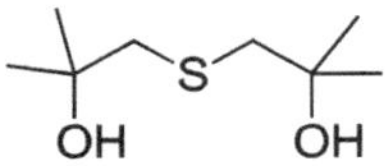

3c

1,1'-Thio*bis*(2-methylpropan-2-ol) (**3c**): 72%

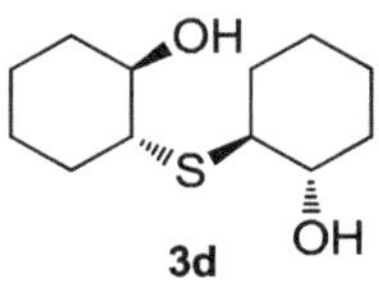

3d

Yield: (1*S*,2*S*)-2-(((1*R*,2*R*)-2-Hydroxycyclohexyl)thio)cyclohexanol (**3d**): Yield: 70%

2.2.4.3 Experimental Procedure

To a stirred solution of epoxides (**1**; 3 mmol) in water (2 mL) in a test tube was added 0.2 g of sodium sulfide (3 mmol), and the resulting mixture was

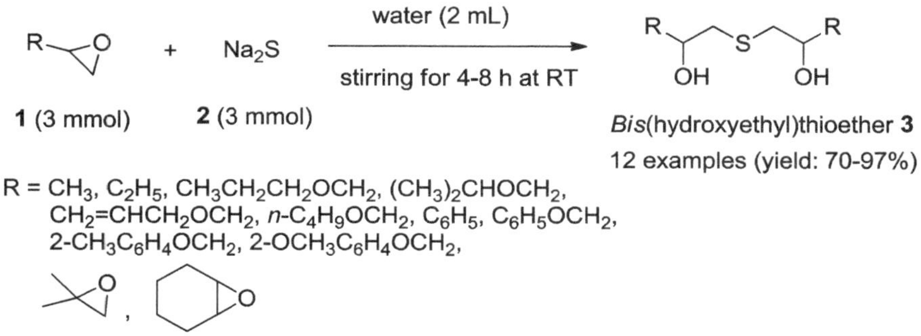

Scheme 2.5 Synthesis of bis(hydroxyethyl)thioethers (**3**).

stirred at room temperature for 4–8 h. On completion of the reaction (monitored by TLC), bis(hydroxyethyl)thioethers (**3**) were obtained as pure by just filtration in the case of solid products. The liquid bis(hydroxyethyl)thioethers (**3**) were extracted with ethyl acetate, dried over anhydrous sodium sulfate, and evaporation of the solvent afforded the desired pure product in most cases. In a few cases, the crude products were further purified by flash column chromatography to provide the corresponding pure products. All compounds were characterized based on their spectroscopic data (IR and NMR) compared with those reported in literature.

2.2.4.4 Critical Views

The investigators developed an easy and simple green protocol for the synthesis of a series of substituted bis(hydroxyethyl)thioethers from their corresponding epoxides on reaction with sodium sulfide in water medium under catalyst-free conditions at ambient temperature.

In an earlier report, Innocenti *et al.*[34] used hexamethyldisilathiane as the sulfur nucleophile for ring opening of epoxides in the presence of tetrabutylammonium fluoride (TBAF) as a catalyst to produce bis(hydroxyethyl)thioethers in tetrahydrofuran (THF). Thus, the present protocol is more advantageous over the earlier method affording good to excellent yields, without the formation of any undesirable side-products. The investigators also observed that the regioselectivity in the reaction of unsymmetrical epoxides is governed by both steric and electronic effects; furthermore, in the case of cyclic epoxides, such as cyclohexene oxide, *trans* products were obtained.

2.2.5 Entry-5: Synthesis of β-Hydroxy Thioesters

Type of reaction: C–S bond formation
Reaction conditions: Water, room temperature
Synthetic strategy: Epoxide ring-opening
Keywords: Epoxides, thioacetic acid, thiobenzoic acid, catalyst-free, water, room temperature, epoxide ring-opening, β-hydroxy thioesters

2.2.5.1 General Reaction Scheme

Halimehjani *et al.*[35] developed a modified chemoselective, ring-opening reaction of epoxides (**1**) employing thioacetic/thiobenzoic acids (**2**) instead of sodium sulfide to produce β-hydroxy thioesters (**3**) in water at room temperature in a regioselective manner (Scheme 2.6).

2.2.5.2 Representative Examples

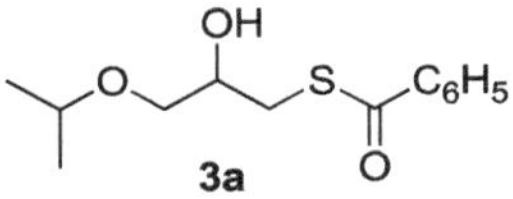

3a

S-(2-Hydroxy-3-isopropoxypropyl) benzothioate (**3a**): yield: 100%

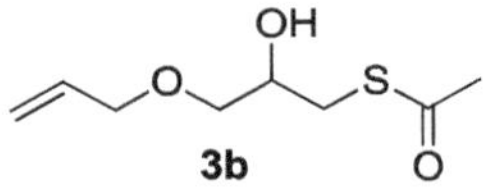

3b

S-(3-(Allyloxy)-2-hydroxypropyl) ethanethioate (**3b**): yield: 100%

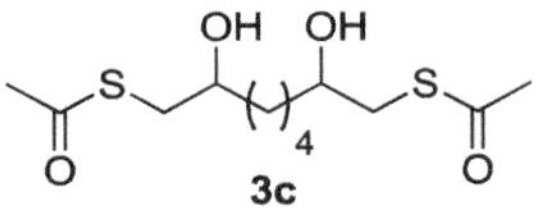

3c

S,S'-(2,4-Dihydroxypentane-1,5-diyl) diethanethioate (**3c**): yield: 96%

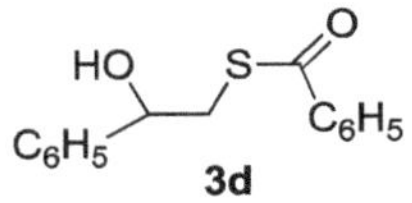

3d

S-(2-Hydroxy-2-phenylethyl) benzothioate (**3d**): yield: 70%

2.2.5.3 Experimental Procedure

A mixture of epoxides (**1**; 5 mmol), water (10 mL), and thioacids (**2**; 5.5 mmol) in a round-bottomed flask was stirred vigorously at room temperature for 10 h. On completion of the reaction, the oily product was extracted by ethyl acetate, washed with 2 M aqueous solution of sodium bicarbonate, and evaporated to give pure β-hydroxy thioesters (**3**) in most of the cases. Further purifications were performed using column chromatography (silica gel; ethyl acetate : *n*-hexane, 2 : 8). The products were characterized by spectral studies and compared with the literature values.

2.2.5.4 Characterization Data of Two Representative Compounds

S-(3-(Allyloxy)-2-hydroxypropyl) ethanethioate (**3b**): yield: 100%; ^{1}H NMR ($CDCl_3$, 300 MHz): δ 5.81 (1H, dd, $J = 10.5$ & 6.2 Hz), 5.08–5.21 (2H, m), 3.93

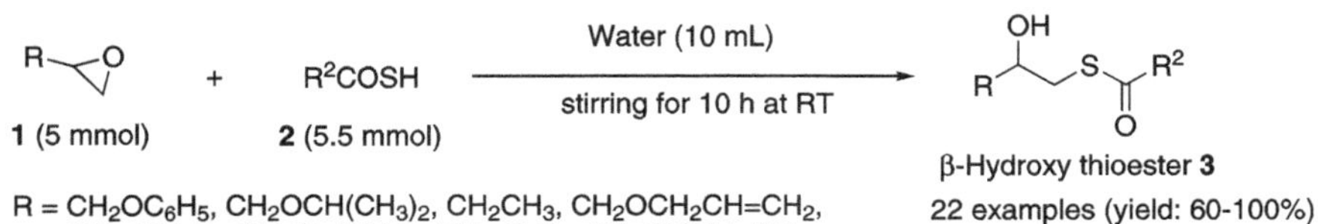

R = $CH_2OC_6H_5$, $CH_2OCH(CH_3)_2$, CH_2CH_3, $CH_2OCH_2CH{=}CH_2$, CH_3, CH_2Br, CH_2Cl, $(CH_2)_{11}CH_3$, C_6H_5, $CH_2O(CH_2)_3CH_3$, $CH_2OCOC(CH_3){=}CH_2$

R^2 = CH_3, C_6H_5

Scheme 2.6 Synthesis of β-hydroxy thioesters **3**.

(2H, d, $J = 4.7$ Hz), 3.81 (1H, m), 3.32–3.41 (2H, m), 3.13 (1H, br s, –OH), 2.91–3.05 (2H, m), 2.27 (3H, s); ^{13}C NMR ($CDCl_3$, 75 MHz): δ 196.2, 134.8, 117.6, 73.1, 72.6, 69.8, 32.8, 30.8; MS (EI): $m/z = 189$ (M − 1), 157, 149, 129, 103, 97, 73, 59, 55, 43 (100), 41.

S,*S*′-(2,4-Dihydroxypentane-1,5-diyl) diethanethioate (**3c**): yield: 96%; ^{1}H NMR ($CDCl_3$, 300 MHz): δ 3.63 (2H, m), 3.04–2.77 (6H, m), 2.26 (6H, s), 1.42–1.33 (8H, m); ^{13}C NMR ($CDCl_3$, 75 MHz): δ 196.4, 69.3, 35.8, 35.7, 30.4, 20.9; MS (EI): $m/z = 295$ (M + 1), 277 (100), 259, 235, 217, 187, 175, 141, 127, 43.

2.2.5.5 Critical Views

The investigators developed a catalyst-free method for the synthesis of β-hydroxyl thioesters with the reaction of epoxides and thioacids in water at room temperature with good to high yields. They demanded that in large-scale synthesis, the product can be extracted by simple decanting. The products can also be purified by distillation only, and it is worthy to mention that the ester and alkyl halide cleavage were not observed in this method.[35]

2.2.6 Entry-6: Synthesis of Thioesters

Type of reaction: C–S bond formation
Reaction conditions: Catalyst-free, water, room temperature
Synthetic strategy: Conjugate addition of thioacids to activated olefins
Keywords: Activated olefins, α,β-unsaturated carbonyl compounds, thioacetic acid, thiobenzoic acid, catalyst-free, water, room temperature, Michael addition, thioesters

2.2.6.1 General Reaction Scheme

Marjani *et al.*[36] reported a catalyst-free room temperature protocol for the conjugate Michael addition of thioacetic/thiobenzoic acids (**2**) to activated olefins (**1**) in aqueous medium affording thioester derivatives (**3**) with excellent yields (Scheme 2.7).

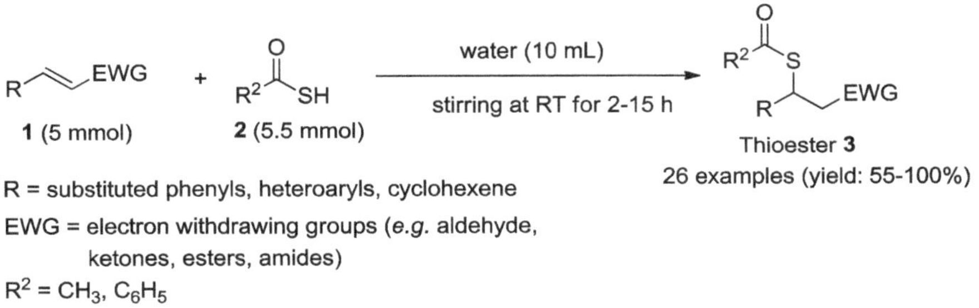

Scheme 2.7 Synthesis of thioesters (**3**).

2.2.6.2 Representative Examples

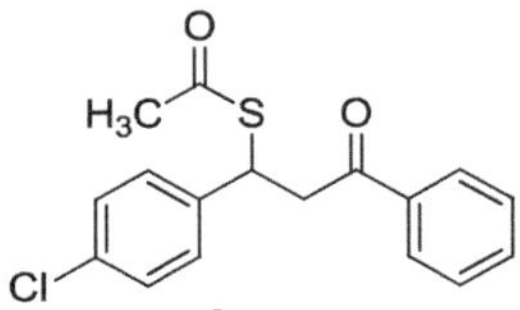

3a

S-(1-(4-Chlorophenyl)-3-oxo-3-phenylpropyl) ethanethioate (**3a**): yield: 100%

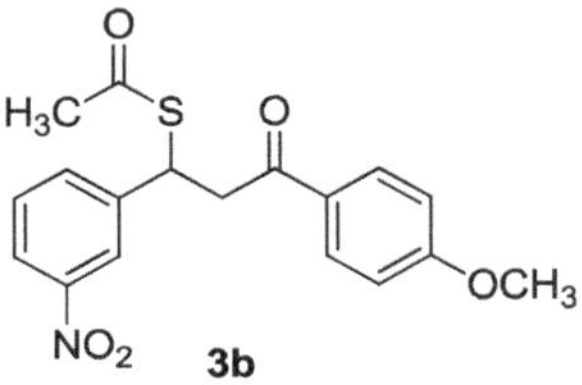

3b

S-(3-(4-Methoxyphenyl)-1-(3-nitrophenyl)-3-oxopropyl) ethanethioate (**3b**): yield: 100%

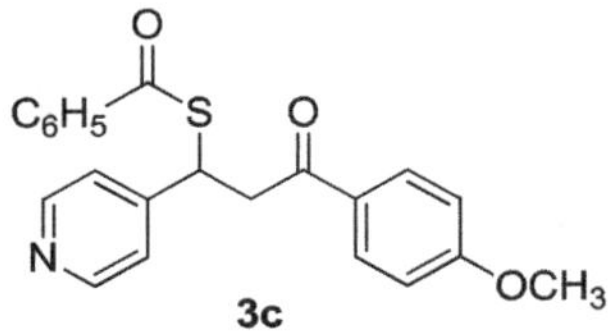

3c

S-(3-(4-Methoxyphenyl)-3-oxo-1-(pyridin-4-yl)propyl) benzothioate (**3c**): yield: 80%

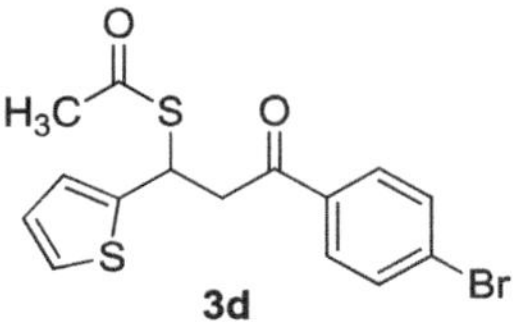

3d

S-(3-(4-Bromophenyl)-3-oxo-1-(thiophen-2-yl)propyl) ethanethioate (**3d**): yield: 80%

2.2.6.3 Experimental Procedure

A round-bottomed flask equipped with a magnetic stirrer was charged with α,β-unsaturated carbonyl compound (**1**; 5 mmol), thioacid (**2**; 5.5 mmol) and water (10 mL), and the reaction mixture was stirred vigorously at room temperature (2 h for thioacetic acid and 15 h for thiobenzoic acid). On completion of the reaction, pure solid products were obtained by filtration in most of the cases; further purification was achieved by crystallization from ethanol or by column chromatography (silica gel; ethyl acetate-petroleum ether = 1 : 5). In the case of oily products, extraction by ethyl acetate followed by washing of the organic layer with 10% aqueous sodium carbonate solution and evaporation of the solvent ultimately afforded pure products. All the products were characterized by spectral studies and compared with the literature values.

2.2.6.4 Characterization Data of Two Representative Compounds

S-(1-(4-Chlorophenyl)-3-oxo-3-phenylpropyl) ethanethioate (**3a**): yield: 100%; ^{1}H NMR ($CDCl_3$, 300 MHz): δ 7.95 (2H, d, $J = 7.1$ Hz), 7.35–7.56 (5H, m), 7.17–7.22 (2H, m), 5.66 (1H, t, $J = 7.1$ Hz), 3.8 (2H, d, $J = 7.2$), 2.32 (3H, s); ^{13}C NMR ($CDCl_3$, 75 MHz): δ 196.1, 194.1, 137.3, 136.2, 133.2, 129.9, 129.5, 128.6, 127.9, 126.8, 43.3, 40.9, 36.1.

S-(3-(4-Methoxyphenyl)-1-(3-nitrophenyl)-3-oxopropyl) ethanethioate (**3b**): yield: 100%; ^{1}H NMR ($CDCl_3$, 300 MHz): δ 8.26 (1H, s), 8.06 (1H, d,

$J = 8.2$ Hz), 7.91 (2H, d, $J = 8.8$ Hz), 7.77 (1H, d, $J = 7.7$ Hz), 7.45 (1H, t, $J = 8.1$ Hz), 6.90 (2H, d, $J = 8.8$ Hz), 5.32 (1H, t, $J = 6.8$ Hz), 3.84 (3H, s), 3.66 (2H, d, $J = 7.1$ Hz), 2.31 (3H, s); ^{13}C NMR ($CDCl_3$, 75 MHz): δ 194.1, 193.7, 163.7, 148.1, 143.3, 134.4, 130.4, 130.2, 129.2, 122.5, 122.2, 113.7, 55.4, 43.3, 42.5, 30.2.

2.2.6.5 Critical Views

The present method was found to proceed smoothly for diverse chalcone derivatives with electron-donating and electron-withdrawing substitutions. In many occasions, complete conversions were observed. Both cyclic and acyclic α,β-unsaturated carbonyl compounds underwent the reaction efficiently. This metal- and catalyst-free water-mediated conjugate addition of thioacids to α,β-unsaturated carbonyl compounds offers an easy access to thioester compounds under eco-friendly and nonhazardous experimental conditions. Room temperature operation, ease of reaction, and excellent yields of products are the added advantages of this method. The investigators were successful in obtaining high-quality products with just a simple filtration technique. It should also be noted that competitive dithiane formation and ester cleavage were not observed. In addition, the investigator applied their present protocol to the synthesis of the thioacetic ester of canrenone efficiently, to underline the industrial potential of the method (Scheme 2.8).

2.2.7 Entry-7: Synthesis of β-Sulfido Carbonyl Compounds

Type of reaction: C–S bond formation
Reaction conditions: Catalyst-free, water, room temperature
Synthetic strategy: Conjugate addition of thiols to α,β-unsaturated carbonyl compounds
Keywords: α,β-Unsaturated carbonyl compounds, thiols, conjugate addition, catalyst-free, water, room temperature, thia-Michael addition, chemoselectivity, β-sulfido carbonyl compounds

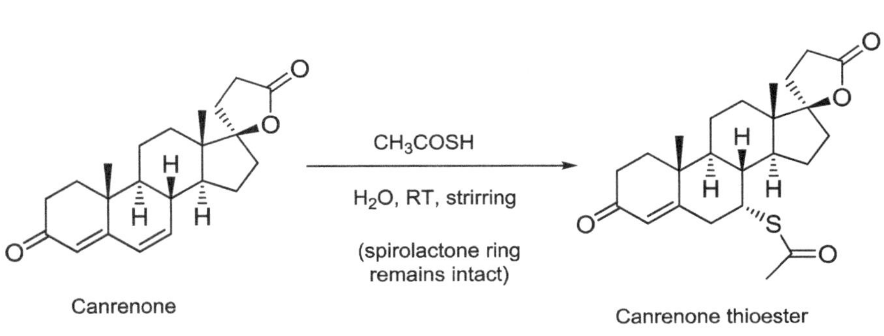

Scheme 2.8 Conversion of canrenone to its thioacetic ester derivative.

2.2.7.1 General Reaction Scheme

Chakraborti and his group[37] reported a catalyst-free room temperature protocol for the conjugate thia-Michael addition of thiols (**2**) to α,β-unsaturated carbonyl compounds (**1**) in aqueous medium for the synthesis of a series of β-sulfido carbonyl compounds (**3**) with excellent yields (Scheme 2.9).

2.2.7.2 Plausible Mechanism

The authors proposed a plausible mechanism for the reaction (Scheme 2.10) underlining the dual activation role of water simultaneously for α,β-unsaturated carbonyl compound and thiol through the formation of hydrogen bonds with them. Thus, hydrogen bond formation between water

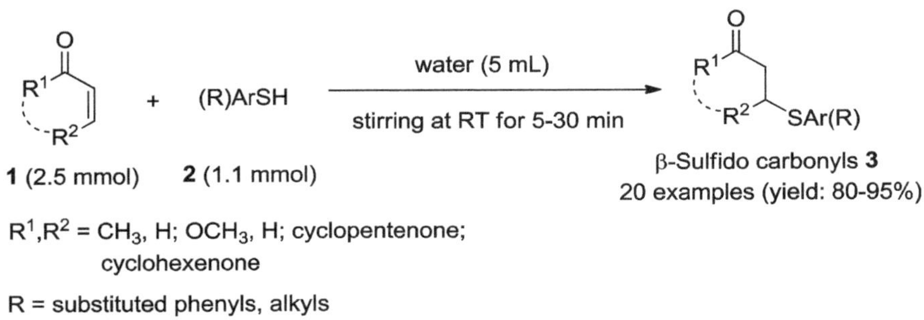

Scheme 2.9 Synthesis of β-sulfido carbonyl compounds (**3**).

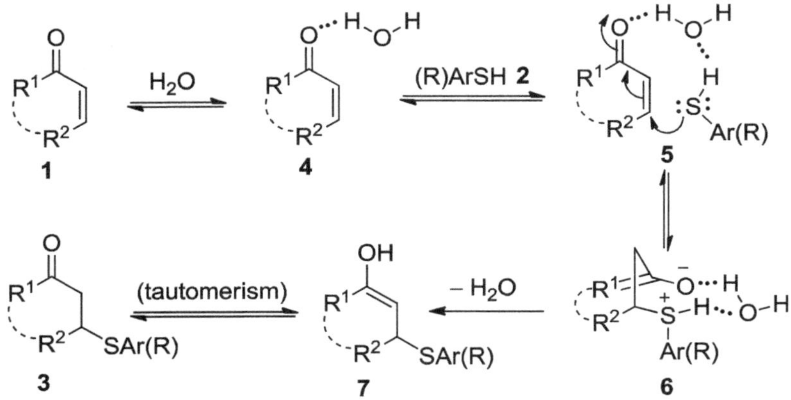

Scheme 2.10 Plausible mechanism for the water-mediated formation of β-sulfido carbonyl compound (**3**).

and the carbonyl oxygen atom of α,β-unsaturated carbonyl compound (**4**) increases the electrophilic character at its β-carbon, and the hydrogen bond formation involving the oxygen atom of water and the sulfhydryl hydrogen of thiol (**5**) enhances nucleophilicity of the sulfur atom of the thiol. Such a simultaneous activation of the substrate molecules through hydrogen bond formation with water facilitates the conjugate addition (thia-Michael addition)—intramolecular nucleophilic attack at the β-carbon atom of the α,β-unsaturated carbonyl compound by the sulfur atom of the thiol *via* the transition state (**6**), followed by intramolecular proton transfer led to the conjugate adduct (**3**). The proposal for '*dual activation by water*' received validity from the fact that the rate of reaction of various thiols with a common α,β-unsaturated carbonyl compound is influenced by the hydrogen bond formation ability of the thiols. The investigators observed that the reaction of aromatic thiols with a common substrate, in general, occurs at a faster rate than that of aryl alkanethiol, alkanethiol, and alkane dithiols.[37]

2.2.7.3 Representative Examples

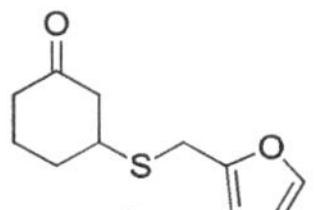

3a

3-((Furan-2-ylmethyl)thio)cyclohexanone (**3a**): colorless oily liquid; yield: 90%

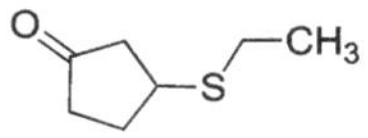

3b

3-(Ethylthio)cyclopentanone (**3b**): oily liquid; yield: 90%

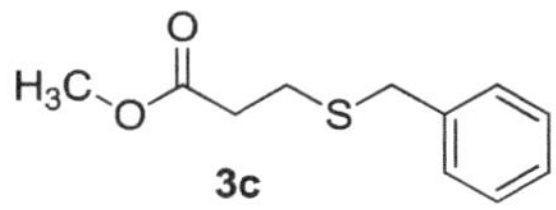

3c

Methyl 3-(benzylthio)propanoate (**3c**): colorless liquid; yield: 95%

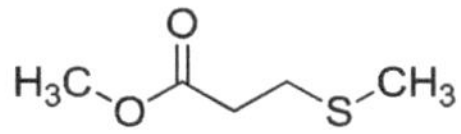

3d

Methyl 3-(methylthio)propanoate (**3d**): colorless liquid; yield: 80%

2.2.7.4 Experimental Procedure

α,β-Unsaturated carbonyl compounds (**1**; 2.5 mmol) in water (5 mL) were treated with thiols (**2**; 1.1 mmol) under magnetic stirring at room temperature (30–35 °C) and, on completion of the reaction (monitored by TLC), the reaction mixture was extracted with ethyl acetate (EtOAC; 3 × 10 mL). The combined EtOAc extracts were dried with sodium sulfate, concentrated under reduced pressure, and the crude product was purified using column chromatography over silica gel (eluent: EtOAc-hexane) to afford β-sulfido carbonyl compounds (**3**). All the products were characterized by their analytical and spectral studies.

2.2.7.5 *Characterization Data of Two Representative Compounds*

3-((Furan-2-ylmethyl)thio)cyclohexanone (**3a**): Colorless oily liquid; yield: 90%; ^{1}H NMR ($CDCl_3$, 300 MHz): δ 1.65–1.73 (2H, m), 2.05–2.16 (2H, m), 2.29–2.41 (3H, m), 2.65–2.71 (1H, m), 3.00–3.05 (1H, m), 3.76 (2H, s), 6.17 (1H, d, $J=3.1$ Hz), 6.29–6.31 (1H, m), 7.34–7.38 (1H, m).

Methyl 3-(benzylthio)propanoate (**3c**): colorless liquid; yield: 95%; IR (neat): 2951, 2923, 1738 cm^{-1}; ^{1}H NMR ($CDCl_3$, 300 MHz): δ 2.31 (3H, s), 2.58 (2H, t, $J=7.41$ Hz), 3.10 (2H, t, $J=7.41$ Hz), 3.66 (3H, s), 7.10 (2H, d, $J=8.0$ Hz), 7.27 (2H, d, $J=8.0$ Hz); ^{13}C NMR ($CDCl_3$, 75 MHz) δ: 20.9, 29.7, 34.2, 51.7, 129.7, 130.9, 131.2, 136.8, 172.1; MS (EI): *m/z* 210 (M^+); calc. for $C_{11}H_{14}O_2S$: C, 62.83; H, 6.71; S, 15.25; found: C, 62.81; H, 6.70; S, 15.23.

2.2.7.6 *Critical Views*

The investigators described an efficient and high-yielding protocol for the green synthesis of β-sulfido carbonyl compounds from a catalyst-free water-mediated conjugate thia-Michael addition of aromatic and aliphatic thiols to α,β-unsaturated carbonyls at ambient conditions. This is the first report on thia-Michael addition reaction in water in the absence of any catalyst. The reaction proceeded with excellent chemoselectivity, and competitive dithiane/dithiolane formation; transesterification, and ester cleavage were not observed. The metal-free and nonhazardous experimental conditions, room temperature operation, ease of reaction, short reaction times, and high yields are the notable features of this environment-friendly method. In addition, water was found to play a dual role in simultaneously activating the α,β-unsaturated carbonyl compound and the thiol, thereby facilitating the conjugate addition.[37]

2.2.8 Entry-8: Synthesis of *S*-alkyl Dithiocarbamates

Type of reaction: C–N and C–S bond formation
Reaction conditions: Catalyst-free, solvent-free (neat condition), room temperature
Synthetic strategy: One-pot synthesis for *S*-alkyl thiocarbamates
Keywords: Secondary amines, carbon disulfide, alkyl halides, solvent-free, catalyst-free, room temperature, dithiocarbamates

2.2.8.1 *General Reaction Scheme*

Azizi *et al.*[38] reported on an efficient and simple synthesis of a wide range of *S*-alkyl dithiocarbamates (**4**) based on a one-pot reaction of amines (**1**), alkyl halides (**2**) and carbon disulfide (**3**) without using a catalyst under solvent-free conditions at room temperature (Scheme 2.11).

Scheme 2.11 Synthesis of *S*-alkyl dithiocarbamates (**4**).

2.2.8.2 Representative Examples

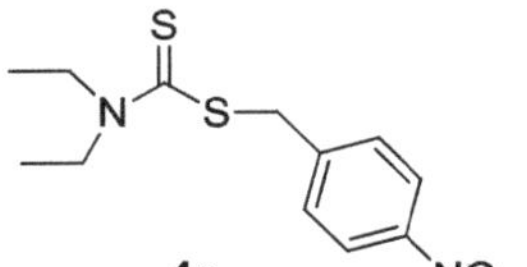

4a

4-Nitrobenzyl diethylcarbamodithioate (**4a**): Yield 76%

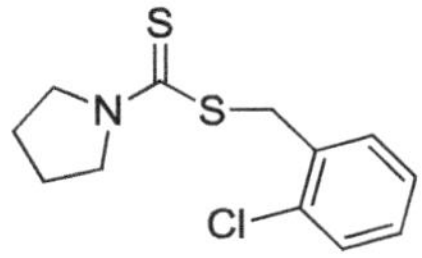

4b

2-Chlorobenzyl pyrrolidine-1-carbodithioate (**4b**): Yield 92%

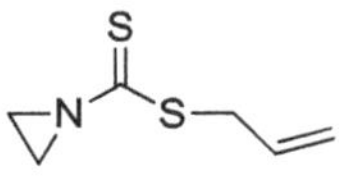

4c

Allyl aziridine-1-carbodithioate (**4c**): Yield 68%

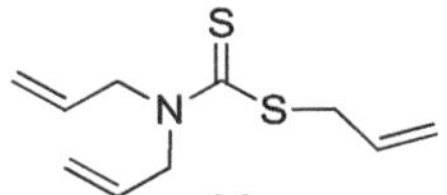

4d

Allyl diallylcarbamodithioate (**4d**): Yield 70%

2.2.8.3 Experimental Procedure

To an ice-cold mixture of organic halides (**2**; 3 mmol) and carbon disulfide (**3**; 6 mmol), secondary amines (**1**: 6 mmol) were diluted and stirred at 0 °C for 30 min, then warmed to room temperature; stirring was continued until the reaction was completed (monitored by TLC). The reaction was quenched by the addition of water and was extracted with ethyl acetate, dried over anhydrous sodium sulfate, and evaporated to give almost pure product. Further purification in some cases was done by silica gel column chromatography eluted with petroleum ether and ethyl acetate or recrystallized to afford pure dithiocarbamate (**4**). All the products were characterized by spectral studies and data were compared with literature values.

2.2.8.4 Critical Views

The present report offers a practical method for the synthesis of diverse *S*-alkyl dithiocarbamates from inexpensive and readily available starting materials in one-pot under mild reaction conditions. It avoids hazardous organic solvents and toxic catalysts and gives nearly quantitative yields without any by-products in most cases. Dithiocarbamates have received considerable attention due to their numerous biological activities and their pivotal role in agriculture and as linkers in solid-phase organic synthesis. Mild reaction conditions, high yields, and broad scope of this protocol may be useful for large-scale synthesis in the pharmaceutical or agrochemical industries.[38]

2.2.9 Entry-9: Synthesis of Densely Substituted Dithiocarbamates

Type of reaction: C–N and C–S bond formation
Reaction conditions: Solvent-free (neat condition), room temperature

Synthetic strategy: One-pot multicomponent reaction
Keywords: Michael acceptors, amines, carbon disulfide (CS_2), neat condition, catalyst-free, room temperature, one-pot multicomponent reaction, dithiocarbamates

2.2.9.1 General Reaction Scheme

Azizi and coworkers further developed a waste-free green technique for the synthesis of a series of densely substituted dithiocarbamates (**4**) *via* a one-pot condensation reaction of amines (**1**), carbon disulfide (CS_2), and a Michael acceptor, under solvent-free conditions at room temperature in good to excellent yields (Scheme 2.12).[39]

2.2.9.2 Representative Examples

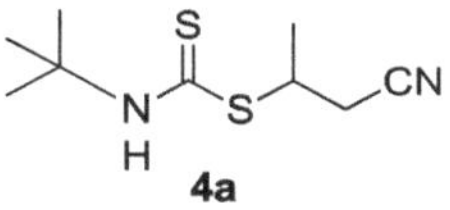

4a

1-Cyanopropan-2-yl *tert*-butylcarbamodithioate (**4a**): yield 84%

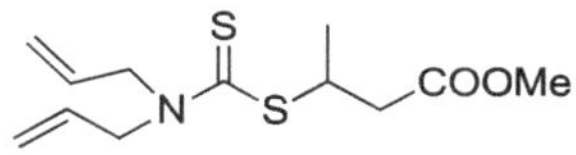

4b

Methyl 3-((diallylcarbamothioyl)thio) butanoate (**4b**): yield 78%

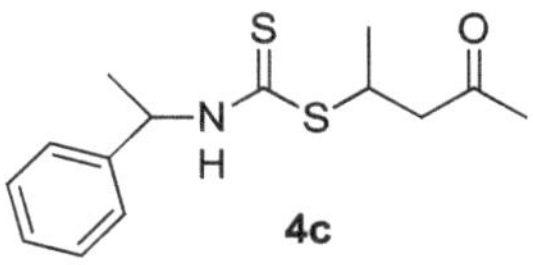

4c

4-Oxopentan-2-yl (1-phenylethyl)carbamodithioate (**4c**): yield 94%

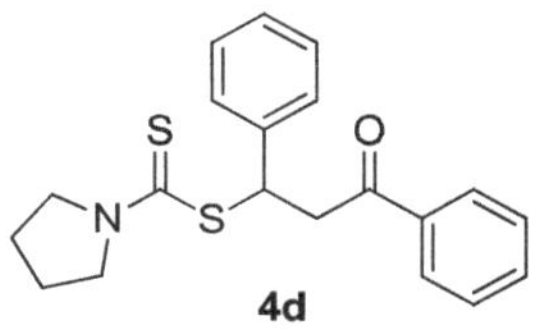

4d

3-Oxo-1,3-diphenylpropyl pyrrolidine-1-carbodithioate (**4d**): yield 92%

2.2.9.3 Experimental Procedure

An amine (**1**; 4.5 mmol) was added slowly to the mixture of carbon disulfide (CS_2, 5 mmol) and the Michael acceptor (**2**; 4 mmol) into a test tube placed in an ice bath. The reaction mixture was stirred at 0 °C for 30 min, and then warmed to room temperature with continued stirring for another 1–12 h to complete the reaction. On completion of the reaction, an excess of CS_2 and

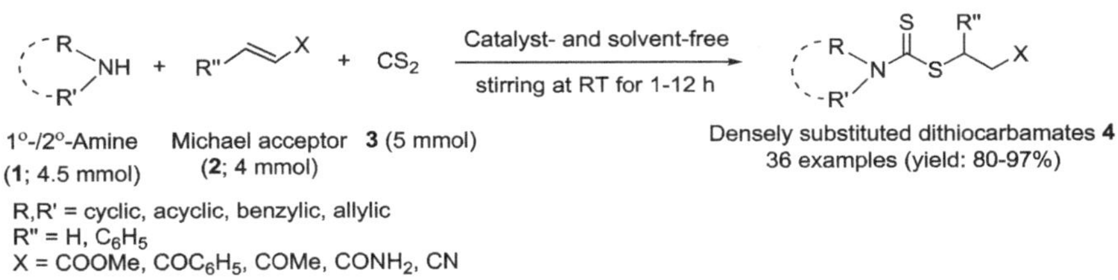

Scheme 2.12 Synthesis of functionalized dithiocarbamates (**4**).

amine was removed under reduced pressure to afford the dithiocarbamate derivative (**4**) in almost pure form; in some cases, further purification was done by recrystallization or short-column chromatography on silica gel using EtOAc–petrol ether mixture as the eluent. All compounds were characterized on the basis of NMR spectral data.

2.2.9.4 Characteristic Data for Two Representative Compounds

4-Oxopentan-2-yl (1-phenylethyl)carbamodithioate (**4c**): ^{1}H NMR (500 MHz, $CDCl_3$–CCl_4,): δ 1.58 (d, $J = 6.8$ Hz, 3H), 2.14 (s, 3H), 2.89 (t, $J = 6.1$ Hz, 2H), 3.40 (t, $J = 6.1$ Hz, 2H), 5.81 (q, $J = 6.8$ Hz, 1H,), 7.26–7.35 (m, 5H), 8.18 (br s, NH, 1H); ^{13}C NMR (125 MHz $CDCl_3$–CCl_4,): δ 21.3, 29.0, 30.4, 43.3, 56.4, 127.0, 128.2, 129.3, 141.7, 196.8, 207.6.

3-Oxo-1,3-diphenylpropyl pyrrolidine-1-carbodithioate (**4d**): ^{1}H NMR (500 MHz, $CDCl_3$): δ 1.93–2.05 (m, 4H), 3.58–4.02 (m, 6H), 5.83 (dd, $J = 4.1$, 9.8 Hz, 1H), 7.27–7.56 (m, 8H), 8.00 (d, $J = 7.3$ Hz, 2H); ^{13}C NMR (125 MHz, $CDCl_3$): δ 24.6, 26.5, 51.0, 54.6, 55.3, 128.0, 128.6, 128.8, 129.0, 129.4, 133.6, 137.0, 140.0, 191.5, 197.4.

2.2.9.5 Critical Views

The investigators developed for the first time a waste-free process for the synthesis of densely substituted dithiocarbamates in the absence of any solvent and catalyst with virtually no by-product. This green protocol possesses considerable merits in terms of mild reaction conditions at room temperature, enhanced reaction rates, clean reaction profiles, operational and experimental simplicity, and with options of further transformations of the resulting dithiocarbamates into synthetically interesting biologically active compounds.[39]

2.2.10 Entry-10: Synthesis of Pivalate Derivatives

Type of reaction: C–O bond formation
Reaction conditions: Solvent- and catalyst-free neat conditions, room temperature
Synthetic strategy: Protection of alcohol through one-step pivaloylation
Keywords: Alcohols (aliphatic and aromatic), pivaloyl chloride, solvent-free, neat condition, catalyst-free, room temperature, protection of alcoholiv group, chemoslectivity, pivalate derivatives

2.2.10.1 General Reaction Scheme

Venkateswarlu and his group[40] developed a simple and efficient protocol for the chemoselective protection of the alcoholic (–OH) group through pivaloylation under solvent- and catalyst-free conditions at room temperature (Scheme 2.13).

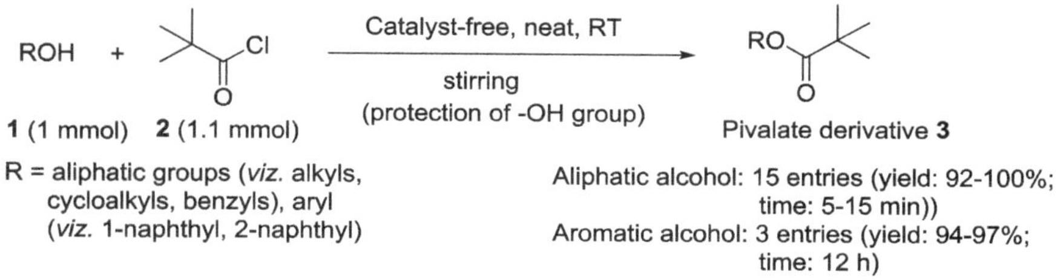

Scheme 2.13 Synthesis of pivalate derivatives (**3**).

2.2.10.2 Representative Examples

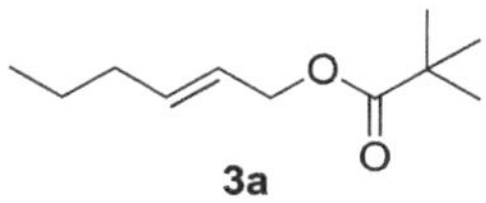

3a

(*E*)-Hex-2-en-1-yl pivalate (**3a**): time, 5 min; yield 100%

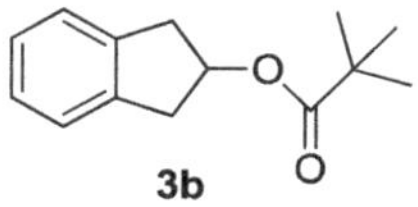

3b

2,3-Dihydro-1*H*-inden-2-yl pivalate (**3b**): time, 15 min; yield 96%

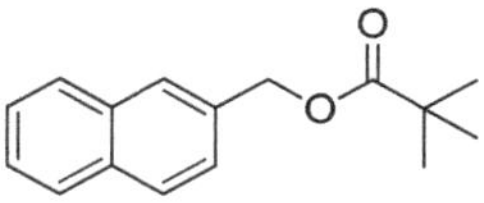

3c

Naphthalen-2-ylmethyl pivalate (**3c**): time, 5 min; yield: 100%

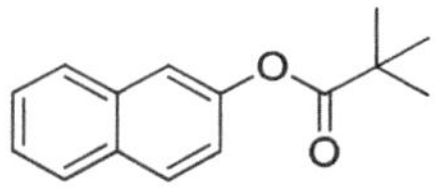

3d

Naphthalen-2-yl pivalate (**3d**): time 12h; yield: 97%

2.2.10.3 Experimental Procedure

Alcohol (**1**; 1 mmol), which was placed in a 5 mL round-bottomed flask, was added to pivaloyl chloride (**2**; 1.1 mmol) and the resulting mixture was stirred at room temperature in the absence of any catalyst and solvent under closed vessel conditions. After completion of the reaction as monitored by TLC, the reaction mixture was dissolved in 5.0 mL of dichloromethane and washed with distilled water; the organic layer was then dried with sodium sulfate and solvent was evaporated under vacuum to yield the corresponding pivalate derivative (**3**) (column chromatography was performed whenever it required). All the products were characterized based on the elemental analyses and spectral studies.

2.2.10.4 Characterization Data of Two Representative Compounds

(*E*)-Hex-2-en-1-yl pivalate (**3a**): IR: 1731 cm^{-1}; ^{1}H NMR (300 MHz, $CDCl_3$): δ 0.91 (t, 3H, $J = 7.0$ Hz), 1.19 (s, 9H), 1.48–1.34 (m, 2H), 2.02 (q, 2H, $J = 6.5$ Hz), 4.47 (d, 2H, $J = 6.5$ Hz), 5.56–5.47 (m, 1H), 5.76–5.67 (m, 1H); Mass (ESI): *m/z* at 185.1 (M + 1). Anal calc. for: $C_{11}H_{20}O_2$ (184) C, 71.70; H, 10.94; found: C, 71.67; H, 10.90.

2,3-Dihydro-1*H*-inden-2-yl pivalate (**3b**): IR: 1715 cm^{-1}; ^{1}H NMR (300 MHz, $CDCl_3$): δ 1.15 (s, 9H), 2.96 (dd, 2H, $J = 3.0$ Hz, 17.3 Hz), 3.30 (dd, 2H,

$J = 6.9$ Hz, 16.6 Hz), 5.48–5.46 (m, 1H), 7.24–7.13 (m, 4H); ^{13}C NMR (75 MHz, $CDCl_3$): δ 178.4, 140.4 (2C), 126.6 (2C), 124.5 (2C), 74.7, 39.5 (2C), 38.5, 27.0 (3C). Mass (ESI): *m/z* at 219.3 (M + 1). Anal calc. for: $C_{14}H_{18}O_2$ (218) C, 77.03; H, 8.31; found: C, 76.90; H, 8.27.

2.2.10.5 *Critical Views*

The present method offers a catalyst- and solvent-free alternative route to protect the alcoholic function under mild reaction conditions through pivaloylation in a chemoselective manner. The key advantages of the reaction are short reaction time, high yields and simple experimental procedure. This protocol may be useful in biologically active natural products synthesis and carbohydrate chemistry.[40]

2.2.11 Entry-11: Synthesis of 2,2′-Arylmethylene bis(3-Hydroxy-5,5-dimethyl-2-cyclohexene-1-one) Derivatives

Type of reaction: C–C bond formation
Reaction conditions: Catalyst-free, water, room temperature
Synthetic strategy: Water-mediated one-pot synthesis
Keywords: Dimedone, aromatic aldehydes, catalyst-free, water, room temperature, Knoevenagel condensation, Michael addition, 2,2′-arylmethylene bis(3-hydroxy-5,5-dimethyl-2-cyclohexene-1-one) derivatives

2.2.11.1 *General Reaction Scheme*

Bayat *et al.*[41] developed an efficient protocol for the synthesis of 2,2′-arylmethylene bis(3-hydroxy-5,5-dimethyl-2-cyclohexene-1-one) derivatives (**3**) from the reaction of dimedone (**1**) with various aromatic aldehydes (**2**) in water at room temperature (Scheme 2.14). The workup procedure is simple and does not require any column chromatographic purification.

2.2.11.2 *Plausible Mechanism*

The authors proposed a plausible mechanism for this transformation (Scheme 2.15). One molecule of dimedone (**1**) undergoes enolizaton in water

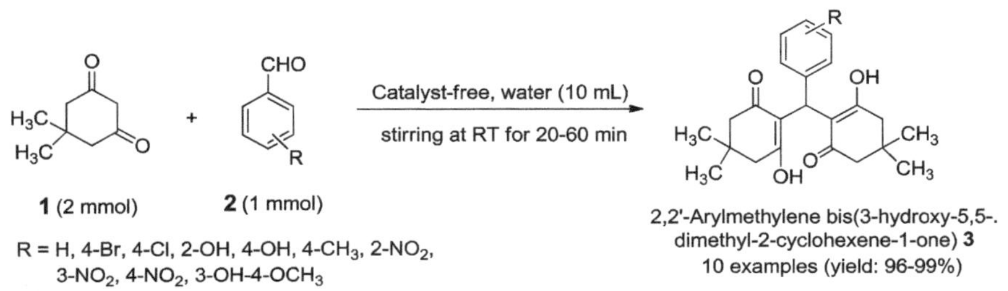

Scheme 2.14 Synthesis of 2,2′-arylmethylene bis(3-hydroxy-5,5-dimethyl-2-cyclohexene-1-one) derivatives (**3**).

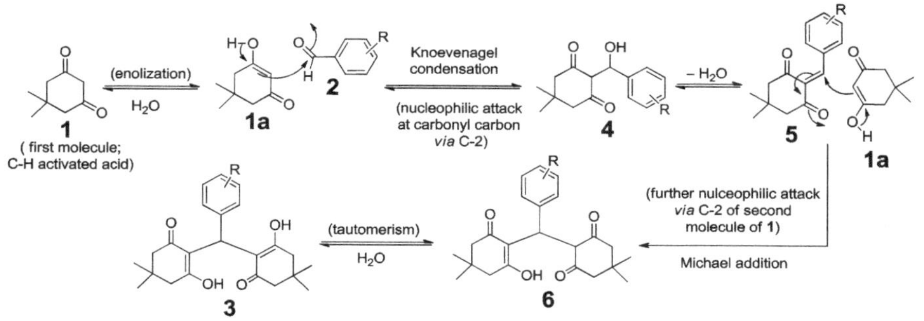

Scheme 2.15 Plausible mechanism of water-mediated synthesis of 2,2′-arylmethylene bis(3-hydroxy-5,5-dimethyl-2-cyclohexene-1-one) derivatives (**3**).

medium and the enolized species (**1a**) participates in Knoevenagel condensation with aldehyde (**2**) generating an intermediate (**5**). Enolized form of a second molecule of dimedone thereafter undergoes Michael addition with the intermediate (**5**) through nucleophilic attack, followed by tautomerization to afford the desired product (**3**).

2.2.11.3 Representative Examples

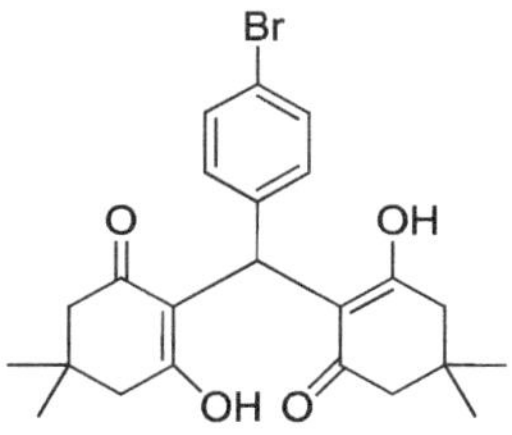

3a

2,2'-((4-Bromophenyl)methylene)bis(3-hydroxy-5,5-dimethylcyclohex-2-enone) (**3a**): solid, mp 161–163 °C; reaction time: 40 min; yield: 98%

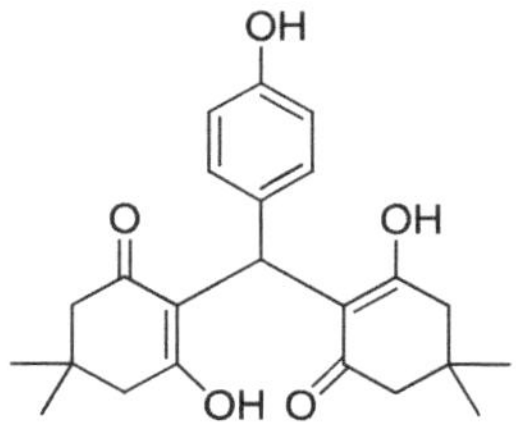

3b

2,2'-((4-Hydroxyphenyl)methylene)bis(3-hydroxy-5,5-dimethylcyclohex-2-enone) (**3b**): solid, mp 192–194 °C; reaction time: 60 min; yield: 98%

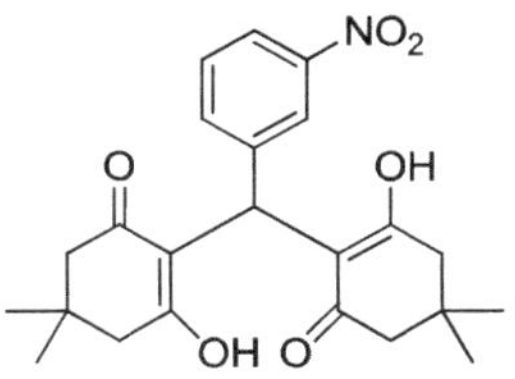

3c

2,2'-((3-Nitrophenyl)methylene)*bis*(3-hydroxy-5,5-dimethylcyclohex-2-enone) (**3c**):solid, mp 120–122 °C; reaction time: 60 min;yield: 98%

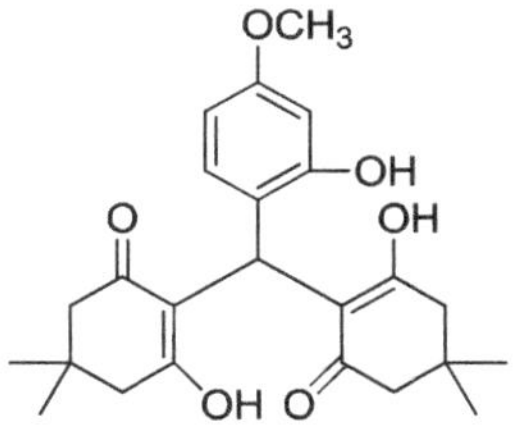

3d

2,2'-((2-Hydroxy-4-methoxyphenyl)methylene)*bis*(3-hydroxy-5,5-dimethylcyclohex-2-enone) (**3d**): solid, mp122–124 °C; reaction time: 25 min; yield: 99%

2.2.11.4 Experimental Procedure

To a stirred solution of dimedone (**1**; 2 mmol) in water (10 mL), aromatic aldehyde (**2**; 1 mmol) was added, all at the same time at room temperature. The reaction was completed within 60 min as monitored by TLC. Upon completion, solid products were precipitated out, filtered, washed with water, and dried to obtain pure products of (**3**). The products were identified based on elemental and spectral analyses.

2.2.11.5 Characterization Data of Two Representative Compounds

2,2′-((4-Bromophenyl)methylene)bis(3-hydroxy-5,5-dimethylcyclohex-2-enone) (**3a**): Solid, mp 161–163 °C; yield: 98%; ^{1}H NMR ($CDCl_3$, 300 MHz): δ 1.11 (6H, s, CMe_2), 1.22 (6H, s, CMe_2), 2.28–2.50 (8H, m, 4CH), 5.45 (1H, s, CH), 6.96 (2H, d, Ar–H), 7.39 (2H, d, Ar–H), 11.58 (1H, s, OH), 11.86 (1H, s, OH); ^{13}C NMR ($CDCl_3$, 75 MHz): δ 27.42 ($2\times CH_3$), 29.60 ($2\times CH_3$), 31.43 ($2\times CMe_2$), 32.47 (CH), 46.42 ($2\times CH_2$), 47.04 ($2\times CH_2$), 115.26 (2×C=C–OH), 119.65, 128.60, 131.29, 137.28 (aromatic carbons), 189.41 (2×C=COH), 190.65 (2×C=O). Anal. calc. for $C_{23}H_{27}BrO_4$: C, 61.75; H, 6.08%; found: C, 61.7; H, 6.1.

2,2′-((2-Hydroxy-4-methoxyphenyl)methylene)bis(3-hydroxy-5,5-dimethyl-cyclohex-2-enone) (**3d**): solid, mp 122–124 °C; yield: 99%; ^{1}H NMR ($CDCl_3$, 300 MHz): δ 1.09 (6H, s, CMe_2), 1.23 (6H, s, CMe_2), 2.27–2.48 (8H, m, 4×CH), 5.46 (1H, s, CH), 5.51 (1H, s, Ar–OH), 6.58 (2H, d, Ar), 6.66 (1H, s, Ar), 6.75 (2H, d, Ar), 11.52 (1H, s, OH), 11.95 (1H, s, OH); ^{13}C NMR ($CDCl_3$, 75 MHz): δ 27.52 ($2\times CH_3$), 29.57 ($2\times CH_3$), 31.46 ($2\times CMe_2$), 32.17 (CH), 46.43 ($2\times CH_2$), 47.09 ($2\times CH_2$), 55.90 (OMe), 110.31 (2×C=C–OH), 113.38, 115.69, 118.25, 131.25 (aromatic carbons), 144.61 (2×C=C–OH), 145.36 (2×C=O). Anal. calc. for $C_{24}H_{30}O_6$: C, 69.54; H, 7.30%; found: C, 69.6; H, 7.2.

2.2.11.6 Critical Views

There are several previous reports on the synthesis of 2,2′-arylmethylene bis(3-hydroxy-5,5-dimethyl-2-cyclohexene-1-one) compounds using a variety of catalysts such as SDS,[42] silica-perchlorate and PPA–SiO_2[43] in water or acetonitrile at reflux condition. Recently, these compounds have also been synthesized in the presence of molecular iodine,[44] solvent-free conditions under heating,[45] ultrasonic,[46] and microwave.[47,48] Hence, the present protocol surpasses all these earlier methods in terms of operational simplicity, avoidance of column chromatographic purification, organic solvents, catalysts and heating, rapid conversion and excellent yields, and obviously use of water as a green solvent and room temperature conditions.

2.2.12 Entry-12: Synthesis of Aryl/Alkyl/Heteroaryl-substituted bis(6-Amino-1,3-dimethyluracil-5-yl)methanes

Type of reaction: C–C bond formation
Reaction conditions: Catalyst-free, water, room temperature

Synthetic strategy: Water-mediated one-pot synthesis
Keywords: 6-Amino-1,3-dimethyluracil, aldehydes, catalyst-free, water, room temperature, aryl/alkyl/heteroaryl-substituted bis(6-amino-1,3-dimethyluracil-5-yl)methanes

2.2.12.1 General Reaction Scheme

Das and Thakur[49] described a highly efficient catalyst-free synthesis of alkyl/aryl/heteroaryl-substituted bis(6-amino-1,3-dimethyluracil-5-yl)methanes (**3**) from a condensation reaction of 6-amino-1,3-dimethyluracil (**2**) with different aldehydes (**1**) at room temperature using water as the reaction medium (Scheme 2.16).

2.2.12.2 Plausible Mechanism

The authors[49] proposed a plausible mechanism (Scheme 2.17) for the water-mediated conversion. Initially, one molecule of 6-amino-1,3-dimethyluracil (**2**) takes part in the nucleophilic attack to the aldehydic carbonyl carbon of **1** *via* its 5-position, followed by water elimination to produce the intermediate (**5**), which is then attacked by a second molecule of **2** through its nucleophilic 5-position to afford the desired product (**3**). The authors presumed

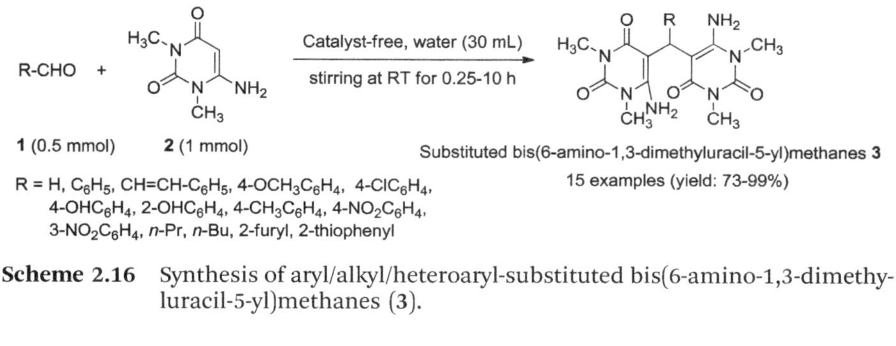

1 (0.5 mmol) **2** (1 mmol) Substituted bis(6-amino-1,3-dimethyluracil-5-yl)methanes **3**

R = H, C_6H_5, CH=CH-C_6H_5, 4-$OCH_3C_6H_4$, 4-ClC_6H_4, 4-OHC_6H_4, 2-OHC_6H_4, 4-$CH_3C_6H_4$, 4-$NO_2C_6H_4$, 3-$NO_2C_6H_4$, *n*-Pr, *n*-Bu, 2-furyl, 2-thiophenyl

15 examples (yield: 73-99%)

Scheme 2.16 Synthesis of aryl/alkyl/heteroaryl-substituted bis(6-amino-1,3-dimethyluracil-5-yl)methanes (**3**).

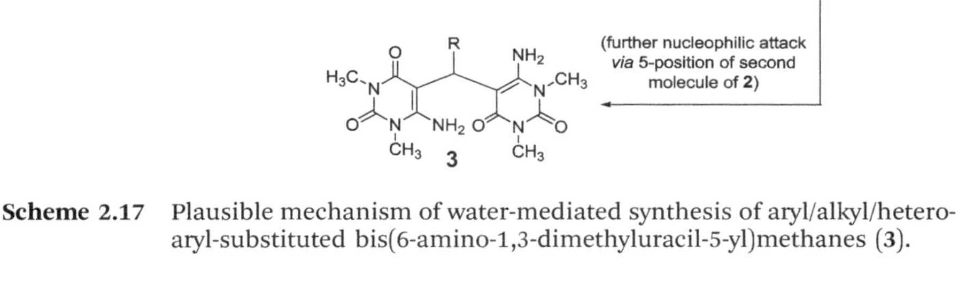

Scheme 2.17 Plausible mechanism of water-mediated synthesis of aryl/alkyl/heteroaryl-substituted bis(6-amino-1,3-dimethyluracil-5-yl)methanes (**3**).

that the product formed from the reaction between the water-soluble 6-amino-1,3-dimethyluracil and water-insoluble aldehyde, immediately precipitates owing to its hydrophobic nature, which might be the driving force for the reaction.[49]

2.2.12.3 Representative Examples

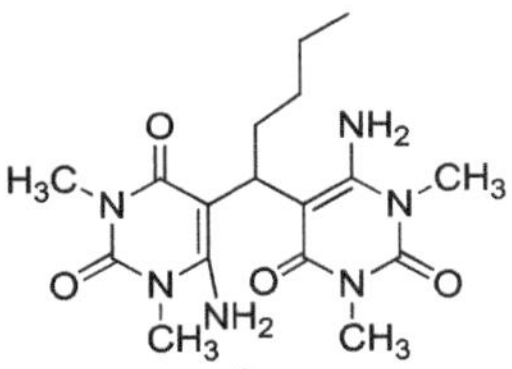

3a

5,5'-(Pentane-1,1-diyl)*bis*(6-amino-1,3-dimethylpyrimidine-2,4(1*H*,3*H*)-dione) (**3a**): white transparent crystalline solid, mp 159–162 °C; reaction time: 3 h; yield: 90%

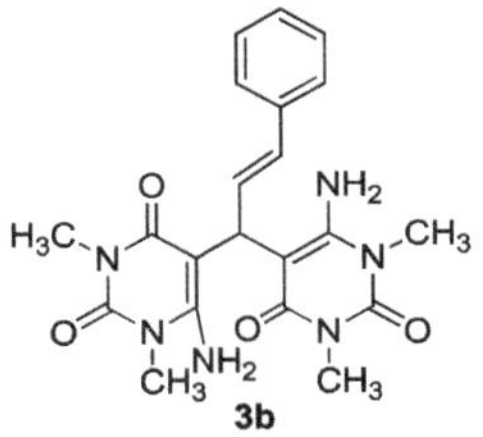

3b

(*E*)-5,5'-(3-Phenylprop-2-ene-1,1-diyl)*bis*(6-amino-1,3-dimethylpyrimidine-2,4(1*H*,3*H*)-dione) (**3b**) : yellowish crystalline solid, mp 265–269 °C; reaction time: 4h; yield: 93%

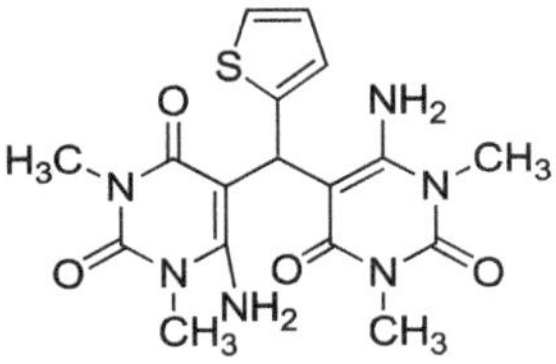

3c

5,5'-(Thiophen-2-ylmethylene)*bis*(6-amino-1,3-dimethylpyrimidine-2,4(1*H*,3*H*)-dione) (**3c**): transparent crystalline solid, mp 309–312 °C; reaction time: 6 h; yield: 87%

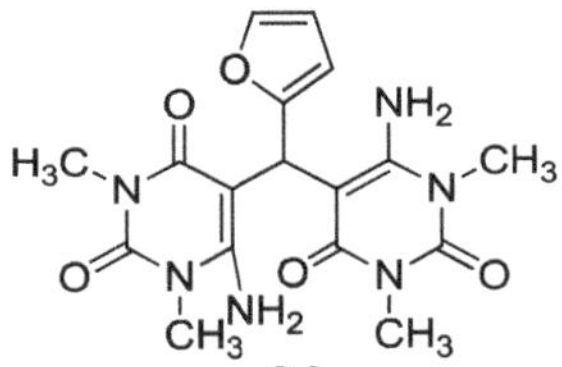

3d

5,5'-(furan-2-ylmethylene)*bis*(6-amino-1,3-dimethylpyrimidine-2,4(1*H*,3*H*)-dione) (**3d**): transparent crystalline solid, mp 246–250 °C; reaction time: 1h; yield: 99%

2.2.12.4 Experimental Procedure

30 mL of distilled water were added to 6-amino-1,3-dimethyluracil (**2**; 155 mg, 1 mmol) placed in a 100 mL round-bottomed flask, and the mixture was stirred at room temperature to dissolve it fully. Then aldehyde (**1**) (0.5 mmol) was added drop-wise to the 6-amino-1,3-dimethyluracil solution with constant stirring. The resulting reaction mixture was continued to stir at room temperature for stipulated time (0.25–10 h) to complete the reaction (TLC monitored). On completion of the reaction, crude product was obtained as solid precipitate, filtered, washed with water, and crystallized from hot ethanol to yield pure bis(6-amino-1,3-dimethyluracil-5-yl)methane derivatives (**3**). All the products were characterized on the basis of analytical and spectral studies.

2.2.12.5 Characterization Data of Two Representative Compounds

5,5′-(Pentane-1,1-diyl)bis(6-amino-1,3-dimethylpyrimidine-2,4(1*H*,3*H*)-dione) (**3a**): white transparent crystalline solid, mp 159–162 °C; yield: 90%; IR (KBr): 3396.53, 3106.74, 2952.08, 2859.75, 1661.83, 1606.21, 1490.77 cm^{-1}; ^{1}H NMR ($CDCl_3$, 400 MHz): δ 0.84–0.88 (3H, m, CH_3), 1.26–1.30 (4H, m, CH_2CH_2), 2.05–2.21 (2H, m, CH_2), 3.34 (3H, s, NCH_3), 3.36 (3H, s, NCH_3), 3.46 (3H, s, NCH_3), 3.47 (3H, s, NCH_3), 4.13–4.18 (1H, m, CH), 6.52 (2H, br s, NH_2), 6.94 (2H, br s, NH_2); ^{13}C NMR ($CDCl_3$, 100 MHz): δ 13.3 (CH_3), 21.8 (CH), 27.4 (CH_2), 27.9 (CH_2), 28.2 (NCH_3), 28.4 (NCH_3), 28.6 (NCH_3), 30.4 (NCH_3), 31.4 (CH_2), 88.2 and 89.2 (C-5 and C-5), 150.1 and 150.2 (C-6 and C′-6), 151.9 and 153.0 (C-2 and C′-2), 163.2 and 163.8 (C-4 and C′-4); MS: *m/z* 378.2 $[M]^{+}$. Anal. calc. for $C_{17}H_{26}N_6O_4$: C, 53.96; H, 6.93; N, 22.21; found C, 53.97; H, 6.93; N, 22.20.

5,5′-(Thiophen-2-ylmethylene)bis(6-amino-1,3-dimethylpyrimidine-2,4(1*H*, 3*H*)-dione) (**3c**): transparent crystalline solid, mp 309–312 °C; yield: 87%; IR (KBr): 3361.56, 3327.33, 3057.73, 2982.55, 1681.87, 1632.89, 1573.09 cm^{-1}; ^{1}H NMR ($CDCl_3$, 400 MHz): δ 3.31 (1H, s, NCH_3), 3.33 (1H, s, NCH_3), 3.42 (1H, s, NCH_3), 3.45 (1H, s, NCH_3), 5.93 (1H, s, CH), 6.54 (2H, br s, NH_2), 6.68 (1H, thiophene), 6.87 (1H, thiophene), 6.91 (2H, br s, NH_2), 7.13 (1H, thiophene); ^{13}C NMR ($CDCl_3$, 100 MHz): δ 29.32 (NCH_3), 29.38 (NCH_3), 29.41 (NCH_3), 29.44 (NCH_3), 33.3 (CH), 89.0 and 89.3 (C-5 and C′-5), 123.4 and 124.0 (C-6 and C′-6), 126.3 (C-3, thiophene), 144.4 (C-4, thiophene), 150.9 (C-5, thiophene), 153.2 and 153.9 (C-2 and C′-2), 163.5 and 164.5 (C-4 and C′-4); MS: *m/z* 404.13 $[M]^{+}$. Anal. calc. for $C_{17}H_{20}N_6O_4S$: C, 50.48; H, 4.98; N, 20.78; found C, 50.46; H, 4.99; N, 20.77.

2.2.12.6 Critical Views

The investigators reported the first catalyst-free synthesis of bis-uracil derivatives in water.[49] The present protocol seems to be efficient and eco-friendly in terms of avoidance of organic solvents, catalyst, and heating. Mild reaction conditions, easy operation, high yields of products, no use of catalyst and organic solvents make this water-mediated strategy more advantageous over the existing methods.[50–55] The investigators observed that this method does not work with ketones, thereby indicating its chemoselective nature with aldehydes only. In addition, the investigators extended their protocol to synthesize tetrakis-uracil adducts—it was evidenced that such an adduct is obtained in case of an aliphatic dicarbaldehyde (glutaraldehyde), but in case of *p*-benzenedicarbaldehyde (terephthaldehyde) bis-adduct is formed instead of tetrakis-adduct.[49]

2.2.13 Entry-13: Synthesis of α-(Acyloxy)-α-(quinolin-4-yl) acetamides

Type of reaction: C–C and C–O bond formation
Reaction conditions: Catalyst-free, water, room temperature

Synthetic strategy: One-pot multicomponent reaction
Keywords: Isocyanides, quinoline-4-carbaldehyde, arylcarboxylic acids, catalyst-free, water, room temperature, one-pot multicomponent synthesis, α-(acyloxy)-α-(quinolin-4-yl)acetamides

2.2.13.1 General Reaction Scheme

Taran *et al.*[56] developed an efficient one-pot three-component synthesis of novel α-(acyloxy)-α-(quinolin-4-yl)acetamides (**4**) from the reaction of isocyanides (**1**), quinoline-4-carbaldehyde (**2**), and arylcarboxylic acids (**3**) at room temperature using water as the reaction medium (Scheme 2.18).

2.2.13.2 Plausible Mechanism

A possible mechanism for the formation of product (**4**) was suggested by the authors (Scheme 2.19).[56] Initially, protonation occurs at the aldehydic carbonyl oxygen by the carboxylic acid (**3**) to generate an electron-deficient species (**2a**), which is then attacked by isocyanide (**1**) through its electron-rich

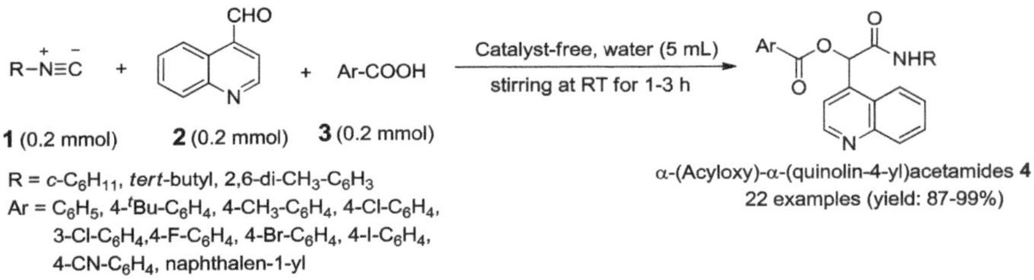

Scheme 2.18 Synthesis of α-(acyloxy)-α-(quinolin-4-yl)acetamides (**4**).

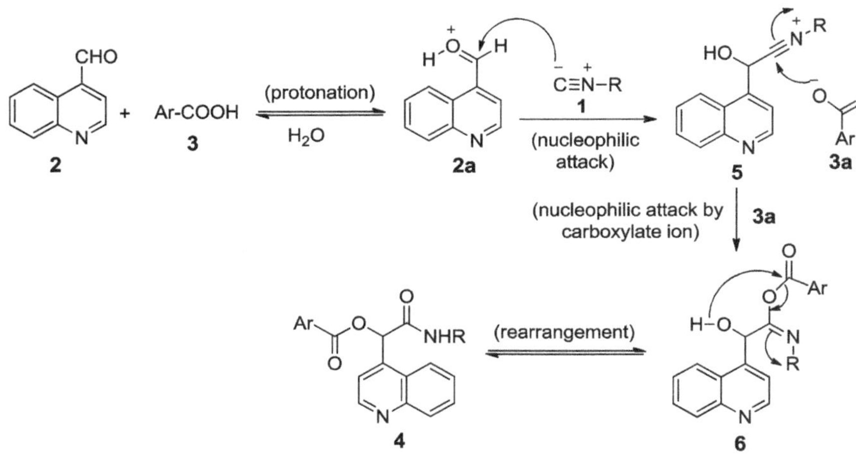

Scheme 2.19 Plausible mechanism for the synthesis of αα-(acyloxy)-α-(quinolin-4-yl)acetamides (**4**).

carbon centre to form intermediate (**5**). The carboxylate ion (**3a**) adds to the intermediate (**5**) in the next step *via* nucleophilic attack leading to the formation of another intermediate (**6**) that undergoes a rearrangement reaction to give ultimately the desired product (**4**).

2.2.13.3 Representative Examples

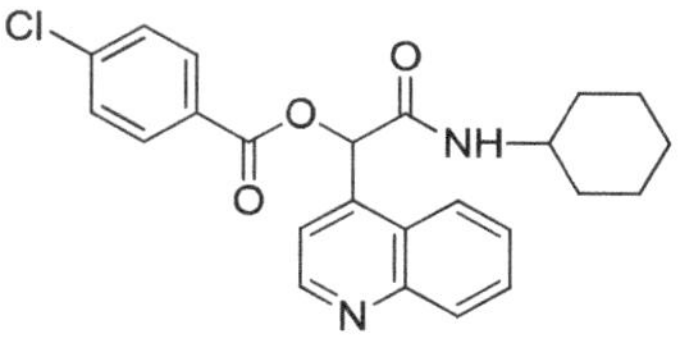

4a

2-(Cyclohexylamino)-2-oxo-1-(quinolin-4-yl)ethyl 4-chlorobenzoate (**4a**): colorless crystals, mp 187–189 °C; reaction time: 1 h; yield: 97%

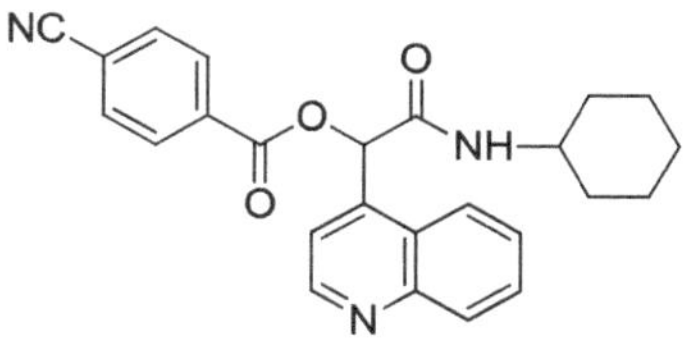

4b

2-(Cyclohexylamino)-2-oxo-1-(quinolin-4-yl)ethyl 4-cyanobenzoate (**4b**): yellow crystals, mp 189–190 °C; reaction time: 2 h; yield: 97%

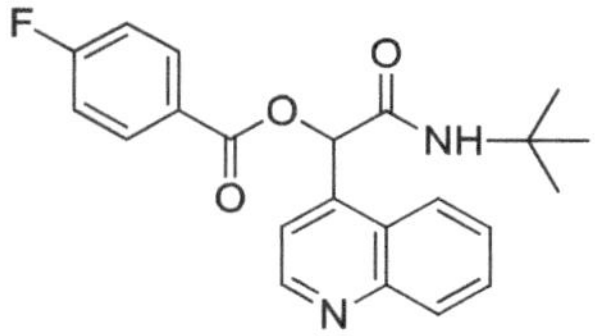

4c

2-(*tert*-Butylamino)-2-oxo-1-(quinolin-4-yl)ethyl 4-fluorobenzoate (**4c**): colorless crystals, mp 178–180 °C; reaction time: 1 h; yield: 98%

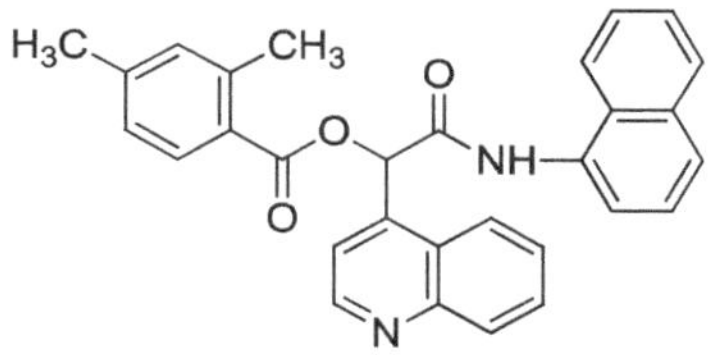

4d

2-(Naphthalen-1-ylamino)-2-oxo-1-(quinolin-4-yl)ethyl 2,4-dimethylbenzoate (**4d**): yellow crystals, mp 212–214 °C; reaction time: 3 h; yield: 94%

2.2.13.4 Experimental Procedure

To a stirred solution of quinoline-4-carbaldehyde (**2**; 0.2 mmol) and aryl carboxylic acids (**3**; 0.2 mmol) in water (5 mL), isocyanides (**1**; 0.2 mmol) were added at room temperature over 5 min. The mixture then continued to be stirred at room temperature for a specified time (1–3 h) to complete the reaction (monitored by TLC). Water was removed under reduced pressure, and the products were purified by TLC and then crystallized to obtain pure α-(acyloxy)-α-(quinolin-4-yl)acetamides (**4**) in excellent yields. Each of the products was characterized by detailed spectral studies.

2.2.13.5 Characterization Data of Two Representative Compounds

2-(Cyclohexylamino)-2-oxo-1-(quinolin-4-yl)ethyl 4-chlorobenzoate (**4a**): colorless crystals, mp 187–189 °C; yield: 97%; IR (KBr): 3445, 3271, 2932, 2851, 1734, 1700, 1569, 1253, 1095 cm^{-1}; ^{1}H NMR ($CDCl_3$, 400 MHz): δ 1.13–1.69

(10 H, m, 5×CH_2), 3.81–3.90 (1H, m, –N–CH), 6.116 (1H, d, $J = 8$ Hz, NH), 6.968 (1H, s, CH–O), 7.29–8.98 (10H, m, Ar– H); ^{13}C NMR ($CDCl_3$, 100 MHz): δ 24.66–32.84 (5×CH_2), 48.75 (NCH), 72.94 (OCH), 120.31, 123.89, 127.61, 129.16, 130.13, 131.26, 149.81, 126.10, 127.20, 140.57, 141.14, 148.43, 164.19 (ester C=O), 165.94 (amide C=O).

2-(Naphthalen-1-ylamino)-2-oxo-1-(quinolin-4-yl)ethyl 2,4-dimethylbenzoate (**4d**): Yellow crystals, mp 212–214 °C; yield: 94%; IR (KBr): 3435, 3260, 3030, 2922, 1712, 1669, 1591, 1243, 1164 cm^{-1}; ^{1}H NMR ($CDCl_3$, 400 MHz): δ 2.12 (6H, s, 2×CH_3), 7.35 (1H, s, CH–O), 7.55 (1H, s, NH), 7.02–9.05 (16H, m, Ar–H) ^{13}C NMR ($CDCl_3$, 100 MHz): δ 18.33, 73.10, 119.84, 124.17, 124.54, 125.46, 126.70, 127.74, 127.84, 128.36, 128.47, 128.80, 130.09, 130.81, 134.71, 135.42, 149.87, 125.29, 126.00, 130.18, 130.49, 132.38, 133.91, 141.27, 148.50, 165.63 (ester C=O), 165.72 (amide C=O).

2.2.13.6 Critical Views

The investigators reported on a new version of water-mediated Passerini reaction under catalyst-free conditions at room temperature. This one-pot three-component protocol has been found highly efficient for the synthesis of a variety of novel α-(acyloxy)-α-quinolin-4-yl)acetamide derivatives with good to excellent yields. Operational simplicity, mild reaction conditions, ease of isolation of products, cleaner reaction profiles, and excellent yields are the key advantages of this method.[56]

2.2.14 Entry-14: Synthesis of Endothiopeptides

Type of reaction: C–N and C–S bond formation
Reaction conditions: Catalyst-free, ethanol or methanol, room temperature
Synthetic strategy: Sequential multicomponent one-pot synthesis
Keywords: Carbonyl compounds, amines, thioacids, alcohol, isonitriles, room temperature, one-pot multicomponent reaction, Ugi reaction, endothiopeptides

2.2.14.1 General Reaction Scheme

Kazmaier and Ackermann[57] successfully utilized Ugi reaction for the easy access of endothiopeptide compounds (**5**) from a sequential multicomponent one-pot reaction of thioacids (**1**), amines (**2**), ketones (**3**) and isonitriles (**4**) in alcohol under ambient conditions (Scheme 2.20).

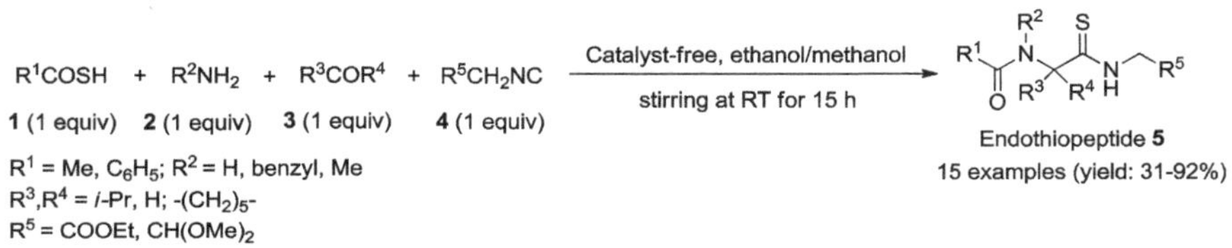

Scheme 2.20 Synthesis of endothiopeptides (**5**).

2.2.14.2 Representative Examples

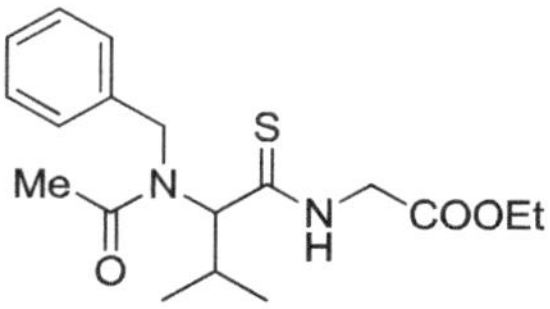

5a

Ethyl 2-(2-(*N*-benzylacetamido)-3-methylbutanethioamido)acetate (**5a**): yellow oil; yield: 68%

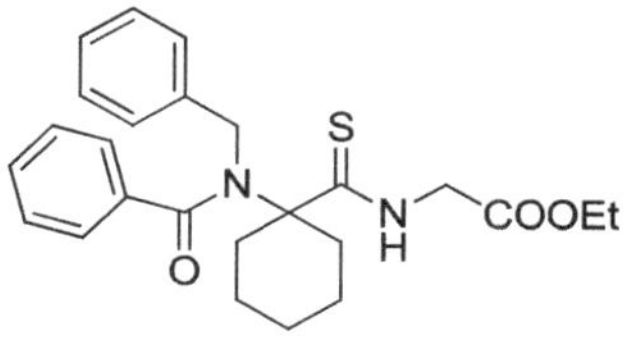

5b

Ethyl 2-(1-(*N*-benzylbenzamido) cyclohexanecarbothioamido)acetate (5b): yield: 55%

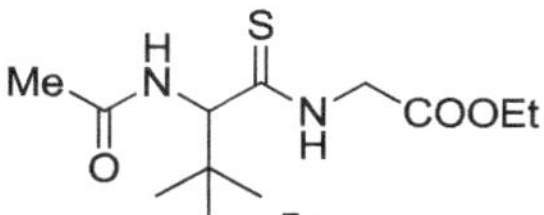

5c

Ethyl 2-(2-acetamido-3,3-dimethylbutanethioamido) acetate (**5c**): yield: 31%

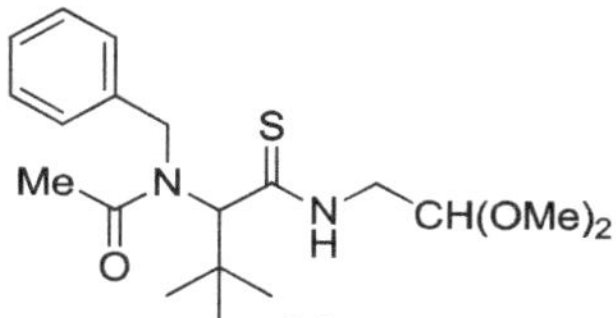

5d

N-Benzyl-*N*-(1-((2,2-dimethoxyethyl)amino)-3,3-dimethyl-1-thioxobutan-2-yl)acetamide (**5d**): white rhombic crystals, mp 110–112 °C; yield: 81%

2.2.14.3 Experimental Procedure

Amine (**2**; 1 equiv.) and carbonyl compound (**3**: 1 equiv.) were dissolved in alcohol (ethanol or methanol) and stirred for 15 min at room temperature. Then the reaction mixture was cooled to 0 °C and the thioacid (**1**; 1 equiv.) was added followed by the isonitrile (**4**) (1 equiv.). The ice bath was removed and the mixture was stirred overnight at room temperature. Upon completion of the reaction, dichloromethane was added and the organic layer was washed twice with saturated sodium carbonate solution and 1 M $KHSO_4$ solution. After drying the organic layer over sodium sulfate, the solvent was removed in *vacuo* and the crude product was purified by column chromatography and/or recrystallization. All the products were characterized on the basis of elemental analyses and detailed spectral studies.

2.2.14.4 Characterization Data of Two Representative Compounds

Ethyl 2-(2-(*N*-benzylacetamido)-3-methylbutanethioamido)acetate (**5a**): yellow oil; ^{1}H NMR ($CDCl_3$, 500 MHz): δ 0.67 (3H, d, $J = 6.1$ Hz), 0.87 (3H, d, $J = 6.7$ Hz), 1.25 (3H, t, $J = 7.0$ Hz), 2.10 (3H, s), 2.81 (1H, br s), 4.18 (2H, q, $J = 7.0$ Hz,), 4.29 (2H, dd, $J = 13.1$, 5.2 Hz), 4.35 (1H, d, $J_{4,5} = 4.9$ Hz), 4.57 (2H, s),

7.16–7.26 (5H, m), 7.97 (1H, br s); ^{13}C NMR ($CDCl_3$,125 MHz): δ 14.1, 20.3, 28.3, 47.2, 61.6, 125.9, 127.5, 128.7, 138.7, 168.6, 173.8, 202.6; HMRS (CI): calc. for $C_{18}H_{26}N_2O_3S$ ($[M]^+$), 350.1664; found: 350.1653.

N-Benzyl-*N*-(1-((2,2-dimethoxyethyl)amino)-3,3-dimethyl-1-thioxobutan-2-yl)acetamide (**5d**): white rhombic crystals, mp 110–112 °C; ^{1}H NMR ($CDCl_3$, 500 MHz): δ =1.06 (9H, s), 2.13 (3H, s), 3.33 (6H, s), 3.70 (2H, m), 4.23–4.54 (4H, m), 7.19–7.26 (5H, m), 10.82 (1H, br s); ^{13}C NMR ($CDCl_3$,125 MHz): δ 23.4, 28.5, 38.2, 54.3, 54.7, 102.2, 129.6, 176.5, 201.5; HRMS (CI): calc. for $C_{19}H_{30}N_2O_3S$ ($[M]^+$), 366.1978; found, 366.1978. Elemental analysis: calc. for $C_{19}H_{30}N_2O_3S$: C, 62.61; H, 7.74; N, 7.69; found: C, 62.10; H, 7.49; N, 7.57.

2.2.14.5 Critical Views

Thioamide unit is found in numerous pharmaceutically potent natural and synthetic molecules. In this report, the investigators developed an energy-efficient and easy protocol to access a series of endothiopeptides in one step.[57] The method offers an option of combinatorial synthesis. In addition, the investigators showed the extension of the present method in synthesizing peptidic thiazoles by using suitable isonitriles, this was, however, under microwave irradiation.

2.2.15 Entry-15: Synthesis of *N*-(*Z*-Alkenyl)imidazole-2-carbothioamides

Type of reaction: C–C and C–N bond formation
Reaction conditions: Catalyst-free, neat, room temperature
Synthetic strategy: One-pot multicomponent nucleophilic addition
Keywords: 1-Substituted imidazoles, isothiocyanates, cyanophenylacetylene, catalyst-free, solvent-free, room temperature, multicomponent reaction, *N*-(*Z*-alkenyl)imidazole-2-carbothioamides, stereoselectivity

2.2.15.1 General Reaction Scheme

Trofimov and coworkers[58] developed a solvent- and catalyst-free room temperature protocol for the stereoselective synthesis of *N*-(*Z*)-(2-cyano-1-phenylethenyl)imidazole-2-carbothioamides (**4**) from one-pot three-component reaction of 1-substituted imidazoles (**1**), isothiocyanates (**2**) and cyanophenylacetylene (**3**) (Scheme 2.21).

2.2.15.2 Plausible Mechanism

The investigators proposed a plausible mechanism for this transformation (Scheme 2.22). Initially, a zwitterionic intermediate (**A**) is formed from the nucleophilic attack of imidazolium nitrogen of (**1**) at the acetylenyl carbon of (**3**) with *E*-configuration. This, in turn, undergoes

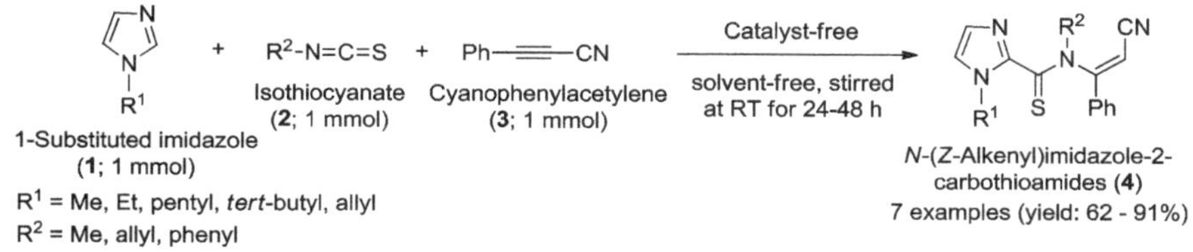

Scheme 2.21 Synthesis of *N*-(*Z*-alkenyl)imidazole-2-carbothioamides (**4**).

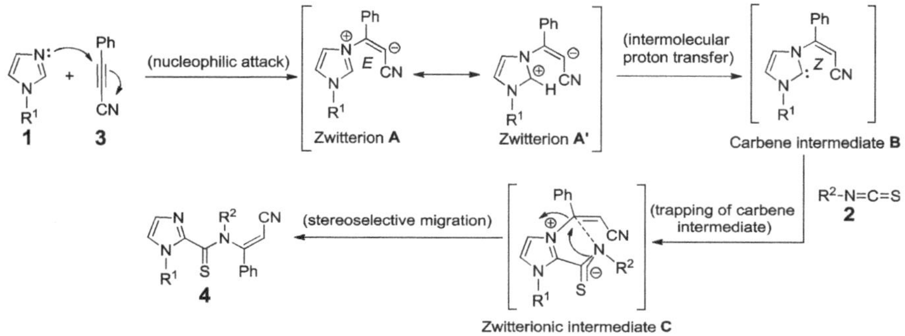

Scheme 2.22 Plausible mechanism for the formation of *N*-(*Z*-alkenyl)imidazole-2-carbothioamides (**4**).

energetically feasible intermolecular proton transfer (intramolecular process is forbidden here due to the adverse configuration of the carbanioic moiety in the zwitterion **A**) to give a carbene intermediate (**B**) with *Z*-configuration, which is then trapped by isothiocyanate (**2**) as a competing electrophile to generate another zwitterion (**C**) having N,S-ambident anionic site. The authors proposed that a stereoselective migration of the (*Z*)-cyanophenylethenyl group onto the nitrogen anionic site then takes place regioselectively to furnish the desired product (**4**) in a concerted manner, since *Z*-configuration of the migrating group is retained in the product.[58]

2.2.15.3 Representative Examples

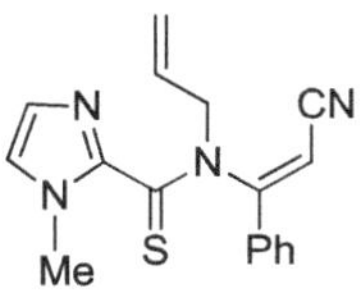

4a

(*Z*)-*N*-Allyl-*N*-(2-cyano-1-phenylvinyl)-1-methyl-1*H*-imidazole-2-carbothioamide (**4a**): dark yellow oil; reaction time: 24 h; yield: 81%

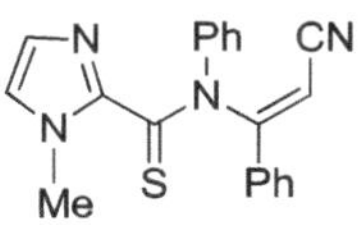

4b

(*Z*)-*N*-(2-Cyano-1-phenylvinyl)-1-methyl-*N*-phenyl-1*H*-imidazole-2-carbothioamide (**4b**): yellow powder, mp167–168 °C; reaction time: 48h; yield: 79%

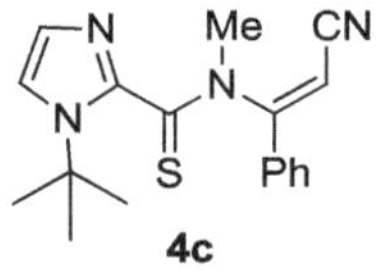

4c

(*Z*)-1-(*tert*-Butyl)-*N*-(2-cyano-1-phenylvinyl)-*N*-methyl-1*H*-imidazole-2-carbothioamide (**4c**): orange oil; reaction time: 24 h; yield: 74%

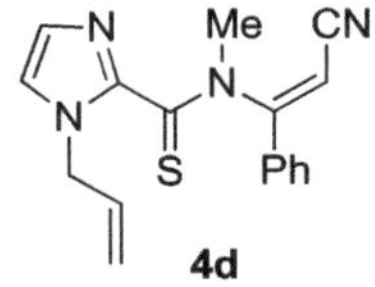

4d

(*Z*)-1-Allyl-*N*-(2-cyano-1-phenylvinyl)-*N*-methyl-1*H*-imidazole-2-carbothioamide (**4d**): orange oil; reaction time: 24 h; yield: 67%

2.2.15.4 Experimental Procedure

To a mixture of isothiocyanate (**2**; 1 mmol) and cyanophenylacetylene (**3**; 1 mmol) was added 1-substituted imidazole (**1**; 1 mmol). The mixture was stirred at room temperature for 24 h (stirring for 48 h is required when arylisothiocyanates are used). After then the reaction mixture was washed with diethyl ether (1×3 mL) to afford *N*-(*Z*-alkenyl)imidazole-2-carbothioamide (**4**) in moderate to good yields (62–91%). All the products were characterized by means of elemental analyses and spectral studies including IR, ^{1}H NMR, and ^{13}C NMR.

2.2.15.5 Characterization Data of Two Representative Compounds

(*Z*)-*N*-Allyl-*N*-(2-cyano-1-phenylvinyl)-1-methyl-1*H*-imidazole-2-carbothioamide (**4a**): Dark yellow oil; IR (microlayer): 2216 (CN), 1607 (C=C), 1382 (C=S) cm^{-1}; ^{1}H NMR ($CDCl_3$, 400.13 MHz): δ 7.60–7.40 (5H, m, C_8-Ph), 6.83 (2H, s, 4-H, 5-H), 6.06 (1H, m, H_X, N_7-allyl]), 5.35 (1H, s, 9-H), 5.21 (1H, d, $^3J_{HA,HX} = 9.2$ Hz, H_A, N_7-allyl), 5.09 (1H, d, $^3J_{HB,HX} = 16.8$ Hz, H_B, N_7-allyl), 4.05 (2H, m, $-CH_2-$, N_7-allyl), 3.88 (3H, s, N_1–CH_3); ^{13}C NMR ($CDCl_3$, 100.62 MHz): δ 189.3 (C-6), 161.9 (C-8), 144.6 (C-2), 137.6 (CH_X, allyl), 133.5 (C_i, C_8-Ph), 131.4 (C_p, C_8-Ph), 129.0 (C_m, C_8-Ph), 127.6 (C-4, C_o, C_8-Ph), 125.1 (C-5), 120.7 ($CH_{A,B}$, allyl), 115.8 (C-10, CN), 92.7 (C-9), 55.3 (N_7–CH_2, allyl), 35.9 (N_1–CH_3). Anal. calc. for $C_{17}H_{16}N_4S$: C, 66.21; H, 5.23; N, 18.17; S, 10.40; found C, 66.59; H, 5.09; N, 18.55; S, 10.01.

(*Z*)-1-Allyl-*N*-(2-cyano-1-phenylvinyl)-*N*-methyl-1*H*-imidazole-2-carbothioamide (**4d**): Orange oil; IR (microlayer): 2217 (CN), 1609 (C=C), 1358 (C=S) cm^{-1}; ^{1}H NMR ($CDCl_3$, 400.13 MHz): δ 7.50–7.30 (5H, m, C_8-Ph), 6.85 (2H, s, 4-H, 5-H), 5.88 (1H, m, H_X, N_1-allyl), 5.38 (1H, s, 9-H), 5.22 (2H, m, H_A, H_B, N_1-allyl), 4.80 (2H, m, CH_2, N_1-allyl), 3.64 (3H, s, N_7–CH_3); ^{13}C NMR ($CDCl_3$, 100.62 MHz): δ 189.2 (C-6), 163.8 (C-8), 144.4 (C-2), 136.0 (C_i, C_8-Ph), 132.8 [CH_X, N_1-allyl], 131.3 (C_p, C_8-Ph), 128.9 (C_m, C_8-Ph), 127.6 (C_o, C_8-Ph), 127.4 (C-4), 122.4 (C-5), 119.1 (*C*$H_{A,B}$, N_1-allyl), 115.6 (C-10, *C*N), 92.0 (C-9), 50.7 (CH_2, N_1-allyl], 42.8 (N_7–CH_3). Anal. calc. for $C_{17}H_{16}N_4S$: C, 66.21; H, 5.23; N, 18.17; S, 10.40; found C, 66.59; H, 5.49; N, 17.79; S, 10.01.

2.2.15.6 Critical Views

The present method offers an access to a new class of stereoselective imidazole-*N*-alkenylcarbothioamide compounds of pharmaceutical interest through a noncatalytic and solvent-free one-pot multicomponent reaction under mild conditions.

2.2.16 Entry-16: Synthesis of Spirooxindole-pyrazolines

Type of reaction: C–C and C–N bond formation
Reaction conditions: Catalyst-free, ethanol, room temperature
Synthetic strategy: One-pot multicomponent reaction
Keywords: Isatin, hydrazine hydrate, 1,1-bis(methylthio)-2-nitroethylene, ethanol, catalyst-free, room temperature, one-pot multicomponent reaction, spirooxindole-pyrazolines

2.2.16.1 General Reaction Scheme

Alizadeh and Zohreh[59] reported a pseudo-five-component synthesis of spirooxindole-pyrazolines (**4**) *via* a one-pot and catalyst-free reaction between isatins (**1**), 1,1-bis(thiomethyl)-2-nitroethylene (**2**) and hydrazine hydrate (**3**) in ethanol at room temperature (Scheme 2.23).

2.2.16.2 Plausible Mechanism

A plausible mechanism for this conversion was suggested by the investigators (Scheme 2.24). Initially, 1,1-bishydrazino-2-nitroethylene (**5**) is formed out of a condensation reaction of two equivalents of hydrazine to 1,1-bis-(methylthio)-2-nitroethylene (**2**) with the loss of two equivalents of methanethiol. This *in situ* generated ketene aminal (**5**) then attacks ketonic carbonyl of isatin through nucleophilic addition *via* both the NH_2 groups followed by loss of two equivalents of water molecules affording bis-hydrazone intermediate (**6**). Finally, intermediate (**6**) undergoes an intramolecular cyclization to furnish the spirooxindole-pyrazoline (**4**).

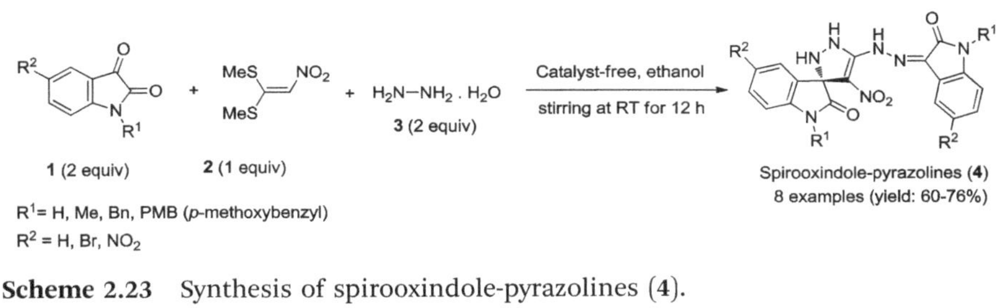

Scheme 2.23 Synthesis of spirooxindole-pyrazolines (**4**).

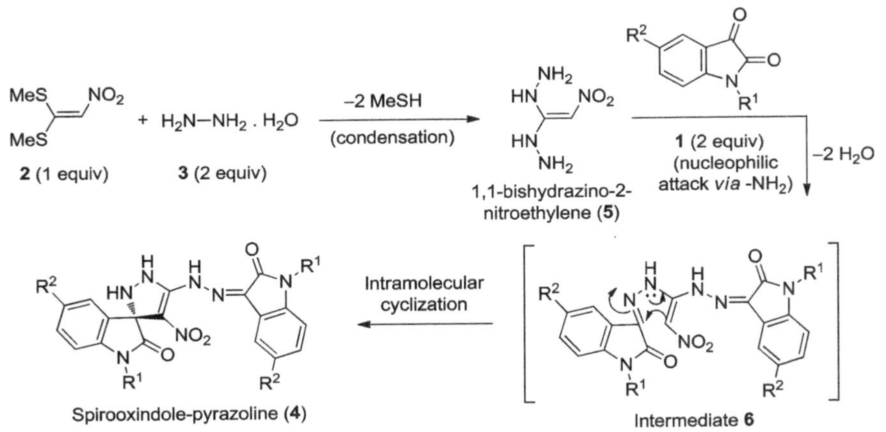

Scheme 2.24 Plausible mechanism for the generation of spirooxindole-pyrazolines (**4**).

2.2.16.3 Representative Examples

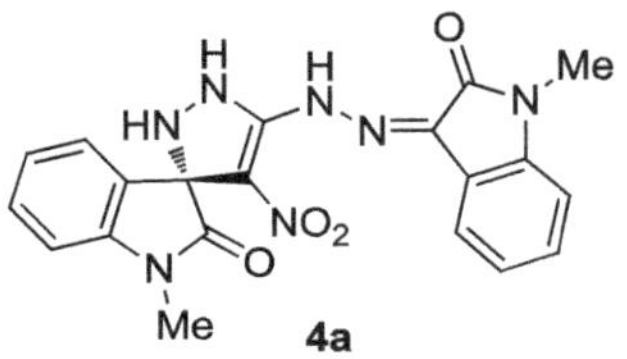

4a

(*R*,*Z*)-1-Methyl-5'-(2-(1-methyl-2-oxoindolin-3-ylidene)hydrazinyl)-4'-nitro-1',2'-dihydrospiro[indoline-3,3'-pyrazol]-2-one (**4a**): orange powder; mp: >300 °C (dec.); yield: 76%

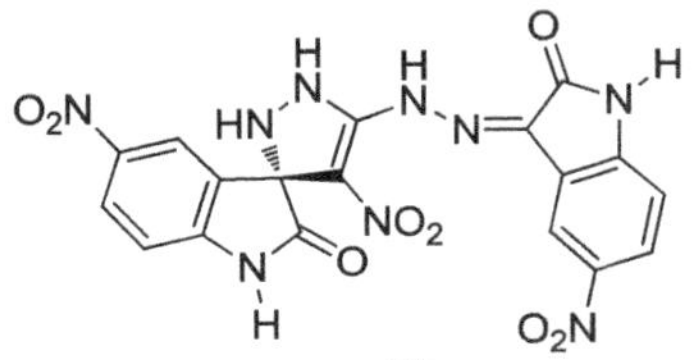

4b

(*R*,*Z*)-4',5-Dinitro-5'-(2-(5-nitro-2-oxoindolin-3-ylidene)hydrazinyl)-1',2'-dihydrospiro[indoline-3,3'-pyrazol]-2-one (**4b**): mustard powder; mp: up 300°C; yield: 64%

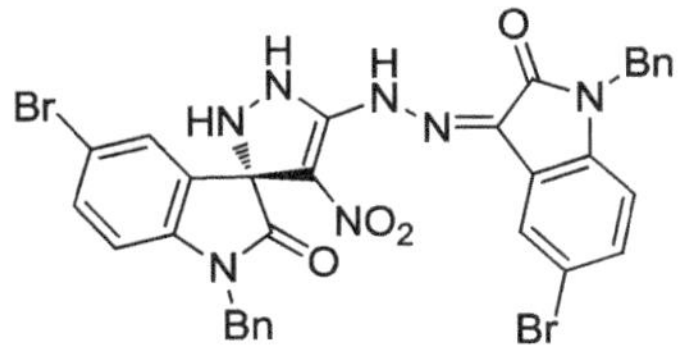

4c

(*R*,*Z*)-1-Benzyl-5'-(2-(1-benzyl-5-bromo-2-oxoindolin-3-ylidene)hydrazinyl)-5-bromo-4'-nitro-1',2'-dihydrospiro[indoline-3,3'-pyrazol]-2-one (**4c**): orange powder; mp: 220 °C; yield: 60%

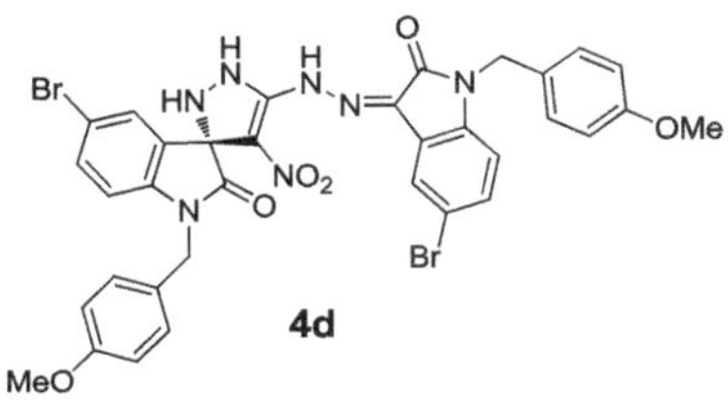

4d

(*R*,*Z*)-5-Bromo-5'-(2-(5-bromo-1-(4-methoxybenzyl)-2-oxoindolin-3-ylidene)hydrazinyl)-1-(4-methoxybenzyl)-4'-nitro-1',2'-dihydrospiro[indoline-3,3'-pyrazol]-2-one (**4d**): dark brown powder; mp: 163 °C; yield: 63%

2.2.16.4 Experimental Procedure

To a solution of 1,1-bis(thiomethyl)-2-nitroethylene (**2**; 1 mmol) in ethanol (4 mL) in a 10 mL flask was added hydrazine hydrate (**3**; 80% aq., 2 mmol) and stirred for 5 h at room temperature. After then, isatins (**1**; 2 mmol) was added to the reaction mixture, and stirring was continued for another 3–7 h. On completion of the reaction, the crude precipitate was filtered and washed with cold ethanol to obtain pure product of spirooxindole-pyrazoline (**4**). All the products were characterized by spectral studies.

2.2.16.5 Characterization Data of Two Representative Compounds

(*R*,*Z*)-1-Benzyl-5′-(2-(1-benzyl-5-bromo-2-oxoindolin-3-ylidene)hydrazinyl)-5-bromo-4′-nitro-1′,2′-dihydrospiro[indoline-3,3′-pyrazol]-2-one (**4c**): orange powder; mp: 220 °C; IR (KBr): 3437, 1721, 1600, 1562, 1342 cm^{-1}; ^1H NMR (DMSO-d_6, 500 MHz): δ 4.97 (4H, d, $^3J_{HH} = 14.6$ Hz, 2×CH_2), 6.27 (2H, s, 2×NH), 6.93–6.96 (2H, m, 2×CH of Ar), 7.27–7.34 (10H, m, 10×CH of Ar), 7.53 (3H, br s, 3×CH of Ar), 8.35 (1H, br s, CH of Ar), 13.41 (1H, s, NH); ^{13}C NMR (DMSO-d_6, 125.7 MHz,): δ 42.7 (CH_2), 42.8 (CH_2), 74.1 (C_{spiro}), 111.9 (C), 112.6 (C), 114.5 (CH), 115.0 (CH), 117.6 (C), 121.1 (C), 122.7 (C), 127.2 (CH), 127.3 (CH), 127.5 (CH), 127.6 (CH), 128.6 (C), 128.7 (2 CH), 130.9 (CH), 133.7 (C), 133.9 (C), 135.3 (CH), 135.7 (CH), 135.9 (CH), 141.8 (C), 144.2 (C), 148.6 (C), 160.5 (C=O), 162.6 (C=O); MS: *m/z* (% rel) 331 (7), 329 (7), 317 (7), 315 (7), 258 (5), 256 (5), 224 (13), 170 (12), 168 (12), 91 (100), 65 (33).

(*R*,*Z*)-5-Bromo-5′-(2-(5-bromo-1-(4-methoxybenzyl)-2-oxoindolin-3-ylidene)-hydrazinyl)-1-(4-methoxybenzyl)-4′-nitro-1′,2′-dihydrospiro[indoline-3,3′-pyrazol]-2-one (**4d**): dark brown powder; mp: 163 °C; IR (KBr): 3308, 3203, 1719, 1604, 1557, 1348 cm^{-1}; ^1H NMR (DMSO-d_6, 500 MHz): δ 2.21 (6H, s, 2×Me), 4.88 (4H, br s, 2×CH_2), 6.23–7.47 (14H, m, 14×CH of Ar), 8.31 (1H, s, NH), 10.17 (1H, s, NH); ^{13}C NMR (DMSO-d_6, 125.7 MHz,): because of insolubility, only some of the resonances are distinguishable including: δ 20.6 (2 Me), 74.1 (C_{spiro}), 127.2 (2 CH of Ar), 127.3 (2 CH of Ar), 129.2 (4 CH of Ar). MS: *m/z* (%rel) 410 (5), 368 (7), 341 (8), 279 (7), 236 (11), 213 (6), 173 (18), 149 (42), 127 (38), 105 (25), 97 (70), 83 (74), 69 (100), 57 (97).

2.2.16.6 Critical Views

The present method offers an operationally simple and catalyst-free one-pot multicomponent reaction for the synthesis of polynitrogeneous spiro-oxindole-pyrazolines of potential synthetic and pharmacological interest under mild reaction conditions. The three other previously reported procedures for the synthesis of such compounds require at least two steps starting from the condensation of isatins and a reactive methylene nucleophile under microwave irradiation or reflux conditions using also catalysts in certain occasions.[60–62] Good yields and easy purification procedures are the added advantages of this present approach.[59]

2.2.17 Entry-17: Synthesis of γ-Aminoethers

Type of reaction: C–N and C–O bond formation
Reaction conditions: Catalyst-free, aliphatic alcohol, room temperature
Synthetic strategy: One-pot four-component reaction
Keywords: Benzyl amines, polyformaldehyde, activated alkenes, aliphatic alcohols, catalyst-free, room temperature, one-pot multicomponent reaction, Mannich-type reaction, γ-aminoethers

2.2.17.1 General Reaction Scheme

Abonia *et al.*[63] reported on a one-pot four-component Mannich-type reaction between benzylamines (**1**), polyformaldehyde (**2**), and activated alkenes (**3**) in aliphatic alcohols (**4**) acting both as solvent and as etherifying agents to furnish novel and diversely substituted γ-aminoethers (**5**) in good yields at room temperature (Scheme 2.25). The investigators confirmed that these reactions proceeded *via* a Mannich-type reaction, where the formation of iminium ions and aminals was identified as the key intermediates to obtain the target products.

2.2.17.2 Representative Examples

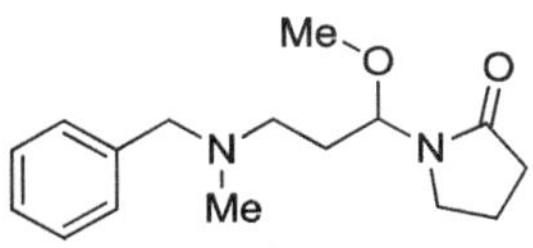

5a

(±)-1-[3-(*N*-Benzyl-*N*-methylamino)-1-methoxypropyl]pyrrolidin-2-one (**5a**): colorless oil; yield: 81%

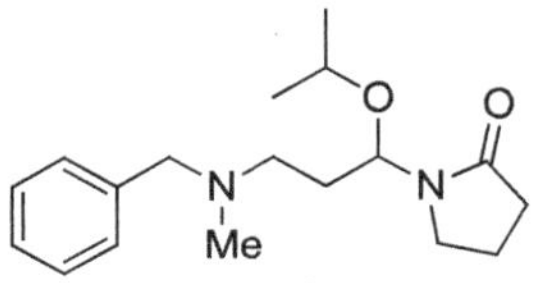

5b

(±)-1-[3-(*N*-Benzyl-*N*-methylamino)-1-isopropoxypropyl]pyrrolidin-2-one (5b): colorless oil; yield: 61%

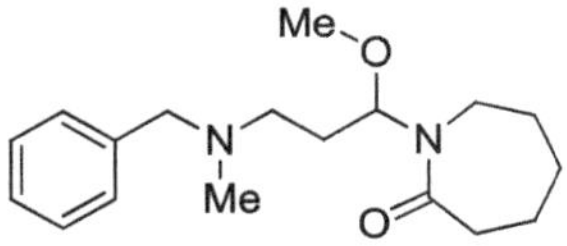

5c

(±)-1-(3-(*N*-Benzyl-*N*-methylamino)-1-methoxypropyl)azepan-2-one (**5c**): colorless oil; yield: 86%

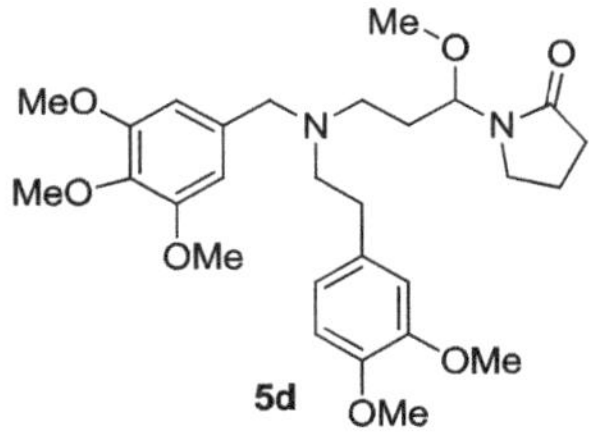

5d

(±)-1-(3-(*N*-(3,4,5-Trimethoxybenzyl)-*N*-(3,4- dimethoxyphenethyl)amino)-1-methoxypropyl)pyrrolidin-2-one (**5d**): colorless oil; yield: 63%

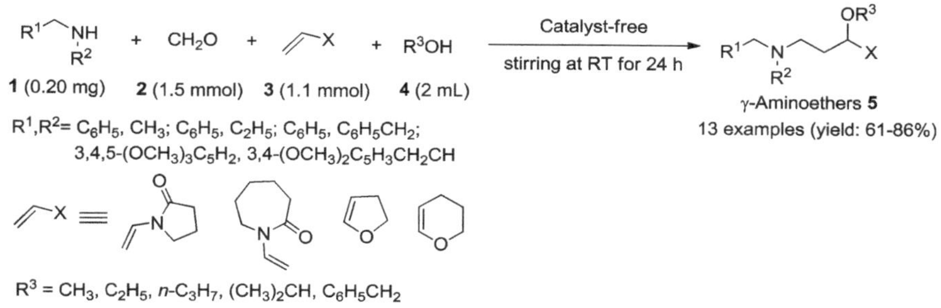

Scheme 2.25 Synthesis of γ-aminoethers (**5**).

2.2.17.3 Experimental Procedure

A mixture of benzylamine (**1**) (0.200 mg each one), polyformaldehyde (**2**; 1.5 mmol) and the activated alkene (**2**) (1.1 mmol) was dissolved in the corresponding aliphatic alcohol (**4**) (2 mL). The solution was then stirred at room temperature for 24 h until the starting benzylamine (**1**) was completely exhausted. On completion of the reaction, excess of the solvent was removed under reduced pressure, and the oily material as obtained was purified by column chromatography on silica gel, using a mixture of CH_2Cl_2–MeOH, $CHCl_3$–MeOH and CH_2Cl_2–EtOAc when a pure product of *gamma*-aminoether (**5**) was yielded. All the products (**5**) were characterized on the basis of elemental analyses and detailed spectral studies.

2.2.17.4 Characterization Data of Two Representative Compounds

(±)-1-[3-(*N*-Benzyl-*N*-methylamino)-1-methoxypropyl]pyrrolidin-2-one (**5a**): colorless oil; yield: 81%; FTIR (film): 3027, 2944, 2835, 1694 (C=O), 1285 and 1078 (C–O) cm^{-1}; ^1H NMR ($CDCl_3$, 300 MHz): δ 1.28–1.70 (1H, m), 1.81–1.96 (3H, m), 2.13 (3H, s, NCH_3), 2.22–2.41 (4H, m,), 3.10–3.28 (5H, m), 3.35 (1H, d, $J = 12.9$ Hz, Bn-H), 3.45 (1H, d, $J = 13.2$ Hz, Bn-H), 5.14 (1H, t, $J = 6.9$ Hz, NCH–O), 7.14–7.25 (5H, m, Ph–H); ^{13}C NMR ($CDCl_3$, 75 MHz): δ 18.1 (CH_2), 30.6 (CH_2), 31.6 (CH_2), 40.9 (CH_2), 42.2 (NCH_3), 52.6 (CH_2), 55.5 (OCH_3), 62.5 ($PhCH_2$), 81.0 (NCH–O), 126.9 (CH), 128.1 (CH), 129.0 (CH), 138.8 (Cq), 176.0 (C=O); MS (70 eV, EI): *m/z* (%) at 277 $[M+1]^+$ (6), 261 (26), 153 (18), 134 (29), 91 (100) $[PhCH_2]^+$. Anal. calc. for $C_{16}H_{24}N_2O_2$: C, 69.53; H, 8.75; N, 10.14; found: C, 69.66; H, 8.90; N, 10.01.

(±)-1-(3-(*N*-(3,4,5-Trimethoxybenzyl)-*N*-(3,4-dimethoxyphenethyl)amino)-1-methoxypropyl)-pyrrolidin-2- one (**5d**): colorless oil; yield: 63%; FTIR (film): 2931, 2833, 1690 (C=O), 1591, 1234 and 1126 (C–O) cm^{-1}; ^1H NMR ($CDCl_3$, 400 MHz): δ 1.62–1.69 (1H, m), 1.84–2.04 (4H, m), 2.30–2.58 (4H, m), 2.64–2.75 (3H, m), 3.14–3.20 (4H, m), 3.27–3.34 (1H, m), 3.47 (1H, d, $J = 13.6$ Hz, Bn–H), 3.61 (1H, d, $J = 13.6$ Hz, Bn–H), 3.82–3.84 (15H, m, $OCH_3 \times 5$), 5.23 (1H, t, $J = 6.4$ Hz, NCH–O), 6.56 (2H, s, Ph-H), 6.66–6.70 (2H, m,

Ph–H), 6.77 (1H, d, $J = 8.0$ Hz, Ph–H); ^{13}C NMR ($CDCl_3$, 100 MHz): δ 18.2 (CH_2), 30.6 (CH_2), 31.5 (CH_2), 33.1 (CH_2), 41.0 (CH_2), 49.3 (CH_2), 55.5 (OCH_3), 55.8 ($CH_2 + OCH_3$), 55.9 (OCH_3), 56.0 ($OCH_3 \times 2$), 58.9 ($PhCH_2$), 60.8 (OCH_3), 80.9 (NCH–O), 105.4 (CH), 111.2 (CH), 112.1 (CH), 120.5 (CH), 133.1 (Cq), 135.5 (Cq), 136.7 (Cq), 147.2 (Cq), 148.7 (Cq), 153.0 (Cq), 175.9 (C=O); MS (70 eV, EI): *m/z* (%) at 365 (9), 254 (7), 181 (100), 151 (8), 124 (11). Anal. calc. for $C_{28}H_{40}N_2O_7$: C, 65.10; H, 7.80; N, 5.42; found: C, 64.93; H, 7.91; N, 5.35.

2.2.17.5 Critical Views

The investigators demonstrated a general one-pot method for the synthesis of novel and diversely functionalized γ-aminoethers of biological significance under mild conditions at room temperature in the absence of any catalyst. This approach involves trapping of the *in situ* formed iminium ion by an activated alkene through a Mannich-type reaction, followed by a subsequent attack of the corresponding alcohol used as solvent also in the next step. In this reaction, three new bonds are formed in only one step and water is released as the unique by-product. Hence, the present method offers an outstanding bond-forming efficiency as well as a considerable environmentally friendly quality.

2.2.18 Entry-18: Synthesis of 1-Substituted-1*H*-pyrazoles

Type of reaction: C–N bond formation
Reaction conditions: Catalyst-free, dichloromethane, room temperature
Synthetic strategy: One-step aza-Michael addition
Keywords: Nitroalkenes, pyrazole, catalyst-free, dichloromethane, room temperature, aza-Michael addition, amination, 1-substituted-1*H*-pyrazoles.

2.2.18.1 General Reaction Scheme

A catalyst-free and efficient aza-Michael addition of azole (**2**) to nitroalkenes (**1**) in dichloromethane has been demonstrated by Wu *et al.* under ambient conditions (Scheme 2.26).[64]

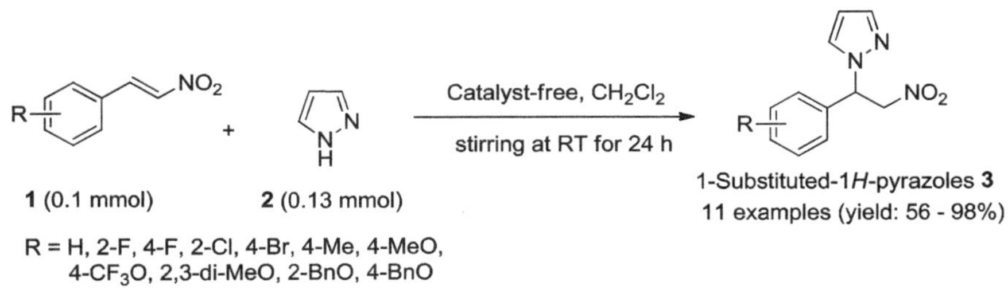

Scheme 2.26 Synthesis of 1-substituted-1*H*-pyrazoles (**3**).

2.2.18.2 Representative Examples

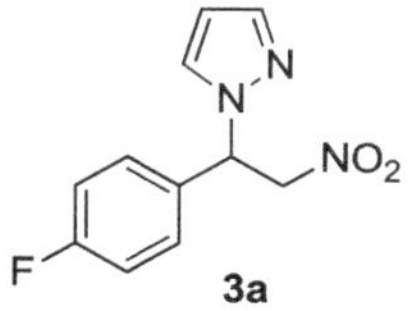

3a

1-(1-(4-Fluorophenyl)-2-nitroethyl)-1*H*-pyrazole (**3a**): yield: 93%

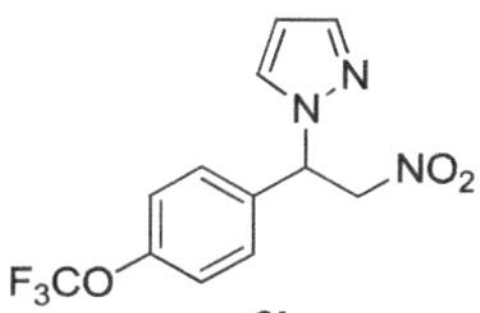

3b

1-(2-Nitro-1-(4-(trifluoromethoxy) phenyl)ethyl)-1*H*-pyrazole (**3b**): yield: 97%

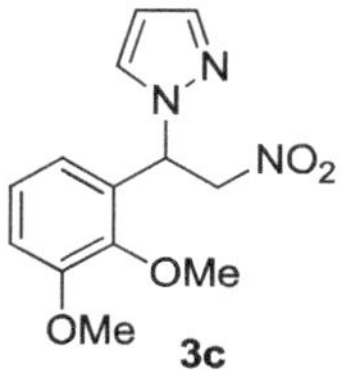

3c

1-(1-(2,3-Dimethoxyphenyl)-2-nitroethyl)-1*H*-pyrazole (**3c**): yield: 81%

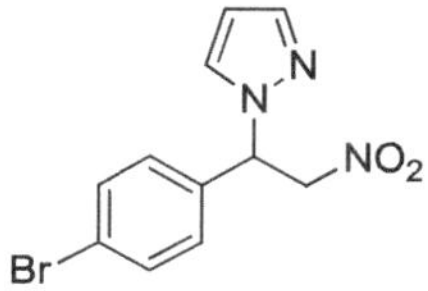

3d

1-(1-(4-Bromophenyl)-2-nitroethyl)-1*H*-pyrazole (**3d**): yield: 82%

2.2.18.3 Experimental Procedure

A mixture of nitroalkenes (**1**; 0.1 mmol) and azole (**2**; 0.13 mmol) in dichloromethane (0.1 mL) was stirred at room temperature for 24 h, and then the solvent was removed under vacuum. The residue was purified by column chromatography on silica gel to yield the Michael adduct (**3**). Each of the products was characterized on the basis of detailed spectral measurements including IR, ^{1}H NMR, ^{13}C NMR and HRMS.

2.2.18.4 Characterization Data of Two Representative Compounds

1-(1-(2,3-Dimethoxyphenyl)-2-nitroethyl)-1*H*-pyrazole (**3c**): IR (KBr): 3120, 2940, 2838, 1556, 1483, 1377, 1273, 1090, 750 cm^{-1}; ^{1}H NMR (CDCl$_3$, 400 MHz): δ 7.57 (1H, d, J = 2.0 Hz), 7.46 (1H, d, J = 2.4 Hz), 7.01 (1H, t, J = 8.0 Hz), 6.91 (1H, dd, J = 1.6, 8.4 Hz), 6.69 (1H, dd, J = 1.2, 8.0 Hz), 6.51 (1H, dd, J = 4.4, 6.0 Hz), 6.28 (1H, t, J = 2.2 Hz), 5.57 (1H, dd, J = 10.0, 14.0 Hz), 4.90 (1H, dd, J = 4.2, 14.2 Hz), 3.91 (3H, s), 3.87 (3H, s); ^{13}CNMR (CDCl$_3$, 100 MHz): δ 152.7, 146.0, 140.1, 130.0, 129.4, 124.7, 119.2, 113.4, 106.4, 76.3, 61.1, 56.9, 56.0; HRMS: $[M + Na]^+$ calc. for $C_{13}H_{15}N_3O_4Na$ at *m/z* 300.0960; found at *m/z* 300.0963.

1-(1-(4-Bromophenyl)-2-nitroethyl)-1*H*-pyrazole (**3d**): IR (KBr): 3117, 3025, 2961, 2917, 1556, 1490, 1377, 1290, 1075, 755 cm^{-1}; ^{1}H NMR (CDCl$_3$, 400 MHz): δ 7.59 (1H, s), 7.50 (2H, d, J = 8.0 Hz), 7.42 (1H, d, J = 1.6 Hz), 7.19 (2H, d, J = 8.4 Hz), 6.29 (1H, s), 6.06 (1H, dd, J = 5.0, 9.4 Hz), 5.59 (1H, dd, J = 5.6, 14.0 Hz), 4.85 (1H, dd, J = 5.0, 14.2 Hz); ^{13}CNMR (CDCl$_3$, 100 MHz): δ

140.6, 134.9, 132.6, 130.0, 128.7, 123.6, 106.8, 77.1, 61.7; HRMS: $[M+H]^+$ calc. for $C_{11}H_{11}N_3O_2Br$ at *m/z* 296.0035; found at *m/z* 296.0032.

2.2.18.5 Critical Views

Azole moiety is commonly found in drug candidates, and hence aza-Michael addition of azole is an important reaction to generate targeted azole derivatives. Reports on catalytic aza-Michael addition of nucleophile containing azole groups are rare but several examples of catalyst-free aza-Michael additions of azole are available in the literature, and they require severe conditions such as high reaction temperature,[65] high pressure[66] and ultraviolet irradiation[67] to yield the adducts. Under this purview, the present catalyst-free protocol offers a general and efficient aza-Michael addition between azoles and a wide spectrum of nitroalkenes with good yields at room temperature condition. The azole serves as both an organocatalyst and *N*-nucleophile in the reaction system.[63]

2.2.19 Entry-19: Synthesis of 2-Thioparabanic Acids

Type of reaction: C–N and C–S bond formation
Reaction conditions: Catalyst-free, dichloromethane, room temperature
Synthetic strategy: One-pot multicomponent reaction
Keywords: Benzoyl thiocyanate, oxalyl chloride, primary amines, dichloromethane, room temperature, multicomponent reaction, 2-thioparabanic acids

2.2.19.1 General Reaction Scheme

Yavari *et al.*[68] developed an efficient protocol for the synthesis of unsymmetrical 2-thioparabanic acids (**4**) *via* a three-component catalyst-free reaction between primary amines (**1**), benzoyl isothiocyanate (**2**) and oxalyl chloride (**3**) at room temperature in dichloromethane (Scheme 2.27).

2.2.19.2 Plausible Mechanism

The investigators proposed a plausible mechanism for this transformation (Scheme 2.28). Initially, the formation of substituted thiourea (**5**) occurs from the reaction of amine (**1**) with benzoyl isothiocyanate (**2**), which is subsequently attacked by oxalyl chloride (**3**) to produce **6**. Intermediate (**6**)

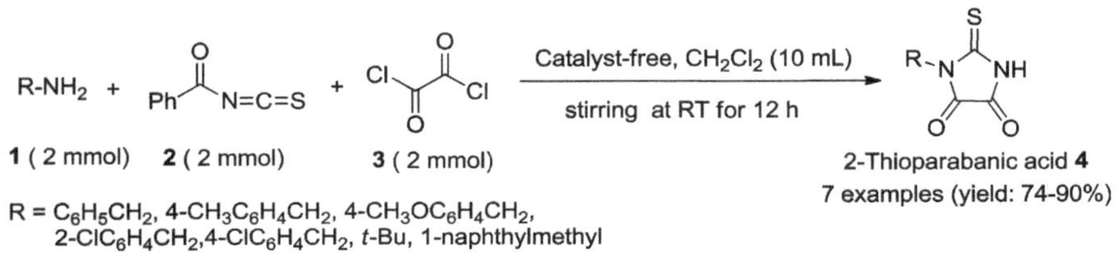

Scheme 2.27 Synthesis of 2-thioparabanic acids (**4**).

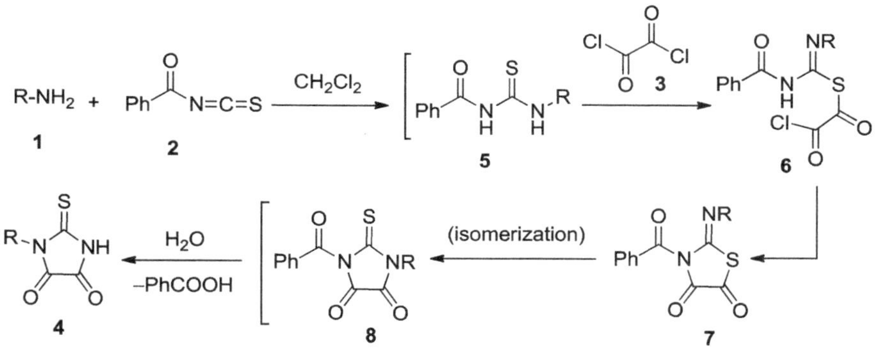

Scheme 2.28 Plausible mechanism for synthesis of 2-thioparabanic acids (**4**).

then undergoes cyclization to produce **7**, followed by isomerization to yield **8**. 2-Thioparabanic acid (**4**) is apparently formed by the addition of adventitious water to **8** and subsequent elimination of benzoic acid.[68]

2.2.19.3 Representative Examples

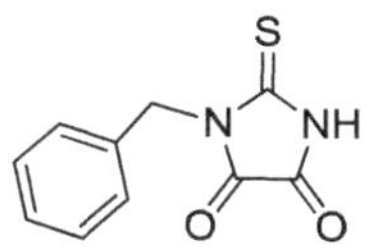

4a

1-Benzyl-2-thioxoimidazolidine-4,5-dione (**4a**): yellow powder, mp 151–153 °C; yield: 78%

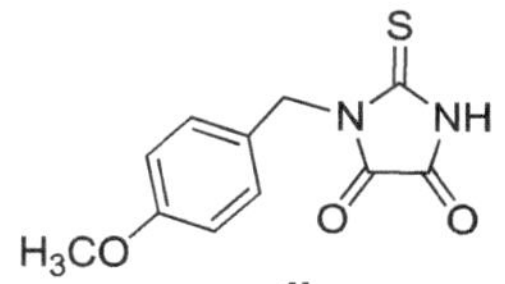

4b

1-(4-Methoxybenzyl)-2-thioxoimidazolidine-4,5-dione (**4b**): yield: 76%

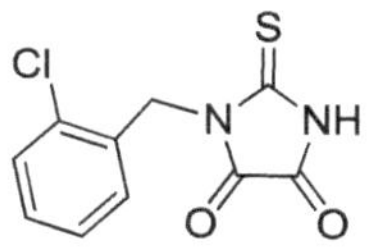

4c

1-(2-Chlorobenzyl)-2-thioxoimidazolidine-4,5-dione (**4c**): yield: 80%

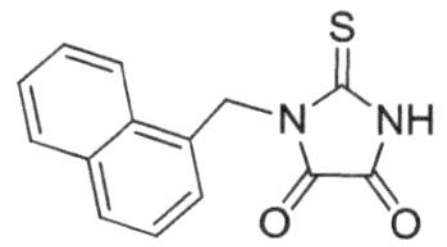

4d

1-(Naphthalen-1-ylmethyl)-2-thioxoimidazolidine-4,5-dione (**4d**): yield: 90%

2.2.19.4 Experimental Procedure

To a stirred solution of primary amine (**1**; 2 mmol) and benzoyl isothiocyanate (**2**; 0.32 g, 2 mmol) in dichloromethane (10 mL) was added oxalyl chloride (**3**; 0.25 g, 2 mmol) at room temperature, and stirring was continued for about 12 h at RT. On completion of the reaction, the solvent was removed under reduced pressure and the viscous residue was purified by flash column chromatography on SiO_2 (Merck 230–400 mesh) using hexane-EtOAc as an eluent to give 2-thioparabanic acid derivatives (**4**), characterized on the basis of spectral and elemental analyses.

2.2.19.5 *Characterization Data of One Representative Compound*

1-Benzyl-2-thioxoimidazolidine-4,5-dione (**4a**): yellow powder, mp 151–153 °C; IR (KBr): 3220, 1764, 1735, 1666, 1441, 1340 cm^{-1}; ^{1}H NMR ($CDCl_3$, 500 MHz): δ 5.10 (2H, s, NCH_2), 7.28 (1H, d, $^{3}J = 7.1$ Hz, CH), 7.32 (2H, t, $^{3}J = 7.6$ Hz, CH), 7.41 (2H, d, $^{3}J = 7.3$ Hz, CH), 11.71 (1H, br s, NH); ^{13}C NMR ($CDCl_3$, 125 MHz): δ 45.0 (NCH_2), 128.6 (CH), 129.0 (CH), 129.2 (CH), 136.2 (CH), 156.8 (C=O), 157.3 (C=O), 182.7 (C=S); MS: *m*/*z* (%rel) 222 (4) [$M^+ + 2$], 221 (10) [$M^+ + 1$], 220 (64) [M^+], 91 (26), 77 (60), 59 (20), 45 (60). Anal. calc. (%) for $C_{10}H_8N_2O_2S$ (220.24): C 54.53, H 3.66, N 12.72; found: C, 54.41, H 3.70, N 12.65.

2.2.19.6 *Critical Views*

The method offers a convenient and practical means to access functionalized 2-thioparabanic acids by simple mixing the starting materials under catalyst-free neutral conditions at ambient conditions.

2.2.20 Entry-20: Synthesis of 5-amino-1,3-aryl-1*H*-pyrazole-4-carbonitriles

Type of reaction: C–C and C–N bond formation
Reaction conditions: Catalyst-free, ethanol-water, room temperature
Synthetic strategy: One-pot tandem Knoevenagel-cyclo condensation reaction
Keywords: Aromatic aldehydes, malononitrile/ethyl cyanoacetate/methyl cyanoacetate, ethanol-water, catalyst-free, room temperature, polysubstituted amino pyrazole analogs, 5-amino-1,3-aryl-1*H*-pyrazole-4-carbonitriles

2.2.20.1 *General Reaction Scheme*

Hasaninejad and Firoozi[69] developed a catalyst-free novel method for the synthesis of a series of polysubstituted amino pyrazole analogs (**4**) in good to excellent yields *via* a one-pot three-component tandem reaction of various aromatic aldehydes (**1**), malono derivatives (**2**), and phenyl hydrazines (**3**) in aqueous ethanol at room temperature (Scheme 2.29).

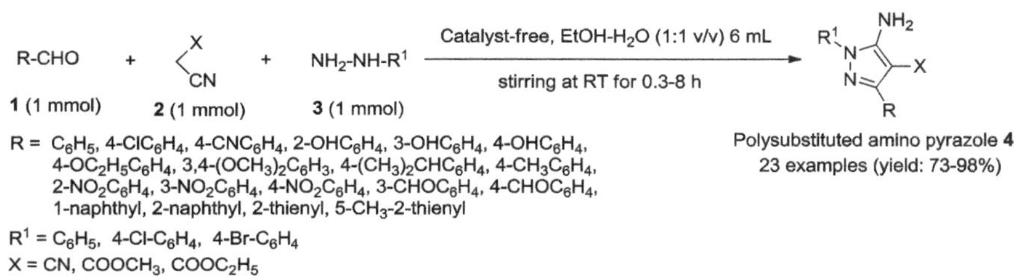

Scheme 2.29 Synthesis of 5-amino-1,3-aryl-1*H*-pyrazole-4-carbonitriles (**4**).

2.2.20.2 Representative Examples

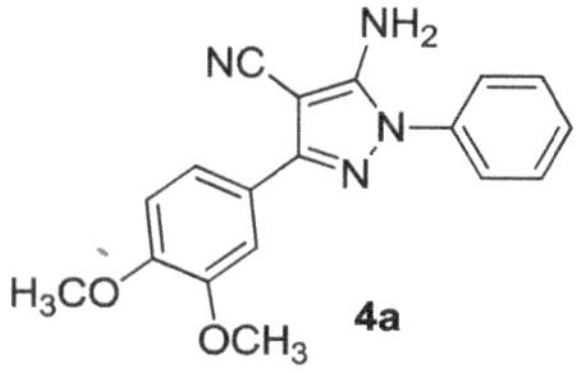

4a

5-Amino-3-(3,4-dimethoxyphenyl)-1-phenyl-1*H*-pyrazole-4-carbonitrile (**4a**): white powder, mp 122–123 °C; yield: 73%

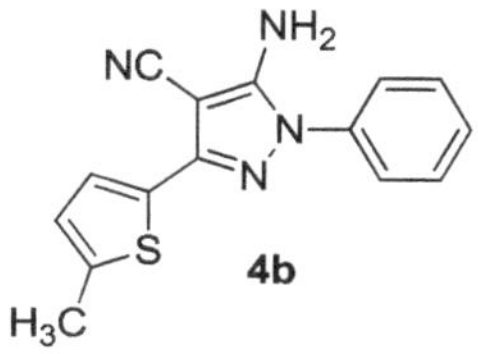

4b

5-Amino-3-(5-methylthiophen-2-yl)-1-phenyl-1*H*-pyrazole-4-carbonitrile (**4b**): yellow powder, mp 131–132 °C; yield: 90%

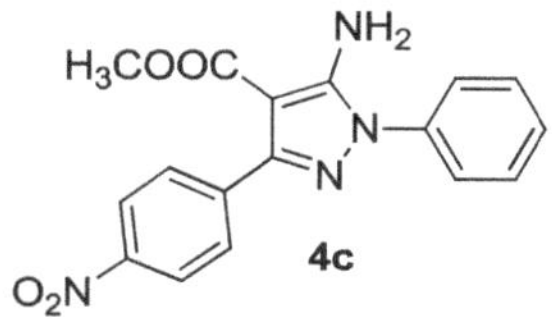

4c

Methyl 5-amino-3-(4-nitrophenyl)-1-phenyl-1*H*-pyrazole-4-carboxylate(**4c**): red powder, mp 162–163 °C; yield: 95%

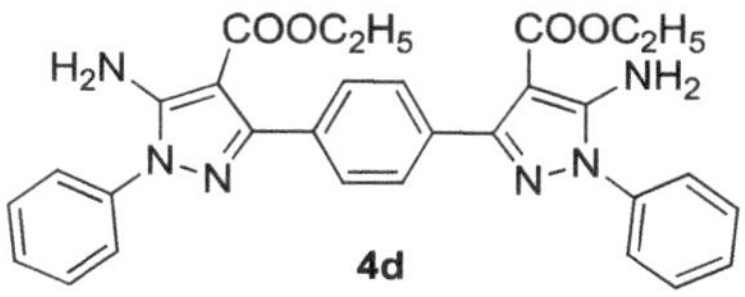

4d

Diethyl 3,3'-(1,4-phenylene)*bis*(5-amino-1-phenyl-1H-pyrazole-4-carboxylate) (**4d**): yellow powder, mp >230 °C; yield: 87%

2.2.20.3 Experimental Procedure

A mixture of aromatic aldehydes (**1**; 1 mmol), malono derivatives (**2**; 1 mmol) and phenyl hydrazine derivatives (**3**; 1 mmol) in aqueous ethanol (1 : 1, v/v) (6 mL) placed within a 25 mL round-bottomed flask was stirred for stipulated time (0.3–8 h) at room temperature. After completion of the reaction (as monitored by TLC), crystals of the product were formed, collected by filtration and then recrystallized from hot ethanol to obtain pure products of 5-amino-1,3-aryl-1*H*-pyrazole-4-carbonitriles (**4**) in good yields. Each of the products was characterized from its detailed spectral studies.

2.2.20.4 Characterization Data of Two Representative Compounds

5-Amino-3-(5-methylthiophen-2-yl)-1-phenyl-1*H*-pyrazole-4-carbonitrile (**4b**): Yellow powder, mp 131–132 °C; yield: 90%; IR (KBr): 3427, 3303, 3102, 2919, 2357, 1600, 1469, 1257, 1230, 1126, 1100, 1070 cm^{-1}; ^{1}H NMR ($CDCl_3$, 500 MHz): δ 2.53 (3H, s), 6.68–6.71 (1H, m), 6.88–6.92 (2H, m), 7.10 (2H, d, $J = 7.7$ Hz), 7.28–7.32 (2H, m), 7.46 (1H, s), 7.76 (1H, s); ^{13} C NMR ($CDCl_3$, 125 MHz): δ 16.08, 113.15, 114.13, 120.42, 125.86, 126.10, 127.16, 129.71, 133.18, 138.66, 141.38, 145.05, 156.73. Anal. calc. for $C_{15}H_{12}N_4S$: C, 64.26; H, 4.31; N, 19.98; found: C, 64.37; H, 4.25; N, 20.05.

Diethyl 3,3′-(1,4-phenylene)bis(5-amino-1-phenyl-1*H*-pyrazole-4-carboxylate) (**4d**): Yellow powder, mp >230 °C; yield: 87%; IR (KBr): 3486, 3348, 3108, 2996, 1722, 1598, 1475, 1251, 1130, 1121, 1079 cm^{-1}; ^{1}H NMR (DMSO-d_6, 500 MHz):

δ 1.43 (6H, t, $J=7.2$ Hz), 4.42 (4H, q, $J=7.2$ Hz), 6.96 (2H, t, $J=7.2$ Hz), 7.18 (4H, d, $J=7.6$ Hz), 7.34 (4H, t, $J=8.0$ Hz), 7.71 (2H, s), 7.79 (2H, d, $J=8.4$ Hz), 8.04 (2H, d, $J=8.4$ Hz), 8.25 (2H, s); MS: *m/z* 536 (M^+). Anal. calc. for $C_{30}H_{28}N_6O_4$: C, 67.15; H, 5.26; N, 15.66; found: C, 67.23; H, 5.33; N, 15.54.

2.2.20.5 Critical Views

In contrast to the majority of the strategies[70–78] those involve either multistep sequences, or expensive catalysts, inert atmosphere or anhydrous conditions and laborious workup, this present protocol offers a convenient and practical catalyst-free alternative route for synthesis of a series of polysubstituted amino pyrazole analogs *via* one-pot multicomponent reaction under ambient conditions. Good to excellent yields, use of aqueous ethanol as green solvent, short reaction time, simple experimental workup procedures, and no toxic by-products are the major advantages of this method.[69]

2.2.21 Entry-21: Synthesis of Functionalized Azole Derivatives

Type of reaction: C–N bond formation
Reaction conditions: Catalyst-free, dichloromethane, room temperature
Synthetic strategy: One-step aza-Michael conjugate addition
Keywords: Substituted azoles, β,γ-unsaturated-α-keto esters, dichloromethane, catalyst-free, room temperature, aza-Michael conjugate addition, azole derivatives

2.2.21.1 General Reaction Scheme

Wang *et al.*[79] developed an efficient aza-Michael addition reaction of azoles (**1**) to a wide spectrum of β,γ-unsaturated-α-keto esters (**2**) in dichloromethane to form the corresponding Michael adduct (**3**) with good to excellent yields under mild reaction reactions (Scheme 2.30).

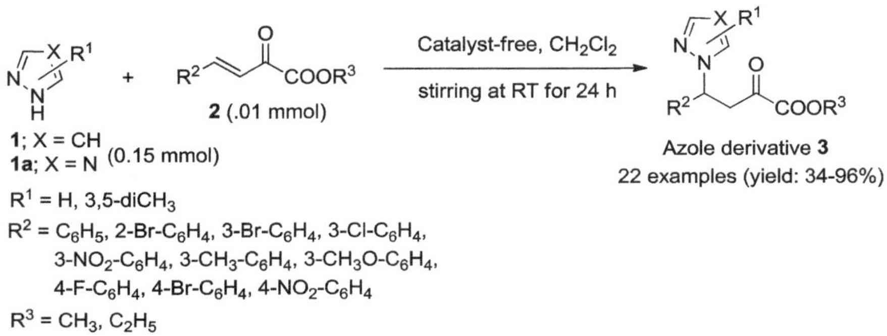

Scheme 2.30 Synthesis of functionalized azole derivatives (**3**).

2.2.21.2 Representative Examples

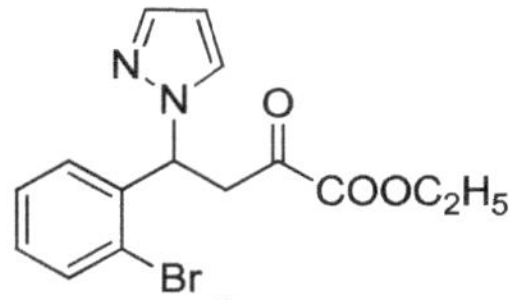

3a

Ethyl 4-(2-bromophenyl)-2-oxo-4-(1*H*-pyrazol-1-yl)butanoate (**3a**): colorless oil; yield: 85%

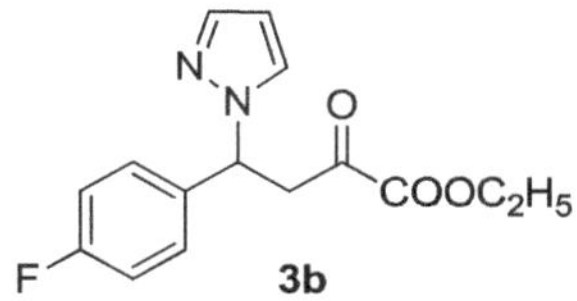

3b

Ethyl 4-(4-flurophenyl)-2-oxo-4-(1*H*-pyrazol-1-yl)butanoate (**3b**): colorless oil; yield: 90%

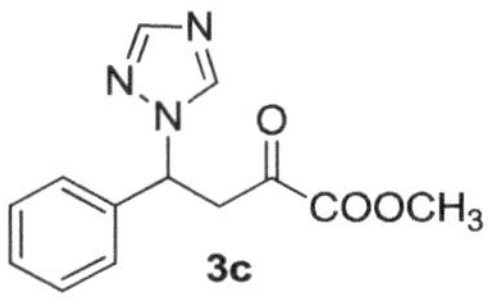

3c

Methyl 2-oxo-4-phenyl-4-(1*H*-1,2,4-triazol-1-yl)butanoate (**3c**): Colorless oil; yield: 42%

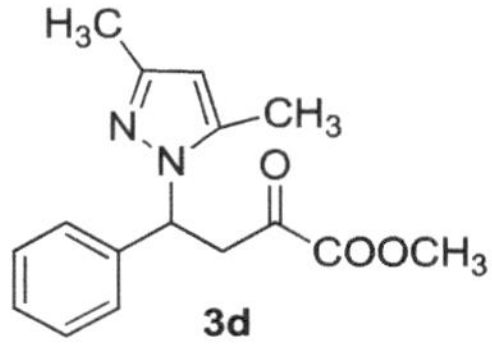

3d

Methyl 4-(3,5-dimethyl-1*H*-pyrazol-1-yl)-2-oxo-4-phenylbutanoate (**3d**): Colorless oil; yield: 84%

2.2.21.3 Experimental Procedure

To dichloromethane (0.1 mL) was added an azole (**1**; 0.15 mmol) and β,γ-unsaturated-α-keto ester (**2**; 0.1 mmol). The reaction mixture was stirred at room temperature for 24 h, and then the solvent was removed under vacuum. The residue was purified by column chromatography on silica gel to yield the desired Michael adduct (**3**). Each of the products was identified by FTIR, NMR and HRMS.

2.2.21.4 Characterization Data of Two Representative Compounds

Ethyl 4-(2-bromophenyl)-2-oxo-4-(1*H*-pyrazol-1-yl)butanoate (**3a**): colorless oil; yield: 85%; IR (KBr): 3063, 2982, 2930, 1728, 1470, 1441, 1394, 1277, 1239, 1079, 1051, 1025, 752 cm^{-1}; ^{1}H NMR ($CDCl_3$, 400 MHz): δ 7.57 (1H, dd, $J = 0.8$, 8.0 Hz), 7.50 (1H, d, $J = 1.2$ Hz), 7.44 (1H, d, $J = 2.4$ Hz), 7.27–7.23 (1H, m), 7.15 (1H, qd, $J = 1.6$, 7.6 Hz), 7.09 (1H, dd, $J = 1.4$, 7.8 Hz), 6.34 (1H, dd, $J = 3.6$, 10.4 Hz), 6.26 (1H, t, $J = 2.0$ Hz), 4.37–4.29 (3H, m), 3.38 (1H, dd, $J = 3.6$, 18.4 Hz), 1.35 (3H, t, $J = 7.0$ Hz); ^{13}C NMR ($CDCl_3$, 100 MHz): δ 190.1, 160.3, 139.2, 133.0, 130.1, 129.7, 128.2, 122.2, 106.3, 62.6, 59.5, 43.1, 14.0; HRMS: *m/z* calc. for $C_{15}H_{16}N_2O_3Br$ [$(M+H)^+$], 351.0344; found: *m/z* 351.0332.

Methyl 2-oxo-4-phenyl-4-(1*H*-1,2,4-triazol-1-yl)butanoate (**3c**): colorless oil; yield: 42%; IR (KBr): 2958, 2920, 1734, 1507, 1461, 1277, 1138, 1071, 702, 679 cm^{-1}; ^{1}H NMR ($CDCl_3$, 400 MHz): δ 8.21 (1H, s), 8.07 (1H, s), 7.92 (1H, s), 7.39–7.33 (5H, m), 5.96 (1H, q, $J = 4.8$ Hz), 4.29 (1H, dd, $J = 9.6$, 18.4 Hz), 3.88 (3H, s), 3.47 (1H, dd, $J = 4.6$, 18.6 Hz); ^{13}C NMR ($CDCl_3$, 100 MHz): δ 189.9,

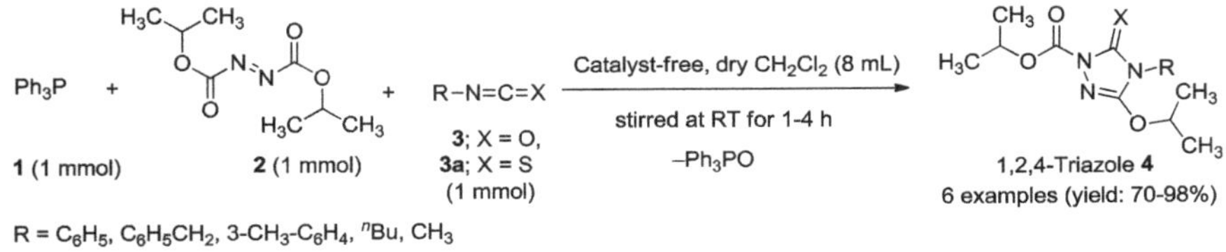

Scheme 2.31 Synthesis of 1,2,4-triazole derivatives (**4**).

160.6, 151.5, 146.9, 143.3, 137.6, 129.4, 129.3, 127.2, 59.0, 53.4, 44.5; HRMS: *m/z* calc. for $C_{13}H_{13}N_3O_3Na$ $[(M+Na)^+]$, 282.0855; found: *m/z* 282.0843.

2.2.21.5 Critical Views

The investigators offered a general and efficient catalyst-free aza-Michael addition of azoles to β,γ-unsaturated-α-keto esters at room temperature with good yields of the desired adducts of useful building blocks for a variety of biologically interesting scaffolds.

2.2.22 Entry-22: Synthesis of 1,2,4-Triazole Derivatives

Type of reaction: C–N bond formation
Reaction conditions: Catalyst-free, dichloromethane, room temperature
Synthetic strategy: One-pot synthesis of 1,2,4-triazoles by using Mitsunobu chemistry
Keywords: Isocyanates, isothiocyanates, diisopropyl azodicarboxylate, dichloromethane, catalyst-free, room temperature, Mitsunobu reaction, one-pot synthesis, 1,2,4-triazole derivatives

2.2.22.1 General Reaction Scheme

An efficient, simple, catalyst-free one-pot three-component method was developed by Alizadeh for the synthesis of 1,2,4-triazole derivatives (**4**) by the reaction of isocyanates (**3**) or isothiocyanates (**3a**) with an *in situ* generated reactive intermediate of triphenylphosphine (**1**) and diisopropyl azodicarboxylate (**2**) in dry dichloromethane under mild conditions (Scheme 2.31).[80]

2.2.22.2 Representative Examples

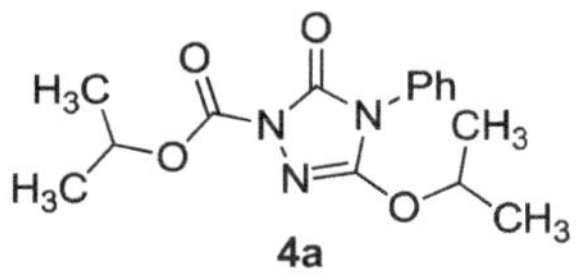

4a

Isopropyl 3-isopropoxy-5-oxo-4-phenyl-4,5-dihydro-1*H*-1,2,4-triazole-1-carboxylate (**4a**): colorless crystals, mp 88–90 °C; yield: 98%

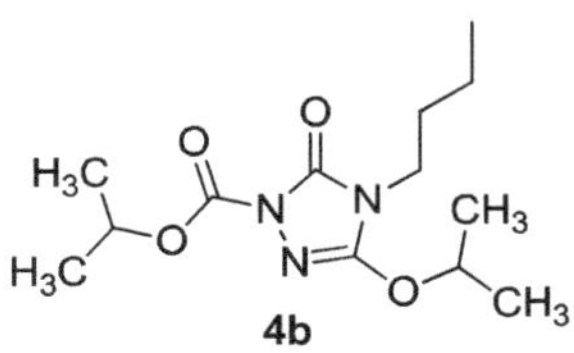

4b

Isopropyl 4-butyl-3-isopropoxy-5-oxo-4,5-dihydro-1*H*-1,2,4-triazole-1-carboxylate (**4b**): colorless wax; yield: 70%

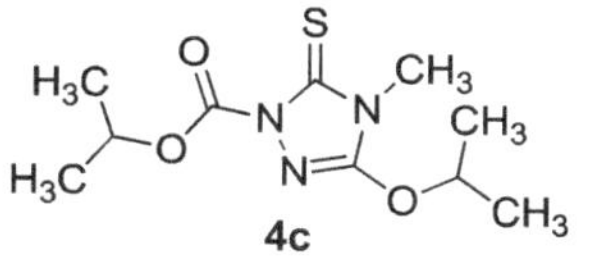

4c

Isopropyl 3-isopropoxy-4-methyl-5-thioxo-4,5-dihydro-1*H*-1,2,4-triazole-1-carboxylate (**4c**): colorless crystals, mp 67–70 °C; yield: 77%

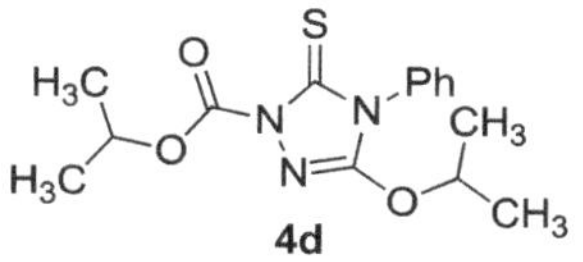

4d

Isopropyl 3-isopropoxy-4-phenyl-5-thioxo-4,5-dihydro-1*H*-1,2,4-triazole-1-carboxylate (**4d**): colorless crystals, mp 121–123 °C; yield: 95%

2.2.22.3 Experimental Procedure

To a magnetically stirred solution of isocyanates (**3**; 1 mmol) or isothiocyanates (**3a**; 1 mmol) and diisopropyl azodicarboxylate (**2**; 0.20 g, 1 mmol) in dry dichloromethane (5 mL) was added drop-wise a solution of triphenyl phosphine (**1**; 0.26 g, 1 mmol) in dichloromethane (3 mL) at room temperature over 10 min. The mixture was then stirred for 1–4 h. On completion of the reaction, the solvent was removed under reduced pressure, and the residue was subjected to column chromatographic resolution using hexane–ethyl acetate mixtures as the eluent to obtain the pure product of 1,2,4-triazoles (**4**). The products were identified based on analytical and spectral studies.

2.2.22.4 Characterization Data of Two Representative Compounds

Isopropyl 4-butyl-3-isopropoxy-5-oxo-4,5-dihydro-1*H*-1,2,4-triazole-1-carboxylate (**4b**): colorless wax; yield: 70%; IR (KBr): 1781 (N–CO–N); 1743 (N–$CO_2{}^i$Pr); 1616 (C=N); 1366, 1313 (C–O) cm^{-1}; ^{1}H NMR ($CDCl_3$, 500.1 MHz): δ 0.86 (3H, t, $J=7.2$), 1.27 (*quint.*, $J=7.3$, 2 H), 1.34 (12H, d, $J=6.1$), 1.55 (2H, *quint.*, $J=7.2$), 3.47 (2H, t, $J=6.9$), 5.13 (1H, *sept.*, $J=6.2$); 5.16 (1H, *sept.*, $J=6.2$); ^{13}C NMR ($CDCl_3$, 125.7 MHz): δ 13.51, 19.63, 21.79, 21.83, 30.17, 40.01, 72.10, 74.42, 148.88, 150.19, 151.77; EIMS (rel.%): 287 (52, $[M+2]^+$), 286 (100, $[M+1]^+$), 285 (10, M^+), 242 (7), 200 (75), 157 (80), 140 (33), 100 (64), 43 (60). Anal. calc. for $C_{13}H_{23}N_3O_4$: C, 54.72; H, 8.12; N, 14.73; found: C, 54.64; H, 8.18; N, 14.70.

Isopropyl 3-isopropoxy-4-phenyl-5-thioxo-4,5-dihydro-1*H*-1,2,4-triazole-1-carboxylate (**4d**): colorless crystals, mp 121–123 °C; yield: 95%; IR (KBr): 1746 (N–$CO_2{}^i$Pr); 1612 (C=N); 1331, 1295 (C–O), 1215 (C=S) cm^{-1}; ^{1}H NMR ($CDCl_3$, 500.1 MHz): δ 1.32 (6H, d, $J=6.2$), 1.43 (6H, d, $J=6.2$), 5.18 (1H, *sept.*, $J=6.2$), 5.23 (1H, *sept.*, $J=6.2$), 7.29 (2H, d, $J=7.4$), 7.44 (1H, t, $J=7.3$), 7.47 (2H, t, $J=6.9$); ^{13}C-NMR ($CDCl_3$, 125.7 MHz): δ 21.43, 21.55, 73.17, 76.46, 127.78, 129.12, 129.45, 131.88, 148.48, 153.56, 169.23; EIMS (rel.%): 323 (42, $[M+2]^+$), 322 (100, $[M+1]^+$), 321 (81, M^+), 278 (6), 235 (50), 193 (100), 136 (20), 43 (60). Anal. calc. for $C_{15}H_{19}N_3O_3S$: C, 56.06; H, 5.96; N, 13.07; found: C, 56.00; H, 5.91; N, 12.95.

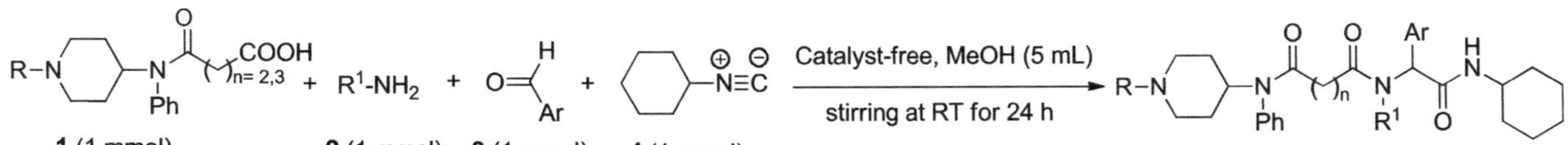

1 (1 mmol) **2** (1 mmol) **3** (1 mmol) **4** (1 mmol)

Amidated fentanyl analog **5**
11 examples (yield: 63-96%)

R = $CH_2CH_2C_6H_5$
R^1 = C_6H_5, $C_6H_5CH_2$, $C_6H_5CH_2CH_2$, 4-BrC_6H_4, 4-IC_6H_4, (thiophen-2-yl)-CH_2CH_2
Ar = C_6H_5, 4-$OCH_3C_6H_4$

Scheme 2.32 Synthesis of amidated fentanyl analogs (5).

2.2.22.5 Critical Views

The present method offers a way to one-pot access of novel 1,2,4-triazole derivatives without the formation of any 1,3,4-oxadiazole or 1,3,4-thiadiazole as a side product, under neutral conditions and at room temperature. The simplicity of the present procedure renders it an interesting alternative to complex multistep approaches.

2.2.23 Entry-23: Synthesis of Amidated Fentanyl Analogs

Type of reaction: C–C and C–N bond formation
Reaction conditions: Catalyst-free, methanol, room temperature
Synthetic strategy: One-pot multicomponent domino addition-amination reaction
Keywords: 4-(*N*-phenylamido)piperidines, amines, aromatic aldehydes, cyclohexylisocyanide, methanol, catalyst-free, room temperature, one-pot multicomponent Ugi reaction, amidated fentanyl analogs

2.2.23.1 General Reaction Scheme

Haghighatnia *et al.*[81] developed an efficient, simple and mild procedure for the generation of amidated fentanyl analogs (**5**) through the Ugi four-component reaction of 4-(*N*-phenylamido)piperidine scaffolds (**1**), amines (**2**), aromatic aldehydes (**3**) and cyclohexylisocyanide (**4**) in methanol at room temperature conditions (Scheme 2.32).

2.2.23.2 Representative Examples

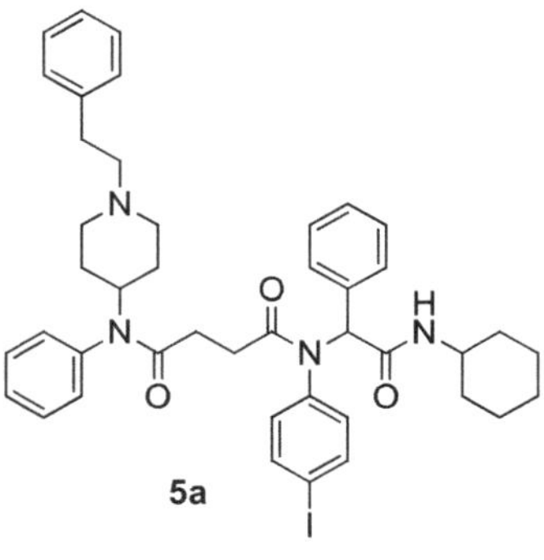

5a

N^1-(2-(cyclohexylamino)-2-oxo-1-phenylethyl)-N^1-(4-iodophenyl)-N^4-(1-phenethylpiperidin-4-yl)-N^4-phenylsuccinamide (**5a**): white solid, mp 81–82 °C; yield: 95%

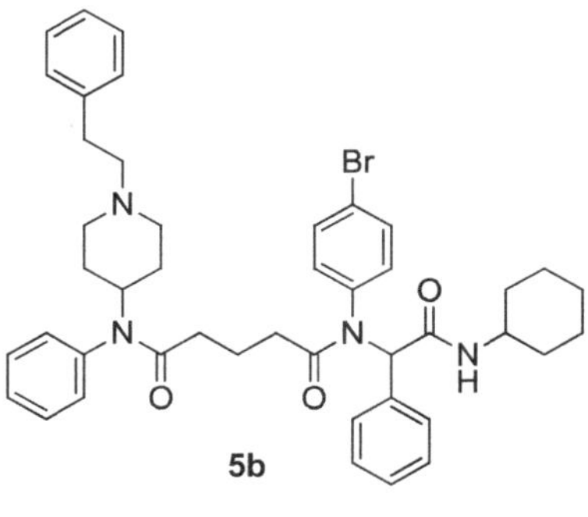

5b

N^1-(4-Bromophenyl)-N^1-[2-(cyclohexylamino)-2-oxo-1-phenylethyl]-N^5-phenyl-N^5-[1-(2-phenylethyl)piperidin-4-yl]pentanediamide (**5b**): white solid, mp 154–155 °C; yield: 77%

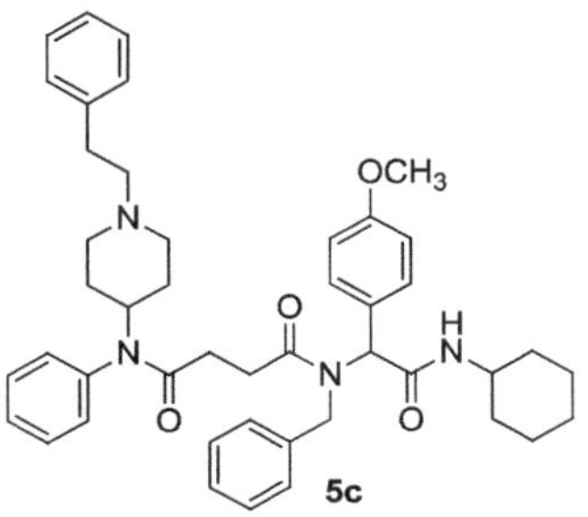

N^1-[2-(Cyclohexylamino)-1-(4-methoxyphenyl)-2-oxoethyl]-N^4-phenyl-N^4-[1-(2-phenylethyl)piperidin-4-yl]-N^1-(phenylmethyl)butanediamide (**5c**): white solid, mp 174–175 °C; yield: 72%

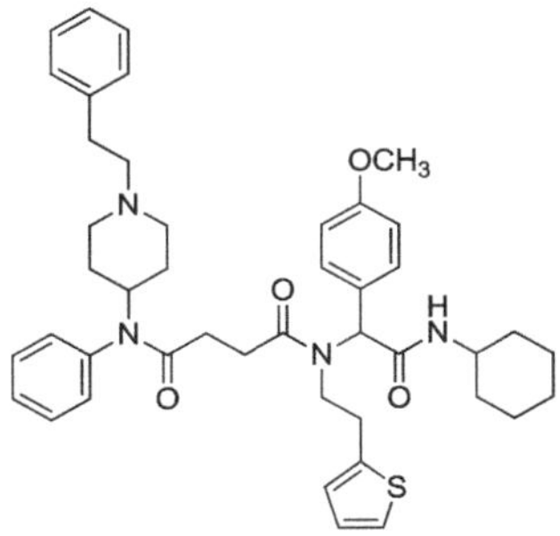

N^1-[2-(Cyclohexylamino)-1-(4-methoxyphenyl)-2-oxoethyl]-N^4-phenyl-N^4-[1-(2-phenylethyl)piperidin-4-yl]-N^1-[2-(thiophen-2-yl)ethyl]butanediamide (**5d**): white solid, mp 150–151 °C; yield: 92%

2.2.23.3 Experimental Procedure

A mixture of amine (**2**; 1 mmol) and aldehyde (**3**; 1 mmol) dissolved in MeOH (5 mL) was first stirred at room temperature for 30 min. Then, carboxylic acid (**1**; 1 mmol) was added, and stirring was continued for another 15 min, followed by addition of cyclohexyl isocyanide (**4**; 1 mmol), and the combined mixture was stirred for 24 h at room temperature. The progress of the reaction was monitored by TLC. Upon completion of the reaction, the solvent was evaporated under reduced pressure, and the resulting crude product was washed with saturated sodium bicarbonate solution (30 mL), followed by crystallization from MeOH to have pure products of amidated fentanyl derivatives (**5**) in good yields. Each of the products was characterized based on spectral studies.

2.2.23.4 Characterization Data of Two Representative Compounds

N^1-(4-Bromophenyl)-N^1-[2-(cyclohexylamino)-2-oxo-1-phenylethyl]-N^5-phenyl-N^5-[1-(2-phenylethyl)piperidin-4-yl]pentanediamide (**5b**): white solid, mp 154–155 °C; yield: 77%; IR (KBr): 3268, 3059, 2927, 2852, 2803, 1650, 1594, 1563 cm^{-1}; ^1H NMR (DMSO-d_6, 500 MHz): δ 0.97–1.94 (20H, m), 2.13 (2H, t, $J = 11.4$ Hz), 2.50–2.53 (2H, m), 2.70–2.73 (2H, m), 2.98 (2H, d, $J = 10.4$ Hz), 3.74–3.83 (1H, m), 4.55–4.65 (1H, m), 5.64 (1H, d, $J = 6.9$ Hz), 6.06 (1H, s), 6.98–7.35 (19H, m, Ar-H); ^{13}C NMR (DMSO-d_6, 125 MHz): δ 21.1, 24.8, 24.9, 25.5, 30.6, 32.8, 32.8, 33.8, 34.1, 34.1, 48.8, 52.2, 53.1, 60.5, 64.3, 122.0, 126.0, 128.3, 128.4, 128.5, 128.6, 129.3, 130.3, 130.4, 131.9, 132.5, 134.6, 138.7, 139.0, 140.3, 168.6, 171.9, 172.7; HRMS (ESI): m/z 763.32081 ($[M+1]^+$, $C_{44}H_{52}{}^{79}BrN_4O_3{}^+$; calc. 763.32069); 765.31801 ($[M+1]^+$, $C_{44}H_{52}{}^{81}BrN_4O_3{}^+$; calc. 765.31789).

N^1-[2-(Cyclohexylamino)-1-(4-methoxyphenyl)-2-oxoethyl]-N^4-phenyl-N^4-[1-(2-phenylethyl)piperidin-4-yl]-N^1-[2-(thiophen-2-yl)ethyl]butanediamide (**5d**): white solid, mp 150–151 °C; yield: 92%; IR (KBr): 3290, 3064, 3028, 2930, 2853, 2804, 1647, 1513 cm^{-1}; ^1H NMR (DMSO-d_6, 500 MHz): δ 1.16–2.74 (26H, m), 3.01–3.05 (2H, m), 3.38–3.47 (1H, m), 3.80 (3H, s), 3.81–3.85 (2H, m), 4.62–4.72 (1H, m), 6.32 (1H, s), 6.57 (1H, d, $J = 3.0$ Hz), 6.78–7.43 (17H, m, Ar-H); ^{13}C NMR (DMSO-d_6, 125 MHz): δ 25.1, 25.2, 25.7, 27.9, 30.1, 30.4, 30.5, 31.8, 32.8, 32.9, 33.1, 33.7, 48.0, 48.8, 52.5, 53.1, 55.3, 60.4, 61.0, 114.1, 123.8, 125.2, 126.1, 126.9, 128.1, 128.4, 128.6, 129.4, 129.6, 130.2, 130.7, 131.5, 138.2, 140.1, 140.3, 159.5, 168.6, 172.0, 173.5; HRMS (ESI): *m/z* 735.39245 ($[M+1]^+$, $C_{44}H_{55}N_4O_4S^+$; calc. 735.39226).

2.2.23.5 Critical Views

The investigators developed a catalyst-free, room temperature procedure for the synthesis of a new series of amidated fentanyl derivatives of potential synthetic and pharmacological applications. Good to excellent yields, high bond-forming efficiency, and tolerance of various groups toward the reaction conditions are advantages of this protocol.

2.2.24 Entry-24: Synthesis of 3-(2-Pyrazolin-5-one) substituted-3-hydroxy-2-oxindoles

Type of reaction: C–C bond formation
Reaction conditions: Catalyst-free, water, room temperature
Synthetic strategy: '*On-water*' direct C–C coupling
Keywords: Isatins, 2-pyrazolin-5-ones, water, catalyst-free, room temperature, direct C–C coupling, atom economy, no column chromatography, 3-pyrazolone substituted-3-hydroxy-2-oxindoles

2.2.24.1 General Reaction Scheme

An '*on-water*' highly efficient synthesis of a novel class of diverse 3-(2-pyrazolin-5-one) substituted-3-hydroxy-2-oxindole scaffolds (**3**) from the reaction of isatins (**1**) with 2-pyrazolin-5-one derivatives (**2**) following a catalyst-free and column chromatography-free protocol at room temperature was demonstrated by Thakur and Meshram (Scheme 2.33).[82]

2.2.24.2 Plausible Mechanism

As proposed by the investigators, 3-substituted-2-pyrazolin-5-one (**2**) first undergoes tautomerization under the influence of water, and the resulting tautomer then takes part in nucleophilic addition-type reaction with isatin (**1**) to form the adduct (**4**), followed by its aromatization to furnish the desired product (**3**) (Scheme 2.34).

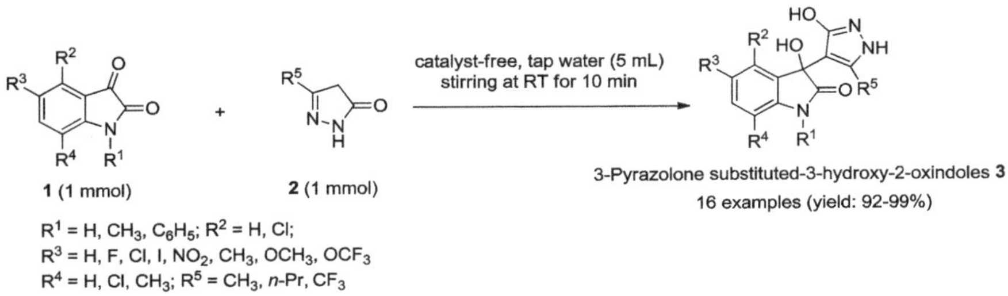

Scheme 2.33 Synthesis of 3-(2-pyrazolin-5-one) substituted-3-hydroxy-2-oxindoles (**3**).

2.2.24.3 Representative Examples

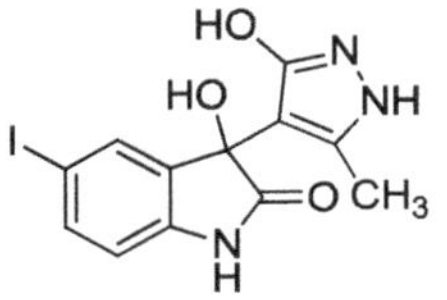

3a

3-Hydroxy-3-(3-hydroxy-5-methyl-1*H*-pyrazol-4-yl)-5-iodoindolin-2-one (**3a**): white solid, mp 218–220 °C; yield: 99%

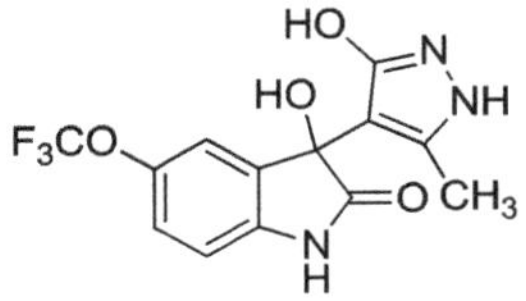

3b

3-Hydroxy-3-(3-hydroxy-5-methyl-1*H*-pyrazol-4-yl)-5-(trifluoromethoxy)indolin-2-one (**3b**): white solid, mp 193–195 °C; yield: 99%

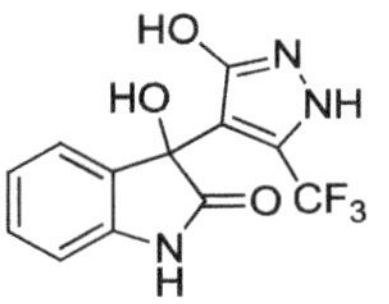

3c

3-Hydroxy-3-(3-hydroxy-5-(trifluoromethyl)-1*H*-pyrazol-4-yl)indolin-2-one (**3c**): white solid, mp 181–183 °C; yield: 92%

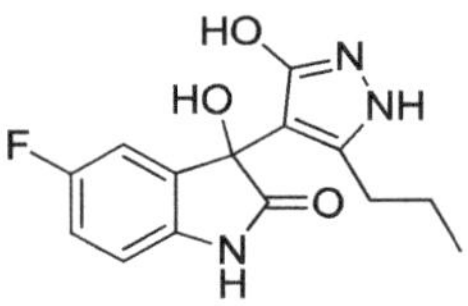

3d

5-Fluoro-3-hydroxy-3-(3-hydroxy-5-propyl-1*H*-pyrazol-4-yl)indolin-2-one (**3d**): white solid, mp 170–172 °C; yield: 99%

2.2.24.4 Experimental Procedure

A mixture of isatins (**1**; 1 mmol) and 2-pyrazolin-5-one derivatives (**2**; 1 mmol) was stirred in 5 mL of tap water at room temperature for 12 h to complete the reaction as monitored by TLC. The crude precipitate, appeared upon completion of the reaction, was filtered, washed with water and dried to afford the desired product (**3**) in very good yield and purity. All products were characterized by IR, ^{1}H NMR, ^{13}C NMR and mass spectral studies.

2.2.24.5 Characterization Data of Two Representative Compounds

3-Hydroxy-3-(3-hydroxy-5-methyl-1*H*-pyrazol-4-yl)-5-(trifluoromethoxy)indolin-2-one (**3b**): white solid, mp 193–195 °C; Yield: 99%; IR (KBr): 3372, 3201, 2569, 1727, 1600, 1533, 1487, 1275, 1186 cm^{-1}; ^{1}H NMR (DMSO-d_6, 300 MHz): δ

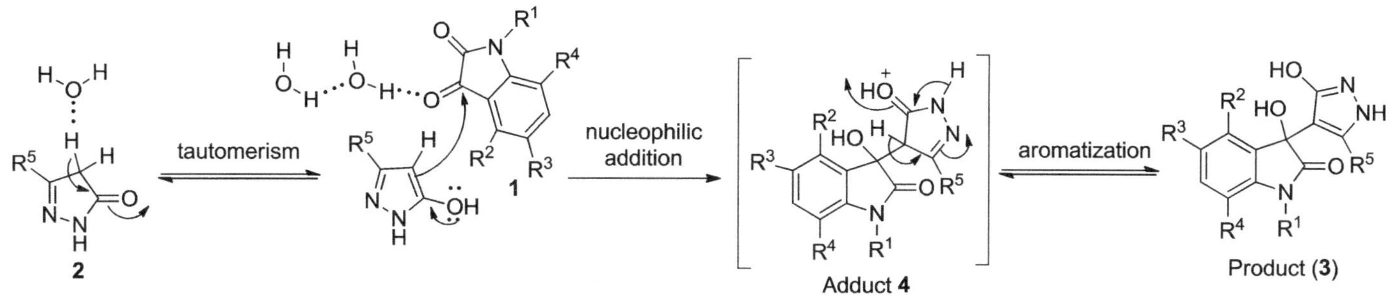

Scheme 2.34 Plausible mechamism for the synthesis of 3-(2-pyrazolin-5-one) substituted-3-hydroxy-2-oxindole derivatives (**3**).

10.43 (1H, s), 7.22–6.99 (2H, m), 6.83 (1H, d, $J = 9.1$ Hz), 6.42 (1H, br s), 2.1 (3H, s); ^{13}C NMR (DMSO-d_6, 75 MHz): δ 178.34, 158.89, 143.33, 140.79, 138.21, 134.77, 130.97, 121.87, 118.12, 110.32, 99.79, 74.02, 11.39; MS (ESI): *m/z* 330 $[M+H]^+$; HRMS (ESI): *m/z* calc. for $C_{13}H_{11}O_4N_3F_3$ $[M+H]^+ = 330.06962$, found 330.06888.

5-Fluoro-3-hydroxy-3-(3-hydroxy-5-propyl-1*H*-pyrazol-4-yl)indolin-2-one (**3d**): white solid, mp 170–172 °C; yield: 99%; IR (KBr): 3371, 3214, 2963, 2875, 2558, 1739, 1696, 1596, 1529, 1489, 1311, 1187, 1147, 815, 701, 596 cm^{-1}; ^{1}H NMR (DMSO-d_6, 300 MHz): δ 10.23 (1H, s), 7.04–6.97 (1H, m), 6.92–6.89 (1H, m), 6.79–6.75 (1H, m), 6.36 (1H, br s), 2.63–2.52 (1H, m), 2.48–2.20 (1H, m), 1.60–1.47 (2H, m), 0.82 (3H, t, $J \frac{1}{4}$ 7.2 Hz); ^{13}C NMR (DMSO-d_6, 75 MHz): 178.12, 159.46, 158.35, 156.32, 142.05, 137.93, 135.05, 134.96, 115.07, 114.76, 112.08, 111.76, δ 110.13, 110.03, 99.34, 74.24, 27.36, 22.54, 13.79; MS (ESI): *m/z* 292 $[M+H]^+$; HRMS (ESI): *m/z* calc. for $C_{14}H_{15}O_3N_3F$ $[M+H]^+ = 292.10920$, found 292.10912.

2.2.24.6 Critical Views

The present protocol is highly efficient and atom-economic for the synthesis of a wide range of functionalized 3-(2-pyrazolin-5-one) substituted-3-hydroxy-2-oxindole scaffolds under catalyst-free and column chromatography-free conditions in water at room temperature with excellent yields. A variety of functionalized isatins as well as 2-pyrazolin-5-one derivatives undergo the reaction smoothly. This method thus offers a rapid and general route for the straightforward preparation of medicinally important 3-hydroxy-2-oxindole structural scaffolds under mild reaction conditions from readily available starting materials. Such a 3-hydroxy-2-oxindole framework with diverse functionality can also provide an additional opportunity for its further transformations in preparing a library of pharmaceutically important compounds.

2.2.25 Entry-25: Synthesis of 4,5-Disubstituted 2-Benzazepines

Type of reaction: C–C and C–N bond formation
Reaction conditions: Catalyst-free, water, room temperature
Synthetic strategy: One-pot domino reaction
Keywords: 4-Chloro-3-formyl coumarin, substituted benzyl amines, catalyst-free, water, room temperature, one-pot domino reaction, no column chromatography, 4,5-disubstituted 2-benzazepines

2.2.25.1 General Reaction Scheme

A novel one-pot eco-friendly protocol for the synthesis of 4,5-disubstituted 2-benzazepine derivatives (**3**) in water under catalyst-free conditions from a domino reaction between 4-chloro-3-formyl coumarin (**1**) and benzyl amines

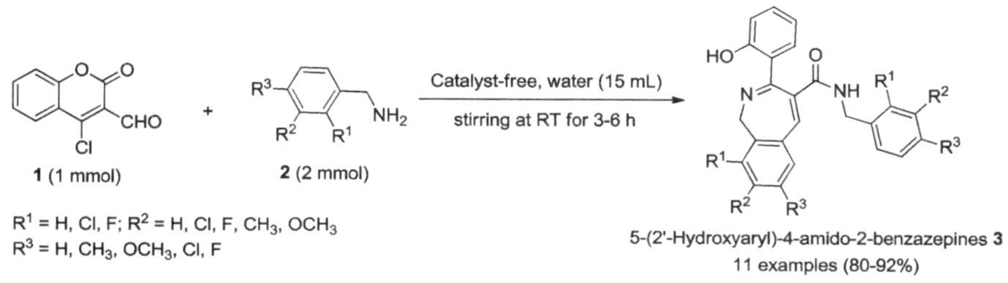

Scheme 2.35 Synthesis of 4,5-disubstituted 2-benzazepines (**3**).

(**2**) at room temperature has been demonstrated by Kumar and his group (Scheme 2.35).[83]

2.2.25.2 Plausible Mechanism

The investigators[83] proposed a plausible mechanism for the synthesis of 4,5-disubstituted 2-benzazepine (**3**) from a domino reaction of 4-chloro-3-formyl coumarin with benzyl amines as outlined in Scheme 2.36.

2.2.25.3 Representative Examples

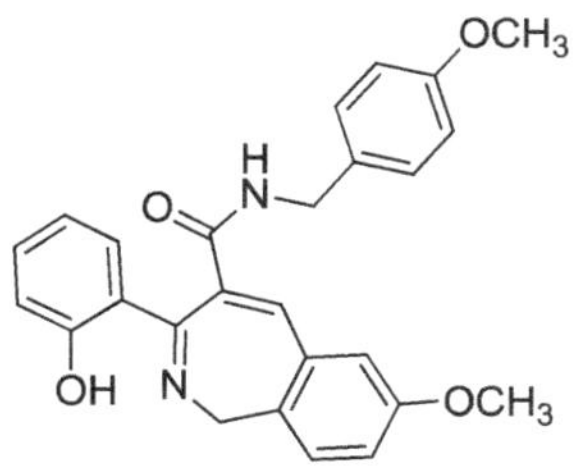

3a

3-(2-Hydroxyphenyl)-7-methoxy-*N*-(4-methoxybenzyl)-1*H*-benzo[*c*]azepine-4-carboxamide (**3a**): solid, reaction time: 5h, yield: 81%

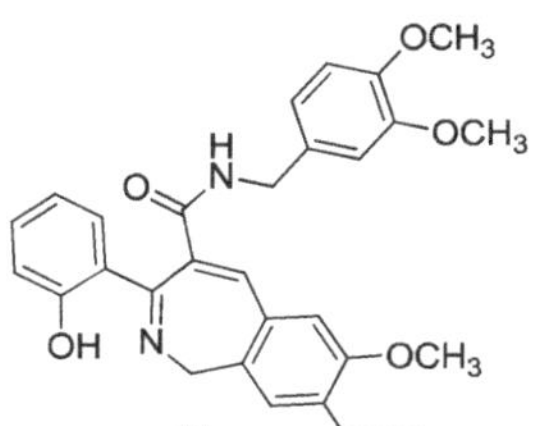

3b

N-(3,4-Dimethoxybenzyl)-3-(2-hydroxyphenyl)-7,8-dimethoxy-1*H*-benzo[*c*]azepine-4-carboxamide (**3b**): solid, reaction time: 6 h, yield: 85%

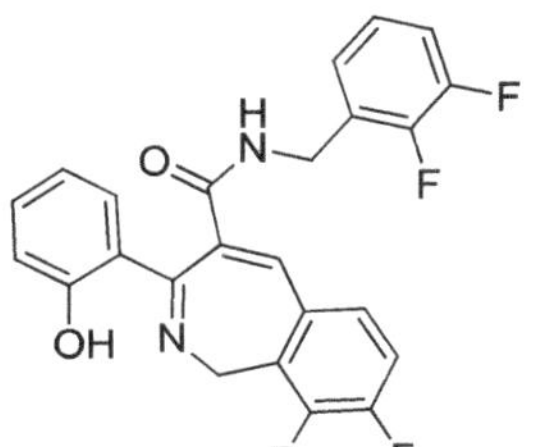

3c

N-(2,3-Difluorobenzyl)-8,9-difluoro-3-(2-hydroxyphenyl)-1*H*-benzo[*c*]azepine-4-carboxamide (**3c**): solid, reaction time: 4h, yield: 90%

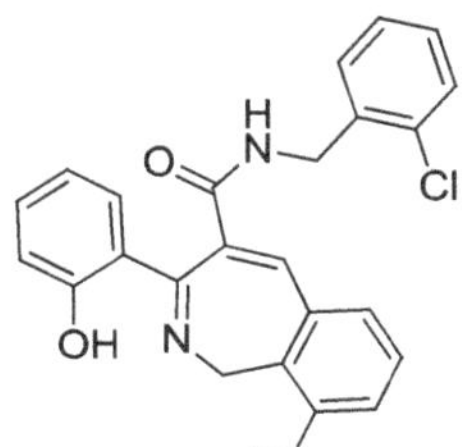

3d

9-Chloro-*N*-(2-chlorobenzyl)-3-(2-hydroxyphenyl)-1*H*-benzo[*c*]azepine-4-carboxamide (**3d**): solid, reaction time: 3h, yield: 92%

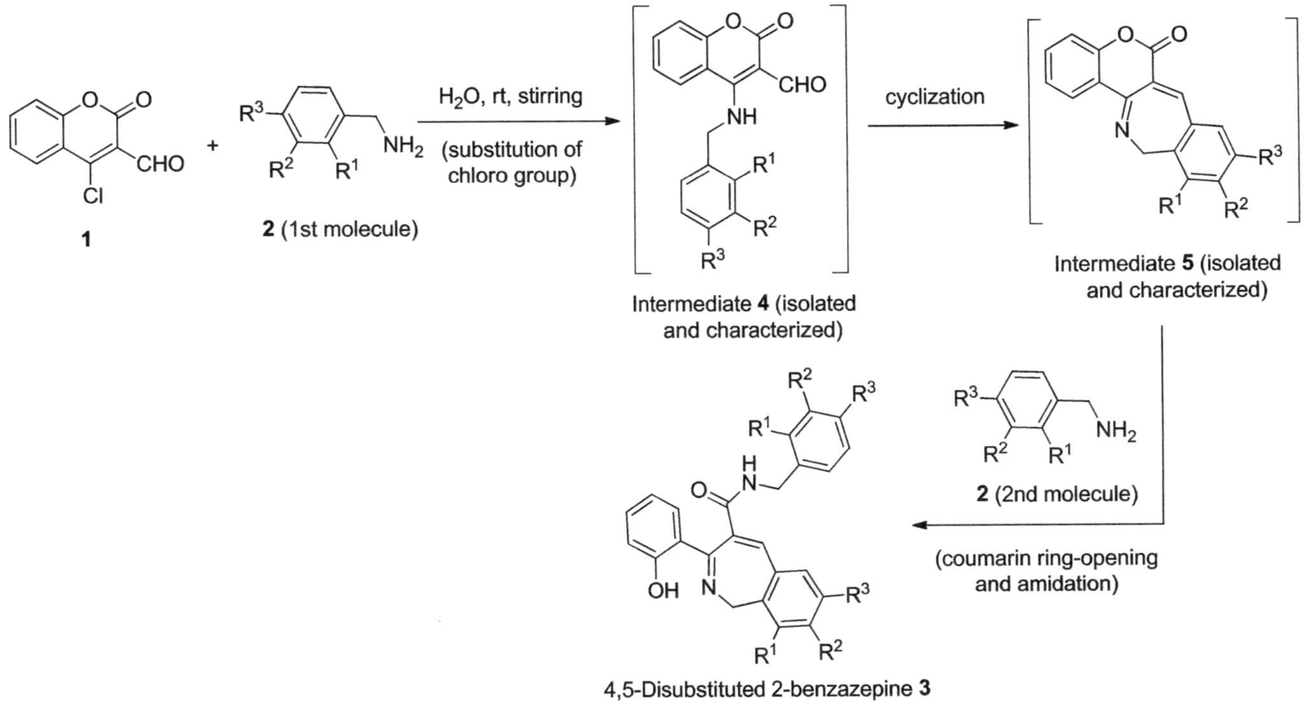

Scheme 2.36 Suggested mechanism for the generation of 4,5-disubstituted 2-benzazepines (**3**).

2.2.25.4 Experimental Procedure

A mixture of 4-chloro-3-formyl coumarin (**1**; 1.0 mmol) and substituted benzyl amines (**2**; 2 mmol) was stirred in water (15 mL) at room temperature for the stipulated time (3–6 h). On completion of the reaction, the solid crude products were filtered and washed with methanol to give pure 2-benzazepines (**3**) in good yields. All the compounds were characterized based on spectral studies compared with literature values.

2.2.25.5 Critical Views

The investigators developed a novel and efficient route for the synthesis of substituted 5-(2′-hydroxyaryl)-4-amido-2-benzazepines from readily available 4-chloro-3-formyl coumarin and benzyl amines in water under catalyst-free conditions at ambient temperature. The method does not involve complicated workup procedures and avoids the use of organic solvents. In addition, the investigators also attempted a mixed substrate combination (use of two different benzyl amines) in one exemplary case with a hope to broadening the scope of the reaction to achieve higher product diversity and they became successful in their attempt. This is also a noteworthy advantage of the present protocol.

2.2.26 Entry-26: Synthesis of Anthranilamide Schiff Bases

Type of reaction: C–N bond formation
Reaction conditions: Catalyst-free, water, room temperature
Synthetic strategy: One-pot multicomponent reaction
Keywords: Isatoic anhydride, amines, aromatic aldehydes, catalyst-free, water, room temperature, one-pot multicomponent reaction, no column chromatography, anthranilamide Schiff bases

2.2.26.1 General Reaction Scheme

A straightforward one-pot three-component catalyst-free protocol for the synthesis of anthranilamide Schiff bases (**4**) from the reaction of isatoic anhydride (**1**), amines (**2**), and aromatic aldehydes (**3**) in water in the absence of any catalyst at room temperature has recently been demonstrated by Ebrahimi *et al.* (Scheme 2.37).[84]

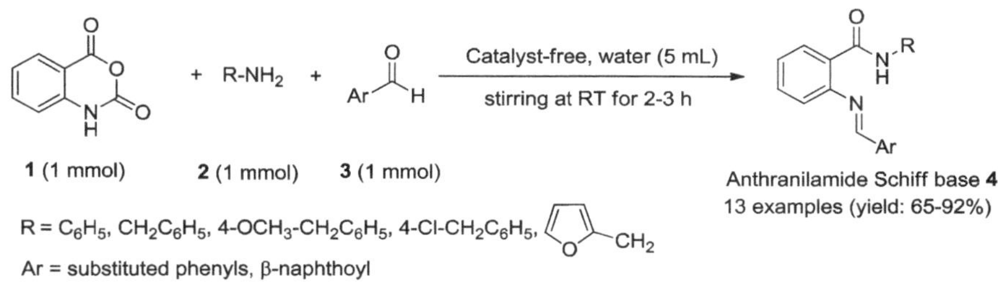

Scheme 2.37 Synthesis of anthranilamide Schiff bases (**4**).

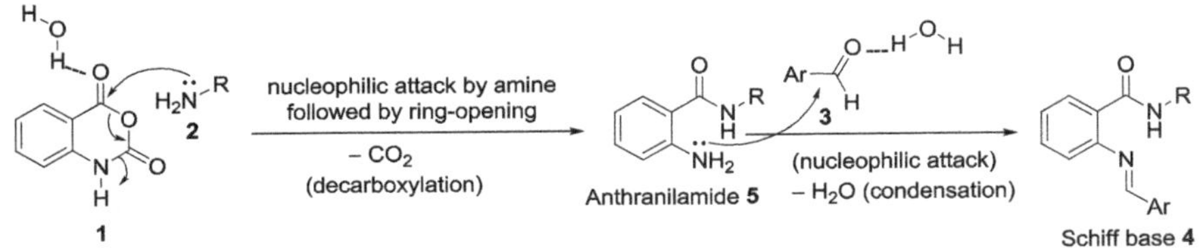

Scheme 2.38 Suggested mechanism for generating anthranilamide Schiff bases (**4**).

2.2.26.2 Plausible Mechanism

The investigators provided a plausible mechanism for the reaction as outlined in Scheme 2.38. Initially an anthranilamide intermediate (**5**) is formed out of nucleophilic attack of amine (**2**) to the carbonyl carbon of isatoic anhydride (**1**) followed by ring opening and decarboxylation. The intermediate (**5**) undergoes Schiff base formation with the aromatic aldehyde (**3**) to afford the desired product (**4**).

2.2.26.3 Representative Examples

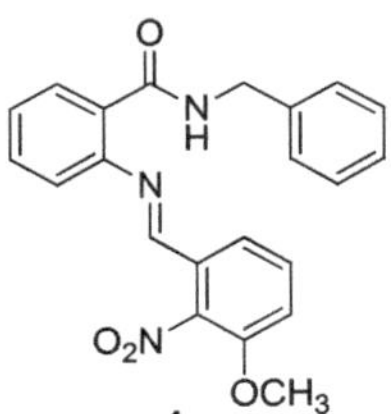

4a

(*E*)-*N*-Benzyl-2-((3-methoxy-2-nitrobenzylidene)amino)benzamide (**4a**): solid, mp 144–145 °C, yield: 86%

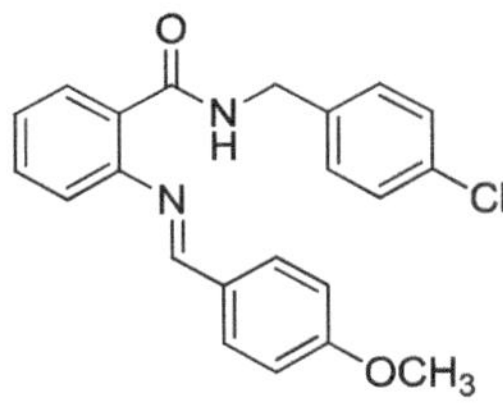

4b

(*E*)-*N*-(4-Chlorobenzyl)-2-((4-methoxybenzylidene)amino)benzamide (**4b**): solid, mp 156–157 °C, yield: 70%

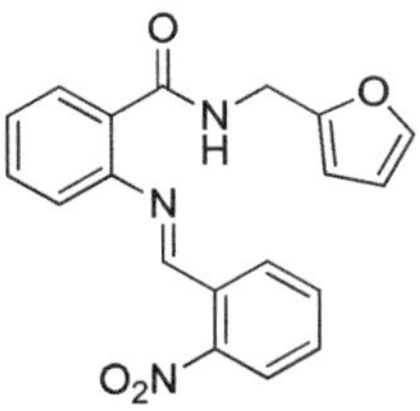

4c

(*E*)-*N*-(Furan-2-ylmethyl)-2-((2-nitrobenzylidene)amino)benzamide (**4c**):solid,mp 112–113°C, yield: 80%

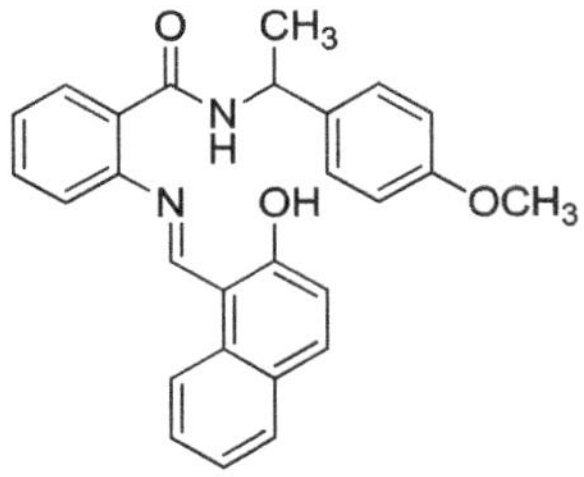

4d (Sirtinol)

(*E*)-2-(((2-Hydroxynaphthalen-1-yl)methylene)amino)-*N*-(1-(4-methoxyphenyl)ethyl)benzamide (**4d**): solid, mp 220–222°C, yield: 87%

2.2.26.4 Experimental Procedure

A mixture of isatoic anhydride (**1**; 1 mmol), amines (**2**; 1 mmol), and aromatic aldehydes (**3**; 1 mmol) in water (5 mL) was stirred at room temperature for 2–3 h (as monitored by TLC). On completion of the reaction, the precipitate was filtered off and recrystallized from ethanol to afford the pure Schiff base (**4**) in good yield. Each of the compounds was fully characterized on the elemental and spectral studies.

2.2.26.5 Characterization Data of Two Representative Compounds

(*E*)-*N*-(Furan-2-ylmethyl)-2-((2-nitrobenzylidene)amino)benzamide (**4c**): solid; yield: 80%; mp 112–113 °C; IR (KBr): 3281, 3068, 2932, 2854, 1658, 1621, 1522, 1347 cm^{-1}; ^1H NMR (DMSO-d_6, 500 MHz): δ 8.96 (1H, t, $J = 5.5$ Hz, NH), 8.88 (1H, s, NCH), 8.16 (1H, dd, $J = 7.5$, 1.0 Hz, ArH), 7.94–7.89 (1H, m, ArH), 7.87–7.79 (3H, m, ArH), 7.58 (1H, td, $J = 7.5$, 1.0 Hz, ArH), 7.54 (1H, m, furan), 7.41 (1H, t, $J = 7.5$ Hz, ArH), 7.26–7.24 (1H, m, ArH), 6.38–6.37 (m, 1H, furan), 6.31 (1H, d, $J = 3.5$ Hz, furan), 4.51 (2H, d, $J = 5.5$ Hz, NCH_2); ^{13}C NMR (DMSO-d_6, 125 MHz): δ 165.7, 158.5, 151.9, 149.2, 148.3, 142.3, 136.9, 134.0, 132.2, 131.8, 129.8, 129.5, 129.0, 126.7, 124.6, 119.1, 115.3, 110.4, 36.1; MS: *m/z* (%) = 350 ($[M+1]^+$, 11), 236 (36), 214 (77), 197 (17), 179 (41), 152 (28), 120 (31), 81 (100), 53 (55); Anal. calc. for $C_{19}H_{15}N_3O_4$: C, 65.32; H, 4.33; N, 12.03; found: C, 65.15; H, 4.11; N, 12.32.

(*E*)-2-(((2-Hydroxynaphthalen-1-yl)methylene)amino)-*N*-(1-(4-methoxyphenyl)-ethyl)benzamide (**4d**): solid; yield: 87%; mp 220–222 °C; IR (KBr): 3407, 3059, 2973, 2849, 1650, 1640, 1529 cm^{-1}; ^1H NMR (DMSO-d_6, 500 MHz): δ 15.0 (1H, s, OH), 9.49 (1H, d, $J = 13.7$ Hz, NH), 8.92 (1H, d, $J = 8.0$ Hz, NCH), 8.46 (1H, d, $J = 8.4$ Hz, ArH), 7.93 (1H, d, $J = 9.2$ Hz, ArH), 7.83–7.80 (2H, m, ArH), 7.61–7.53 (3H, m, ArH), 7.40–7.7.34 (4H, m, ArH), 7.28–7.25 (2H, m, ArH), 7.19 (1H, t, $J = 6.9$ Hz, ArH), 6.98 (1H, d, $J = 8.9$ Hz, ArH), 5.17 (1H, m, NCH), 1.43 (3H, d, $J = 7.1$ Hz, CH_3); ^{13}C NMR (DMSO-d_6, 125 MHz,): δ 170.0, 166.3, 155.6, 155.1, 144.5, 142.7, 136.9, 133.3, 131.0, 129.0, 128.3, 128.2, 126.8, 126.5, 126.0, 125.7, 123.5, 122.2, 120.4, 119.2, 108.9, 48.3, 22.2; MS: *m/z* (%) = 394 ($[M]^+$, 61), 289 (33), 273 (17), 245 (50), 230 (16), 217 (23), 171 (28), 120 (43), 105 (100), 77 (56), 51 (23); Anal. calc. for $C_{26}H_{22}N_2O_2$: C, 79.16; H, 5.62; N, 7.10; found: C, 79.02; H, 5.83; N, 6.94.

2.2.26.6 Critical Views

The investigators developed an efficient and straightforward route for an exclusive synthesis of anthranilamide Schiff bases in water at ambient temperature without using any catalyst. The present protocol offers several advantages such as simple and mild experimental procedure, excellent yields, low-cost and easy workup procedure that make it an attractive strategy for the synthesis of such pharmaceutically potent compounds like sirtinol and its analogs.

2.2.27 Entry-27: Synthesis of 1,6-Dihydropyrazine-2,3-dicarbonitriles

Type of reaction: C–C and C–N bond formation
Reaction conditions: Catalyst-free, water, room temperature

Synthetic strategy: One-pot multicomponent reaction
Keywords: Isocyanides, 2,3-diaminomaleonitrile, 3-oxopentanedioic acid, catalyst-free, water, room temperature, one-pot multicomponent synthesis, 1,6-dihydropyrazine-2,3-dicarbonitriles

2.2.27.1 General Reaction Scheme

A novel and efficient one-pot multicomponent reaction was developed by Shaabani *et al.*[85] for the synthesis of 1,6-dihydro-6,6-dimethylpyrazine-2,3-dicarbonitriles (**4**) from the reaction between alkyl or aryl isocyanides (**1**), 2,3-diaminomaleonitrile (**2**) and 3-oxopentanedioic acid (**3**) in water at room temperature (Scheme 2.39).

2.2.27.2 Plausible Mechanism

A possible mechanism for the formation of product (**4**) was suggested by the authors (Scheme 2.40). Initially, a condensation reaction occurs between 2,3-diaminomaleonitrile (**2**) and 3-oxopentanedioic acid (**3**) to form an imine derivative (**5**) which immediately undergoes decarboxylation leading to the formation of another imine derivative (**6**). In the next step, an intermediate (**7**) is produced by a nucleophilic attack of isocyanide (**1**) on **6**, followed by an intramolecular nucleophilic attack by the $-NH_2$ group at the activated nitrile moiety to give intermediate (**8**). Finally, imine-enamine tautomerization of intermediate (**8**) affords the 1,6-dihydropyrazine-2,3-dicarbonitrile derivative (**4**).

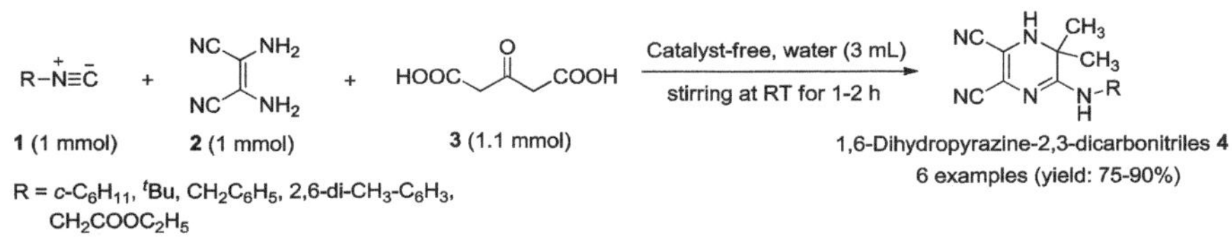

Scheme 2.39 Synthesis of 1,6-dihydropyrazine-2,3-dicarbonitriles (**4**).

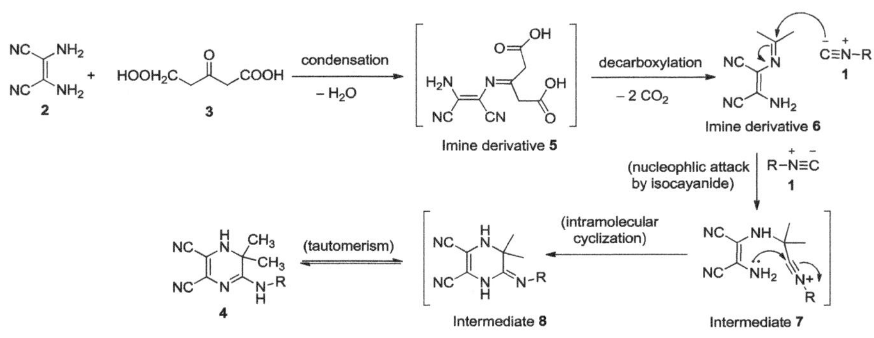

Scheme 2.40 Proposed mechanism for the synthesis of 1,6-dihydropyrazine-2,3-dicarbonitriles (**4**).

2.2.27.3 Representative Examples

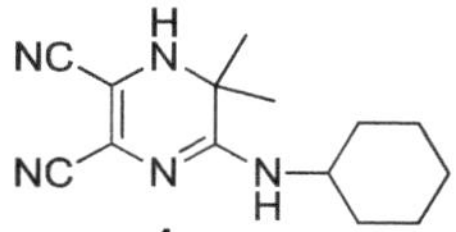

4a

5-(Cyclohexylamino)-6,6-dimethyl-1,6-dihydropyrazine-2,3-dicarbonitrile (**4a**): colorless crystals, mp 252–255 °C; yield: 90%

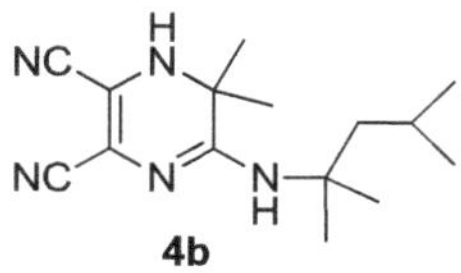

4b

5-((2,4-Dimethylpentan-2-yl)amino)-6,6-dimethyl-1,6-dihydropyrazine-2,3-dicarbonitrile (**4b**): white powder, mp 149–151 °C; yield: 80%

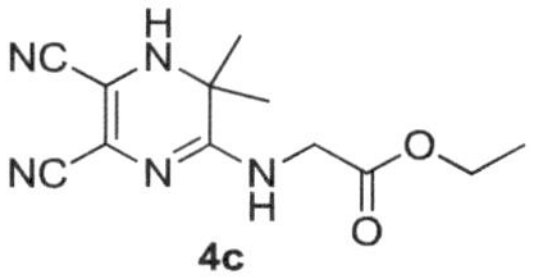

4c

Ethyl 2-((5,6-dicyano-3,3-dimethyl-3,4-dihydropyrazin-2-yl)amino)acetate (**4c**): colorless crystals, mp 190–192°C; yield:78%

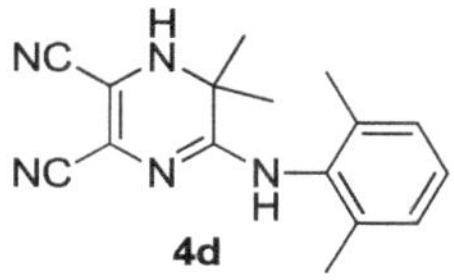

4d

5-((2,6-Dimethylphenyl)amino)-6,6-dimethyl-1,6-dihydropyrazine-2,3-dicarbonitrile (**4d**): colorless crystals, mp 260–262 °C; yield: 75%

2.2.27.4 Experimental Procedure

A mixture of alkyl or aryl isocyanides (**1**; 1.0 mmol), 2,3-diaminomaleonitrile (**2**; 1.0 mmol) and 3-oxopentanedioic acid (**3**; 1.1 mmol) in water (3 mL) was stirred for 1–2 h at room temperature. After completion of the reaction, as monitored by TLC (ethyl acetate-hexane = 3 : 1), the crude precipitate was filtered off and then purified by crystallization from acetone to obtain the desired product, 1,6-dihydro-6,6-dimethylpyrazine-2,3-dicarbonitriles (**4**). All the products were characterized by their analytical and spectral studies.

2.2.27.5 Characterization Data of Two Representative Compounds

5-((2,4-Dimethylpentan-2-yl)amino)-6,6-dimethyl-1,6-dihydropyrazine-2,3-dicarbonitrile (**4b**): white powder, mp 149–151 °C; yield: 80%; IR(KBr): 3412, 3344, 2981, 2966, 2950, 2903, 2861, 2212, 1577, 1559, 1542, 1448, 1393, 1372 cm^{-1}; ^1H NMR (DMSO-d_6, 300 MHz): δ 0.93 (9H, s, 3×CH_3), 1.17 (6H, s, 2×CH_3), 1.36 (6H, br. s, 2×CH_3), 1.82 (2H, br. s, CH_2), 6.04 (1H, s, NH), 7.12 (1H, s, NH); ^{13}C NMR (DMSO-d_6, 75 MHz): δ 23.9, 29.3, 31.6, 31.8, 49.2, 49.8, 56.4, 109.8, 110.1, 114.9, 118.4, 154.0; EIMS (%rel): 287 (17, M^+), 176 (48), 160 (100), 133 (15), 97 (17), 57 (60), 41 (63). Anal. calc. for $C_{16}H_{25}N_5$: C, 66.86; H, 8.77; N, 24.37; found: C, 66.75; H, 8.71; N, 24.27.

Ethyl 2-((5,6-dicyano-3,3-dimethyl-3,4-dihydropyrazin-2-yl)amino)acetate (**4c**): colorless crystals, mp 190–192 °C; yield: 78%; IR(KBr): 3379, 3293, 3080, 3049, 2997, 2960, 2934, 2217, 1744, 1579, 1502, 1461, 1406, 1322 cm^{-1}; ^1H

NMR (DMSO-d_6, 300 MHz): δ 1.17 (3H, t, $J=7.2$, CH_3), 1.24 (6H, s, 2×CH_3), 3.90 (2H, d, $J=5.4$, CH_2), 4.09 (2H, q, $J=6.9$, CH_2), 7.37 (1H, s, NH), 7.77 (1H, br s, NH); ^{13}C NMR (DMSO-d_6, 75 MHz): δ 14.5, 24.4, 43.0, 50.0, 60.8, 106.5, 109.6, 111.7, 114.5, 118.2, 156.6, 169.8; EIMS (%rel): 262 (50, $[M+1]^+$), 246 (45), 172 (100), 133 (25), 42 (27). Anal. calc. for $C_{12}H_{15}N_5O_2$: C, 55.16; H, 5.79; N, 26.80; found: C, 55.11; H, 5.72; N, 26.70.

2.2.27.6 Critical Views

The investigators[85] reported on an isocyanide-based multicomponent reaction (IMCR) for the one-pot catalyst-free synthesis of biologically relevant highly-substituted 1,6-dihydropyrazine-2,3-dicarbonitriles in water from the reaction of commercially available isocyanides, 2,3-diaminomaleonitrile and 3-oxopentanedioic acid at ambient temperature and pressure. Easy reaction set-up, facile workup procedure, catalyst-free and mild reaction conditions, no column chromatographic purification, and high yields are the notable features of this present protocol; however, the investigators synthesized just six compounds of this series. Scope of this present protocol is thus limited.

2.2.28 Entry-28: Synthesis of Polyhydroquinolines

Types of reaction: C–C and C–N bond formation
Reaction conditions: Catalyst- and solvent-free, grinding, room temperature
Synthetic strategy: One-pot multicomponent
Keywords: Aldehydes, dimedone, active methylene compounds, ammonium acetate, solvent-free, catalyst-free, room temperature, grinding, one-pot multicomponent synthesis, Hantzsch condensation, polyhydroquinolines

2.2.28.1 General Reaction Scheme

Kumar *et al.*[86] developed a facile method for the synthesis of polyhydroquinoline derivatives (**5**) *via* a four-component reaction of aldehydes (**1**), dimedone (**2**), active methylene compounds (**3**), and ammonium acetate (**4**) in one-pot under solvent-free conditions at room temperature on grinding (Scheme 2.41). Using malononitrile/ethyl cyanoacetate (**6**) in place of

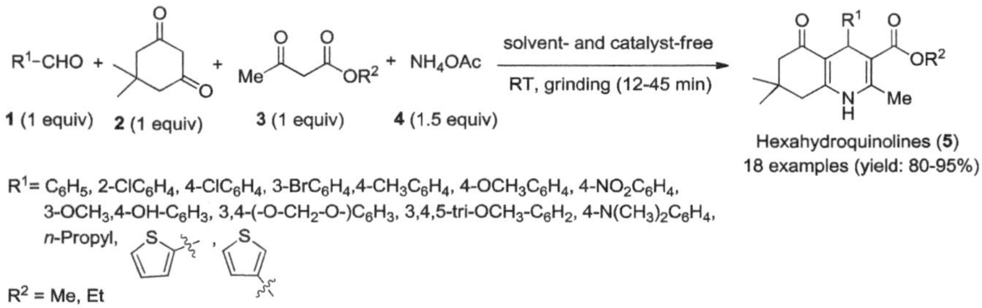

Scheme 2.41 Synthesis of polyhydroquinolines (5).

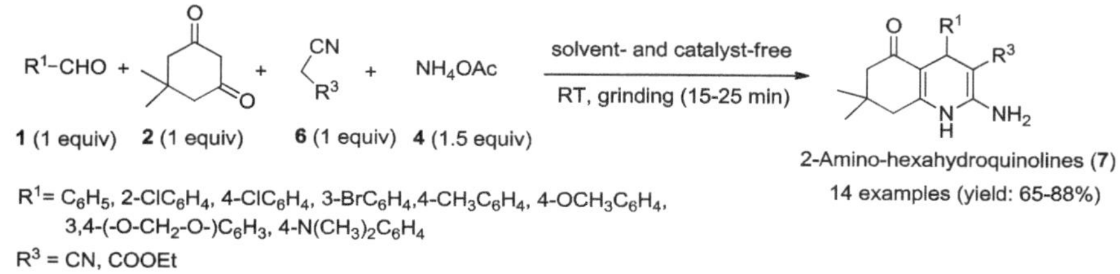

Scheme 2.42 Synthesis of polyhydroquinolines (7).

β-ketoesters, the investigators observed that the reaction follows the expected route to yield 2-amino-hexahydroquinolines (7) in good to excellent yields under the same reaction conditions (Scheme 2.42).[86]

2.2.28.2 Representative Examples

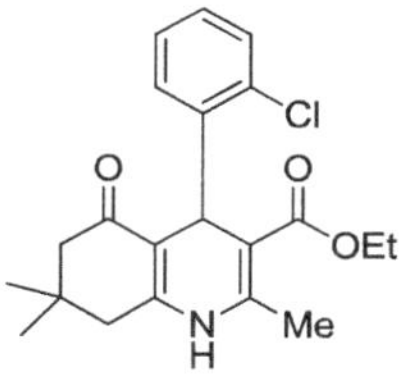

5a

Ethyl-1,4,7,8-tetrahydro-2,7,7-trimethyl-4-(2-chlorolphenyl)-5(6*H*)-oxoquinolin-3-carboxylate (**5a**): yellow solid, mp 209–211°C; reaction time: 15 min; yield: 93%

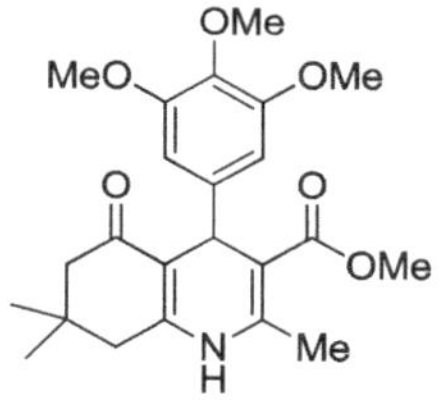

5b

Methyl-1,4,7,8-tetrahydro-2,7,7-trimethyl-4-(3,4,5-trimethoxylphenyl)-5(6*H*)-oxoquinolin-3-carboxylate (**5b**): yellow solid, mp 203–206 °C; reaction time: 20 min; yield: 88%

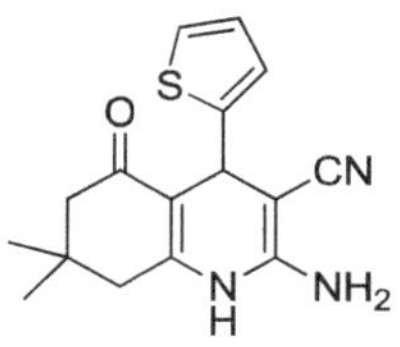

7a

2-Amino-4-(2′-thienyl)-3-cyano-7,7-dimethyl-5-oxo-1,4,5,6,7,8-hexahydroquinoline (**7a**): yellow solid, mp 253–255 °C; reaction time: 20 min; yield: 74%

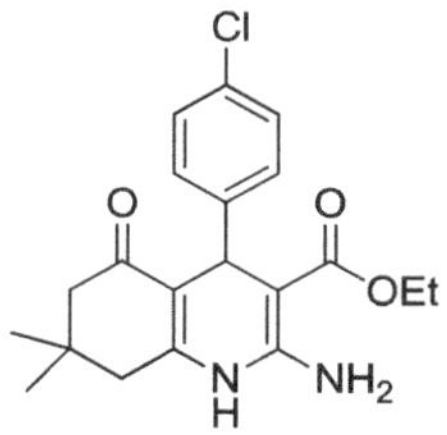

7b

2-Amino-7,7-dimethyl-5-oxo-4-(4′-chlorophenyl)-1,4,5,6,7,8-hexahydroquinoline-3-carboxylic acid ethyl ester (**7b**): yellow solid, mp 174–176 °C; reactiontime: 15min; yield: 75%

2.2.28.3 Experimental Procedure

General procedure for the synthesis of 1,4,7,8-tetrahydro-2,7,7-trimethyl-4-aryl-5(6H)-oxoquinolin-3-carboxylic acid alkyl ester (5). A mixture of aldehydes

(**1**; 2 mmol), dimedone (**2**; 2 mmol), β-ketoesters (**3**; 2 mmol), and ammonium acetate (**4**; 3 mmol) was thoroughly mixed in a mortar with a pestle, followed by grinding, until the completion of the reaction, as indicated by TLC (12–45 min). The resultant mass was then washed with water to remove any unreacted ammonium acetate and was air-dried to give a crude product which on recrystallization from ethanol afforded the pure desired compound (**5**) in excellent yield (80–95%).

General procedure for the synthesis of 2-amino-4-phenyl-3-cyano-7,7-dimethyl-5-oxo-1,4,5,6,7,8-hexahydroquinoline (**7**). A mixture of aldehydes (**1**; 2 mmol), dimedone (**2**; 2 mmol), malononitrile or ethyl cyanoacetate (**6**; 2 mmol), and ammonium acetate (**4**; 3 mmol) was mixed thoroughly in a mortar with a pestle, followed by grinding, until the completion of the reaction, as indicated by TLC (15–25 min). The resultant material was then washed with water to remove any unreacted ammonium acetate and was air-dried to furnish the crude product. The pure product (**7**) was obtained by recrystallization from ethyl alcohol.

All the products were characterized from elemental analyses as well as their detailed spectral studies including IR, ^{1}H NMR, ^{13}C NMR and MS.

2.2.28.4 Characterization Data of Two Representative Compounds

Methyl-1,4,7,8-tetrahydro-2,7,7-trimethyl-4-(3,4,5-trimethoxylphenyl)-5(6*H*)-oxoquinolin-3-carboxylate (**5b**): yellow solid, mp 203–206 °C; IR (KBr): 3280, 3185, 3058, 2931, 1686, 1644, 1603, 1490, 1382, 1333, 1227, 1114, 1004, 787 cm^{-1}; ^{1}H NMR (DMSO-d_6, 200 MHz): δ 0.94 (3H, s, CH_3), 1.03 (3H, s, CH_3), 2.00–2.51 (4H, m, 2×CH_2), 2.27 (3H, s, CH_3), 3.47 (3H, s, CO_2Me), 3.59 (3H, s, OMe), 3.67 (6H, s, 2×OMe), 4.83 (1H, s, CH), 6.41 (2H, s, ArH), 8.39 (1H, s, NH); ^{13}C NMR (DMSO-d_6, 50 MHz): δ 18.6, 26.7, 29.6, 32.5, 35.9, 50.6, 55.9, 60.2, 103.0, 104.7, 110.0, 136.1, 143.5, 145.4, 150.3, 152.7, 167.8, 194.9; MS (ESI): *m/z* 416.0 $(M+H)^+$. Anal. calc. for $C_{23}H_{29}NO_6$: C, 66.49; H, 7.04; N, 3.37; found: C, 66.57; H, 7.11; N, 3.46.

2-Amino-7,7-dimethyl-5-oxo-4-(4′-chlorophenyl)-1,4,5,6,7,8-hexahydroquinoline-3-carboxylic acid ethyl ester (**7b**): yellow solid, mp 174–176 °C; IR (KBr): 3378, 3319, 3286, 2198, 1679, 1607 cm^{-1}; ^{1}H NMR (DMSO-d_6, 200 MHz): δ 0.92 (3H, s, CH_3), 1.01 (3H, s, CH_3), 2.06–2.29 (4H, m, 2×CH_2), 4.51 (1H, s, CH), 5.85 (2H, s, NH_2), 6.85–7.12 (3H, m, ArH), 8.98 (1H, s, NH); MS (ESI): *m/z* 300.2 $(M+H)^+$. Anal. calc. for $C_{16}H_{17}N_3OS$: C, 64.19; H, 5.72; N, 14.04; found: C, 64.15; H, 5.68; N, 13.99.

2.2.28.5 Critical Views

Due to considerable biological importance of the polyhydroquinolines, several earlier methods involving a wide range of homogeneous and heterogeneous catalysts and varying techniques such as heating, refluxing and microwave irradiation, were reported.[87–94] The present method[86] seems to be quite advantageous over the earlier ones in terms of avoidance of toxic organic solvent, room temperature condition, short reaction time, high yields, simple workup and easy purification without the use of column chromatography.

2.2.29 Entry-29: Synthesis of Functionalized 1,3,5-Trisubstituted Hydantoins

Type of reaction: C–N bond formation
Reaction conditions: Ethanol, room temperature
Synthetic strategy: Sequential three-component one-pot reaction
Keywords: 1,2-Diaza-1,3-dienes, amines, isocyanates, ethanol, room temperature, sequential three-component reaction, aza-Michael reaction, one-pot synthesis, functionalized 1,3,5-trisubstituted hydantoins

2.2.29.1 General Reaction Scheme

One-pot synthesis of functionalized 1,3,5-trisubstituted hydantoins (**4**) from the sequential aza-Michael addition/condensation reactions of 1,2-diaza-1,3-dienes (**1**) with primary amines (**2**) and isocyanates (**3**) in ethanol under ambient conditions was reported for the first time by Attanasi *et al.* (Scheme 2.43).[95]

2.2.29.2 Plausible Mechanism

A plausible mechanism for such transformation is suggested in Scheme 2.44. 1,2-Diaza-1,3-diene (**1**) acts as a Michael acceptor and produces corresponding α-aminohydrazone (**5**) *via* an aza-Michael reaction with the primary amine (**2**). The intermediate (**5**) undergoes subsequent acylation with isocyanate (**3**), thereby generating the requisite asymmetric urea (**6**) that provides spontaneous regioselective hetero-ring closure owing to the nucleophilic attack of the amidic NH at the terminal ester function of the azoene system, affording the hydantoin derivative (**4**) by loss of an alcohol molecule.

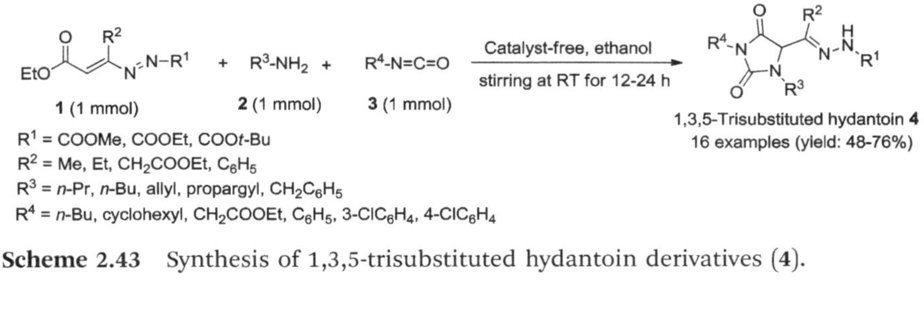

Scheme 2.43 Synthesis of 1,3,5-trisubstituted hydantoin derivatives (**4**).

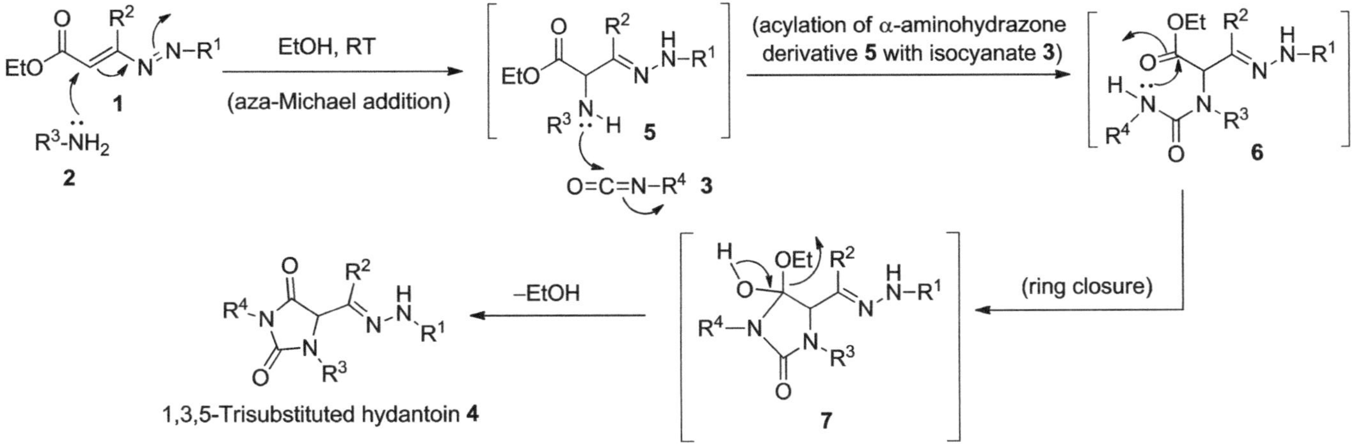

Scheme 2.44 Proposed mechanism for the synthesis of 1,3,5-trisubstituted hydantoins (**4**).

2.2.29.3 Representative Examples

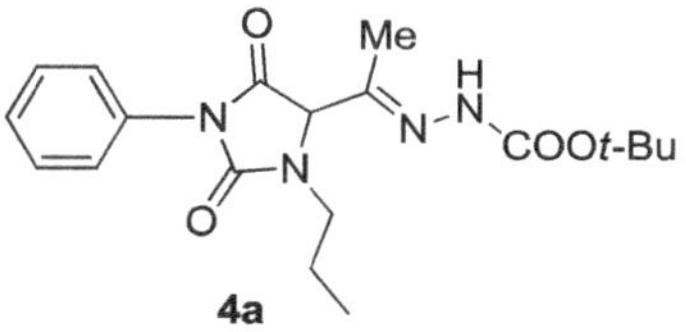

4a

(*E*)-*tert*-Butyl 2-(1-(2,5-dioxo-1-phenyl-3-propylimidazolidin-4-yl)ethylidene)hydrazinecarboxylate (**4a**): white powder, mp 179–180°C; yield: 76%

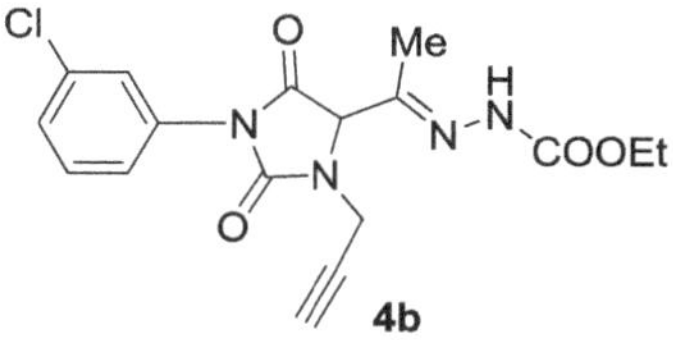

4b

(*E*)-Ethyl 2-(1-(1-(3-chlorophenyl)-2,5-dioxo-3-(prop-2-yn-1-yl)imidazolidin-4-yl)ethylidene)hydrazinecarboxylate (**4b**): white powder, mp 151–153°C; yield: 63%

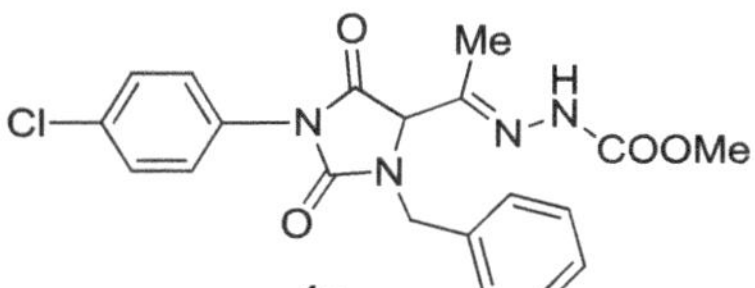

4c

(*E*)-Methyl 2-(1-(3-benzyl-1-(4-chlorophenyl)-2,5-dioxoimidazolidin-4-yl)ethylidene)hydrazinecarboxylate (**4c**): white powder, mp 192–194 °C; yield: 73%

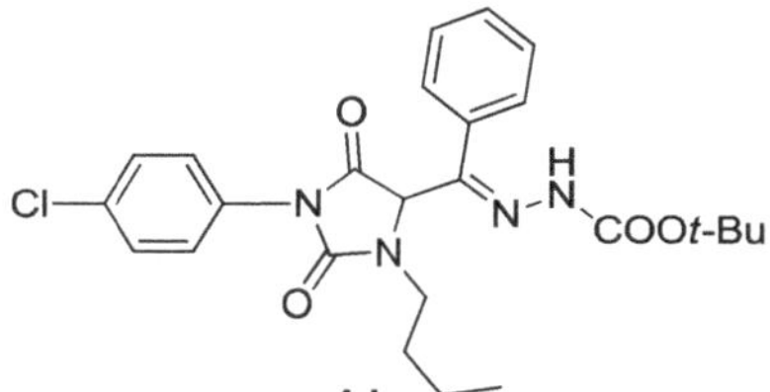

4d

(*E*)-*tert*-Butyl 2-((3-butyl-1-(4-chlorophenyl)-2,5-dioxoimidazolidin-4-yl)(phenyl)methylene)hydrazinecarboxylate (**4d**): white powder, mp 172–176 °C; yield: 48%

2.2.29.4 Experimental Procedure

To a stirred solution of 1,2-diaza-1,3-diene (**1**; 1 mmol) in ethanol, an amine (**2**; 1 mmol) was added, and stirring was continued at room temperature. On dissolution of **1** (within 0.1–0.5 h as checked by TLC), isocyanate (**3**) was added and the reaction mixture was allowed to stand at room temperature (12–24 h). In some cases, solid hydantoin derivative (**4**) was obtained by precipitation from the reaction medium or by crystallization from appropriate solvents, while, in other cases, it was obtained after removal of ethanol *in vacuo* and subsequent column chromatographic purification. All the products were characterized based on elemental analyses and detailed spectral studies.

2.2.29.5 Characterization Data of Two Representative Compounds

(*E*)-Ethyl 2-(1-(1-(3-chlorophenyl)-2,5-dioxo-3-(prop-2-yn-1-yl)imidazolidin-4-yl)ethylidene)hydrazinecarboxylate (**4b**): white powder, mp 151–153 °C; IR (nujol): 3244, 3209, 3038, 1780, 1722, 1709, 1645, 1595, 1583, 1538 cm^{-1}; ^1H NMR (DMSO-d_6, 400 MHz): δ 1.85 (3H, s, CH_3), 3.36 (1H, s, CH), 3.70 (3H, s, OCH_3), 4.21 (2H, s, CH_2), 4.88 (1H, s, CH), 7.42 (1H, s, ArH), 7.51–7.57 (3H, m, ArH), 10.37 (1H, s, NH); ^{13}C NMR (DMSO-d_6, 100 MHz): δ 12.4 (q), 31.4 (t), 52.0 (q), 66.5 (d), 75.6 (d), 77.6 (s), 125.4 (d), 126.5 (d), 128.2 (d), 130.5 (d), 132.9 (s), 133.0 (s), 144.5 (s), 154.2 (s), 154.4 (s), 168.0 (s); MS (EI): *m/z* (%) = 364 ([M + 2]$^+$, 1), 362 (M$^+$, 3), 288 (10), 153 (22), 135 (27), 111 (100). Anal. calc. for $C_{16}H_{15}ClN_4O_4$: C, 52.97; H, 4.17; N, 15.44; found: C, 52.84; H, 4.26; N, 15.35.

(*E*)-Methyl 2-(1-(3-benzyl-1-(4-chlorophenyl)-2,5-dioxoimidazolidin-4-yl)ethylidene) hydrazinecarboxylate (**4c**): white powder, mp 192–194 °C; IR (nujol): 3211, 1780, 1732, 1713, 1704, 1637, 1542 cm^{-1}; ^{1}H NMR (DMSO-d_6, 400 MHz): δ 1.84 (3H, s, CH_3), 3.69 (3H, s, OCH_3), 4.44 (1H, d, J = 15.2 Hz, NCH_aH_bPh), 4.54 (1H, d, J = 15.2 Hz, NCH_aH_bPh), 4.79 (1H, s, CH), 7.27–7.36 (5H, m, ArH), 7.47–7.50 (2H, m, ArH), 7.57–7.60 (2H, m, ArH), 10.22 (1H, s, NH); ^{13}C NMR (DMSO-d_6, 100 MHz): δ 12.1 (q), 45.4 (t), 52.0 (q), 66.8 (d), 127.6 (d), 128.3 (d), 128.5 (d), 128.6 (d), 128.9 (d), 130.8 (s), 132.6 (s), 135.9 (s), 144.6 (s), 154.3 (s), 154.9 (s), 168.3 (s); MS (EI): *m/z* (%) = 416 ([M + 2]$^+$, 1), 414 (M$^+$, 3), 340 (10), 324 (89), 209 (4), 153 (57), 132 (100). Anal. calc. for $C_{20}H_{19}ClN_4O_4$: C, 57.90; H, 4.62; N, 13.51; found: C, 58.02; H, 4.72; N, 13.47.

2.2.29.6 Critical Views

The present work offers a synthetic strategy for regioselective 1,3,5-trisubstituted hydantoin derivatives under mild reaction conditions in the absence of any catalyst for the first time. The hydrazone side chain introduced by the conjugated heterodiene system at the 5-position of the heterocycle represents a valuable functionality for accessing novel 5-acyl derivatives difficult to obtain by other methods.[96–104]

2.2.30 Entry-30: Synthesis of 1,3,5-Trisubstituted-2-thiohydantoins

Type of reaction: C–N bond formation
Reaction conditions: Catalyst-free, chloroform, room temperature
Synthetic strategy: Sequential three-component one-pot reaction
Keywords: 1,2-Diaza-1,3-dienes, amines, isothiocyanates, chloroform, room temperature, sequential three-component reaction, one-pot synthesis, 1,3,5-trisubstituted-2-thiohydantoins

2.2.30.1 General Reaction Scheme

One-pot and catalyst-free efficient synthesis of functionalized 1,3,5-trisubstituted 2-thiohydantoins (**4**) from the sequential aza-Michael addition/condensation reactions of 1,2-diaza-1,3-dienes (**1**) with primary amines (**2**) and isothiocyanates (**3**), under ambient conditions, was also accomplished for the first time by the same group of investigators (Attanasi *et al.*) (Scheme 2.45).[105]

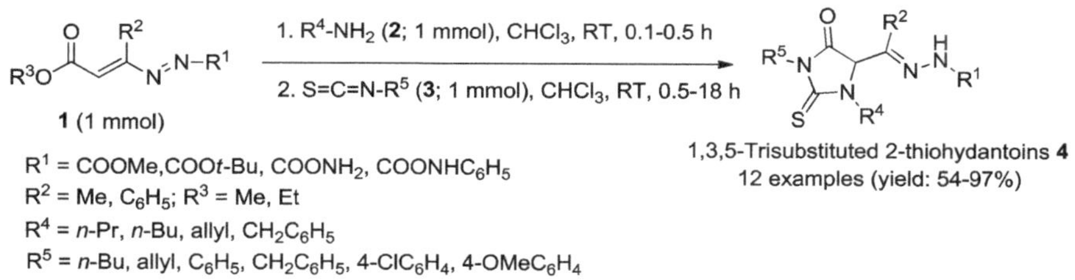

Scheme 2.45 Synthesis of 1,3,5-trisubstituted-2-thiohydantoins (**4**).

2.2.30.2 Plausible Mechanism

The investigators suggested a plausible mechanism for this transformation (Scheme 2.46). 1,2-Diaza-1,3-diene (**1**) undergoes aza-Michael condensation with the primary amine (**2**) to produce the corresponding α-aminohydrazone (**5**) which then couples with isothiocyanate (**3**); such an acylation reaction generates the requisite asymmetric isothiourea (**6**), which provides spontaneous regioselective hetero-ring closure owing to the nucleophilic attack of the amidic NH at the terminal ester function, thus affording the 2-thiohydantoin derivative (**4**) by loss of an alcohol molecule.

2.2.30.3 Representative Examples

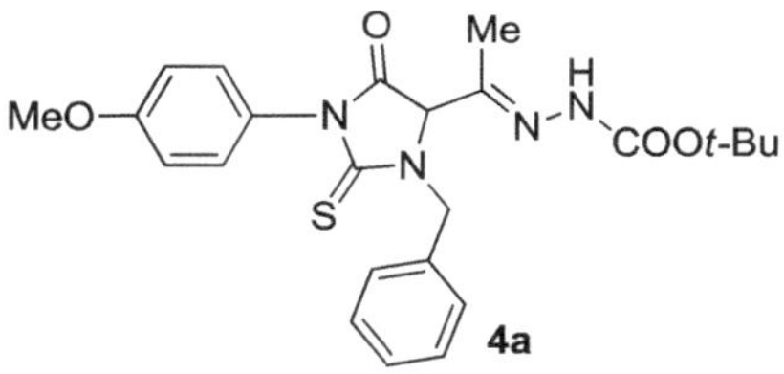

4a

(*E*)-*tert*-Butyl 2-(1-(3-benzyl-1-(4-methoxyphenyl)-5-oxo-2-thioxoimidazolidin-4-yl)ethylidene)hydrazinecarboxylate (**4a**): white powder, mp 193–195 °C; yield: 86%

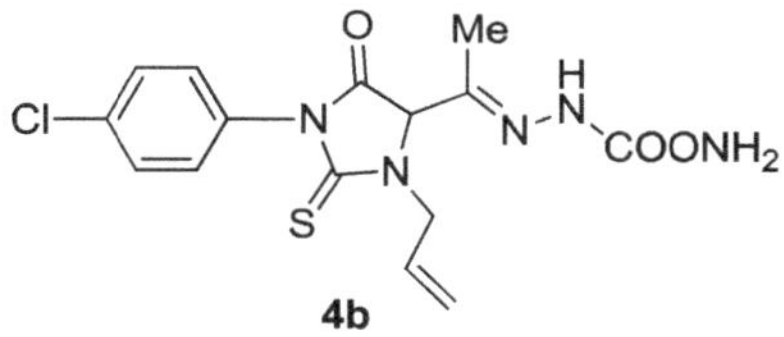

4b

(*E*)-1-Allyl-5-(1-(2-((aminooxy)carbonyl)hydrazono)ethyl)-3-(4-chlorophenyl)-2-thioxoimidazolidin-4-one (**4b**): white powder, mp 182–184 °C; yield: 97%

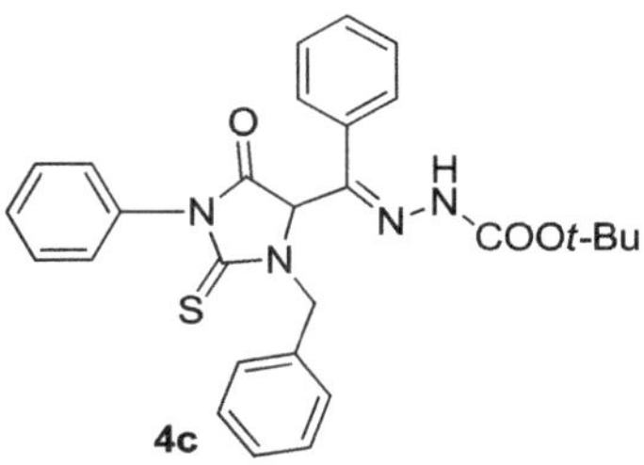

4c

(*E*)-*tert*-Butyl 2-((3-benzyl-5-oxo-1-phenyl-2-thioxoimidazolidin-4-yl)(phenyl)methylene)hydrazinecarboxylate (**4c**): white powder, mp 181–183 °C; yield: 75%

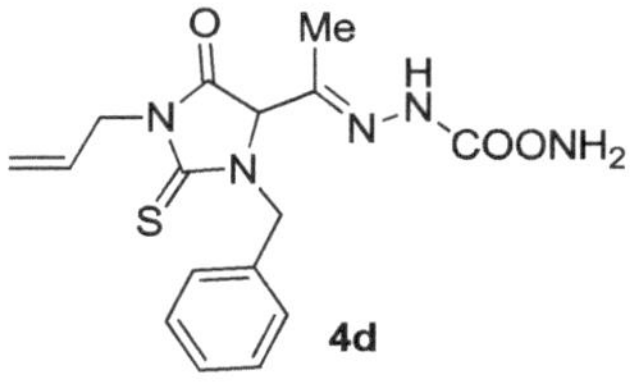

4d

(*E*)-3-Allyl-5-(1-(2-((aminooxy)carbonyl)hydrazono)ethyl)-1-benzyl-2-thioxoimidazolidin-4-one (**4d**): white powder, mp 163–166 °C; yield: 54%

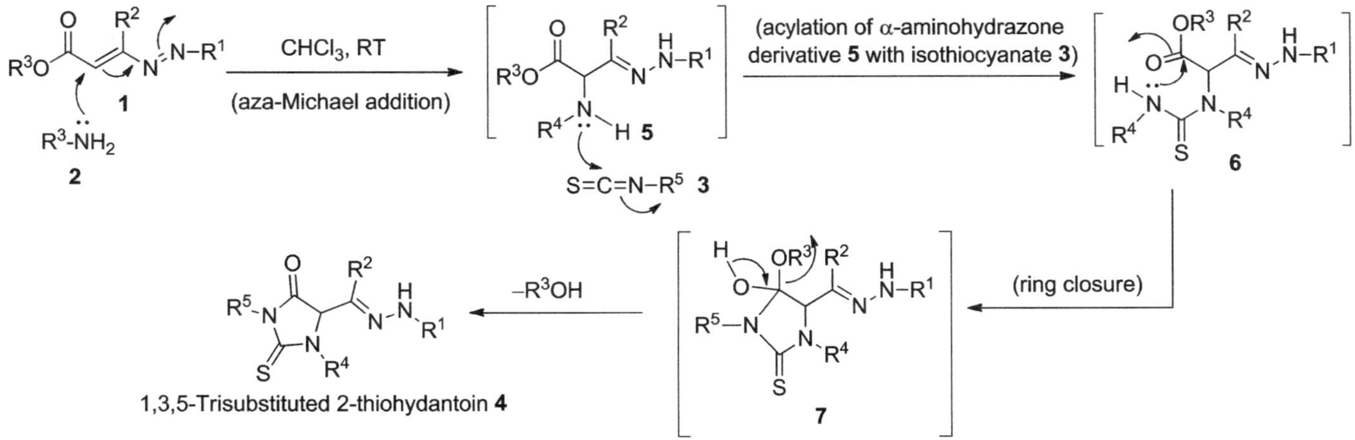

Scheme 2.46 Proposed mechanism for the synthesis of 1,3,5-trisubstituted-2-thiohydantoins (**4**).

2.2.30.4 *Experimental Procedure*

To a stirred solution of 1,2-diaza-1,3-diene (**1**: 1 mmol) in chloroform, amine (**2**; 1 mmol) was added and stirred at room temperature. On disappearance of the reagents (within 0.1–0.5 h as checked by TLC), isothiocyanate (**3**) was added to the reaction vessel and the reaction mixture was allowed to stand at room temperature until completion of the reaction (0.5–18 h). In some cases solid thiohydantoin derivative (**4**) was obtained by precipitation from the reaction medium or by crystallization from appropriate solvents, while, in other cases, it was obtained after removal of chloroform *in vacuo* and subsequent column chromatographic purification (cyclohexane–ethyl acetate mixtures). All the products were characterized based on elemental analyses and detailed spectral studies.

2.2.30.5 *Characterization Data of Two Representative Compounds*

(*E*)-*tert*-Butyl 2-(1-(3-benzyl-1-(4-methoxyphenyl)-5-oxo-2-thioxoimidazolidin-4-yl)ethylidene)hydrazinecarboxylate (**4a**): white powder, mp 193–195 °C; IR (nujol): 3230, 1755, 1715, 1638, 1620, 1592 cm^{-1}; ^{1}H NMR (DMSO-d_6, 400 MHz): δ 1.48 (9H, s), 1.55 (3H, s), 3.80 (3H, s), 4.84 (1H, d, J=15.2 Hz), 5.01 (1H, s), 5.11 (1H, d, J=15.2 Hz), 7.04 (2H, d, J=6.8 Hz), 7.29–7.39 (7H, m), 9.98 (1H, s); ^{13}C NMR (DMSO-d_6, 100 MHz): δ 12.5 (q), 28.0 (q), 48.8 (t), 55.4 (q), 69.1 (d), 79.8 (s), 114.1 (d), 126.3 (s), 127.7 (d), 128.4 (d), 128.5 (d), 129.9 (d), 135.3 (s), 142.9 (s), 152.7 (s), 159.4 (s), 170.1(s), 183.1 (s). MS (EI): *m/z* at 468 (M^+, 5), 412 (15), 368 (17), 352 (25), 311 (9), 277 (12), 165 (100), 106 (40). Anal. calc. for $C_{24}H_{28}N_4O_4S$ (468.56): C, 61.52; H, 6.02; N, 11.96; found: C, 61.38; H, 6.14; N, 11.89.

(*E*)-3-Allyl-5-(1-(2-((aminooxy)carbonyl)hydrazono)ethyl)-1-benzyl-2-thioxoimidazolidin-4-one (**4d**): white powder, mp 163–166 °C; IR (nujol): 3414, 3287, 3285, 3157, 1750, 1702, 1671, 1628, 1584 cm^{-1}; ^{1}H NMR (DMSO-d_6, 400 MHz): δ 1.58 (3H, s), 4.09 (1H, dd, J=15.4, 6.0 Hz), 4.49 (1H, dd, J=15.4, 6.0 Hz), 4.89–5.01 (3H, m), 5.16–5.20 (2H, m), 5.75–5.82 (1H, m), 6.39 (2H, br s), 7.26–7.35 (5H, m), 9.59 (1H, s); ^{13}C NMR (DMSO-d_6, 100 MHz): δ 12.5 (q), 44.7 (t), 47.8 (t), 68.3 (d), 119.0 (t), 127.4 (d), 128.4 (d), 130.9 (d), 136.0 (s), 138.3 (s), 156.5 (s), 170.6 (s), 182.0 (s); MS (EI): *m/z* (%) at 345 (M^+, 2),328 (7), 286 (100), 270 (43), 245 (17), 205 (52), 106 (27), 100 (58). Anal. calc. for $C_{16}H_{19}N_5O_2S$ (345.42): C, 55.63; H, 5.54; N, 20.27; found: C, 55.50; H, 5.66; N, 20.34.

2.2.30.6 *Critical Views*

The present work offers a synthetic strategy for regioselective substituted 2-thiohydantoin derivatives from a sequential addition of 1,2-diaza-1,3-diene, primary amine and isothiocyanate, respectively under mild reaction conditions in the absence of any catalyst for the first time. The hydrazone side chain introduced by the conjugated heterodiene system at the 5-position of the heterocycle represents a valuable functionality for accessing novel 5-acyl derivatives difficult to obtain by other methods.[106,107] The mild and simple reaction conditions make the protocol suitable for the generation of a library of functionalized 1,3,5-thiohydantoin derivatives.

2.2.31 Entry-31: Synthesis of 1,5-Disubstituted 1*H*-Tetrazoles

Type of reaction: C–C and C–N bond formation
Reaction conditions: Catalyst-free, methanol, room temperature
Synthetic strategy: One-pot multicomponent reaction
Keywords: Isocyanides, carbodiimides, trimethylsilyl azide, methanol, room temperature, one-pot multicomponent synthesis, 1,5-disubstituted 1*H*-tetrazoles

2.2.31.1 General Reaction Scheme

Kazemizadeh *et al.*[108] demonstrated a one-pot three-component synthesis of 1,5-disubstituted 1*H*-tetrazoles (**4**) from the reaction of isocyanides (**1**), carbodiimides (**2**) and trimethylsilyl azide (**3**) in methanol at room temperature in the absence of any catalyst (Scheme 2.47).

2.2.31.2 Proposed Mechanism

The investigators proposed a plausible mechanism for the transformation (Scheme 2.48). In the first step, hydrazoic acid (HN_3) is generated *in situ* from

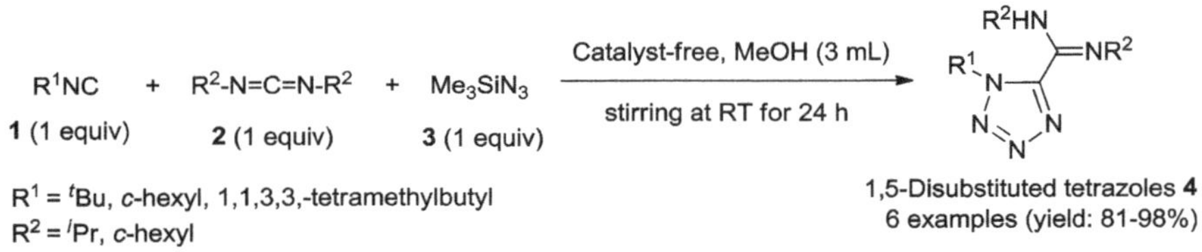

Scheme 2.47 Synthesis of 1,5-disubstituted 1*H*-tetrazoles (**4**).

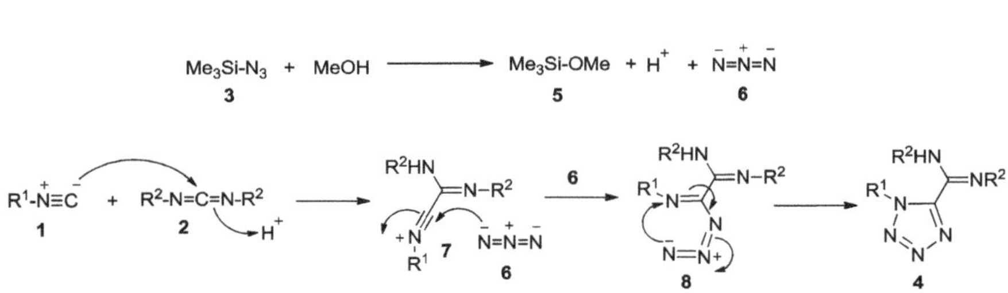

Scheme 2.48 Proposed mechanism for the synthesis of 1,5-disubstituted 1*H*-tetrazoles (**4**).

the reaction between trimethylsilyl azide (**3**) and methanol. In another step, isocyanide (**1**) undergoes a nucleophilic addition to carbodiimide (**2**) and protonation of the resulting adduct leads to the nitrilium intermediate (**7**). This intermediate may then be attacked by the azide anion (**6**) to form the adduct (**8**), and finally ring closing of the latter affords 1,5-disubstituted 1*H*-tetrazole derivative (**4**).

2.2.31.3 *Representative Examples*

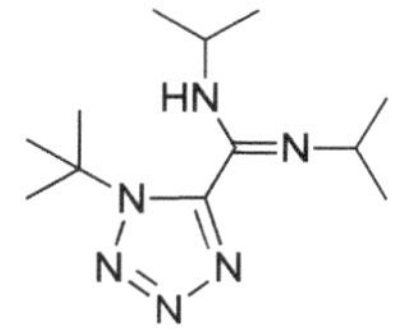

4a

1-(*tert*-Butyl)-*N*,*N*′-diisopropyl-1*H*-tetrazole-5-carboximidamide (**4a**): white solid, mp 105.4–107.9°C; yield: 93%

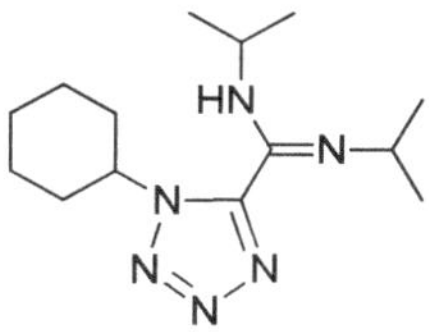

4b

1-Cyclohexyl-*N*,*N*′-diisopropyl-1*H*-tetrazole-5-carboximidamide (**4b**): white solid, mp 111.9–113.4 °C; yield: 82%

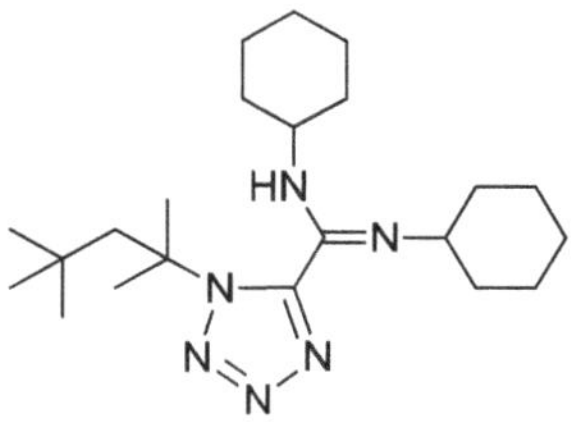

4c

N,*N*′-Dicyclohexyl-1-(2,4,4-trimethylpentan-2-yl)-1*H*-tetrazole-5-carboximidamide (**4c**): white solid, mp 65.8–66.2 °C; yield: 97%

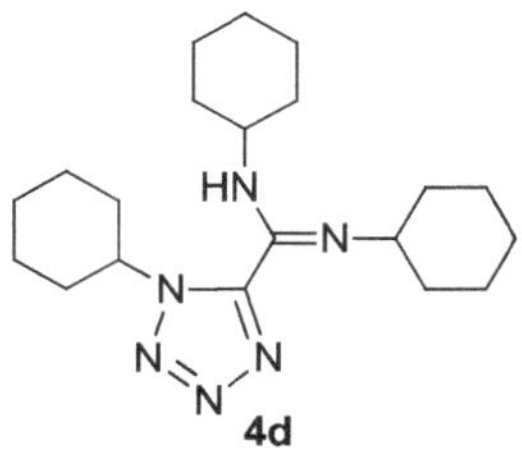

4d

N,*N*′,1-Tricyclohexyl-1*H*-tetrazole-5-carboximidamide (**4d**): white solid, mp 72.4–74.1 °C; yield: 81%

2.2.31.4 *Experimental Procedure*

To a magnetically stirred solution of isocyanide (**1**; 0.2 mmol) and carbodiimide (**2**; 0.2 mmol) in methanol (3 mL), trimethylsilyl azide (**3**; 0.2 mmol) was added, and the resulting mixture was then stirred at room temperature for 24 h. The solvent was evaporated to give the pure product of 1,5-disubstituted tetrazole (**4**), characterized based on spectral and analytical studies.

2.2.31.5 *Characterization Data of One Representative Compound*

N,*N*′-Dicyclohexyl-1-(2,4,4-trimethylpentan-2-yl)-1*H*-tetrazole-5-carboximidamide (**4c**): white solid, mp 65.8–66.2 °C; IR (KBr): 3313, 2926, 1594, 1445, 1100 cm^{-1};

^{1}H NMR ($CDCl_3$, 250 MHz): δ 0.69 (9H, s, Me_3C), 1.07–2.04 (20H, m, 10 CH_2 of *c*-hex), 1.71 (6H, s, Me_2C), 1.99 (2H, s, CH_2), 3.51–3.68 (1H, m, CH of *c*-hex), 3.93–4.08 (1H, m, CH of *c*-hex), 5.15 (1H, d, $J=7.5$, NH); ^{13}C NMR ($CDCl_3$, 62.5 MHz): δ 24.88, 24.98, 25.02, 25.47, 31.86, 33.14 (10 CH_2 of *c*-hex); 29.82 (Me_2C); 30.50 (Me_3C); 31.46 (Me_3C); 53.52, 55.17 (2 CH of *c*-hex); 54.11 (CH_2); 62.90 (Me_2C); 140.74, 154.11 (2 C=N). Anal. calc. for $C_{22}H_{40}N_6$ (388.59): C 68.00, H 10.38, N 21.63; found: C 68.07, H 10.34, N 21.60.

2.2.31.6 Critical Views

The present method offers mild reaction conditions, ease of workup, good to excellent yields and a clean reaction profile for the synthesis of biologically relevant 1,5-disubstituted 1*H*-tetrazole derivatives.[108]

2.2.32 Entry-32: Synthesis of 2-Thioxo-tetrahydropyrimidines

Type of reaction: C–C and C–N bond formation
Reaction conditions: Catalyst-free, acetonitrile, room temperature
Synthetic strategy: One-pot three-component reaction
Keywords: Alkyl isocyanaides, alkyl chloroglyoxalate, α-ketoimidoyl chloride, thiourea derivatives, acetonitrile, catalyst-free, room temperature, 2-thioxo-tetrahydropyrimidines

2.2.32.1 General Reaction Scheme

Moradi *et al.*[109] reported a simple and facile catalyst-free one-pot entry for the synthesis of novel 2-thioxo-tetrahydropyrimidine derivatives (**4**) of potential synthetic and pharmacological interest *via* the three-component reaction of alkyl isocyanides (**1**), alkyl chloroglyoxalates (**2**) and thiourea derivatives (**3**) in acetonitrile at ambient conditions (Scheme 2.49). This is the first report of the synthesis of 2-thioxo-tetrahydropyrimidine derivatives using the Nef reaction.

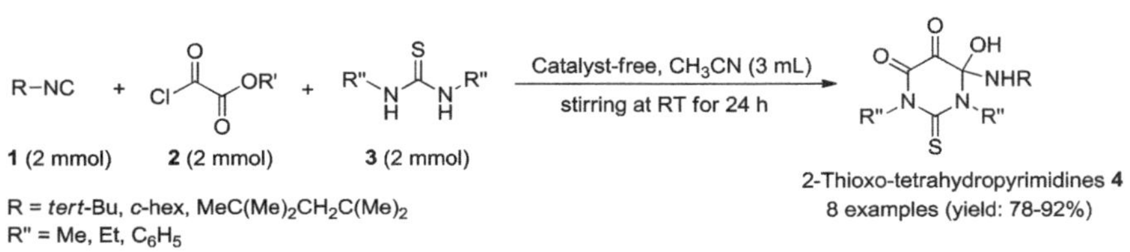

Scheme 2.49 Synthesis of 2-thioxo-tetrahydropyrimidines (**4**).

2.2.32.2 Representative Examples

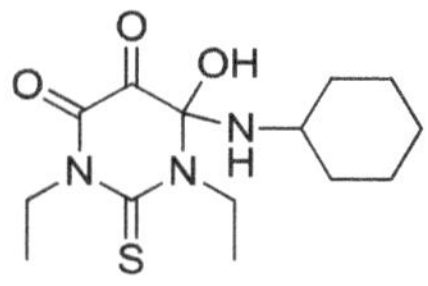

4a

6-(Cyclohexylamino)-1,3-diethyl-6-hydroxyl-2-thioxo-tetrahydropyrimidine-4,5-dione (**4a**): white powder, mp 201–203 °C; yield: 92%

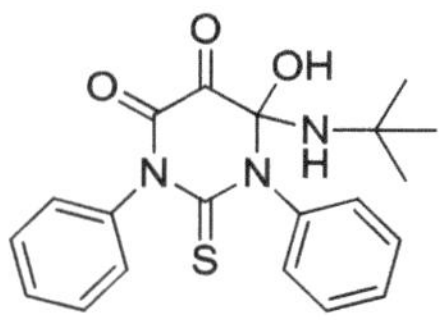

4b

6-(*tert*-Butylamino)-6-hydroxy-1,3-diphenyl-2-thioxotetrahydropyrimidine-4,5-dione (**4b**): white powder, mp 230–232 °C; yield: 95%

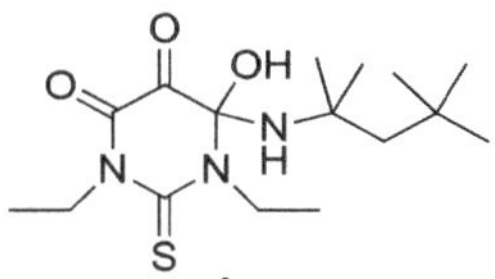

4c

1,3-Diethyl-6-hydroxy-2-thioxo-6-((2,4,4-trimethylpentan-2-yl)amino)tetrahydropyrimidine-4,5-dione (**4c**): white powder, mp194–196°C; yield: 82%

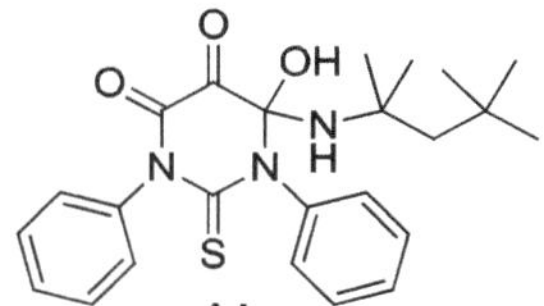

4d

6-Hydroxy-1,3-diphenyl-2-thioxo-6-((2,4,4-trimethylpentan-2-yl)amino)tetrahydropyrimidine-4,5-dione (**4d**): white powder, mp 210–212°C; yield: 78%

2.2.32.3 Experimental Procedure

To a stirred solution of alkyl isocyanide (**1**; 2 mmol) and alkyl chloroglyoxalate (**2**; 2 mmol) in acetonitrile (2 mL) was added a solution of thiourea derivative (**3**; 2 mmol) in acetonitrile (1 mL) at room temperature. The mixture was then allowed to stir for 24 h to complete the reaction when a solid product (**4**) was precipitated out. The crude product was filtered and washed with a mixture of hexane-diethyl ether (1 : 2), followed by neutralization with triethylamine in a mixture of water-methanol (1 : 2). The powder obtained was dried and used for further analysis without recrystallization. All the products (**4**) were characterized based on elemental analyses and detailed spectral studies.

2.2.32.4 Characterization Data of Two Representative Compounds

6-(*tert*-Butylamino)-6-hydroxy-1,3-diphenyl-2-thioxotetrahydropyrimidine-4,5-dione (**4b**): white powder, mp 230–232 °C; yield: 95%; IR (KBr): 3390 (NH), 1762 (C=O), 1655 (C=O), 1537, 1492, 1397 cm^{-1}; ^{1}H NMR

(DMSO-d_6, 250.1 MHz): δ 1.04–1.37 (9H, m, $C(CH_3)_3$), 7.13 (1H, s, NH), 7.29–7.54 (10H, m, 2×C_6H_5) 8.97 (1H, s, OH); ^{13}C NMR (DMSO-d_6, 62.9 MHz): δ 28.4 ($C(CH_3)_3$), 51.6 (C–N), 88.5 (C), 129.0 (2 CH) 129.3 (CH), 129.4 (2 CH), 129.5 (2 CH), 129.6 (3 CH), 133.6 (C), 135.7 (C), 163.5 (C=O), 170.1 (C=O), 183.2 (C=S); MS: *m/z* (%): 383 (M^+, 2), 284 (100), 268 (25), 136 (17), 120 (15), 93 (14), 77 (19). Anal. calc. for $C_{20}H_{21}N_3O_3S$ (383.47): C, 62.64; H, 5.52; N, 10.96. found: C, 62.60; H, 5.40; N, 10.98.

1,3-Diethyl-6-hydroxy-2-thioxo-6-((2,4,4-trimethylpentan-2-yl)amino)-tetrahydropyrimidine-4,5-dione (**4c**): white powder, mp 194–196 °C; yield: 82%; IR (KBr): 3211 (NH), 1741 (C=O), 1642 (C=O), 1510, 1441, 1401, 1383 cm^{-1}; ^1H NMR (DMSO-d_6, 250.1 MHz): δ 0.98 (s, 9H, $C(CH_3)_3$), 1.01 (s, 3H, CH_3), 1.05 (s, 3H, CH_3), 1.07–1.11 (m, 6H, 2×CH_3), 1.37–1.38 (m, 2H, CH_2), 3.50–3.52 (m, 4H, 2×CH_2), 7.41 (br s, 1H, NH), 8.35 (br s, 1H, OH); ^{13}C NMR (DMSO-d_6, 62.9 MHz): δ 13.2 (CH_3), 13.3 (CH_3), 28.5 ($C(CH_3)_3$), 29.0 (C), 31.3 (CH_2), 35.6 (2×CH_3), 42.2 (C), 36.4 (CH_2), 38.4 (CH_2), 86.8 (C), 163.4 (C=O), 170.2 (C=O), 182.5 (C=S). Anal. calc. for $C_{16}H_{29}N_3O_3S$ (343.49): C, 55.95; H, 8.51; N, 12.23. found: C, 55.54; H, 8.90; N, 12.46.

2.2.32.5 *Critical Views*

The investigators offered a simple alternative method for the synthesis of biologically interesting novel 2-thioxo-tetrahydropyrimidine derivatives at room temperature using the Nef reaction for the first time. The simplicity of the present procedure without using an acid catalyst makes it an interesting alternative to other approaches. However, this method was found not to work when thiourea is replaced with urea.[109]

2.2.33 Entry-33: Synthesis of Polycyclic Spiroindolines

Type of reaction: C–C and C–N bond formation
Reaction conditions: Catalyst-free, ethanol, room temperature
Synthetic strategy: One-pot three-component reaction
Keywords: 2-Isocyanoethylindole, malononitrile, aromatic aldehydes, ethanol, catalyst-free, room temperature, one-pot multicomponent reaction, stereo-selectivity, polycyclic spiroindolines

2.2.33.1 *General Reaction Scheme*

Wang *et al.*[110] developed a novel catalyst-free room temperature method for the one-pot synthesis of diversely functionalized polycyclic spiroindolines (**4**) from a three-component reaction between 2-isocyanoethylindole (**1**), malononitrile (**2**) and substituted aromatic aldehydes (**3**) in ethanol (Scheme 2.50).

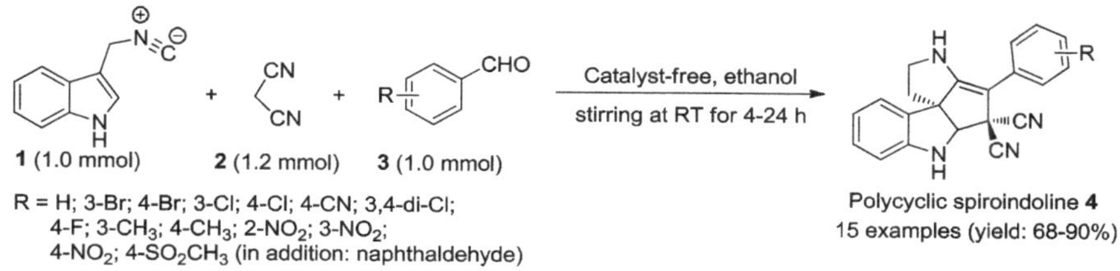

Scheme 2.50 Synthesis of polycyclic spiroindolines (**4**).

2.2.33.2 Representative Examples

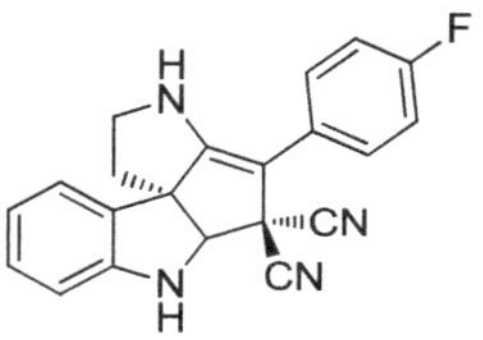

4a

4-Fluorophenyl-2,3,5a,6-tetrahydropyrrolo[3',2':2,3]cyclopenta[1,2-*b*]indole-5,5(1*H*)-dicarbonitrile (**4a**): white solid, mp 191–192 °C; yield: 78%

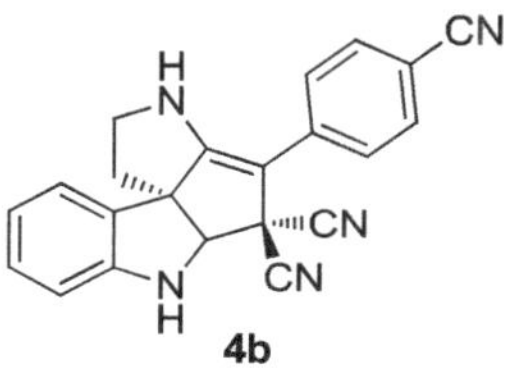

4b

4-Cyanophenyl-2,3,5a,6-tetrahydropyrrolo[3',2':2,3]cyclopenta[1,2-b]indole-5,5(1*H*)-dicarbonitrile (**4b**): white solid, mp 249–250 °C; yield: 88%

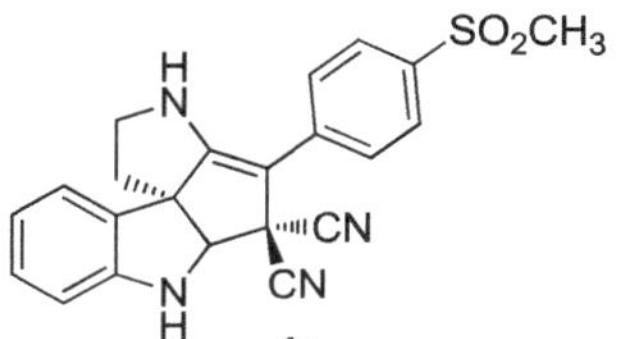

4c

4-(Methylsulfonyl)phenyl-2,3,5a,6-tetrahydropyrrolo[3',2':2,3]cyclopenta[1,2-*b*]indole-5,5(1*H*)-dicarbonitrile (**4c**): white solid, mp 209–210 °C; yield: 90%

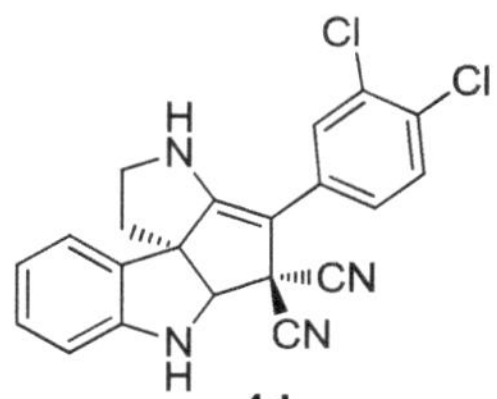

4d

3,4-Dichlorophenyl-2,3,5a,6-tetrahydropyrrolo[3',2':2,3]cyclopenta[1,2-*b*]indole-5,5(1*H*)-dicarbonitrile (**4d**): white solid, mp 207–208 °C; yield: 78%

2.2.33.3 Experimental Procedure

In a 25 mL round bottom flask, aromatic aldehyde (**3**; 1.0 mmol) was first added with malononitrile (**2**; 1.2 mmol) in EtOH and the mixture was stirred for 10 minutes at room temperature. Then 2-isocyanoethylindole (**1**; 1.0 mmol) was added to the reaction system and the stirring was continued for the stipulated time (4–24 h). Upon completion of the reaction (as monitored by TLC), the residue was directly purified by flash column chromatography by using ethyl acetate and petroleum ether as eluents to afford the pure product of diversely functionalized polycyclic spiroindolines (**4**). All the products were characterized based on detailed spectral studies.

2.2.33.4 *Characterization Data of Two Representative Compounds*

4-Fluorophenyl-2,3,5a,6-tetrahydropyrrolo[3′,2′ : 2,3]cyclopenta[1,2-*b*]indole-5, 5(1*H*)-dicarbonitrile (**4a**): white solid, mp 191–192 °C; yield: 78%; IR (KBr): 3384, 2974, 2890, 2236, 1646, 1516, 1465, 1304, 1230, 1087, 1018, 919 cm^{-1}; ^1H NMR (DMSO-d_6, 300 MHz): δ 7.45 (1H, s, NH), 7.38 (2H, t, $J = 8.4$ Hz, Ar–H), 7.31 (1H, s, NH), 7.22 (2H, t, $J = 8.8$ Hz, Ar–H), 7.11 (2H, dd, $J = 7.6$ Hz, Ar–H), 6.71–6.66 (2H, m, Ar–H), 4.86 (1H, s, CH), 4.04–3.98 (1H, m, CH_2), 3.89–3.85 (1H, m, CH_2), 2.41–2.33 (1H, m, CH_2), 1.78–1.73 (1H, m, CH_2); ^{13}C NMR (DMSO-d_6, 75 MHz): δ 161.2, 158.3, 158.0, 148.6, 131.0, 130.5, 129.6, 126.1, 126.0, 122.8, 119.2, 117.6, 116.1, 115.9, 114.7, 110.2, 87.5, 73.7, 67.6, 54.6, 51.2, 35.9; HRMS (ESI): *m/z* calc. for: 341.1208 $[M+H]^+$, found: 341.1214.

4-(Methylsulfonyl)phenyl-2,3,5a,6-tetrahydropyrrolo[3′,2′ : 2,3]cyclopenta [1,2-*b*]indole-5,5(1*H*)-dicarbonitrile (**4c**): white solid, mp 209–210 °C; yield: 90%; IR (KBr): 3409, 2907, 2240, 1639, 1578, 1474, 1405, 1315, 1146, 1027, 954, 829 cm^{-1}; ^1H NMR (DMSO-d_6, 300 MHz): δ 8.08 (1H, s, NH), 7.86 (2H, d, $J = 6.3$ Hz, Ar–H), 7.51 (2H, d, $J = 6.3$ Hz, Ar–H), 7.37 (1H, s, NH), 7.16–7.09 (2H, m, Ar–H), 6.72–6.67 (2H, m, Ar–H), 4.93 (1H, s, CH), 4.15–4.08 (1H, m, CH_2), 3.98–3.93 (1H, m, CH_2), 2.48–2.40 (1H, m, CH_2), 1.81–1.77 (1H, m, CH_2); ^{13}C NMR (DMSO-d_6, 75 MHz): δ 161.95, 1486, 139.1, 135.3, 130.6, 129.8, 128.0, 123.9, 122.8, 119.3, 117.4, 114.3, 110.4, 87.5, 73.4, 68.3, 53.9, 51.9, 44.2, 35.6; HRMS (ESI): *m/z* calc. for: 401.1078 $[M+H]^+$, found: 401.1085.

2.2.33.5 *Critical Views*

This protocol offers an easy and practical one-pot access to diversely substituted polycyclic spiroindoline scaffolds of biological promise in the absence of any catalyst under mild conditions. This is very advantageous over earlier methods[111–114] due to the avoidance of any kind of catalyst including transition metals and high temperature; moreover, the method is highly diastereoselective and good yielding.

2.2.34 Entry-34: Synthesis of Fused Polyhalogeno-7α-hydroxy-[1,2-*a*]indol-5-one Derivatives

Type of reaction: C–C and C–N bond formation
Reaction conditions: Catalyst-free, acetone, room temperature
Synthetic strategy: Cyclization based on Nenitzescu reaction
Keywords: Heterocyclic ketene aminals (HKAs), halogenated quinones, acetone, catalyst-free, room temperature, Nenitzescu reaction, fused polyhalogeno-7α-hydroxy-[1,2-*a*]indol-5-one derivatives

2.2.34.1 *General Reaction Scheme*

A series of novel fused polyhalogeno-7α-hydroxy-[1,2-*a*]indol-5-one derivatives (**3**) has recently been synthesized with good yields based on the

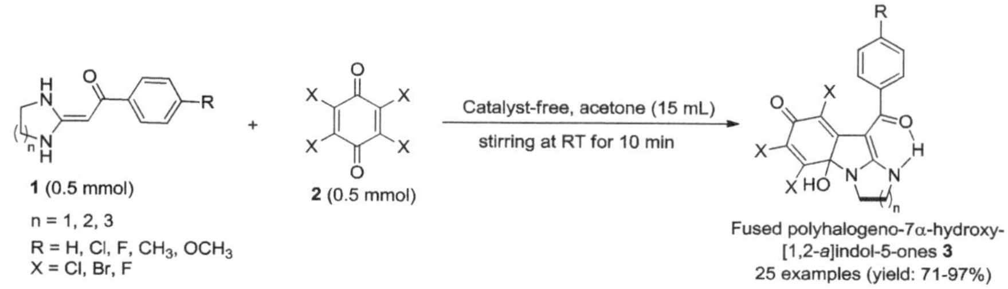

Scheme 2.51 Synthesis of fused polyhalogeno-7α-hydroxy-[1,2-*a*]indol-5-ones (**3**).

Nenitzescu reaction of heterocyclic ketene aminals (**1**) with halogenated quinones (**2**) under catalyst-free condition in acetone at room temperature by Yu *et al.*[115] (Scheme 2.51).

2.2.34.2 Representative Examples

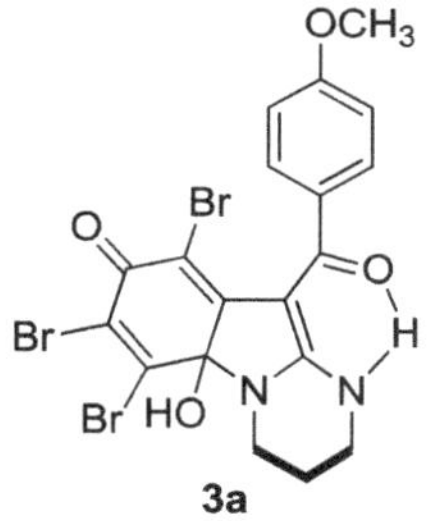

3a

6,7,9-Tribromo-5α-hydroxy-10-(4-methoxybenzoyl)-1,2,3,4-tetrahydropyrimido[1,2-*a*]indol-8(5a*H*)-one (**3a**): blood red solid, mp 164–166 °C; yield: 84%

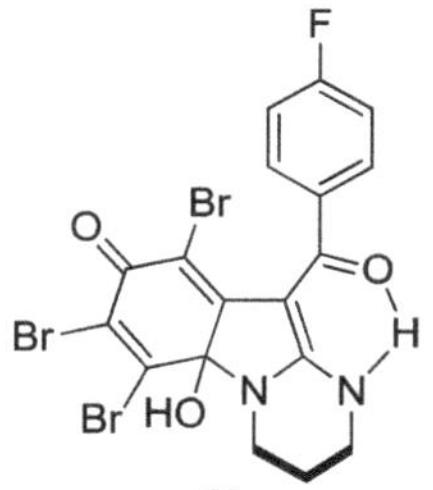

3b

6,7,9-Tribromo-10-(4-fluorobenzoyl)-5α-hydroxy-1,2,3,4-tetrahydropyrimido[1,2-*a*]indol-8(5a*H*)-one (**3b**): saffron yellow solid, mp 167–169 °C; yield: 91%

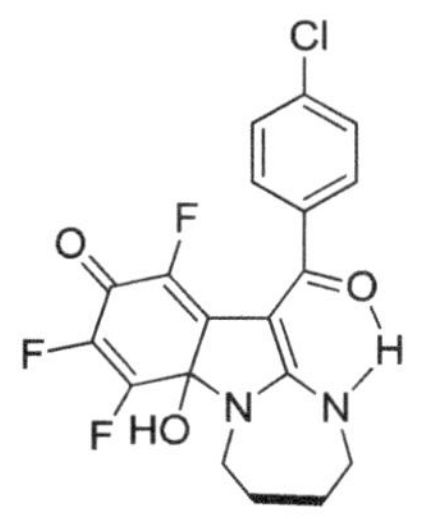

3c

11-(4-Chlorobenzoyl)-7,8,10-trifluoro-6α-hydroxy-2,3,4, 5-tetrahydro-1*H*-[1,3]diazepino[1,2-*a*]indol-9(6a*H*)-one (**3c**): orange solid, mp 233–235 °C; yield: 97%

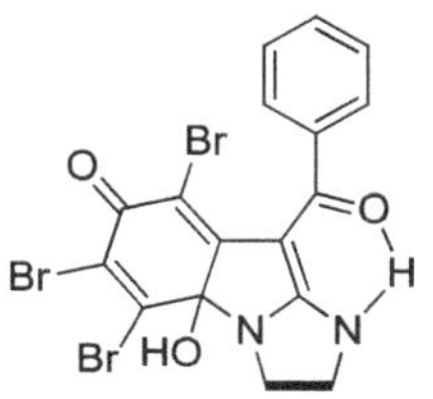

3d

9-Benzoyl-5,6,8-tribromo-4α-hydroxy-2,3-dihydro-1*H*-imidazo-[1,2-*a*]indol-7(4a*H*)-one (**3d**): orange solid, mp 160–162 °C; yield: 72%

2.2.34.3 Experimental Procedure

In a 25 mL round bottom flask, a heterocyclic ketene aminal derivative (**1**; 0.5 mmol) was first dissolved in acetone (5 mL). To this solution, an acetone solution (10 mL) of halogenated quinone derivatives (**2**) (0.5 mmol) was added slowly and drop-wise over a period of 10 min with the help of a 25 mL dropping funnel. The whole mixture was then stirred for 10 min at room temperature. On completion of the reaction as monitored by TLC, the crude product was filtered, followed by washing with 95% ethanol to obtain pure product of polyhalogeno-7α-hydroxy-[1,2-*a*]indol-5-one derivatives (**3**) in good yields. Each of the products was identified by FTIR, NMR and HRMS.

2.2.34.4 Characterization Data of Two Representative Compounds

6,7,9-Tribromo-5α-hydroxy-10-(4-methoxybenzoyl)-1,2,3,4-tetrahydropyrimido [1,2-*a*]indol-8(5a*H*)-one (**3a**): blood red solid, mp 164–166 °C; yield: 84%; IR (KBr): 3415, 1636, 1595, 1549,1497, 1408, 1310, 1254, 1171, 1032, 670 cm^{-1}; ^1H NMR (DMSO-d_6, 500 MHz): δ 1.76–1.78 (1H, m, CH_2), 2.06–2.08 (1H, m, CH_2), 3.30–3.31 (2H, m, CH_2N), 3.81 (3H, s, Ar–OCH_3), 3.83–3.85 (2H, m, N–CH_2), 6.94–6.95 (2H, m, Ar–H), 7.64–7.65 (2H, m, Ar–H), 8.32 (1H, br s, OH), 9.24 (1H, br s, NH); ^{13}C NMR (125 MHz, DMSO-d_6): δ 21.5, 38.8, 41.8, 55.7, 91.8, 96.0, 98.3, 113.8, 130.4, 132.5, 135.5, 138.2, 160.9, 162.3, 163.9, 169.4, 186.7; HRMS (TOF ES^+): *m/z* calc. for $C_{19}H_{16}Br_3N_2O_4$ ($[M+H]^+$), 572.8655; found, 572.8681.

11-(4-Chlorobenzoyl)-7,8,10-trifluoro-6α-hydroxy-2,3,4, 5-tetrahydro-1*H*-[1,3]diazepino[1,2-*a*]indol-9(6a*H*)-one (**3c**): orange solid, mp 233–235 °C; yield: 97%; IR (KBr): 3426, 1608, 1437, 1355, 1285, 1242, 1087, 1042, 782 cm^{-1}; ^1H NMR (DMSO-d_6, 500 MHz): δ 1.64–1.93 (4H, m, CH_2CH_2), 3.53–3.57 (2H, m, CH_2N), 3.74–3.76 (2H, m, NCH_2), 7.44–6.61 (4H, m, Ar–H), 8.48 (1H, br s, OH), 10.21 (1H, br s, NH); ^{13}C NMR (DMSO-d_6, 125 MHz): δ 26.1, 26.3, 43.9, 45.5, 88.0, 93.0, 128.8, 130.1, 136.2, 138.3 (d, $J=268.8$ Hz), 139.3, 139.6 (d, $J=246.3$ Hz), 141.3, 153.4 (d, $J=278.8$ Hz), 169.9, 171.0, 186.5; HRMS (TOF ES^+): *m/z* calc. for $C_{19}H_{15}ClF_3N_2O_3$ ($[M+H]^+$) , 411.0718; found, 411.0734.

2.2.34.5 Critical Views

The investigators[115] have offered direct access to a series of pharmaceutically interesting and novel fused polyhalogeno-7α-hydroxy-[1,2-*a*]indol-5-ones in good yields at room temperature under catalyst-free conditions. The method satisfies a number of green aspects such as energy efficiency, avoidance of any catalyst and easy purification without using any column chromatography. In addition, this novel Nenitzescu reaction is somewhat different from conventional Nenitzescu reaction.[116–118]

2.2.35 Entry-35: Synthesis of Dialkyl-1,5-dihydro-5-oxo-1-phenyl-2*H*-[1]benzopyrano[2,3-*b*]pyridine-2,3-dicarboxylates

Type of reaction: C–C and C–N bond formation
Reaction conditions: Catalyst-free, dichloromethane, room temperature
Synthetic strategy: One-pot multicomponent reaction
Keywords: Dialkyl acetylenedicarboxylates, triphenylphosphine, 2-amino-4-oxo-4*H*-1-benzopyran-3-carboxaldehydes, dichloromethane, catalyst-free, room temperature, one-pot multicomponent reaction, dialkyl-1,5-dihydro-5-oxo-1-phenyl-2*H*-[1]benzopyrano[2,3-*b*]pyridine-2,3-dicarboxylates

2.2.35.1 General reaction scheme

Esmaeil and Ghareghani[119] developed a simple and clean protocol for the synthesis of highly functionalized chromenone fused dialkyl-1,5-dihydro-5-oxo-1-phenyl-2*H*-[1]benzopyrano[2,3-*b*]pyridine-2,3-dicarboxylates (**4**) of biological relevance from the one-pot three-component reaction of triphenylphosphine (**1**), dialkyl acetylenedicarboxylate (**2**) and 2-amino-4-oxo-4*H*-1-benzopyran-3-carboxaldehydes (**3**) in dichloromethane in high yield (Scheme 2.52).

2.2.35.2 Representative Examples

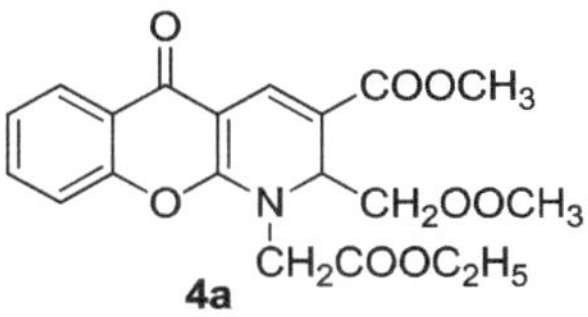

4a

Dimethyl 1-(2-ethoxy-2-oxoethyl)-5-oxo-2,5-dihydro-1*H*-chromeno[2,3-*b*]pyridine-2,3-dicarboxylate (**4a**): solid, mp 206–208 °C; yield: 93%

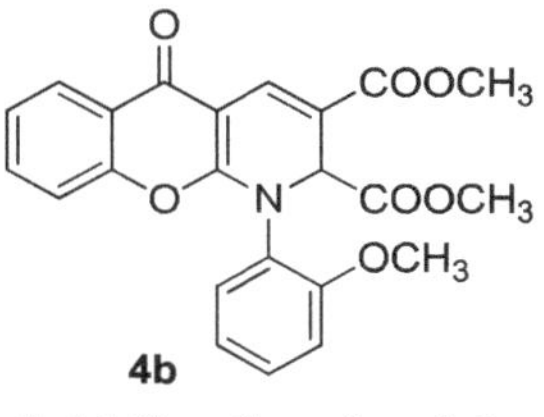

4b

Dimethyl 1-(2-methoxyphenyl)-5-oxo-2,5-dihydro-1*H*-chromeno[2,3-*b*]pyridine-2,3-dicarboxylate (**4b**): solid, mp 210–212 °C; yield: 96%

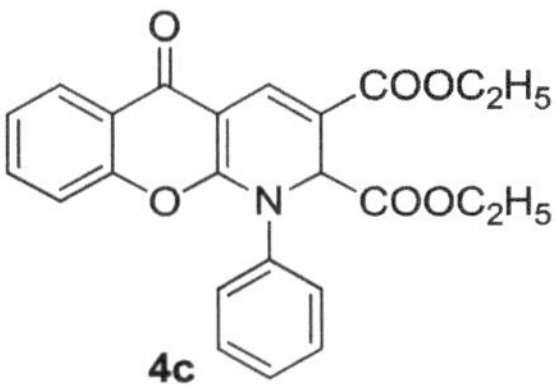

4c

Diethyl 5-oxo-1-phenyl-2,5-dihydro-1*H*-chromeno[2,3-*b*]pyridine-2,3-dicarboxylate (**4c**): solid, mp 190–191 °C; yield: 96%

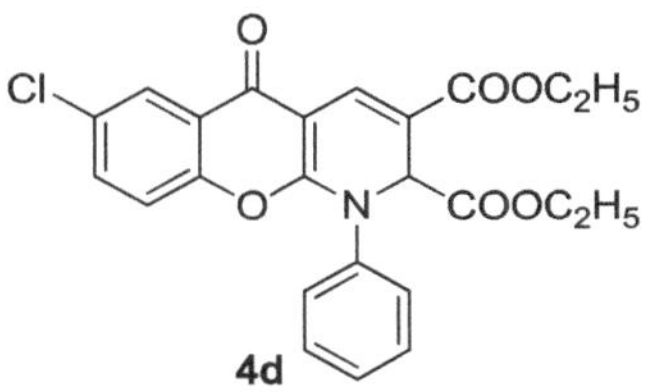

4d

Diethyl 7-chloro-5-oxo-1-phenyl-2,5-dihydro-1*H*-chromeno[2,3-*b*]pyridine-2,3-dicarboxylate (**4d**): solid, mp 210–212°C; yield: 96%

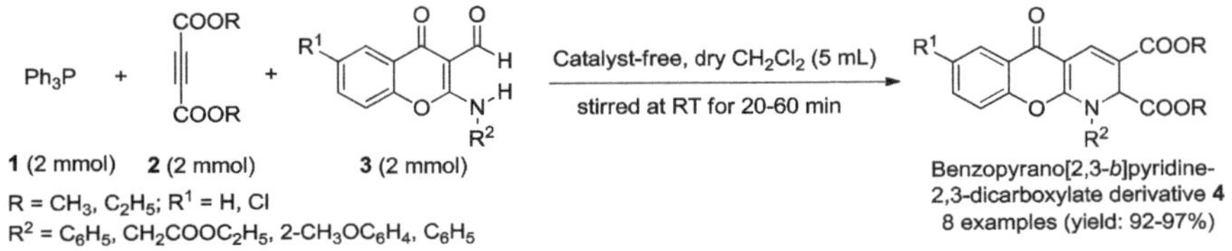

Scheme 2.52 Synthesis of dialkyl-1,5-dihydro-5-oxo-1-phenyl-2*H*-[1]benzopyrano[2,3-*b*]-pyridine-2,3-dicarboxylates (**4**).

2.2.35.3 *Experimental Procedure*

To a magnetically stirred solution of 4-oxo-2-amino-4*H*-1-benzopyran-3-carboxaldehydes (**3**; 2 mmol) and triphenylphosphine (**1**; 2 mmol) in dichloromethane (3 mL), dialkyl acetylenedicarboxylate (**2**; 2 mmol) was added drop-wise at −5 °C within 10 min. The mixture was then allowed to warm up to room temperature and stirred for another 10–50 min. On completion of the reaction, the solvent was removed under reduced pressure and the residue was crystallized from ethanol to afford pure products of dialkyl-1,5-dihydro-5-oxo-1-phenyl-2*H*-[1]benzopyrano[2,3-*b*]pyridine-2,3-dicarboxylates (**4**), characterized using analytical and spectral techniques.

2.2.35.4 *Characterization Data of Two Representative Compounds*

Dimethyl 1-(2-ethoxy-2-oxoethyl)-5-oxo-2,5-dihydro-1*H*-chromeno[2,3-*b*]pyridine-2,3-dicarboxylate (**4a**): solid, mp 206–208 °C; yield: 93%; IR (KBr): 3090 (arom. C–H), 2974, 2924 (aliph. C–H), 1750, 1675 (2 CO_2Me), 1650 (C=O), 1500 (arom. C=C) cm^{-1}; ^1H NMR ($CDCl_3$, 500 MHz): δ 1.21 (3H, t, $^3J = 5.7$, Me), 3.65 (3H, s, MeO), 3.74 (3H, s, MeO), 4.21 (2H, q, $^3J = 7.13$, CH_2O); 4.34 (2H, s, CH_2N), 5.33 (1H, s, CHN), 7.59 (1H, t, $^3J = 7.5$, Ar–H), 7.33–7.45 (2H, m, Ar–H); 7.96 (s, 1 olef. H); 8.1 (1H, d, $^3J = 7.1$, Ar–H); ^{13}C NMR ($CDCl_3$, 125.7 MHz): δ 14.02 (Me), 49.57 (CH_2N), 51.82 (MeO), 52.86 (MeO), 61.29 (CH_2O), 61.97 (*C*HN), 96.96 (C-4a), 110.70 (C-3), 116.56 (C-9), 122.40 (C-5a), 125.60 (C-6), 126.26 (C-7), 131.56 (C-8), 132.78 (C-4), 153.03 (C-9a), 160.38 (C-10a), 165.17 (ester), 169.15 (ester), 167.31 (ester), 172.09 (C-5); EIMS (70 eV, rel.%): 401 (1, M^+), 402 (1, [M +1]), 370 (1, $[M\text{–}MeO]^+$), 343 (21), 342 (100, $[M\text{–}CO_2Me]^+$), 328 (2), 315 (11), 314 (61), 300 (1, $[M\text{–}NCH_2CO_2Et]^+$), 196 (5, $[M\text{–}2CO_2Me\text{–}CH_2CO_2Et]^+$), 59 (6, CO_2Me^+). Anal. calc. for $C_{20}H_{19}NO_8$: C, 59.85; H, 4.77; N, 3.49; found: C, 60.0; H, 4.8; N, 3.5.

Diethyl 7-chloro-5-oxo-1-phenyl-2,5-dihydro-1*H*-chromeno[2,3-*b*]pyridine-2,3-dicarboxylate (**4d**): solid, mp 210–212 °C; yield: 96%; IR (KBr): 3050 (arom. C–H), 2950 (aliph. C–H), 1745 and 1700 (2 CO_2Me), 1625 (C=O) cm^{-1};

^{1}H NMR ($CDCl_3$, 500 MHz): δ 3.64 (3H, s, MeO), 3.74 (3H, s, MeO), 5.61 (1H, s, CHN), 6.98 (1H d, $^3J = 8.8$, Ar–H), 7.92 (1H, s, olefinic H), 7.34–7.44 (6H, m, Ar–H), 7.98 (1H, s, Ar–H); ^{13}C NMR ($CDCl_3$, 125.7 MHz): δ 51.93 (MeO), 52.94 (MeO), 62.96 (CHN), 98.53 (C-4a), 111.41 (C-3), 118.59 (C-9), 123.52 (C-5a), 125.50 (C-6), 126.42 (arom. CH), 129.57 (arom. CH), 128.37 (arom. CH), 130.99 (C-8), 131.2 (C-7), 132.73 (C-4), 139.55 (arom. C), 151.49 (C-9a), 159.49 (C-10a), 169.57 (CO_2Me), 164.77 (CO_2Me), 171.35 (C-5); EIMS (70 eV, rel.%): 366 (83, $[M{-}CO_2Me]^+$), 367 (17), 258 (7), 230 (4), 77 (100, $C_6H_5^+$), 59 (25, CO_2Me^+). Anal. calc. for $C_{22}H_{16}ClNO_6$: C, 62.05; H, 3.79; N, 3.29; found: C, 63.1; H, 3.8; N, 3.3.

2.2.35.5 Critical Views

The present method offers a catalyst-free and clean one-pot alternative to multistep approaches to chromenone derivatives fused to a heterocycle, and demonstrated the synthesis of a series of highly functionalized dialkyl 1-aryl- and 1-alkyl-1,5-dihydro-5-oxo-2*H*-[1]benzopyrano[2,3-*b*]pyridine-2,3-dicarboxylates of potential synthetic and pharmaceutical interest with excellent yields under neutral conditions without any activation or modification of the starting materials.[119]

2.2.36 Entry-36: Synthesis of 2-Aryl-2-(2,3,4,5-tetrahydro-2,4-dioxo-1*H*-1,5-benzodiazepin-3-yl)acetamides

Type of reaction: C–C and C–N bond formation
Reaction conditions: Catalyst-free, dichloromethane, room temperature
Synthetic strategy: One-pot multicomponent reaction
Keywords: Benzene-1,2-diamine, Meldrum's acid, aromatic aldehydes, isocyanides, water, dichloromethane, catalyst-free, room temperature, one-pot multicomponent reaction, Knoevenagel condensation, Michael-type addition, 2-aryl-2-(2,3,4,5-tetrahydro-2,4-dioxo-1*H*-1,5-benzodiazepin-3-yl)acetamides

2.2.36.1 General Reaction Scheme

A catalyst-free, simple and efficient five-component condensation reaction of readily available benzene-1,2-diamine (**1**), Meldrum's acid (**2**), aromatic aldehydes (**3**), isocyanides (**4**) and H_2O in dichloromethane at room temperature was reported by Akbarzadeh *et al.*[120] for the synthesis of a series of new 2-aryl-2-(2,3,4,5-tetrahydro-2,4-dioxo-1*H*-1,5-benzodiazepin-3-yl)acetamides (**5**) (Scheme 2.53).

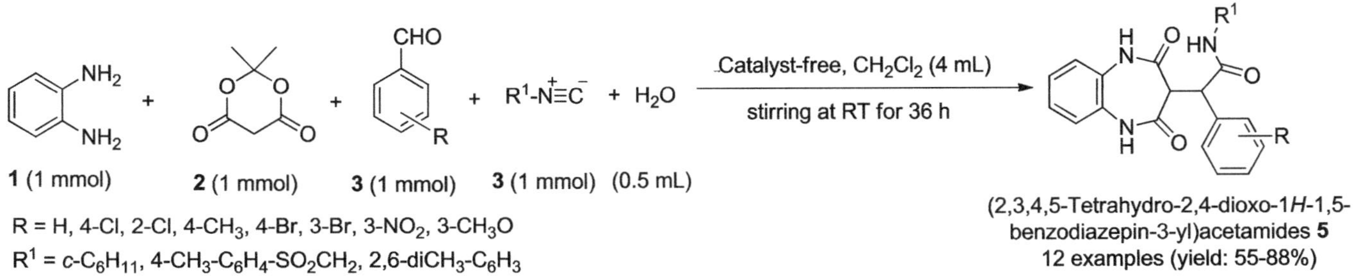

Scheme 2.53 Synthesis of 2-aryl-2-(2,3,4,5-tetrahydro-2,4-dioxo-1*H*-1,5-benzodiazepin-3-yl)acetamides (5).

2.2.36.2 Representative Examples

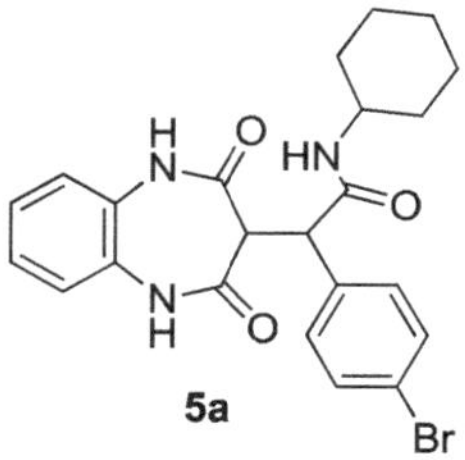

5a

2-(4-Bromophenyl)-*N*-cyclohexyl-2-(2,4-dioxo-2,3,4,5-tetrahydro-1*H*-benzo[*b*][1,4]diazepin-3-yl)acetamide (**5a**): white powder, mp >270 °C; yield: 78%

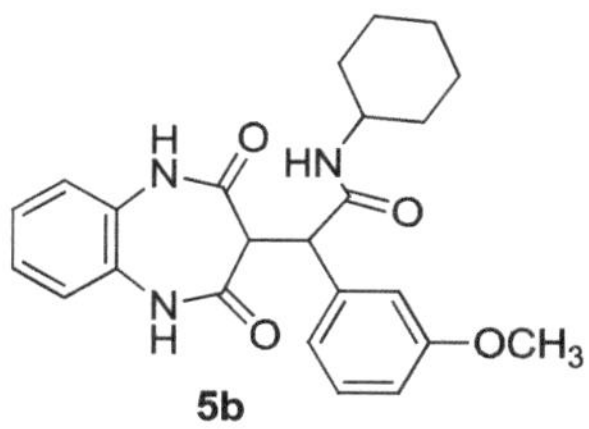

5b

N-Cyclohexyl-2-(2,4-dioxo-2,3,4,5-tetrahydro-1*H*-benzo[*b*][1,4]diazepin-3-yl)-2-(3-methoxyphenyl)acetamide (**5b**): white powder, mp >260 °C; yield: 88%

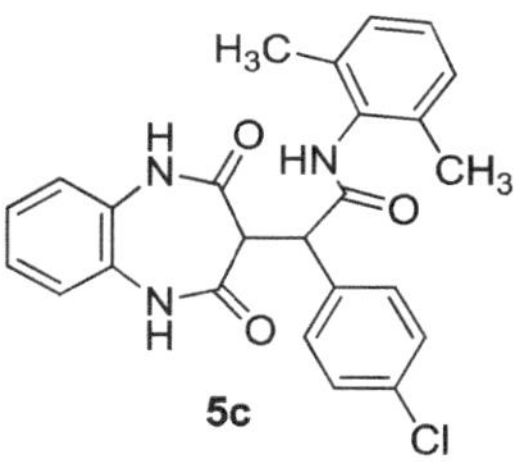

5c

2-(4-Chlorophenyl)-*N*-(2,6-dimethylphenyl)-2-(2,4-dioxo-2,3,4,5-tetrahydro-1*H*-benzo[*b*][1,4]diazepin-3-yl)acetamide (**5c**): white powder, mp >260 °C; yield: 69%

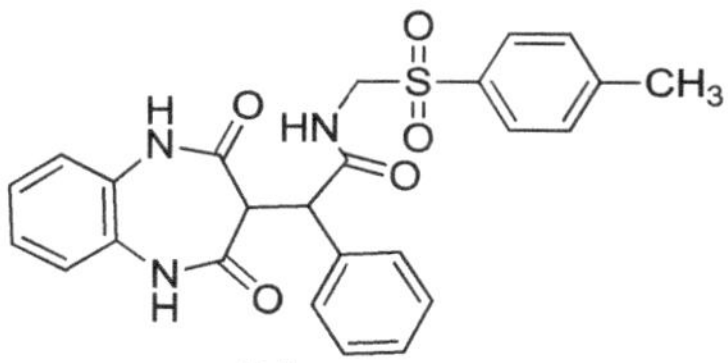

5d

2-(2,4-Dioxo-2,3,4,5-tetrahydro-1*H*-benzo[*b*][1,4]diazepin-3-yl)-2-phenyl-*N*-(tosylmethyl)acetamide (**5d**): white powder, mp 228–232 °C; yield: 55%

2.2.36.3 Experimental Procedure

A mixture of benzene-1,2-diamine (**1**; 1 mmol), Meldrum's acid (**2**; 1 mmol), aromatic aldehydes (**3**; 1 mmol), cyclohexyl isocyanides (**4**; 1 mmol) and H_2O (0.5 mL) in dichloromethane (4 mL) was stirred for 36 h at room temperature. After completion of the reaction (monitored by TLC), hexane (3 mL) was added, the mixture filtered, and the precipitate washed with ethanol (5 mL) to afford the pure products of 2-aryl-2-(2,3,4,5-tetrahydro-2,4-dioxo-1*H*-1,5-benzodiazepin-3-yl)acetamides (**5**) with good to excellent yields. The synthesized compounds were identified based on elemental and spectral studies.

2.2.36.4 Characterization Data of Two Representative Compounds

2-(4-Chlorophenyl)-*N*-(2,6-dimethylphenyl)-2-(2,4-dioxo-2,3,4,5-tetrahydro-1*H*-benzo[*b*][1,4]diazepin-3-yl)acetamide (**5c**): white powder, mp >260 °C; yield: 69%; IR (KBr): 3450, 3300, 3290 (NH), 1699, 1662, 1600 (C=O) cm^{-1}; ^1H NMR (DMSO-d_6, 300.13 MHz): δ 1.94 (6H, s, 2×Me), 3.81 (1H, d,

$J = 12.9$, CH), 4.65 (1H, d, $J = 12.9$, CH), 6.97–7.54 (11H, m, Ar–H), 9.65 (1H, s, NH), 10.39 (1H, s, NH), 10.67 (1H, s, NH); EIMS (70 eV): *m/z* 447 (M^+). Anal. calc. for $C_{25}H_{22}ClN_3O_3$: C, 67.04; H, 4.95; N, 9.38; found: C, 67.14; H, 4.87; N, 9.31.

2-(2,4-Dioxo-2,3,4,5-tetrahydro-1*H*-benzo[*b*][1,4]diazepin-3-yl)-2-phenyl-*N*-(tosylmethyl)acetamide (**5d**): white powder, mp 228–232 °C; yield: 55%; IR (KBr): 3555, 3300, 3266 (NH), 1687, 1630, 1601 (C=O) cm^{-1}; ^1H NMR (DMSO-d_6, 300.13 MHz): δ 2.34 (3H, s, Me), 4.29–4.71 (4H, m, 2×CH, CH_2), 6.41–7.32 (14 H, m, 13 Ar–H and NH), 8.95 (1H, br s, NH), 9.18 (1H, br s, NH); EIMS (70 eV): *m/z* 477 (M^+). Anal. calc. for $C_{25}H_{23}N_3O_5S$: C, 62.88; H, 4.85; N, 8.80; found: C, 62.74; H, 4.73; N, 8.71.

2.2.36.5 Critical Views

The present protocol offers an easy and efficient practical route to synthesize a series of 2-aryl-2-(2,3,4,5-tetrahydro-2,4-dioxo-1*H*-1,5-benzodiazepin-3-yl)-acetamides with *in vitro* antibacterial activities from easily available starting materials just at ambient conditions. The method does not need a catalyst, no column chromatographic purification of the products, and, more importantly, has no need of activation or modifications. The simplicity of the presented procedure makes it an interesting alternative to the complex multistep approaches[121–126] for the synthesis of 1,5-benzodiazepines of potential synthetic and pharmacological interest. However, the reaction involves a long reaction time (36 h).

2.2.37 Entry-37: Synthesis of Functionalized Tetrahydro-4-oxoindeno[1,2-*b*]pyrroles

Type of reaction: C–C and C–N bond formation
Reaction conditions: Catalyst-free, dichloromethane, room temperature
Synthetic strategy: One-pot multicomponent reaction
Keywords: Indan-1,2,3-trione hydrate (ninhydrin), dialkyl acetylenedicarboxylates, primary amines, dichloromethane, catalyst-free, room temperature, one-pot multicomponent reaction, functionalized tetrahydro-4-oxoindeno[1,2-*b*]pyrroles

2.2.37.1 General Reaction Scheme

A series of highly functionalized dialkyl 1,3a,4,8b-tetrahydro-3a,8b-dihydroxy-1-alkyl-4-oxoindeno[1,2-*b*]pyrrole-2,3-dicarboxylates (**4**) were synthesized in good yields by Yavari *et al.*[127] *via* a one-pot three-component reaction of indan-1,2,3-trione hydrate (ninhydrin) (**1**), dialkyl acetylenedicarboxylates (**2**), and primary amines (**3**) in dichloromethane under ambient conditions (Scheme 2.54).

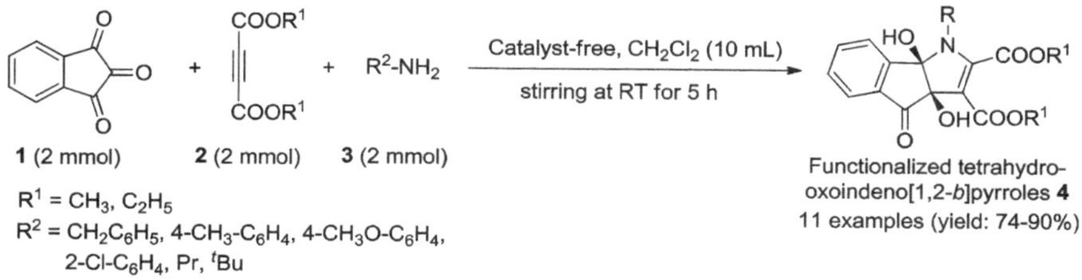

Scheme 2.54 Synthesis of functionalized tetrahydro-4-oxoindeno[1,2-*b*]pyrroles (**4**).

2.2.37.2 Representative Examples

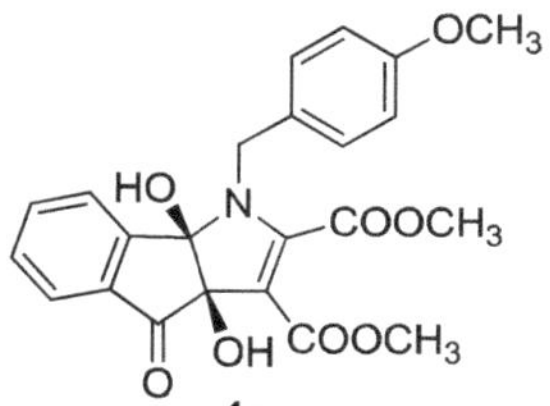

4a

(3a*R*,8b*R*)-Dimethyl 3a,8b-dihydroxy-1-(4-methoxybenzyl)-4-oxo-1,3a,4,8b-tetrahydroindeno[1,2-*b*]pyrrole-2,3-dicarboxylate (**4a**): yellow powder, mp 127–129 °C; yield: 90%

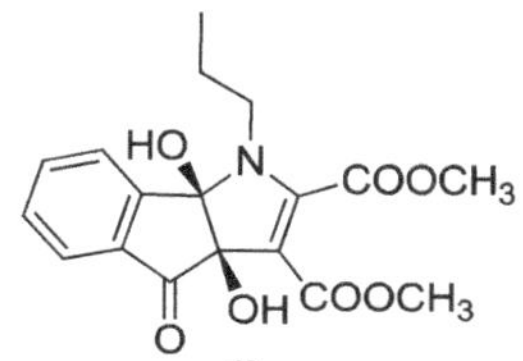

4b

(3a*R*,8b*R*)-Dimethyl 3a,8b-dihydroxy-4-oxo-1-propyl-1,3a,4,8b-tetrahydroindeno[1,2-*b*]pyrrole-2,3-dicarboxylate (**4b**): white powder, mp 125–127 °C; yield: 78%

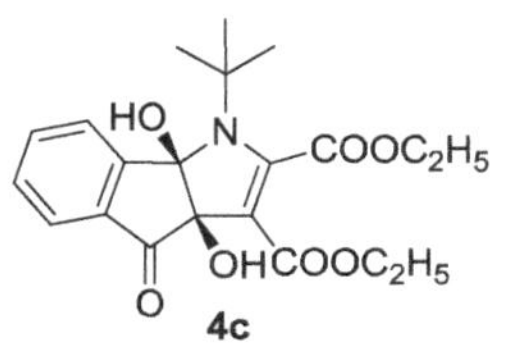

4c

(3a*R*,8b*R*)-Diethyl 1-(tert-butyl)-3a,8b-dihydroxy-4-oxo-1,3a,4,8b-tetrahydroindeno[1,2-*b*]pyrrole-2,3-dicarboxylate (**4c**): pale yellow powder, mp 137–139 °C; yield: 78%

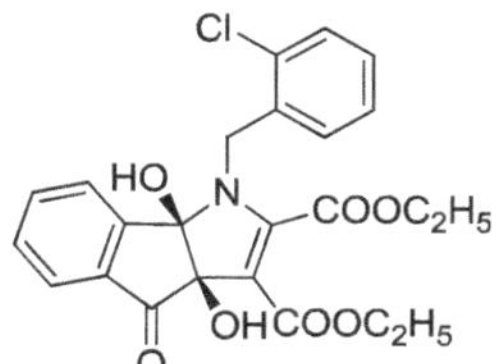

4d

(3a*R*,8b*R*)-Diethyl 1-(2-chlorobenzyl)-3a,8b-dihydroxy-4-oxo-1,3a,4,8b-tetrahydroindeno[1,2-*b*]pyrrole-2,3-dicarboxylate (**4d**): pale yellow powder, mp 133–135 °C; yield: 74%

2.2.37.3 Experimental Procedure

To a magnetically stirred solution of primary amines (**3**; 2 mmol) and dialkyl acetylenedicarboxylates (**2**; 2 mmol) in dichloromethane (5 mL), a solution of ninhydrin (**1**; 0.32 g, 2 mmol) in dichloromethane (5 mL) was added, at room temperature for 5 h. Upon completion of the reaction as indicated by TLC, the solvent was removed under reduced pressure to leave a residue that was purified by column chromatography (SiO_2; hexane–ethyl acetate 8 : 1) to afford the pure products of functionalized tetrahydrooxoindeno[1,2-*b*]pyrroles (**4**), fully characterized by means elemental analyses and detailed spectral studies.

2.2.37.4 Characterization Data of Two Representative Compounds

(3a*R*,8b*R*)-Dimethyl 3a,8b-dihydroxy-1-(4-methoxybenzyl)-4-oxo-1,3a,4,8b-tetrahydroindeno[1,2-*b*]pyrrole-2,3-dicarboxylate (**4a**): yellow powder, mp 127–129 °C; yield: 90%; IR (KBr): 3445 (br.), 1742, 1712, 1662, 1569, 1468, 1240, 1180 cm^{-1}; ^{1}H NMR ($CDCl_3$, 500.1 MHz): δ 3.47 (3H, s, MeO), 3.66 (3H, s, MeO), 3.77 (3H, s, MeO), 4.67 (1H, d, $^2J=15.5$ Hz, CH), 4.87 (1H, d, $^2J=15.5$ Hz, CH), 4.92 (s, 1H, OH), 5.41 (1H, s, OH), 7.05 (d, $^3J=8.3$ Hz, 2H, 2×CH), 7.12 (d, $^3J=8.3$ Hz, 2×CH), 7.55 (t, $^3J=7.4$, 1H, CH), 7.67 (1H, t, $^3J=7.7$ Hz, CH), 8.36 (2H, d, $^3J=8.0$ Hz, 2×CH); ^{13}C-NMR ($CDCl_3$, 125.7 MHz): δ 46.4 (CH_2N), 51.2 (MeO), 52.7 (MeO), 55.3 (MeO), 83.7 (C), 95.3 (C), 114.0 (2 CH), 124.5 (CH), 124.7 (CH), 128.4 (C), 129.5 (2 CH), 130.6 (CH), 135.2 (2C), 136.2 (CH), 147.4 (C), 151.1 (C), 159.4 (C), 162.2 (C=O), 164.2 (C=O), 197.1 (C=O). Anal. calc. for $C_{23}H_{21}NO_8$: C, 62.87; H, 4.82; N, 3.19; found: C, 62.74; H, 4.76; N, 3.08.

(3a*R*,8b*R*)-Diethyl 1-(*tert*-butyl)-3a,8b-dihydroxy-4-oxo-1,3a,4,8b-tetrahydroindeno[1,2-*b*]pyrrole-2,3-dicarboxylate (**4c**): pale yellow powder, mp 137–139 °C; yield: 78%; IR: 3435 (br.), 1737, 1710, 1664, 1578, 1460, 1279, 1125 cm^{-1}; ^{1}H NMR ($CDCl_3$, 500.1 MHz): δ 1.12 (3H, t, $^3J=7.2$ Hz, Me), 1.18 (3H, t, $^3J=7.2$ Hz, Me), 1.45 (9H, s, ^{t}Bu), 4.15 (2H, q, $^3J=7.2$ Hz, CH_2), 4.18 (2H, q, $^3J=7.2$ Hz, CH_2), 4.98 (1H, s, OH), 5.10 (1H, s, OH), 7.20 (1H, t, $^3J=7.3$ Hz, CH), 7.60–7.63 (2H, m, 2×CH); 7.72 (1H, d, $^3J=7.3$ Hz, CH); ^{13}C-NMR ($CDCl_3$, 125.7 MHz): δ 14.0 (Me), 14.2 (Me), 28.7 (Me_3C), 57.2 (Me_3C), 60.4 (CH_2O), 61.3 (CH_2O), 85.4 (C), 96.2 (C), 123.8 (CH), 124.5 (C), 124.8 (CH), 130.8 (CH), 134.6 (C), 137.2 (CH), 148.5 (C), 151.6 (C), 161.9 (C=O), 165.7 (C=O), 196.7 (C=O). Anal. calc. for $C_{21}H_{25}NO_7$: C, 62.52; H, 6.25; N, 3.47; found: C, 62.48; H, 6.18; N, 3.38.

2.2.37.5 Critical Views

The investigators demonstrated a simple, efficient and convenient route to functionalized tetrahydrooxoindeno[1,2-*b*]pyrroles by simple mixing of the starting materials without the use of any catalyst or activation.

2.2.38 Entry-38: Synthesis of 4-(alkylamino)-1-(arylsulfonyl)-3-benzoyl-1,5-dihydro-5-hydroxy-5-phenyl-2*H*-pyrrol-2-ones

Type of reaction: C–C, C–N and C–S bond formation
Reaction conditions: Catalyst-free, dichloromethane, room temperature
Synthetic strategy: One-pot multicomponent reaction
Keywords: Primary alkylamines, phenyl isothiocyanate/benzoyl isothiocyanate, itaconic anhydride, dichloromethane, catalyst-free, room temperature, one-pot multicomponent reaction, 1,5-dihydro-2*H*-pyrrol-2-ones

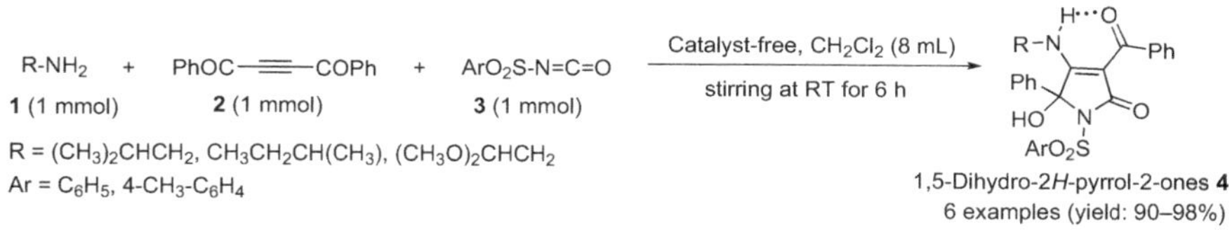

Scheme 2.55 Synthesis of 4-(alkylamino)-1-(arylsulfonyl)-3-benzoyl-1,5-dihydro-5-hydroxy-5-phenyl-2*H*-pyrrol-2-ones (**4**).

2.2.38.1 General Reaction Scheme

Alizadeh and his group[128] reported a facile, straightforward route for the synthesis of pharmaceutically interesting novel 4-(alkylamino)-1-(arylsulfonyl)-3-benzoyl-1,5-dihydro-5-hydroxy-5-phenyl-2*H*-pyrrol-2-ones (**4**) *via* one-pot three-component reaction of primary amines (**1**), 1,4-diphenylbut-2-yne-1,4-dione (**2**) and arenesulfonyl isocyanate (**3**) in dichloromethane under ambient conditions (Scheme 2.55).

2.2.38.2 Representative Examples

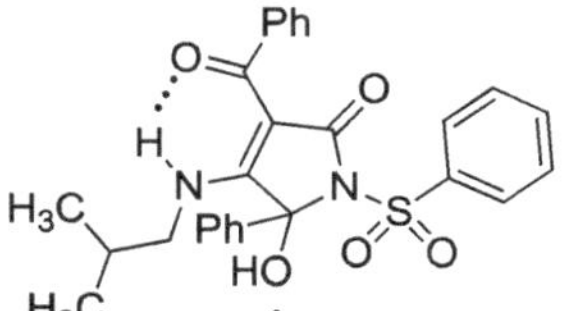

4a

3-Benzoyl-5-hydroxy-4-(isobutylamino)-5-phenyl-1-(phenylsulfonyl)-1*H*-pyrrol-2(5*H*)-one (**4a**): colorless crystals, mp 170–172 °C; yield: 95%

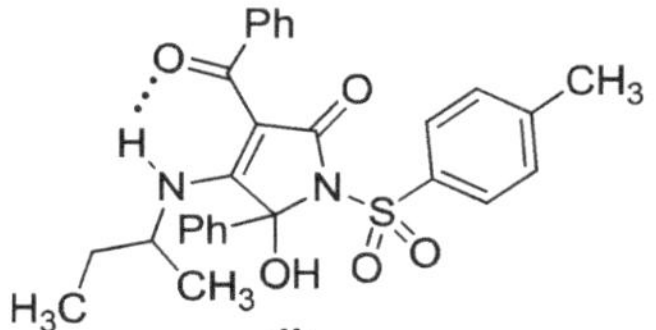

4b

3-Benzoyl-4-(*sec*-butylamino)-5-hydroxy-5-phenyl-1-tosyl-1*H*-pyrrol-2(5*H*)-one (**4b**): colorless crystals, mp 177–179 °C; yield: 95%

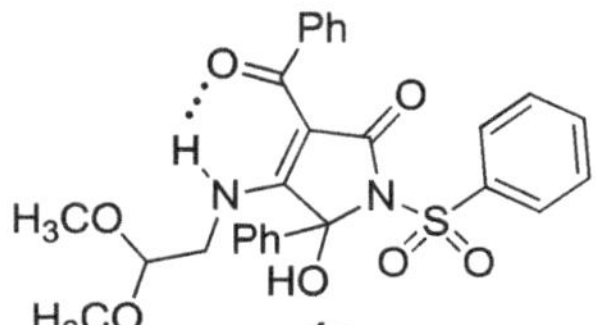

4c

3-Benzoyl-4-((2,2-dimethoxyethyl)amino)-5-hydroxy-5-phenyl-1-(phenylsulfonyl)-1*H*-pyrrol-2(5*H*)-one (**4c**): colorless crystals, mp 150–152 °C; yield: 90%

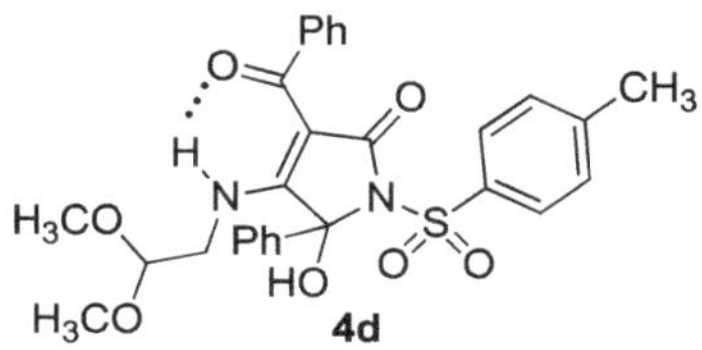

4d

3-Benzoyl-4-((2,2-dimethoxyethyl)amino)-5-hydroxy-5-phenyl-1-tosyl-1*H*-pyrrol-2(5*H*)-one (**4d**): colorless crystals, mp 145–147 °C; yield: 98%

2.2.38.3 Experimental Procedure

A solution of primary amines (**1**; 1 mmol) and 1,4-diphenylbut-2-yne-1,4-dione (**2**; 1 mmol) in dry dichloromethane (5 mL) was stirred for 1 h. Then a solution of arenesulfonyl isocyanate (**3**; 1 mmol) in 3 mL of dichloromethane was added drop-wise to the reaction mixture at room temperature, and stirring was continued for another 5 h. Upon completion of the reaction as indicated by TLC, the solvent was removed under reduced pressure to leave a residue that was purified by column chromatography (SiO_2; hexane–ethyl acetate mixtures) to afford the pure products of 4-(alkylamino)-1-(arylsulfonyl)-3-benzoyl-1,5-dihydro-5-hydroxy-5-phenyl-2*H*-pyrrol-2-ones (**4**). All the products were characterized by means of elemental analyses and detailed spectral studies.

2.2.38.4 Characterization Data of Two Representative Compounds

3-Benzoyl-5-hydroxy-4-(isobutylamino)-5-phenyl-1-(phenylsulfonyl)-1*H*-pyrrol-2(5*H*)-one (**4a**): colorless crystals, mp 170–172 °C; yield: 95%; IR (KBr): 3410 (OH), 3030 (NH), 1754 (C=O), 1626 (NC=C), 1575 and 1440 (Ar), 1375 and 1185 (SO_2), 1212, 1082 cm^{-1}; ^1H NMR ($CDCl_3$, 500.1 MHz): δ 1.03 (3H, d, $J=6.5$ Hz), 1.05 (3H, d, $J=6.5$ Hz), 2.19–2.22 (1H, m), 3.41 (1H, dd, $J=13.7$, 7.0 Hz), 3.60 (1H, dd, $J=13.8$, 7.5 Hz), 6.14 (1H, s), 7.29–7.56 (11H, m), 7.69 (2H, d, $J=7.2$ Hz), 7.80 (2H, d, $J=7.3$ Hz), 8.35 (1H, s); ^{13}C NMR ($CDCl_3$, 125.7 MHz): δ 19.54, 19.68, 25.91, 48.18, 90.32, 94.85, 125.77, 126.97, 127.49, 127.84, 127.94, 128.29, 128.31, 132.49, 133.04, 137.21, 137.34, 138.40, 150.76, 160.90, 189.58; EIMS (20 eV, rel.%): *m/z* 490 (3, M^+), 333 (6), 264 (5), 228 (5), 202 (6), 146 (19), 105 (100), 91 (9), 57 (14), 51 (19), 41 (17). Anal. calc. for $C_{27}H_{26}N_2O_5S$: C, 66.11; H, 5.34; N, 5.71; found: C, 67.00; H, 5.26; N, 5.63.

3-Benzoyl-4-((2,2-dimethoxyethyl)amino)-5-hydroxy-5-phenyl-1-tosyl-1*H*-pyrrol-2(5*H*)-one (**4d**): colorless crystals, mp 145–147 °C; yield: 98%; IR (KBr): 3390 (OH), 3160 (NH), 1754 (C=O), 1636 (NC=C), 1570, 1435 (Ar), 1373, 1187 (SO_2), 1212, 1070 cm^{-1}; ^1H NMR ($CDCl_3$, 500.1 MHz): δ 2.41 (3H, s), 3.49 (3H, s), 3.50 (3H, s), 3.59 (1H, dd, $J=14.6$, 5.6 Hz), 3.99 (1H, dd, $J=14.6$, 4.4 Hz), 4.65 (1H, t, $J=4.7$ Hz), 6.47 (1H, s), 7.21 (2H, d, $J=7.9$ Hz), 7.28–7.29 (3H, m), 7.38 (2H, t, $J=7.3$ Hz), 7.51 (1H, t, $J=7.7$ Hz), 7.58–7.59 (2H, m), 7.69 (2H, d, $J=7.5$ Hz), 7.72 (2H, d, $J=7.9$ Hz), 8.36 (1H, s); ^{13}C NMR ($CDCl_3$, 125.7 MHz): δ 21.67, 43.62, 55.28, 55.77, 91.21, 96.70, 102.50, 126.47, 127.53, 128.30, 128.55, 128.74, 128.89, 129.16, 133.16, 136.19, 137.95, 138.11, 144.84, 151.27, 161.06, 190.42; EIMS (20 eV, rel.%): *m/z* 308 (6), 307 (12), 105 (23), 91 (11), 75 (100), 47 (8). Anal. calc. for $C_{28}H_{28}N_2O_7S$: C, 62.67; H, 5.26; N, 5.22; found: C, 61.52; H, 5.10; N, 5.11.

2.2.38.5 Critical Views

A batch of new 1,5-dihydro-5-hydroxy-2*H*-pyrrol-2-one derivatives, of promising synthetic and pharmaceutical interest, were synthesized by a simple one-pot catalyst-free protocol under neutral conditions and requiring no activation or modification of the starting materials.

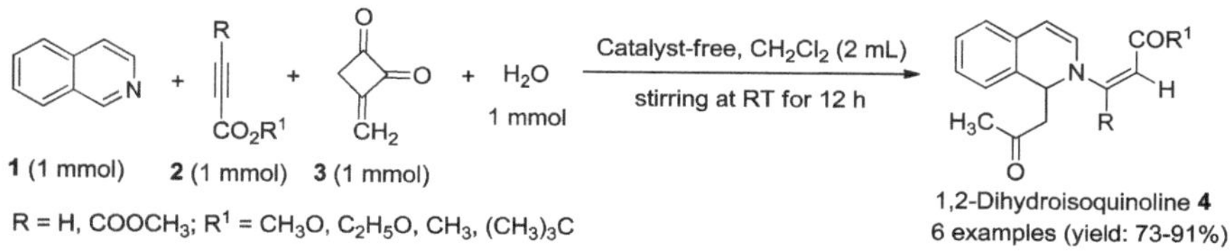

Scheme 2.56 Synthesis of 1,2-dihydroisoquinoline derivatives (**4**).

2.2.39 Entry-39: Synthesis of 1,2-Dihydroisoquinoline Derivatives

Type of reaction: C–C and C–N bond formation
Reaction conditions: Catalyst-free, dichloromethane, room temperature
Synthetic strategy: One-pot multicomponent reaction
Keywords: Isoquinoline, electron-deficient acetylene derivatives, diketene, water, dichloromethane, catalyst-free, room temperature, one-pot multicomponent reaction, 1,2-dihydroisoquinoline derivatives

2.2.39.1 General Reaction Scheme

Alizadeh and Zohreh[129] designed a convenient, one-pot four-component protocol for the synthesis of 1,2-dihydroisoquinoline derivatives (**4**) in good yields *via* reaction between a 1,4-dipolar intermediate generated by the reaction of isoquinoline (**1**) and an electron-deficient acetylene compound (**2**), and water in the presence of diketene (**3**) in dichloromethane at room temperature (Scheme 2.56).

2.2.39.2 Representative Examples

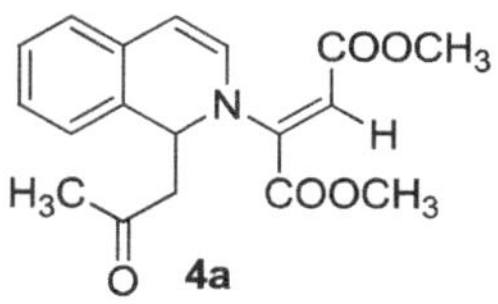

4a

Dimethyl 2-(1-(2-oxopropyl)isoquinolin-2(1*H*)-yl)fumarate (**4a**): pale yellow powder, mp 123–125 °C; yield: 95%

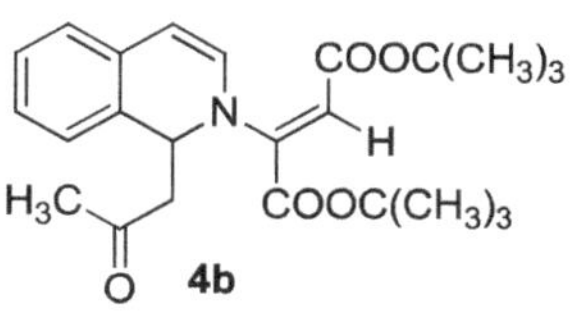

4b

Di-*tert*-butyl 2-(1-(2-oxopropyl)isoquinolin-2(1*H*)-yl)fumarate (**4b**): pale yellow powder, mp 120–122 °C; yield: 90%

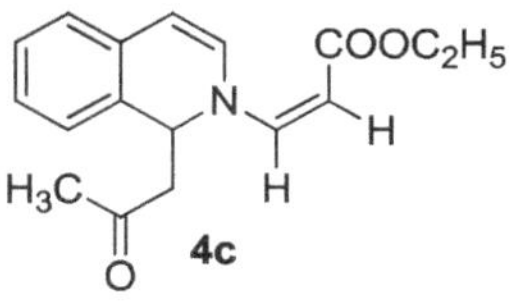

4c

(*Z*)-Ethyl 3-(1-(2-oxopropyl)isoquinolin-2(1*H*)-yl)acrylate (**4c**): pale yellow powder, mp 90–92 °C; yield: 85%

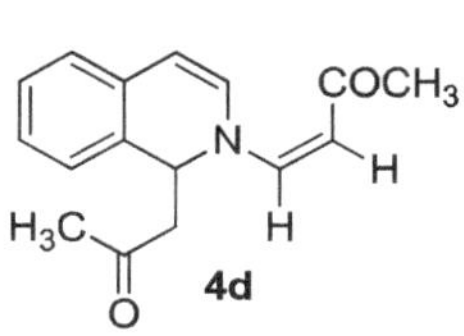

4d

(*Z*)-4-(1-(2-Oxopropyl)isoquinolin-2(1*H*)-yl)but-3-en-2-one (**4d**): pale yellow crystals, mp 190 °C; yield: 75%

2.2.39.3 Experimental Procedure

To a magnetically stirred solution of isoquinoline (**1**; 0.13 g, 1 mmol) and water (0.02 g, 1 mmol) in dichloromethane (2 mL) was added drop-wise a solution of the diketene (**3**; 0.08 g, 1 mmol) and acetylene derivative (**2**; 1 mmol) at 0 °C. The mixture was then stirred for 10–12 h at room temperature to complete the reaction. Upon completion, the solvent was evaporated, and the residue was subjected to column chromatography (silica gel 230–240 mesh; hexane–ethyl acetate mixtures as eluent) to obtain pure products of 1,2-dihydroisoquinoline derivatives (**4**). All the products were characterized by means of elemental analyses and detailed spectral studies.

2.2.39.4 Characterization Data of Two Representative Compounds

Di-*tert*-butyl 2-(1-(2-oxopropyl)isoquinolin-2(1*H*)-yl)fumarate (**4b**): pale yellow powder, mp 120–122 °C; yield: 90%; IR (KBr): 1719 (COMe), 1690 ($CO_2{}^tBu$), 1577 (C=C), 1556 and 1509 (arom.), 1223 and 1151 (C–O, ester) cm^{-1}; ^1H NMR ($CDCl_3$, 500.1 MHz): δ 1.46 (9H, s), 1.60 (9H, s), 1.97 (3H, s), 2.73 (1H, dd, J = 16.8, 2.7 Hz), 3.15 (1H, dd, J = 16.8, 10.2 Hz), 5.08 (1H, s,), 5.32 (1H, d, J = 9.7 Hz), 5.92 (1H, d, J = 7.6 Hz), 6.35 (1H, d, J = 7.6 Hz), 7.05 (1H, d, J = 7.5 Hz), 7.12–7.20 (3H, m); ^{13}C NMR ($CDCl_3$, 125.7 MHz): δ 27.9, 28.3, 31.1, 46.9, 53.9, 79.5, 83.9, 93.2, 109.2, 124.7, 125.8, 126.7, 127.0, 128.1, 129.6, 131.3, 149.0, 163.5, 165.9, 205.6; EIMS (20 eV, %rel.): *m/z* 413 (4, M^+), 356 (12), 300 (19), 244 (51), 200 (25), 154 (13), 57 (100), 43 (32). Anal. calc. for $C_{24}H_{31}NO_5$: C, 69.71; H, 7.56; N, 3.39; found: C, 69.53; H, 7.32; N, 3.46.

(*Z*)-4-(1-(2-Oxopropyl)isoquinolin-2(1*H*)-yl)but-3-en-2-one (**4d**): pale yellow crystals, mp 190 °C; yield: 75%; IR (KBr): 3490 (OH), 1726 (COMe), 1702 (COPh), 1580 (C=C), 1555 and 1498 (arom.), 1165 (C–OH) cm^{-1}; ^1H NMR ($CDCl_3$, 500.1 MHz): δ 2.35 (3H, s), 4.14 (1H, d, J = 9.4 Hz), 5.73 (1H, d, J = 9.4 Hz), 6.05 (1H, s), 6.37(1H, d, J = 7.5 Hz), 6.69 (1H, d, J = 7.5 Hz), 6.89 (1H, d, J = 7.5 Hz), 7.17 (1H, d, J = 7.2 Hz), 7.21–7.28 (3H, m), 7.32 (2H, t, J = 7.7 Hz), 7.36 (2H, t, J = 7.6 Hz), 7.47 (1H, t, J = 7.6 Hz), 7.52 (2H, d, J = 6.6 Hz), 7.79 (2H, d, J = 7.6 Hz), 9.01 (1H, s); ^{13}C NMR ($CDCl_3$, 125.7 MHz): δ 31.2, 59.9, 61.3, 67.8, 84.6, 89.1, 116.5, 122.1, 123.8, 124.8, 125.7, 127.7, 128.1, 128.2, 128.3, 128.7, 130.6, 131.6, 131.9, 138.9, 140.2, 166.8, 188.9, 204.9; EIMS (20 eV, %rel.): *m/z* 421 (2, M^+), 360 (16), 277(23), 236 (10), 167(17), 149 (33), 129 (51), 105 (100), 94 (24), 77 (72), 57 (32), 43 (44). Anal. calc. for $C_{28}H_{23}NO_3$: C, 79.79; H, 5.50; N, 3.32; found: C, 79.61; H, 5.68; N, 3.41.

2.2.39.5 Critical Views

The present method offers a convenient one-pot route for preparing isoquinoline derivatives under neutral and aqueous conditions without any prior activation or modification of the starting materials.

2.2.40 Entry-40: Synthesis of Arylsulfonamido-substituted 1,5-Benzodiazepines

Type of reaction: C–C and C–N bond formation
Reaction conditions: Catalyst-free, dichloromethane, room temperature
Synthetic strategy: One-pot multicomponent reaction
Keywords: Benzene-1,2-diamine, diketene, arylsulfonyl isocyanate, dichloromethane, catalyst-free, room temperature, one-pot multicomponent reaction, arylsulfonamido-substituted 1,5-benzodiazepines

2.2.40.1 General Reaction Scheme

A facile and efficient catalyst-free method for the synthesis of pharmacologically interesting arylsulfonamido-substituted 1,5-benzodiazepines (**4**) *via* a one-pot three-component condensation reaction of benzene-1,2-diamine (**1**), diketene (**2**), and arylsulfonyl isocyanate (**3**) in dichloromethane at room temperature was reported by Alizadeh and Zohreh[130] (Scheme 2.57).

2.2.40.2 Representative Examples

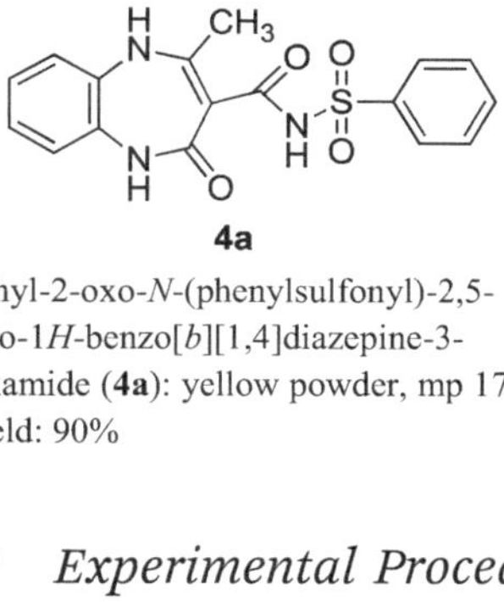

4a

4-Methyl-2-oxo-*N*-(phenylsulfonyl)-2,5-dihydro-1*H*-benzo[*b*][1,4]diazepine-3-carboxamide (**4a**): yellow powder, mp 173–175 °C; yield: 90%

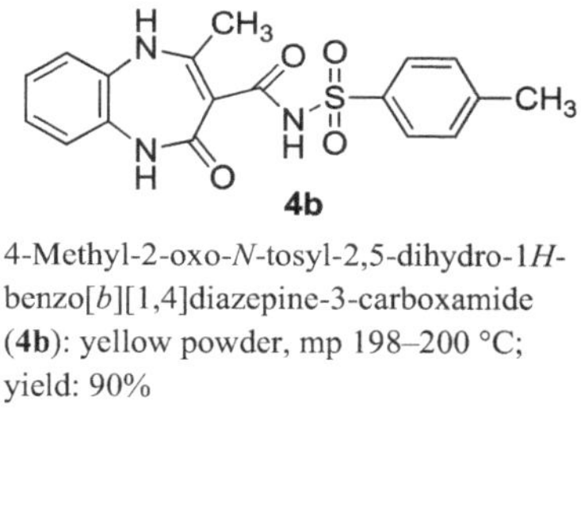

4b

4-Methyl-2-oxo-*N*-tosyl-2,5-dihydro-1*H*-benzo[*b*][1,4]diazepine-3-carboxamide (**4b**): yellow powder, mp 198–200 °C; yield: 90%

2.2.40.3 Experimental Procedure

To a magnetically stirred solution of benzene-1,2-diamine (**1**; 0.11 g, 1 mmol) and diketene (**2**; 0.08 g, 1 mmol) in dichloromethane (5 mL) was added after 1 h arylsulfonyl isocyanate (**3**; 1 mmol) at room temperature, and the mixture was continued for another 5 h to complete the conversion. After completion of the reaction, the solvent was removed, and the crude product of arylsulfonamido-substituted 1,5-benzodiazepine (**4**) was precipitated in diethyl

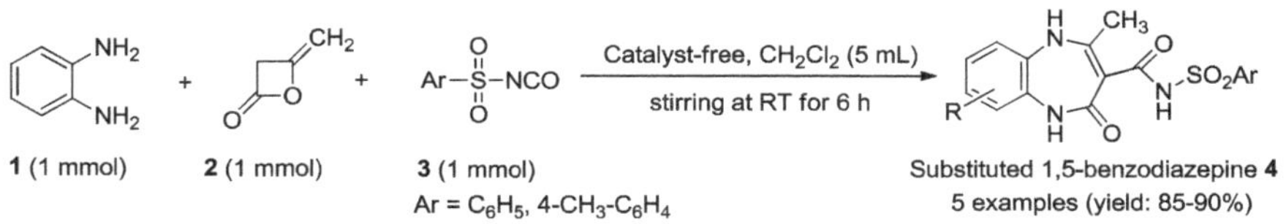

Scheme 2.57 Synthesis of arylsulfonamido-substituted 1,5-benzodiazepines (**4**).

ether. The product was filtered and washed with cold dichloromethane for further purification. All the products were characterized by means of elemental analyses and detailed spectral studies.

2.2.40.4 Characterization Data of One Representative Compound

4-Methyl-2-oxo-*N*-tosyl-2,5-dihydro-1*H*-benzo[*b*][1,4]diazepine-3-carboxamide (**4b**): yellow powder, mp 198–200 °C; yield: 90%; IR (KBr): 3290, 3180, 3020 (3×NH), 1699, 1646 (2×NC=O), 1522, 1408 (2×Ar), 1334, 1149 (2×SO_2) cm^{-1}; ^{1}H NMR (CD_3COCD_3, 500 MHz): δ 2.31 (3H, s), 2.39 (3H, s), 7.01–7.05 (3H, m), 7.13–7.15 (1H, m), 7.35 (2H, d, J=7.9), 7.90 (2H, d, J=7.8), 8.41 (1H, s), 9.10 (1H, s), 12.58 (1H, s); ^{13}C NMR (CD_3COCD_3, 125 MHz): δ 21.4, 25.1, 105.4, 122.3, 125.5, 126.9, 128.0, 128.9, 130.0, 133.3, 138.8, 144.5, 165.8, 172.9, 173.0; EIMS (20 eV, %rel.): *m/z* 371 (5, M^+), 311 (10), 197 (23), 174 (22), 155 (39), 132 (98), 131 (50), 91 (100), 65 (35), 39 (40). Anal. calc. for $C_{18}H_{17}N_3O_4S$: C, 58.21; H, 4.61; N, 11.31; found: C, 58.29; H, 4.64; N, 11.33.

2.2.40.5 Critical Views

Arylsulfonamido-substituted 1,5-benzodiazepines were synthesized following a catalyst-free protocol under ambient conditions. The developed method bears several unique advantages, such as relatively short reaction times, simplicity of operation under neutral conditions, high yields of products and no need for a catalyst.

2.2.41 Entry-41: Synthesis of *N*-(1,7-Dioxotetrahydropyrazolo [1,2-*a*]pyrazol-2-yl)-Benzamides

Type of reaction: C–C and C–N bond formation
Reaction conditions: Catalyst-free, dichloromethane, room temperature
Synthetic strategy: One-pot multicomponent reaction
Keywords: Azomethine imines, azalactones, dichloromethane, catalyst-free, room temperature, *N*-(1,7-dioxotetrahydropyrazolo[1,2-*a*]pyrazol-2-yl)-benzamides

2.2.41.1 General Reaction Scheme

A series of *N*-(1,7-dioxotetrahydropyrazolo[1,2-*a*]pyrazol-2-yl)-benzamide derivatives (**3**), a novel class of bioactive compounds, were synthesized in moderate to high yields by Liu *et al.*[131] *via* the 1,3-dipolar cycloaddition of azomethine imines (**1**) with azalactones (**2**) in dichloromethane. This was followed by subsequent rearrangement under ambient conditions (Scheme 2.58).

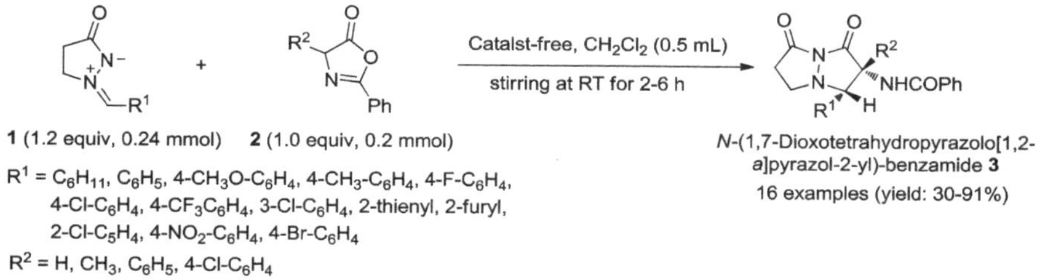

Scheme 2.58 Synthesis of *N*-(1,7-dioxotetrahydropyrazolo[1,2-*a*]pyrazol-2-yl)-benzamides (**3**).

2.2.41.2 Representative Examples

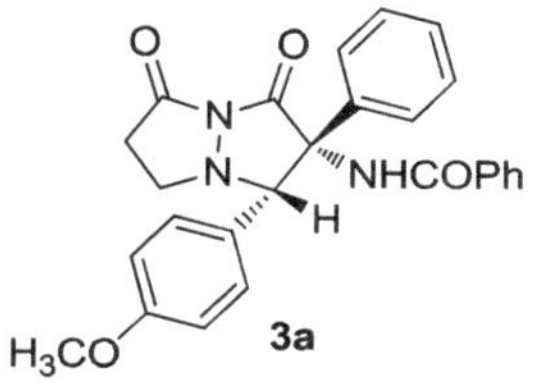

3a

N-((1*S*,2*R*)-1-(4-methoxyphenyl)-3,5-dioxo-2-phenylhexahydropyrazolo[1,2-*a*]pyrazol-2-yl)benzamide (**3a**): white solid, mp 159–160 °C; yield: 70%

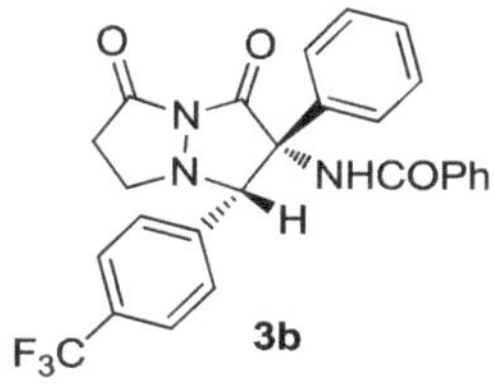

3b

N-((1*S*,2*R*)-3,5-dioxo-2-phenyl-1-(4-(trifluoromethyl)phenyl)hexahydropyrazolo[1,2-*a*]pyrazol-2-yl)benzamide (**3b**): yield: 67%

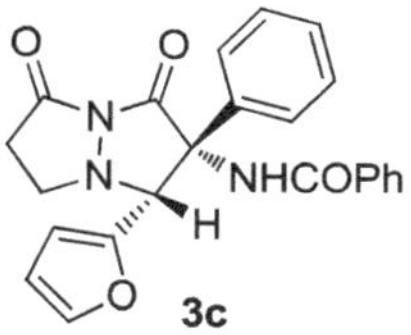

3c

N-((1*R*,2*R*)-1-(furan-2-yl)-3,5-dioxo-2-phenylhexahydropyrazolo[1,2-*a*]pyrazol-2-yl)benzamide (**3c**): white solid, mp 166–168 °C; yield: 69%

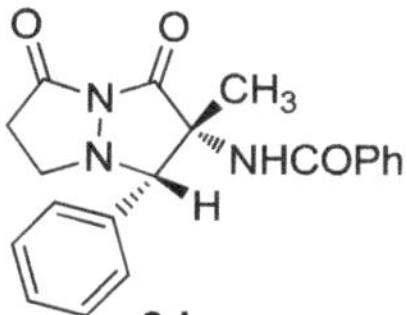

3d

N-((1*S*,2*R*)-2-methyl-3,5-dioxo-1-phenylhexahydropyrazolo[1,2-*a*]pyrazol-2-yl)benzamide (**3d**): yield: 68%

2.2.41.3 Experimental Procedure

A mixture of azomethine imines (**1**; 1.2 equiv, 0.24 mmol) and azalactones (**2**; 0.2 mmol) in dichloromethane (0.5 mL) was stirred in a reaction tube containing a magnetic stirrer at room temperature for 2–6 h. The crude product as obtained upon completion of the reaction (monitored by TLC) was then purified by column chromatography (silica gel, petrol ether-ethyl-acetate 1 : 1) to obtain pure products of *N*-(1,7-dioxotetrahydropyrazolo[1,2-*a*]pyrazol-2-yl)-benzamides (**3**). Each of the products was characterized by means of detailed spectral studies.

2.2.41.4 Characterization Data of Two Representative Compounds

N-((1*S*,2*R*)-1-(4-methoxyphenyl)-3,5-dioxo-2-phenylhexahydropyrazolo[1,2-*a*] pyrazol-2-yl)benzamide (**3a**): white solid, mp 159–160 °C; yield: 70%; IR (KBr): 3413, 3061, 2932, 1783, 1713, 1597, 1503, 1267, 1173, 1045, 916, 835, 716 cm^{-1}; ^{1}H NMR ($CDCl_3$, 400 MHz): δ 7.54 (2H, d, $J = 8.4$ Hz), 7.45 (3H, t, $J = 6.9$ Hz), 7.38–7.33 (5H, m), 7.13 (2H, d, $J = 8.7$ Hz), 6.87 (2H, d, $J = 8.8$ Hz), 6.75 (1H, s), 4.68 (1H, s), 3.78 (3H, s), 3.39 (1H, s), 2.99–2.84 (3H, m); ^{13}C NMR ($CDCl_3$, 101 MHz): δ 190.77, 165.57, 165.10, 160.22, 135.65, 133.49, 131.91, 131.82, 128.68, 128.59, 128.52, 126.90, 126.56, 122.56, 114.28, 114.24, 70.86, 55.52, 55.02, 35.69; HRMS (ESI): *m/z* $[M+Na]^+$ calc. for $C_{26}H_{23}N_3O_4Na$: 464.1581; found: 464.1573.

N-((1*R*,2*R*)-1-(furan-2-yl)-3,5-dioxo-2-phenylhexahydropyrazolo[1,2-*a*]pyrazol-2-yl)benzamide (**3c**): white solid, mp 166–168 °C; yield: 69%; IR (KBr): 3424, 2932, 1783, 1713, 1666, 1502, 1314, 904, 730 cm^{-1}; ^{1}H NMR ($CDCl_3$, 400 MHz): δ 7.83 (2H, d, $J = 7.1$ Hz), 7.48–7.31 (10H, m), 6.52 (1H, dd, $J = 3.2$, 0.5 Hz), 6.29 (1H, dd, $J = 3.3$, 1.9 Hz), 5.90 (1H, s), 3.39–3.34 (1H, m), 2.94 (1H, ddd, $J = 16.1$, 10.9, 9.0 Hz), 2.80–2.65 (2H, m); ^{13}C NMR ($CDCl_3$, 101 MHz): δ 166.05, 165.50, 163.88, 147.86, 143.87, 135.51, 133.24, 131.71, 128.93, 128.74, 128.39, 126.79, 126.56, 113.92, 110.33, 69.23, 46.41, 35.05; HRMS (ESI): *m/z* $[M+Na]^+$ calc. for $C_{23}H_{19}N_3O_4Na$: 424.1268; found: 424.1261.

2.2.41.5 Critical Views

The investigators developed a new and efficient catalyst-free approach for the 1,3-dipolar cycloaddition reaction between azomethine imines and azalactones to synthesize a series of biologically promising *N*,*N*-bicyclo compounds, namely *N*-(1,7-dioxotetrahydropyrazolo[1,2-*a*]pyrazol-2-yl)benzamides at room temperature. This method may find extended uses to synthesize *N*,*N*-bicyclic derivatives with important bioactivities, without any further activation or modification of the starting materials.

2.2.42 Entry-42: Synthesis of Substituted Pyridin-2(1*H*)-ones

Type of reaction: C–C and C–N bond formation
Reaction conditions: Catalyst-free, dichloromethane, room temperature
Synthetic strategy: One-pot multicomponent reaction
Keywords: Isocyanides, dimethyl acetylenedicarboxylate, arylidene-2-cyanoacetohydrazides, dichloromethane, catalyst-free, room temperature, one-pot multicomponent, regioselectivity, highly-substituted pyridin-2(1*H*)-ones

2.2.42.1 General Reaction Scheme

Li and his group[132] reported a catalyst-free, efficient and regioselective one-pot three-component reaction protocol for the synthesis of highly-substituted

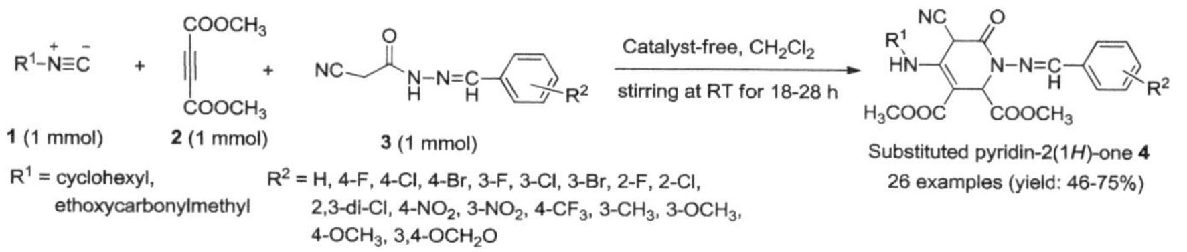

Scheme 2.59 Synthesis of substituted pyridin-2(1*H*)-ones (**4**).

novel bioactive pyridin-2(1*H*)-one derivatives (**4**) by the reaction of isocyanides (**1**), dimethyl acetylenedicarboxylate (**2**) and a series of *N*-arylidene-2-cyanoacetohydrazides (**3**) in dichloromethane at room temperature (Scheme 2.59). Easy workup, the readily available starting materials, good functional group tolerance, and moderate to good yields are some of the major advantages of this present protocol.

2.2.42.2 Representative Examples

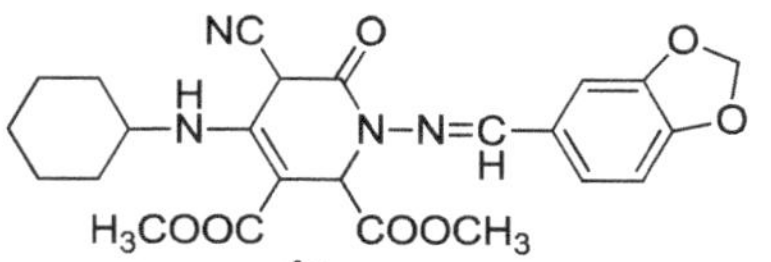

4a

Dimethyl 1-((benzo[*d*][1,3]dioxol-5-ylmethylene)amino)-5-cyano-4-(cyclohexylamino)-6-oxo-1,2,5,6-tetrahydropyridine-2,3-dicarboxylate (**4a**): white powder, mp 138–140 °C; yield: 65%

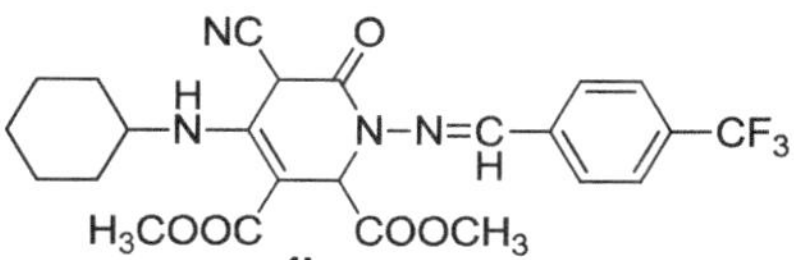

4b

Dimethyl 5-cyano-4-(cyclohexylamino)-6-oxo-1-((4-(trifluoromethyl)benzylidene) amino)-1,2,5,6-tetrahydropyridine-2,3-dicarboxylate (**4b**): white powder, mp 196–198 °C; yield: 72%

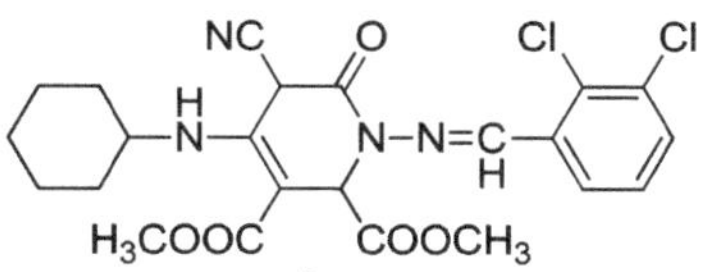

4c

Dimethyl 5-cyano-4-(cyclohexylamino)-1-((2,3-dichlorobenzylidene)amino)-6-oxo-1,2,5,6-tetrahydropyridine-2,3-dicarboxylate (**4c**): white powder, mp 190–193 °C; yield: 70%

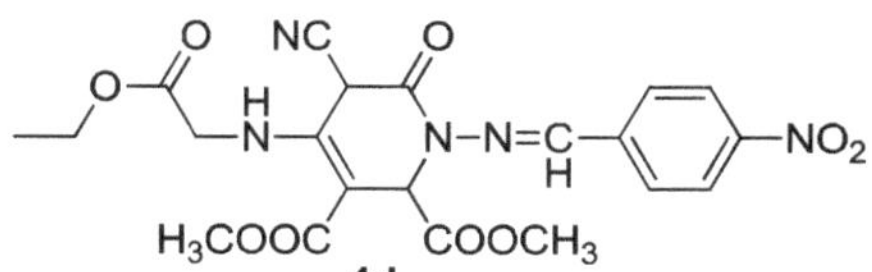

4d

Dimethyl 5-cyano-4-((2-ethoxy-2-oxoethyl)amino)-1-((4-nitrobenzylidene)amino)-6-oxo-1,2,5,6-tetrahydropyridine-2,3-dicarboxylate (**4d**): yellow powder, mp 197–199 °C; yield: 73%

2.2.42.3 Experimental Procedure

A mixture of isocyanides (**1**; 1.0 mmol), dimethyl acetylenedicarboxylate (**2**; DMAD; 1.0 mmol), and *N*-arylidene-2-cyanoacetohydrazides (**3**; 1.0 mmol) was stirred in dichloromethane at room temperature for 18–28 h. After completion of the reaction, as indicated by TLC, the solvent was removed under vacuum, and the residue was subjected to column chromatography (silica gel, petroleum ether–ethyl acetate, 8:1, v/v) to afford pure products of highly-substituted pyridin-2(1*H*)-one derivatives (**4**) in moderate to good yields. Each of the products was identified based on spectral studies including FTIR, NMR and HRMS.

2.2.42.4 Characterization Data of Two Representative Compounds

Dimethyl 1-((benzo[*d*][1,3]dioxol-5-ylmethylene)amino)-5-cyano-4-(cyclohexylamino)-6-oxo-1,2,5,6-tetrahydropyridine-2,3-dicarboxylate (**4a**): white powder, mp 138–140 °C; yield: 65%; IR (KBr): 3442, 2950, 2859, 2204, 1742, 1652, 1607, 1452, 1400, 1256, 1151, 1033, 818, 742 cm^{-1}; ^{1}H NMR ($CDCl_3$, 500 MHz): δ 1.19–1.97 (10H, m, 5×CH_2 of cyclohexyl), 3.35 (1H, m, CH of cyclohexyl), 3.77 (3H, s, OCH_3), 3.81 (3H, s, OCH_3), 4.38 (1H, s, CHCN), 5.18 (1H, s, CH–COO), 6.85–6.87 (3H, m, ArH, CH_2), 6.78–6.79 (1H, d, $J=8.0$ Hz, ArH), 7.03–7.04 (1H, dd, $J=8.0$, 1.5 Hz, ArH), 7.29 (1H, s, CH=N), 9.12 (1H, s, NH); ^{13}C NMR ($CDCl_3$, 125 MHz): δ 24.6, 24.8, 32.9, 34.2, 53.6, 53.7, 60.3, 63.6, 101.3, 105.6, 108.1, 115.5, 124.1, 129.4, 148.1, 149.5, 153.1, 158.5, 160.5, 166.2, 169.7; HRMS (ESI-TOF, $[M+H]^+$): calc. for $C_{24}H_{27}N_4O_7$, 483.1880; found, 483.1879.

Dimethyl 5-cyano-4-((2-ethoxy-2-oxoethyl)amino)-1-((4-nitrobenzylidene)amino)-6-oxo-1,2,5,6-tetrahydropyridine-2,3-dicarboxylate (**4d**): yellow powder, mp 197–199 °C; yield: 73%; IR (KBr): 3434, 2925, 2853, 2209, 1734, 1700, 1637, 1522, 1472, 1408, 1345, 1216, 1109, 845 cm^{-1}; ^{1}H NMR ($CDCl_3$, 500 MHz): δ 1.31 (3H, t, $J=7.0$ Hz, CH_3), 3.72 (3H, s, OCH_3), 3.81 (3H, s, OCH_3), 4.13–4.18 (1H, dd, $J=18.5$, 6.0, Hz, C(O)CH_2N), 4.28 (2H, q, $J=7.0$ Hz, OCH_2), 4.33–4.38 (1H, dd, $J=18.5$, 6.0 Hz, C(O)CH_2N), 4.77 (1H, s, CHCN), 5.55 (1H, s, CH–COO), 7.82–7.84 (2H, d, $J=8.5$ Hz, ArH), 7.87 (1H, s, CH=N), 8.26–8.28 (2H, d, $J=8.5$ Hz, ArH), 8.61 (1H, s, NH); ^{13}C NMR ($CDCl_3$, 125 MHz): δ 14.1, 42.9, 51.7, 52.4, 53.5, 57.2, 62.1, 72.7, 116.4, 124.2, 128.1, 139.1, 139.7, 148.7, 157.9, 168.3, 168.9; HRMS (ESI-TOF, $[M+H]^+$): calc. for $C_{21}H_{22}N_5O_9$, 488.1418; found, 488.1425.

2.2.42.5 Critical Views

Due to their diverse range of biological activities, pyridin-2(1*H*)-ones are recognized as privileged structures making these structural motifs attractive targets for library preparation. The present investigators outlined a new synthetic approach to design the synthesis of such a new series of highly functionalized pyridine-2(1*H*)-one scaffolds utilizing *N*-arylidene-2-cyanoacetohydrazides, for the first time, along with isocyanides and DMAD *via* a one-pot multicomponent reaction under ambient conditions in the absence of any catalyst. The present procedure has several advantages, such as the

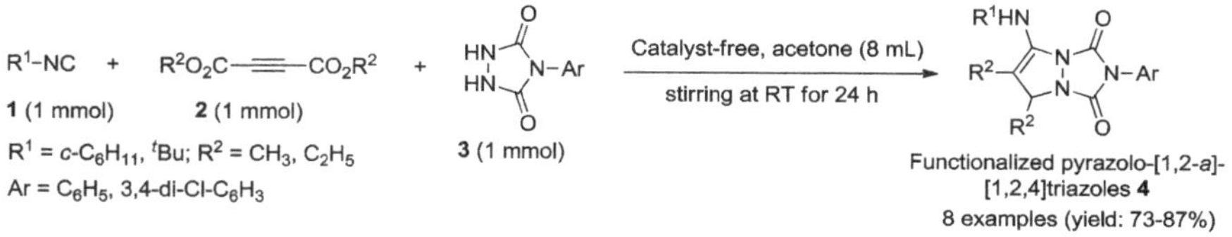

Scheme 2.60 Synthesis of functionalized pyrazolo[1,2-*a*][1,2,4]triazoles (**4**).

ready availability of the starting materials, easy workup procedure, mild reaction conditions, good functional group tolerance, highly regioselectivity, and moderate to good yields. This synthetic strategy opens a convenient way to construct a pyridin-2-one core for many synthetic fragments with unique properties for medicinal chemistry.

2.2.43 Entry-43: Synthesis of Functionalized Pyrazolo[1,2-*a*][1,2,4]triazoles

Type of reaction: C–C and C–N bond formation
Reaction conditions: Catalyst-free, ethanol-water, room temperature
Synthetic strategy: One-pot multicomponent reaction
Keywords: Isocyanides, dialkyl acetylenedicarboxylates, 4-arylurazoles, acetone, catalyst-free, room temperature, one-pot multicomponent reaction, functionalized pyrazolo[1,2-*a*][1,2,4]triazoles

2.2.43.1 General Reaction Scheme

Adib and his group[133] reported a straightforward one-pot three-component protocol for the synthesis of densely functionalized pyrazolo[1,2-*a*][1,2,4]triazoles (**4**) in good yields from the reaction of 4-arylurazoles (**3**) with highly reactive, generated *in situ*, 1 : 1 zwitterionic intermediates of isocyanides (**1**) and dialkyl acetylenedicarboxylates (**2**) in acetone at room temperature (Scheme 2.60).

2.2.43.2 Representative Examples

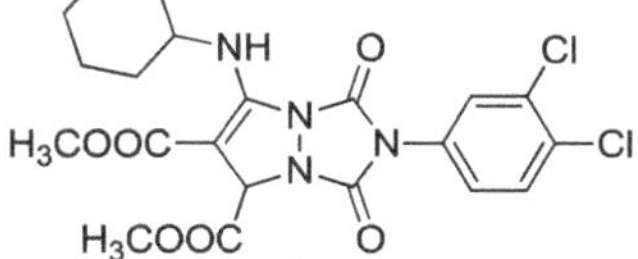

4a

Dimethyl 7-(cyclohexylamino)-2-(3,4-dichlorophenyl)-1,3-dioxo-1,2,3,5-tetrahydropyrazolo[1,2-*a*][1,2,4]triazole-5,6-dicarboxylate (**4a**): pale yellow crystals, mp 147–151 °C; yield: 85%

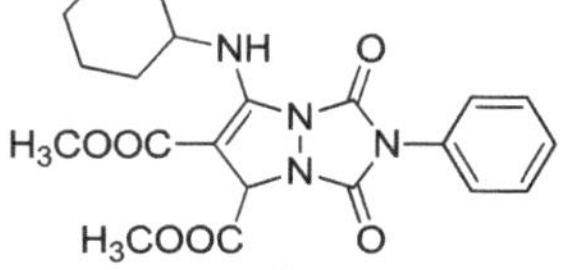

4b

Dimethyl 7-(cyclohexylamino)-1,3-dioxo-2-phenyl-1,2,3,5-tetrahydropyrazolo[1,2-*a*][1,2,4]triazole-5,6-dicarboxylate (**4b**): colorless crystals, mp 85–88 °C; yield: 84%

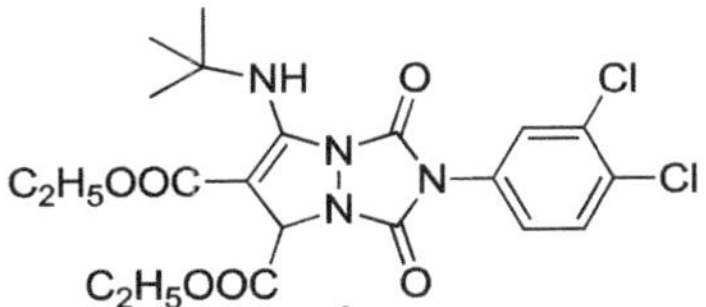

4c

Diethyl 7-(*tert*-butylamino)-2-(3,4-dichlorophenyl)-1,3-dioxo-1,2,3,5-tetrahydropyrazolo[1,2-*a*][1,2,4]triazole-5,6-dicarboxylate (**4c**): pale yellow crystals, mp 90–102 °C; yield: 75%

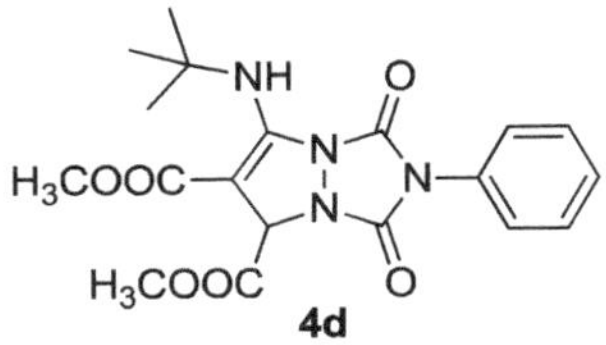

4d

Dimethyl 7-(*tert*-butylamino)-1,3-dioxo-2-phenyl-1,2,3,5-tetrahydropyrazolo[1,2-*a*][1,2,4]triazole-5,6-dicarboxylate (**4d**): colorless crystals, mp 129–131 °C; yield: 87%

2.2.43.3 Experimental Procedure

To a magnetically stirred solution of dialkyl acetylenedicarboxylates (**2**; 1 mmol) and 4-arylurazoles (**3**; 1 mmol) in anhydrous acetone (6 mL), a solution of isocyanides (**1**; 1 mmol) in anhydrous acetone (2 mL) was added drop-wise, at room temperature over 10 min. The mixture was then stirred for 24 h. Upon completion of the reaction, the solvent was removed, and the crude product was purified using column chromatography (silica gel; hexane–ethyl acetate 3 : 1), followed by recrystallization from hexane–ethyl acetate (1 : 1) mixture to obtain pure functionalized pyrazolo[1,2-*a*][1,2,4]triazoles (**4**) in good yields. All the synthesized compounds were characterized by means of spectral and elemental analyses.

2.2.43.4 Characterization Data of Two Representative Compounds

Dimethyl 7-(cyclohexylamino)-1,3-dioxo-2-phenyl-1,2,3,5-tetrahydropyrazolo-[1,2-*a*][1,2,4]triazole-5,6-dicarboxylate (**4b**): colorless crystals, mp 85–88 °C; yield: 84%; IR (KBr): 3290 (NH); 1783m, 1734s, 1663 (C=O); 1598, 1487, 1453, 1399, 1218, 1094, 1023, 751 cm^{-1}; ^{1}H NMR ($CDCl_3$, 500 MHz): δ 1.20–2.05 (10H, m), 3.65 (3H, s), 3.70 (3H, s), 4.05 (H, m), 5.26 (1H, s), 7.20–7.25 (2H, m), 7.39–7.42 (4H, m); ^{13}C NMR ($CDCl_3$, 125 MHz): δ 24.34, 24.38, 25.10, 33.69, 34.07, 50.75, 52.70, 55.81, 61.78, 80.52, 126.01, 128.75, 129.11, 130.76, 148.62, 150.27, 153.02, 165.43, 169.17; EIMS (20 eV): *m/z* (%rel.) 428 (5, M^+), 397 (10), 370 (65), 288 (100), 216 (56), 168 (70), 140 (66), 111 (80), 86 (95), 55 (82), 41 (25). Anal. calc. for $C_{21}H_{24}N_4O_6$: C, 58.87; H, 5.65; N, 13.08; found: C, 59.0; H, 5.7; N, 13.0.

Diethyl 7-(*tert*-butylamino)-2-(3,4-dichlorophenyl)-1,3-dioxo-1,2,3,5-tetrahydropyrazolo[1,2-*a*][1,2,4]triazole-5,6-dicarboxylate (**4c**): pale yellow crystals, mp 90–102 °C; yield: 75%; IR (KBr): 3250 (NH); 1801m, 1753s, 1668 (C=O); 1614, 1477, 1380, 1335, 1221, 1096, 1028, 874, 646 cm^{-1}; ^{1}H NMR ($CDCl_3$, 500 MHz): δ 1.16 (3H, t, J =7.1 Hz), 1.18 (3H, t, J = 7.1), 1.37 (9H, s), 4.05 (1H, dq, J = 10.8 and 7.1 Hz), 4.12 (2H, q, J = 7.1 Hz), 4.14 (1H, dq, J = 10.8 and 7.1 Hz), 5.19 (1H, s), 6.97 (1H, br s), 7.31 (1H, dd, J = 2.4 and 8.6 Hz), 7.43

(1H, d, $J=8.6$ Hz), 7.57 (1H, d, $J=2.4$ Hz); ^{13}C NMR ($CDCl_3$,125 MHz): δ 14.05, 14.33, 30.28, 57.85, 59.97, 62.28, 62.41, 86.13, 124.96, 127.58, 130.24, 130.81, 132.92, 133.17, 148.46, 149.81, 153.13, 164.93, 168.69; EIMS (20 eV): *m/z* (%rel.) 499 (11, M^+), 443 (10), 425 (59), 369 (100), 341 (6), 182 (7), 154 (13), 57 (60), 41 (31), 29 (74). Anal. calc. for $C_{21}H_{24}Cl_2N_4O_6$: C, 50.51; H, 4.84; N, 11.22; found: C, 50.8; H, 5.0; N, 11.0.

2.2.43.5 Critical Views

The present method offers a new route for the generation of synthetically and biologically potent densely functionalized pyrazolo[1,2-*a*][1,2,4]triazoles under neutral conditions at room temperature, requiring no activation or modification of the starting materials.

2.2.44 Entry-44: Synthesis of bis(Indolyl)-1,4-quinones

Type of reaction: C–C bond formation
Reaction conditions: Catalyst-free, water, room temperature
Synthetic strategy: '*On-water*' direct C–C coupling
Keywords: Indoles, 1,4-benzoquinones, water, catalyst-free, room temperature, direct C–C coupling, 3-indolylquinones, bis(indolyl)-1,4-quinones

2.2.44.1 General Reaction Scheme

A catalyst-free '*on-water*' direct C–C coupling between indoles (**1**) and 1,4-benzoquinones (**2**) affording both 3-indolyl-1,4-benzoquinones (**3**) and bis-(indolyl)-1,4-benzoquinones (**4**), at room temperature, has been developed by Li and his group[134] (Scheme 2.61).

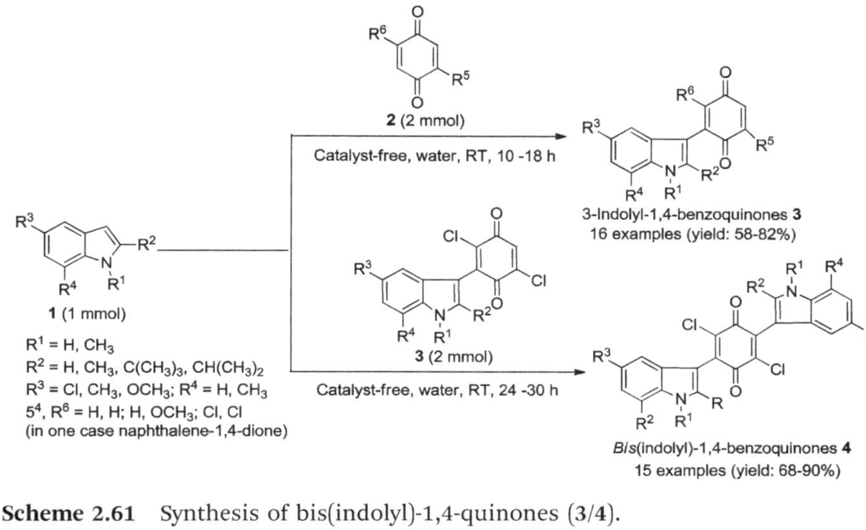

Scheme 2.61 Synthesis of bis(indolyl)-1,4-quinones (3/4).

2.2.44.2 Representative Examples

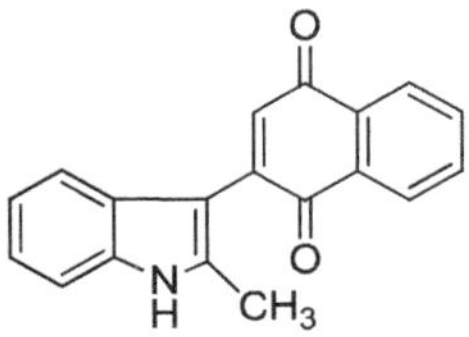

3a

2-(2-Methyl-1*H*-indol-3-yl)naphthalene-1,4-dione (**3a**): solid; yield: 84%

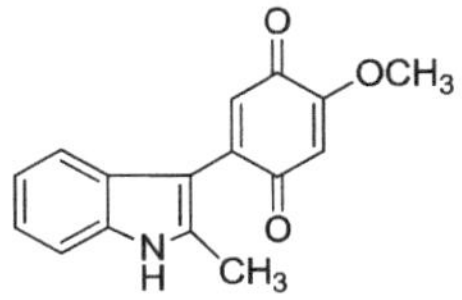

3b

2-Methoxy-5-(2-methyl-1*H*-indol-3-yl)cyclohexa-2,5-diene-1,4-dione (**3b**): solid; yield: 72%

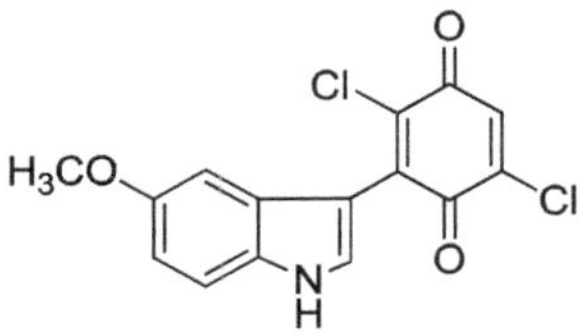

3c

2,5-Dichloro-3-(5-methoxy-1*H*-indol-3-yl)cyclohexa-2,5-diene-1,4-dione (**3c**): solid; yield: 65%

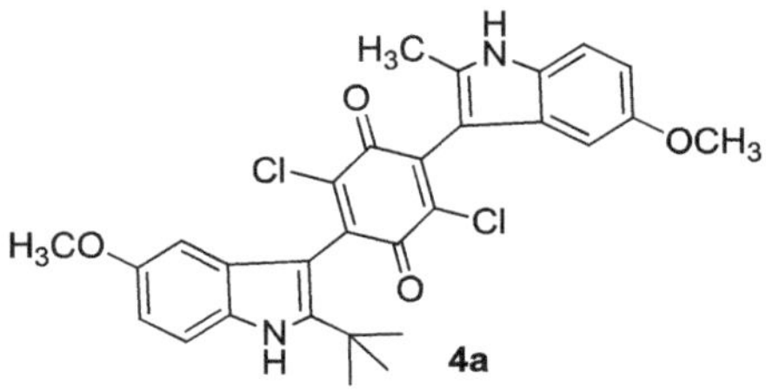

4a

2-(2-(*tert*-Butyl)-5-methoxy-1*H*-indol-3-yl)-3,6-dichloro-5-(5-methoxy-2-methyl-1H-indol-3-yl)cyclohexa-2,5-diene-1,4-dione (**4d**): white solid; yield: 83%

2.2.44.3 Experimental Procedure

A mixture of indole (**1**; 1 mmol) and 1,4-benzoquinone (**2**; 2 mmol) was stirred in distilled water at room temperature for a stipulated time (10–48 h) to complete the reaction as monitored by TLC. The solid precipitate was collected and purified to obtain the desired product of 3-indolylquinone (**3**) with good yields. The bis(indolyl)-1,4-benzoquinones (**4**) were prepared following the similar technique from the coupling between indoles (**1**; 1 mmol) and the synthesized 2,5-dichloro-3-indolyl-1,4-benzoquinones (**3**; 2 mmol). Physical and spectral data of all the compounds were compared with those reported in literature.[134]

2.2.44.4 Critical Views

The present '*on-water*'-promoted direct coupling reaction of indoles with 1,4-benzoquinones has been reported to occur efficiently with good to excellent yields in the absence of any catalyst, organic co-solvent, or additives, and thus appears to be much more beneficial and greener compared to the earlier methods involving the use of protic (HCl, H_2SO_4 and CH_3CO_2H)

acids[135,136], montmorillonite clay[137] and Lewis acids, such as $InBr_3$[138] and $Bi(OTf)_3$,[139] in an organic solvent. The synthesis of bis(indolyl)-1,4-benzoquinone was also previously reported under relatively intense reaction conditions that require refluxing in the presence of $Zn(OTf)_2$ catalyst in tetrahydrofuran for 16 h under nitrogen followed by oxidation with DDQ[140] or by coupling with mercury reagents catalyzed by $Pd^{II}/Cu{-}(OAc)_2$.[141]

2.2.45 Entry-45: Synthesis of Substituted 3-Hydroxy-2-oxindoles

Type of reaction: C–C bond formation
Reaction conditions: Catalyst-free, water, room temperature
Synthetic strategy: '*On-water*' direct C–C coupling
Keywords: Isatins, thiazolidinedione, oxindoles, water, catalyst-free, room temperature, direct C–C coupling, 3- thiazolidinedione substituted, 3-hydroxy-2-oxindoles, 3-oxindole substituted, 3-hydroxy-2-oxindoles, diastereoselectivity, no column chromatographic purification

2.2.45.1 General Reaction Scheme

An '*on-water*' diastereoselective synthesis of a novel class of diverse 3-(thiazolidinedione or oxindole) substituted, 3-hydroxy-2-oxindole scaffolds (**3**/**5**) from the reaction of isatins (**1**) with thiazolidinediones (**2**) or oxindoles (**4**) following a catalyst-free and column chromatography-free protocol at room temperature has recently been demonstrated by Thakur and Meshram[142] (Scheme 2.62).

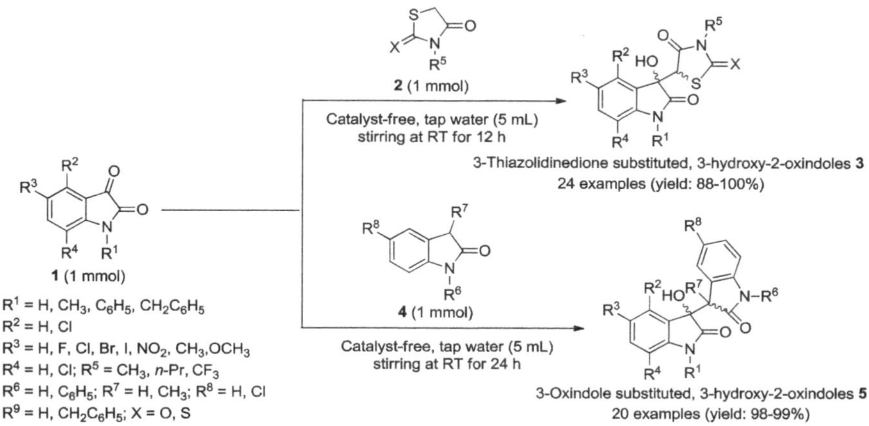

Scheme 2.62 Synthesis of substituted 3-hydroxy-2-oxindoles (3/5).

2.2.45.2 Representative Examples

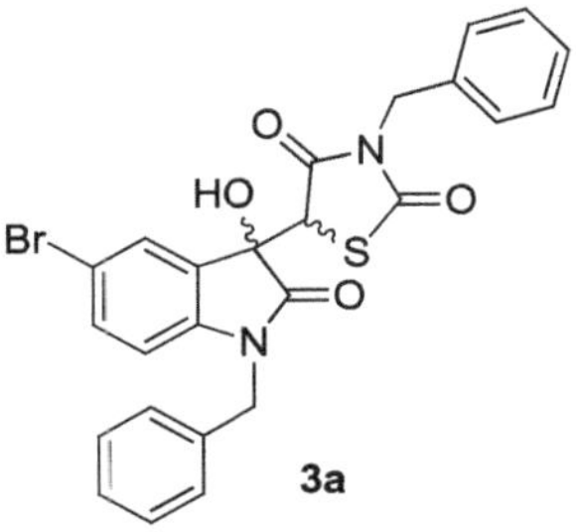

3a

3-Benzyl-5-(1-benzyl-5-bromo-3-hydroxy-2-oxoindolin-3-yl)thiazolidine-2,4-dione (**3a**): white solid, mp 164–166 °C; yield: 90% (dr 100:0)

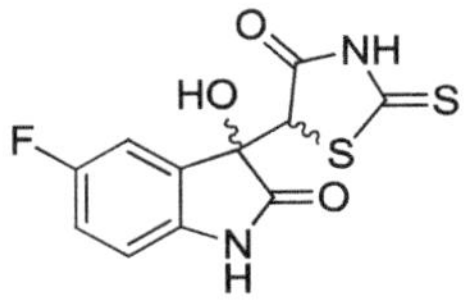

3b

5-(5-Fluoro-3-hydroxy-2-oxoindolin-3-yl)-2-thioxothiazolidin-4-one (**3b**): white solid, mp 184–186°C; yield: 99% (dr 100:0)

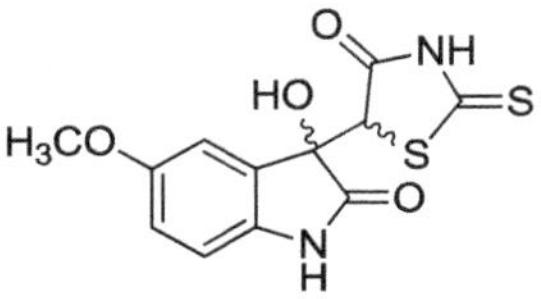

3c

5-(3-Hydroxy-5-methoxy-2-oxoindolin-3-yl)-2-thioxothiazolidin-4-one (**3c**): white solid, mp 140–142°C; yield: 99% (dr 100:0)

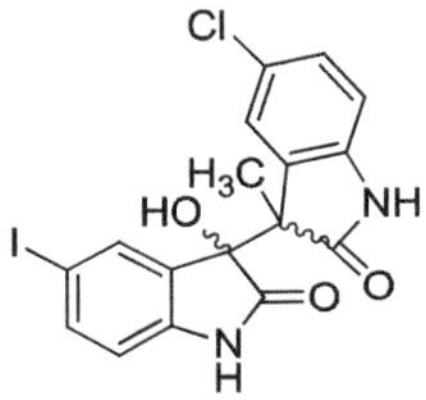

5a

5-Chloro-3'-hydroxy-5'-iodo-3-methyl-[3,3'-biindoline]-2,2'-dione (**5a**): white solid, mp >350 °C; yield: 99% (dr 96:4)

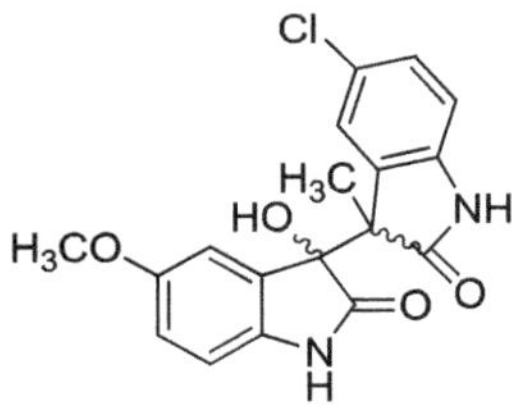

5b

5-Chloro-3'-hydroxy-5'-methoxy-3-methyl-[3,3'-biindoline]-2,2'-dione (**5b**): white solid, mp >350 °C; yield: 99% (dr 97:3)

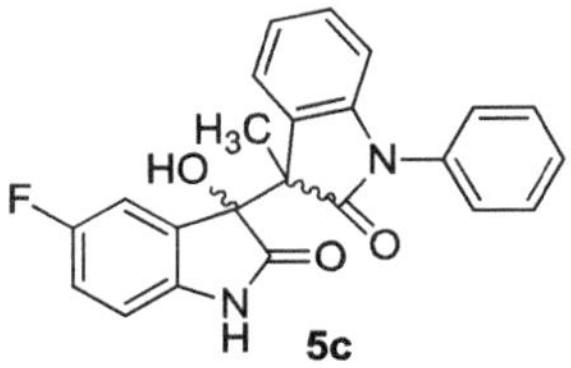

5c

5'-Fluoro-3'-hydroxy-3-methyl-1-phenyl-[3,3'-biindoline]-2,2'-dione (**5c**): white solid, mp 194–196 °C; yield: 99% (dr 97:3)

2.2.45.3 Experimental Procedures

Synthesis of 3-thiazolidinedione substituted, 3-hydroxy-2-oxindoles (**3**). A mixture of isatin (**1**; 1 mmol) and thiazolidinedione derivative (**2**; 1 mmol) was stirred in 5 mL of tap water at room temperature for 12 h until the reaction had gone to completion, as monitored by TLC. The crude precipitate, which appeared upon completion, was filtered, washed with water and dried to afford the desired product (**3**) in very good yield and purity. All products were characterized by IR, ^{1}H NMR, ^{13}C NMR and mass spectral studies.

Synthesis of 3-oxindole substituted, 3-hydroxy-2-oxindoles (**5**): A mixture of isatin (**1**; 1 mmol) and oxindole derivative (**4**; 1 mmol) was stirred in 5 mL of tap water at room temperature for 24 h until the reaction had gone to completion, as monitored by TLC. The crude precipitate, which appeared upon completion, was filtered, washed with water and dried to afford the desired product (**3**) in very good yield and purity. All products were characterized by IR, ^{1}H NMR, ^{13}C NMR and mass spectral studies.

2.2.45.4 Characterization Data of Four Representative Compounds

3-Benzyl-5-(1-benzyl-5-bromo-3-hydroxy-2-oxoindolin-3-yl)thiazolidine-2,4-dione (**3a**): white solid, mp 164–166 °C; yield: 90% (dr 100:0); IR (KBr): 3328, 2927, 1718, 1678, 1612, 1430, 1334, 1150, 1080, 698 cm^{-1}; ^{1}H NMR ($CDCl_3$ + DMSO-d_6, 300 MHz): δ 7.66 (1H, s), 7.43–7.31 (6H, m), 7.19–7.08 (4H, m), 6.86 (2H, d, J = 5.9 Hz,), 6.39 (1H, d, J = 8.1 Hz), 5.17 (1H, s), 4.91 (2H, s), 4.53 (2H, s); ^{13}C NMR ($CDCl_3$ + DMSO-d_6, 75 MHz): δ 174.8, 170.6, 169.7, 142.4, 134.8, 134.3, 133.3, 128.7, 128.6, 128.5, 127.7, 127.6, 127.3, 115.2, 111.1, 74.8, 57.1, 44.7, 33.8; MS (ESI): *m/z* 542 $[M + NH_4]^+$; HRMS (ESI): *m/z* calc. for $C_{25}H_{23}O_4N_3BrS$ $[M + NH_4]^+$ = 542.05722; found 542.05736.

5-(5-Fluoro-3-hydroxy-2-oxoindolin-3-yl)-2-thioxothiazolidin-4-one (**3b**): white solid, mp 184–186 °C; yield: 99% (dr 100:0); IR (KBr): 3241, 3064, 2950, 2950, 2881, 1702, 1487, 1455, 1313, 1220, 1189, 1083, 824, 782, 665, 581, 512 cm^{-1}; ^{1}H NMR (DMSO-d_6, 300 MHz): δ 13.04 (1H, br s), 10.65 (1H, s), 7.28–7.24 (1H, m), 7.18–7.14 (1H, m), 7.13 (1H, s), 6.86–6.22 (1H, m), 5.21 (1H, s); ^{13}C NMR (DMSO-d_6, 75 MHz): δ 203.3, 175.5, 173.9, 159.1, 156.0, 138.8, 129.0, 128.8, 116.9, 116.6, 111.9, 111.6, 111.0, 110.9, 74.7, 61.6; MS (ESI): *m/z* 376 $[M + H]^+$; HRMS (ESI): *m/z* calc. for $C_{11}H_7O_3N_2FS_2Na$ $[M + Na]^+$ = 320.97743; found 320.97757.

5-Chloro-3′-hydroxy-5′-iodo-3-methyl-[3,3′-biindoline]-2,2′-dione (**5a**): white solid, mp >350 °C; yield: 99% (dr 96:4); IR (KBr): 3285, 3058, 2856, 1732, 1694, 1617, 1473, 1445, 1326, 1196, 1167, 816, 760, 657, 617, 560 cm^{-1}; ^{1}H NMR (DMSO-d_6, 300 MHz): δ 11.21 (1H, s), 10.36 (1H, s), 8.14 (1H, dd, J = 8.4, 2.2 Hz), 7.63 (1H, s), 7.40 (1H, dd, J = 8.4, 1.9 Hz), 7.11 (1H, s), 6.99 (1H, d, J = 8.4 Hz), 6.88 (1H, d, J = 2.5 Hz), 6.80 (1H, d, J = 8.1 Hz), 4.14 1H, (1H, s); ^{13}C NMR (DMSO-d_6, 75 MHz): δ 176.1, 173.4, 142.4, 142.3, 138.1, 132.1, 130.5, 128.4, 127.6, 126.4, 125.2, 112.2, 110.3, 83.6, 75.2, 51.6; MS (ESI): *m/z* 440 $[M]^+$; HRMS (ESI): *m/z* calc. for $C_{16}H_{10}O_3N_2ClI$ $[M + H]^+$ = 439.94246; found 439.94261.

5′-Fluoro-3′-hydroxy-3-methyl-1-phenyl-[3,3′-biindoline]-2,2′-dione (**5c**): white solid, mp 194–196 °C; yield: 99% (dr 97:3); IR (KBr): 3424, 3226, 2929, 2879, 1731, 1689, 1605, 1474, 1375, 1324, 1200, 755, 701, 629, 488 cm^{-1}; ^{1}H NMR (DMSO-d_6, 300 MHz): δ 10.34 (1H, s), 7.67–7.49 (1H, m), 7.46–7.32

(4H, m), 7.19 (1H, t, $J=7.4$ Hz), 7.02 (1H, dt, $J=8.6$, 2.6 Hz), 6.84–6.69 (4H, m), 6.59 (1H, d, $J=7.7$ Hz), 5.45 (1H, br s), 1.74 (3H, s); ^{13}C NMR (DMSO-d_6, 75 MHz): δ 176.1, 175.2, 158.4, 155.2, 143.4, 138.4, 133.8, 130.0, 129.4, 128.7, 128.1, 126.3, 125.6, 122.5, 115.8, 115.5, 111.8, 111.5, 110.0, 109.9, 108.3, 76.6, 53.7, 15.1; MS (ESI): *m/z* 411 $[M+Na]^+$; HRMS (ESI): *m/z* calc. for $C_{23}H_{17}O_3N_2FNa$ $[M+Na]^+=411.11154$; found 411.11133.

2.2.45.5 Critical Views

The present protocol is highly efficient and atom-economic in generating a wide range of functionalized 3-(thiazolidinedione or oxindole)-substituted, 3-hydroxy-2-oxindole frameworks from readily available starting materials under catalyst-free conditions, in water at ambient temperature, with excellent yields. A variety of functionalized isatins, as well as thiazolidinedione and oxindole derivatives, underwent the reaction smoothly. There was no need for column chromatographic purification of products, thus avoiding toxic organic solvents. This method offers an easy access to the green synthesis of a novel class of medicinally important 3-substituted, 3-hydroxy-2-oxindole molecular frameworks with diverse functionalities.

2.2.46 Entry-46: Synthesis of Pyrano[3,2-*c*]pyridines

Type of reaction: C–C and C–O bond formation
Reaction conditions: Catalyst-free, ethanol, room temperature
Synthetic strategy: One-pot addition-cyclization reaction
Keywords: (*E*)-3,5-bis(benzylidene)-4-piperidones, malononitrile, ethanol, catalyst-free, room temperature, pyrano[3,2-*c*]pyridines

2.2.46.1 General Reaction Scheme

An eco-friendly, simple, waste- and catalyst-free protocol was developed by Rostamizadeh *et al.*[143] for the synthesis of biologically relavent pyrano[3,2-*c*]pyridines (3), in good to excellent yields, from the reaction of (*E*)-3,5-bis(benzylidene)-4-piperidones (**1**) and malononitrile (**2**) at room temperature (Scheme 2.63).

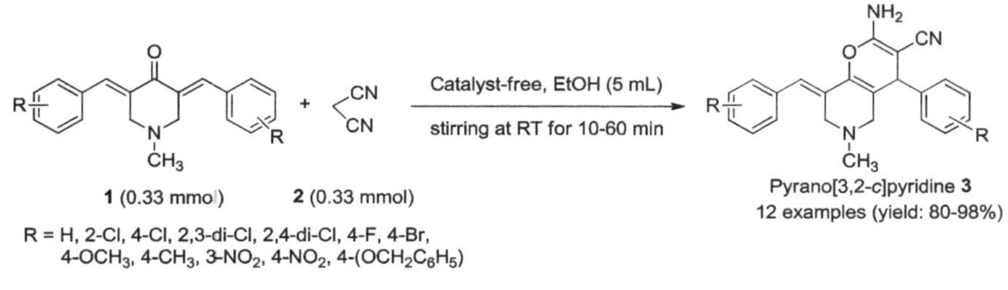

Scheme 2.63 Synthesis of pyrano[3,2-*c*]pyridines (3).

2.2.46.2 Representative Examples

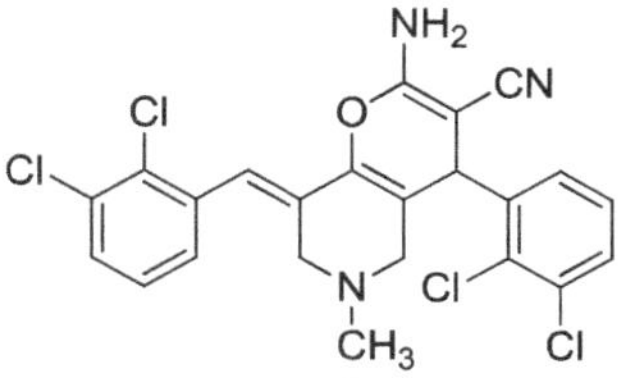

3a

(8*E*)-2-Amino-8-(2,3-dichlorobenzylidene)-4-(2,3-dichlorophenyl)-6-methyl-5,6,7,8-tetrahydro-4*H*-pyrano[3,2-*c*]pyridine-3-carbonitrile (**3a**): solid, mp 213–215 °C; yield: 98%

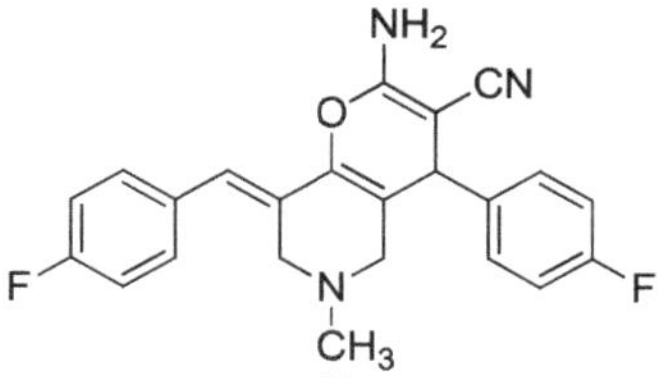

3b

(8*E*)-2-Amino-8-(4-fluorobenzylidene)-4-(4-fluorophenyl)-6-methyl-5,6,7,8-tetrahydro-4*H*-pyrano[3,2-*c*]pyridine-3-carbonitrile (**3b**): solid; mp 204–205 °C; yield: 98%

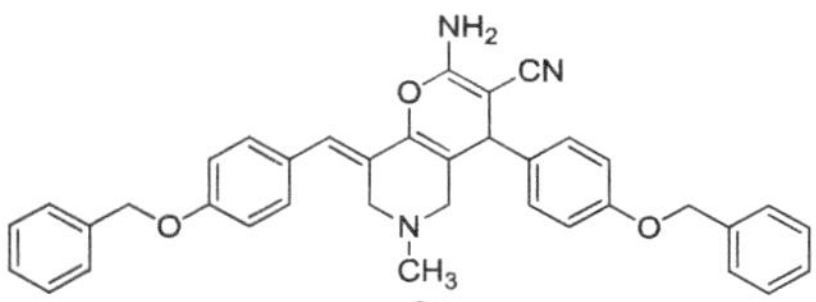

3c

(8*E*)-2-Amino-8-(4-(benzyloxy)benzylidene)-4-(4-(benzyloxy)phenyl)-6-methyl-5,6,7,8-tetrahydro-4*H*-pyrano[3,2-*c*]pyridine-3-carbonitrile (**3c**): solid, mp 193–195 °C; yield: 89%

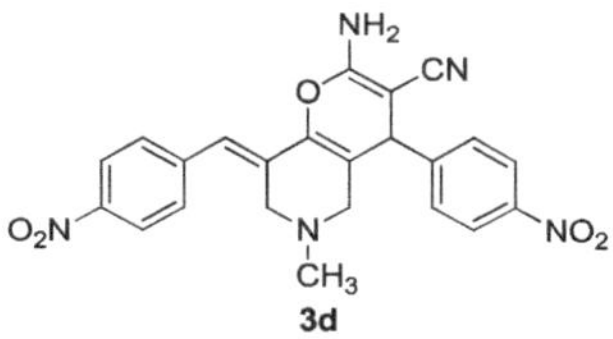

3d

(8*E*)-2-amino-6-methyl-8-(4-nitrobenzylidene)-4-(4-nitrophenyl)-5,6,7,8-tetrahydro-4*H*-pyrano[3,2-*c*]pyridine-3-carbonitrile (**3d**): solid, mp 238–240 °C; yield: 96%

2.2.46.3 Experimental Procedure

A mixture of (*E*)-3,5-bis(benzylidene)-4-piperidones (**1**; 0.33 mmol) and malononitrile (**2**; 0.33 mmol) in ethanol (5 mL) was stirred at room temperature for 10–60 min and the progress of the reaction was monitored by TLC. After completion of the reaction, the crude product was collected and recrystallized from 95% EtOH to have the pure products of pyrano[3,2-*c*]pyridines (**3**), characterized based on spectral studies.

2.2.46.4 Characterization Data of Two Representative Compounds

(8*E*)-2-Amino-8-(2,3-dichlorobenzylidene)-4-(2,3-dichlorophenyl)-6-methyl-5,6,7,8-tetrahydro-4*H*-pyrano[3,2-*c*]pyridine-3-carbonitrile (**3a**): molecular formula $C_{23}H_{17}Cl_4N_3O$; solid, mp 213–215 °C; yield: 98%; IR (KBr): 3322, 3260, 2202, 1678, 1637, 1601 cm^{-1}; ^1H NMR (DMSO-d_6, 300 MHz): δ 2.09 (3H, s, N-CH_3), 2.51 (1H, d, $J = 14.0$ Hz), 3.02 (1H, d, $J = 16.2$ Hz), 3.15 (1H, d,

$J = 14.0$ Hz), 3.31 (1H, d, $J = 14.8$ Hz), 4.7 (1H, s, CH), 6.94 (1H, s, C=CH), 7.05 (2H, s, NH_2), 7.24–7.44 (4H, m), 7.58 (2H, d, $J = 8.0$ Hz); ^{13}C NMR (DMSO-d_6, 75 MHz): δ 44.2, 53.7, 54.2, 118.8, 119.9, 128.0, 128.9, 129.2, 129.4, 129.6, 130.3, 130.6, 132.0, 136.3, 139.5, 142.7, 160.1; MS (EI): *m/z* (rel.%) 42 (100), 81 (46), 113 (30), 149 (70), 181 (60), 216 (53), 250 (91), 293 (12), 346 (11), 390 (14), 449 (19), 492 (22).

(8*E*)-2-Amino-8-(4-(benzyloxy)benzylidene)-4-(4-(benzyloxy)phenyl)-6-methyl-5,6,7,8-tetrahydro-4*H*-pyrano[3,2-*c*]pyridine-3-carbonitrile (**3c**): molecular formula $C_{37}H_{33}N_3O_3$; solid, mp 193–195 °C; yield: 89%; IR (KBr): 3388, 3301, 2182, 1683, 1642, 1606 cm^{-1}; ^1H NMR (DMSO-d_6, 300 MHz): δ 2.13 (3H, s, N–CH_3), 2.54 (1H, d, $J = 14.4$ Hz), 2.94 (1H, d, $J = 14.0$ Hz), 3.25 (1H, d, $J = 14.4$ Hz), 3.45 (1H, d, $J = 14.0$ Hz), 3.97 (1H, s, CH), 5.06 (2H, s, OCH_2), 5.10 (2H, s, OCH_2), 6.75 (2H, s, NH_2), 6.83 (1H, s, C=CH), 7.00 (2H, d, $J = 8.7$ Hz), 7.03 (2H, d, $J = 8.7$ Hz), 7.12 (2H, d, $J = 8.7$ Hz), 7.18 (2H, d, $J = 8.7$ Hz), 7.29–7.46 (10H, m); ^{13}C NMR (DMSO-d_6, 75 MHz): δ 44.4, 53.9, 54.4, 55.7, 113.0, 115.3, 115.5, 120.3, 120.4, 127.4, 129.4, 129.5, 130.9, 131.0, 132.3, 132.4, 139.1, 139.6, 139.7, 159.5, 159.6, 159.7, 162.8; MS (EI): *m/z* (rel.%) 42 (16), 65 (63), 91 (100), 131 (40), 174 (44), 198 (19), 222 (23), 250 (63), 382 (86), 410 (54), 473 (30), 501 (75), 567 (16).

2.2.46.5 Critical Views

The present methodology is advantageous over previous reports,[144–151] where either a catalyst and/or microwave was employed; besides, operational simplicity, short reaction time, nontoxicity of the solvent, as well as economic feasibility are also notable benefits of this chemically waste-free process.

2.2.47 Entry-47: Synthesis of Iminofuranones

Type of reaction: C–C and C–O bond formation
Reaction conditions: Catalyst-free, dichloromethane, room temperature
Synthetic strategy: One-pot multicomponent reaction
Keywords: Alkyl isocyanides, alkylidene-substituted Meldrum's acid, arylcarboxylic acids, dichloromethane, catalyst-free, room temperature, one-pot multicomponent, iminofuranones

2.2.47.1 General Reaction Scheme

Habibi and Rahmani[152] demonstrated a simple and catalyst-free new route for the synthesis of novel derivatives of iminofuranones (**4**) *via* a one-pot pseudo-four-component reaction between two molecules of alkyl isocyanides (**1**), alkylidene-substituted Meldrum's acid (**2**), and arylcarboxylic acids (**3**) at room temperature (Scheme 2.64).

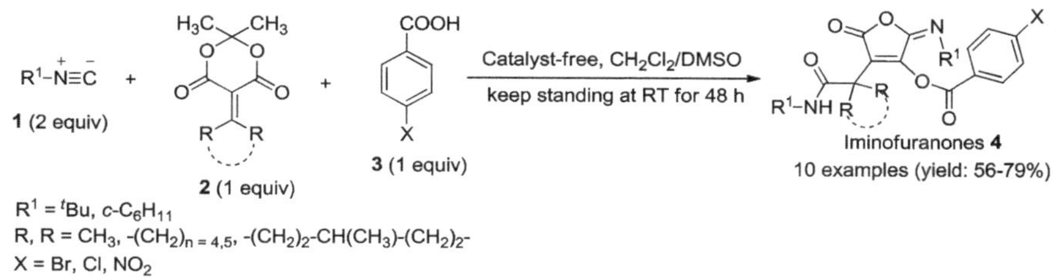

Scheme 2.64 Synthesis of iminofuranones (**4**).

2.2.47.2 Representative Examples

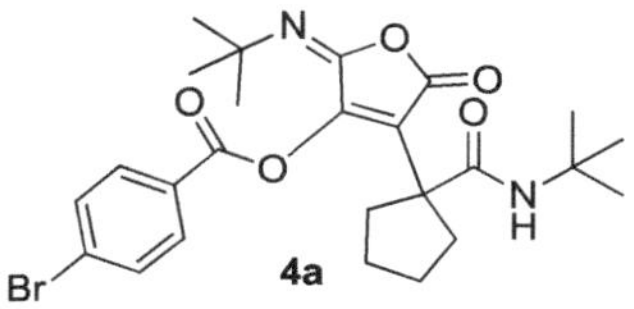

4a

(2*E*)-4-{1-[(*tert*-Butyl)carbamoyl]cyclopentyl}-2-[(*tert*-butyl)imino]-2,5-dihydro-5-oxofuran-3-yl 4-bromobenzoate (**4a**): white powder, mp 201–203 °C; yield: 75%

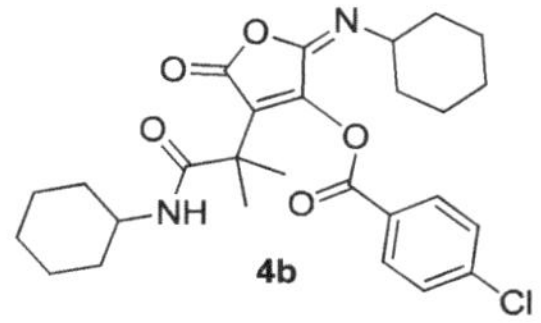

4b

(2*E*)-4-[1-(Cyclohexylamino)-2-methyl-1-oxopropan-2-yl]-2-(cyclohexylimino)-2,5-dihydro-5-oxofuran-3-yl 4-chlorobenzoate (**4b**): white powder, mp 196–198 °C; yield: 75%

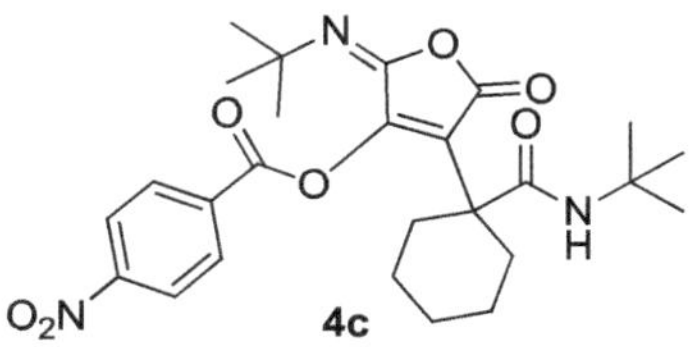

4c

(2*E*)-4-{1-[(*tert*-Butyl)carbamoyl]cyclohexyl}-2-[(*tert*-butyl)imino]-2,5-dihydro-5-oxofuran-3-yl 4-nitrobenzoate (**4c**): white powder, mp 173–175 °C; yield: 73%

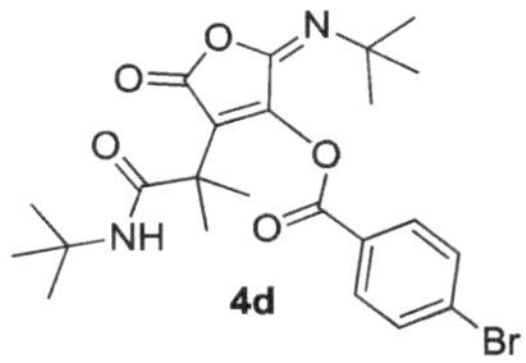

4d

(2*E*)-4-{1-[(*tert*-Butyl)amino]-2-methyl-1-oxopropan-2-yl}-2-[(*tert*-butyl)imino]-2,5-dihydro-5-oxofuran-3-yl 4-bromobenzoate (**4d**): white powder, mp 174–176 °C; yield: 65%

2.2.47.3 Experimental Procedure

To a magnetically stirred solution of alkylidene-substituted Meldrum's acid (**2**, 1 mmol) and arylcarboxylic acids (**3**, 1 mmol) in dichloromethane (10 mL), plus a few drops of DMSO, a solution of alkyl isocyanides (**1**; 2 mmol) in dichloromethane was added drop-wise at room temperature over 5 min. The mixture was kept at room temperature for 48 h. The solvent was then removed under reduced pressure, and the residue was extracted with ethyl acetate and cold methanol, and then filtered. Removal of this solvent afforded pure products of iminofuranones (**4**). All the products were characterized based on spectral and elemental analyses.

2.2.47.4 Characterization Data of Two Representative Compounds

(2*E*)-4-{1-[(*tert*-Butyl)carbamoyl]cyclopentyl}-2-[(*tert*-butyl)imino]-2,5-dihydro-5-oxofuran-3-yl 4-bromobenzoate (**4a**): white powder, mp 201–203 °C; yield: 75%; IR (KBr): 3222, 3060, 2974, 1759, 1741 1690, 1667, 1589, 1579, 1485, 1361, 1261, 1219, 1167, 1124, 1068, 997, 848, 739 cm^{-1}; ^1H NMR ($CDCl_3$, 300 MHz): δ 1.42 (9H, s, ^{t}Bu), 1.65 (9H, s, ^{t}Bu), 1.46–2.06 (8H, m, cyclopentyl), 7.64 and7.97 (AA′BB′, $J = 9.2$ Hz, 4×CH), 9.78 (s, 1H, NH); ^{13}C NMR ($CDCl_3$, 75 MHz): δ 28.0, 28.3, 28.4, 28.7, 37.4, 53.0, 59.4, 127.3, 129.5, 129.7, 132.0, 132.2, 149.2, 157.3, 164.1, 169.9, 181.6; EIMS (70 eV): *m/z* (%rel.) 520 (2, M^+), 518 (2, M^+), 518 (2), 420 (8), 418 (8), 318 (10), 262 (16), 206 (10), 185 (100), 183 (97), 180 (15), 157 (11), 153 (10), 76 (7), 57 (18), 41 (10). Anal. calc. for $C_{25}H_{31}BrN_2O_5$: C, 57.81; H, 6.02; N, 5.39; found: C, 57.73; H, 6.05; N, 5.33.

(2*E*)-4-{1-[(*tert*-Butyl)carbamoyl]cyclohexyl}-2-[(*tert*-butyl)imino]-2,5-dihydro-5-oxofuran-3-yl 4-nitrobenzoate (**4c**): white powder, mp 173–175 °C; yield: 73%; IR (KBr): 3218, 3063, 2975, 2935, 1761, 1742, 1694, 1668, 1619, 1566, 1529, 1459, 1362, 1260, 1183, 1118, 1050, 1010, 865, 847, 708 cm^{-1}; ^1H NMR ($CDCl_3$, 300 MHz): δ 1.42 (9H, s, ^{t}Bu), 1.63 (9H, s, ^{t}Bu), 1.56–1.95 (10H, m, cyclohexyl), 8.30–8.39 (4H, m, 4×CH), 9.48 (1H, s, NH); ^{13}C NMR ($CDCl_3$, 75 MHz): δ 21.4, 24.8, 29.6, 28.4, 28.6, 46.5, 52.1, 59.3, 123.9, 128.8, 131.6, 133.7, 149.5, 151.1, 157.5, 162.8, 169.0, 178.3; EIMS (70 eV): *m/z* (rel.%) 499 (5, M^+), 427 (8), 399 (8), 371 (5), 343 (18), 276 (7), 220 (7), 194 (14), 150 (100), 120 (5), 104 (19), 92 (6), 76 (6), 57 (29), 41 (13). Anal. calc. for $C_{26}H_{33}N_3O_7$: C, 62.51; H, 6.66; N, 8.41; found: C, 63.07; H, 7.00; N, 8.09.

2.2.47.5 Critical Views

The present method furnished the synthesis of some new derivatives of iminofuranones under neutral conditions at room temperature without any activation or modification of the starting materials. Operational simplicity, mild reaction conditions and good yields make this protocol a suitable alternative to complex multistep approaches.[153–156]

2.2.48 Entry-48: Synthesis of Functionalized 5-Pyridylfuran-2-amines

Type of reaction: C–C and C–O bond formation
Reaction conditions: Catalyst-free, dichloromethane, room temperature
Synthetic strategy: One-pot multicomponent reaction
Keywords: Pyridine carbaldehydes, isocyanides, diaroylacetylenes, dichloromethane, catalyst-free, room temperature, one-pot multicomponent, functionalized 5-pyridylfuran-2-amines

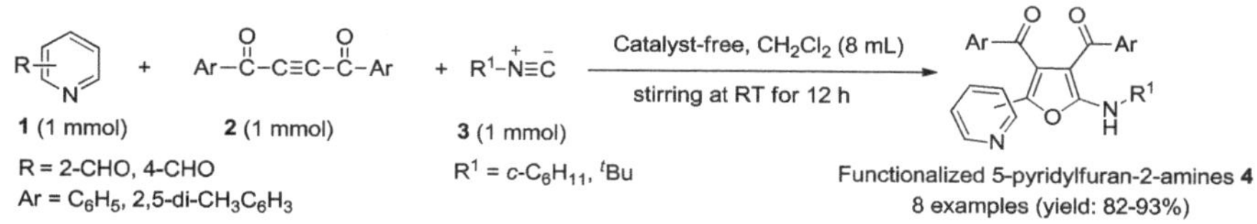

Scheme 2.65 Synthesis of functionalized 5-pyridylfuran-2-amines (**4**).

2.2.48.1 General Reaction Scheme

Adib *et al.*[157] developed a simple and convenient one-pot three-component method for the synthesis of functionalized 5-pyridylfuran-2-amines (**4**) in good yields from the reaction of pyridine carbaldehydes (**1**) with 1:1 zwitterionic intermediates. These intermediates were generated *in situ* from a series of isocyanides (**3**) and diaroylacetylenes (**2**) in dichloromethane under ambient conditions (Scheme 2.65).

2.2.48.2 Representative Examples

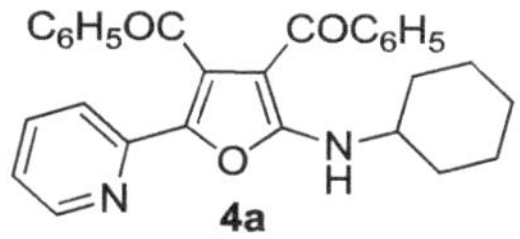

4a

(2-(Cyclohexylamino)-5-(pyridin-2-yl)furan-3,4-diyl)*bis*(phenylmethanone) (**4a**): yellow crystals, mp 138–139 °C; yield: 93%

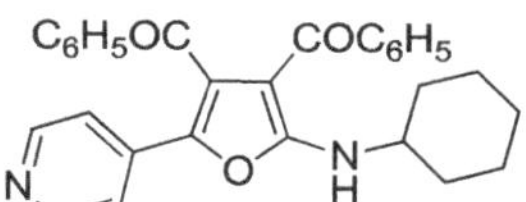

4b

(2-(Cyclohexylamino)-5-(pyridin-4-yl)furan-3,4-diyl)*bis*(phenylmethanone) (**4b**): yellowish crystals, mp 114–115 °C; yield: 92%

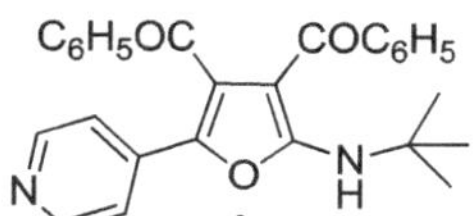

4c

(2-(*tert*-butylamino)-5-(pyridin-4-yl)furan-3,4-diyl)*bis*(phenylmethanone) (**4c**): yellow crystals, mp 125–129 °C; yield: 89%

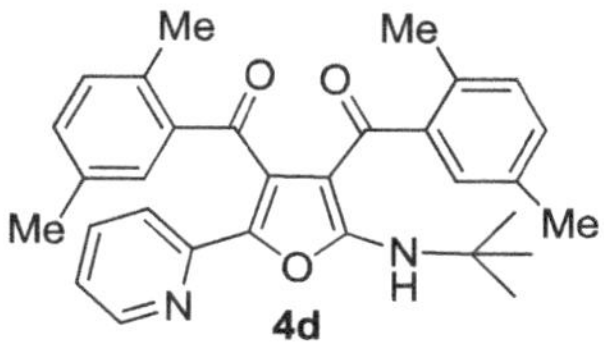

4d

(2-(*tert*-butylamino)-5-(pyridin-2-yl)furan-3,4-diyl)*bis*((2,5-dimethylphenyl) methanone) (**4d**): yellow crystals, mp 111–114 °C; yield: 91%

2.2.48.3 Experimental Procedure

To a magnetically stirred solution of pyridine carbaldehydes (**1**; 1 mmol) and diaroylacetylenes (**2**; 1 mmol) in dichloromethane (6 mL), a solution of isocyanides (**3**; 1 mmol) in dichloromethane (2 mL) was added drop-wise at −5 °C over 10 min. The mixture was then allowed to warm to room temperature and was stirred for 12 h. Upon completion of the reaction, the solvent was removed, and the residue was purified by column chromatography (silica gel; hexane–ethyl acetate 2 : 1) and recrystallization from hexane-ethyl acetate (1 : 1) to afford pure products of functionalized 5-pyridylfuran-2-amines (**3**) in good yields. All the products were characterized based on spectral and elemental analyses.

2.2.48.4 Characterization Data of Two Representative Compounds

(2-(Cyclohexylamino)-5-(pyridin-4-yl)furan-3,4-diyl)bis(phenylmethanone) (**4b**): yellowish crystals, mp 114–115 °C; yield: 92%; IR (KBr): 3265 (NH), 1670, 1634 (C=O), 1593, 1554, 1460, 1404, 1336, 1234, 1045, 889, 692 cm^{-1}; ^1H NMR ($CDCl_3$, 500 MHz): δ 1.29–2.20 (10 H, m), 3.93 (1 H, m), 7.06 (2 H, t, J = 7.6 Hz), 7.13 (2 H, d, J = 7.0 Hz), 7.17 (2 H, t, J = 7.5 Hz), 7.22 (1 H, t, J = 7.3 Hz), 7.37 (1 H, t, J = 7.4 Hz), 7.38–7.43 (4 H, m), 8.39 (1 H, d, J = 7.9 Hz), 8.50 (2 H, d, J = 6.0); ^{13}C NMR ($CDCl_3$, 125 MHz): δ 24.43, 25.30, 33.20, 51.70, 100.40, 118.50, 124.32, 127.12, 127.84, 128.12, 128.79, 130.43, 133.26, 136.22, 137.08, 139.24, 140.27, 149.77, 162.56, 190.06, 192.06; EIMS (20 eV): *m/z* (rel.%) 450 (33, M^+), 406 (22), 105 (100), 77 (20), 55 (41), 40 (25), 29 (30). Anal. calc. for $C_{29}H_{26}N_2O_3$: C, 77.31; H, 5.82; N, 6.22; found: C, 77.4; H, 5.7; N, 6.2.

(2-(*tert*-butylamino)-5-(pyridin-2-yl)furan-3,4-diyl)bis((2,5-dimethylphenyl) methanone) (**4d**): yellow crystals, mp 111–114 °C; yield: 91%; IR (KBr): 3298 (NH), 1674, 1626 (C=O), 1551, 1463, 1373, 1265, 1207, 1175, 1105, 960, 789 cm^{-1}; ^1H NMR ($CDCl_3$, 500 MHz): δ 1.64 (9H, s), 2.01 (3H, s), 2.02 (3H, s), 2.21 (3H, s), 2.23 (3H, s), 6.71 (1H, d, J = 1.2 Hz), 6.75 (1H, d, J = 7.7 Hz), 6.83 (1H, dd, J = 1.6 & 7.6 Hz), 6.91 (1H, d, J = 7.7 Hz), 6.96 (1H, dd, J = 1.1 & 7.4 Hz), 7.05 (1H, dd, J = 1.4 & 7.7 Hz), 7.32 (1H, d, J = 1.2 Hz), 7.50 (1H, d, J = 8.0 Hz), 7.59 (1H, dt, J = 1.7 & 7.5 Hz), 8.28 (1H, dd, J = 1.1 & 4.9 Hz), 8.85 (1H, s); ^{13}C NMR ($CDCl_3$, 125.8 MHz): δ 18.54, 20.53, 20.81, 21.52, 29.72, 53.10, 101.66, 118.52, 121.17, 124.35, 127.67, 129.65, 129.88, 131.63, 131.67, 132.58, 133.01, 134.25, 134.34, 135.31, 136.13, 137.70, 139.66, 140.52, 148.13, 149.56, 163.00, 191.77, 192.57; EIMS (20 eV): *m/z* (rel.%) 480 (50, M^+), 409 (28), 380 (13), 291 (27), 133 (100), 105 (47), 79 (20), 57 (29), 41 (16), 29 (7). Anal. calc. for $C_{31}H_{32}N_2O_3$: C, 77.47; H, 6.71; N, 5.83; found: C, 77.6; H, 6.7; N, 5.7.

2.2.48.5 Critical Views

The present method offers a one-pot route alternative to multistep approaches[158–163] for the synthesis of highly functionalized pyridylfurans of potential synthetic and pharmaceutical interest. This is achieved from easily available starting materials without prior activation or modification under neutral and mild conditions.

2.2.49 Entry-49: Synthesis of Functionalized γ-Iminolactones

Type of reaction: C–C and C–O bond formation
Reaction conditions: Catalyst-free, water, room temperature

Synthetic strategy: One-pot multicomponent reaction
Keywords: Isocyanides, dialkyl acetylenedicarboxylates, phenacyl halides, catalyst-free, water, room temperature, one-pot multicomponent synthesis, functionalized γ-iminolactones

2.2.49.1 General Reaction Scheme

Ramazani *et al.*[164] reported an operationally simple, mild and water-mediated synthesis of highly functionalized γ-iminolactone derivatives (**4**) *via* a one-pot three-component reaction of alkyl isocyanides (**1**), dialkyl acetylenedicarboxylates (**2**) and phenacyl halides (**3**) at room temperature (Scheme 2.66).

2.2.49.2 Plausible Mechanism

The authors proposed a plausible mechanism for the condensation reaction (Scheme 2.67).[166] Initially, a zwitterionic species (**5**) is formed from the reaction between isocyanide (**1**) and dialkyl carboxylate (**2**). In the next step, the carbanion part of this zwitterion intermediate attacks the electro-deficient carbonyl carbon of phenacyl halide (**3**), leading to a dipolar specie (**6**). Cyclization of (**6**) eventually yields the γ-iminolactone (**4**).

2.2.49.3 Representative Examples

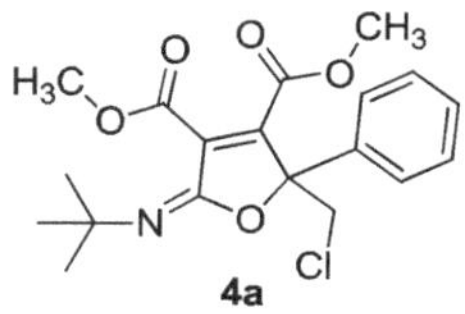

4a

Dimethyl 5-(*tert*-butylimino)-2-(chloromethyl)-2-phenyl-2,5-dihydrofuran-3,4-dicarboxylate (**4a**): yield: 97%

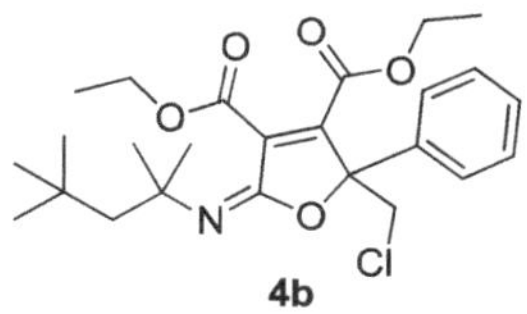

4b

Diethyl 2-(chloromethyl)-2-phenyl-5-[(1,1,3,3-tetramethylbutyl)imino]-2,5-dihydrofuran-3,4-dicarboxylate (**4b**): yield: 84%

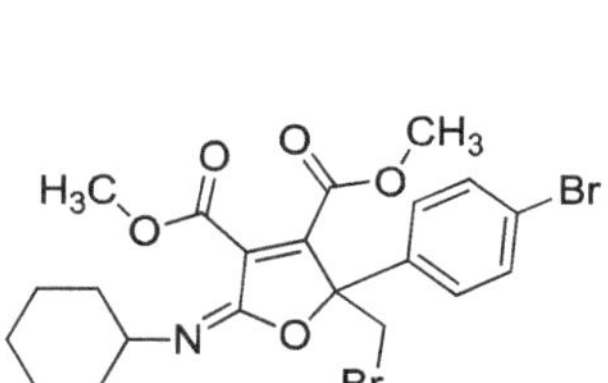

4c

Dimethyl 2-(bromomethyl)-2-(4-bromophenyl)-5-(cyclohexylimino)-2,5-dihydrofuran-3,4-dicarboxylate (**4c**): yield: 97%

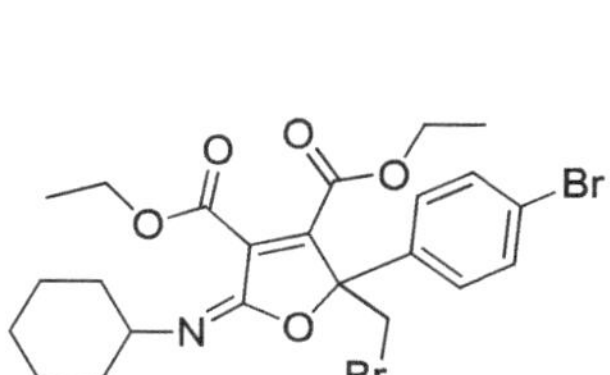

4d

Diethyl 2-(bromomethyl)-2-(4-bromophenyl)-5-(cyclohexylimino)-2,5-dihydrofuran-3,4-dicarboxylate (**4d**): yield: 92%

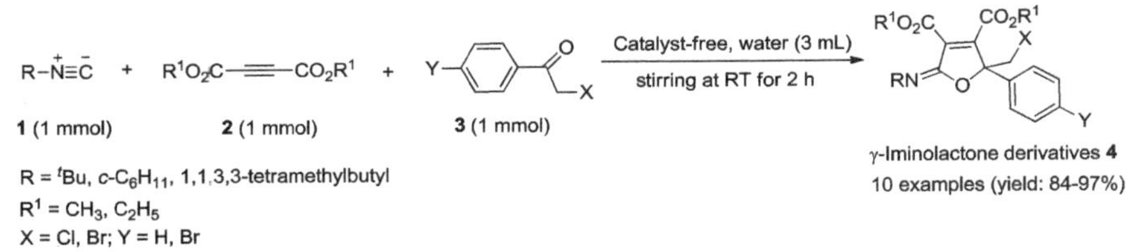

Scheme 2.66 Synthesis of functionalized γ-iminolactones (**4**).

2.2.49.4 Experimental Procedure

To a magnetically stirred solution of phenacyl halide (**3**; 1 mmol) and dialkyl acetylenedicarboxylate (**2**; 1 mmol) in water (3 mL), a solution of isocyanide (**1**; 1 mmol) in water (1 mL)was added drop-wise at room temperature over 5 min, followed by vigorous stirring of the whole mixture for 2 h to complete the reaction. The solvent was then removed under reduced pressure, and the viscous residue was purified by flash column chromatography (silica gel; petroleum ether-ethyl acetate) to obtain pure product of γ-iminolactone (**4**). All the products were characterized by their analytical and spectral studies.

2.2.49.5 Characterization Data of Two Representative Compounds

Dimethyl 5-(*tert*-butylimino)-2-(chloromethyl)-2-phenyl-2,5-dihydrofuran-3,4-dicarboxylate (**4a**): yield: 97%; IR (KBr): 2937, 2870, 1748, 1715, 1680, 1452, 1244 cm^{-1}; ^{1}H NMR ($CDCl_3$, 250 MHz): δ 1.38 (9H, s, Me_3C), 3.78 (3H, s, OCH_3), 3.90 (3H, s, OCH_3); 4.20 (1H, d, $J=11.8$, CH_2Cl), 4.64 (1H, d, $J=11.8$, CH_2Cl), 7.26–7.40 (5H, m, Ar–H); ^{13}C NMR ($CDCl_3$, 62.5 MHz): δ 29.54, 48.08, 52.84, 52.95, 54.96, 91.35, 125.75, 128.82, 129.07, 136.73, 139.04, 141.20, 152.31, 161.13, 162.37; EIMS (rel.%): *m/z* 380 (6), 356 (100), 302 (83), 250 (38), 236 (41), 157 (9), 105 (23), 57 (23), 41 (15). Anal. calc. for $C_{19}H_{22}ClNO_5$: C, 60.08; H, 5.84; N, 3.69; found: C, 60.00; H, 5.78; N, 3.58.

Diethyl 2-(chloromethyl)-2-phenyl-5-[(1,1,3,3-tetramethylbutyl)imino]-2,5-dihydrofuran-3,4-dicarboxylate (**4b**): yield: 84%; IR (KBr): 2967, 2878, 1746, 1715, 1685, 1457, 1222 cm^{-1}; ^{1}H NMR ($CDCl_3$, 250 MHz): δ 1.00 (9H, s, Me_3C), 1.27 (1H, t, $J=7.3$ Hz, $MeCH_2$), 1.36 (1H, t, $J=7.3$, $MeCH_2$), 1.42 (6H, s, Me_2C), 1.66 (2H, s, CH_2CMe_3), 4.20 (1H, d, $J=11.8$ Hz, CH_2Cl), 4.59 (1H, d, $J=11.8$ Hz, CH_2Cl); 4.22 (1H, q, $J=7.3$ Hz, $MeCH_2$), 4.33 (1H, q, $J=7.3$ Hz, $MeCH_2$); 7.26–7.49 (5H, m, Ar–H); ^{13}C NMR ($CDCl_3$, 62.5 MHz): δ 13.74, 14.08, 29.75, 31.68, 31.93, 48.18, 55.27, 58.41, 61.96, 62.03, 91.39, 125.90, 128.70, 128.93, 137.11, 139.62, 140.95, 153.86, 160.84, 162.10; EIMS (rel.%): *m/z* 464 (13), 428 (38), 408 (100), 392 (90), 362 (36), 296 (32), 250 (80), 233

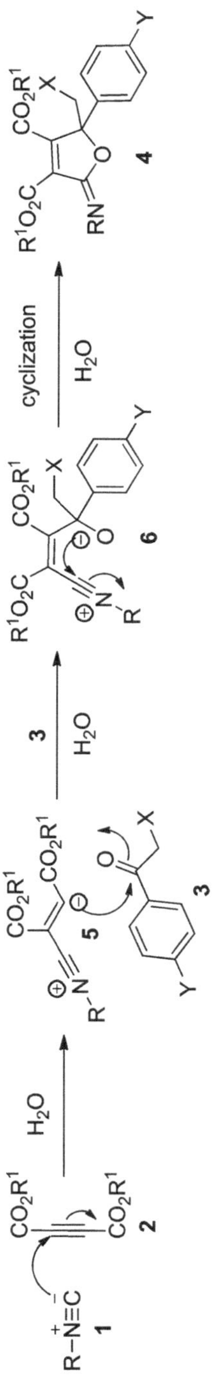

Scheme 2.67 Suggested mechanism for the formation of γ-iminolactones (4).

(40), 105 (37), 57 (45), 41 (20). Anal. calc. for $C_{25}H_{34}ClNO_5$: C, 64.71, H, 7.39, N, 3.02; found: C, 64.09, H, 7.22, N, 2.99.

2.2.49.6 *Critical Views*

The investigators reported a simple and practical procedure to prepare highly functionalized γ-iminolactones by a one-pot three-component reaction of isocyanides, acetylenedicarboxylates, and phenacyl chloride or bromide in the presence of water at room temperature. In their report, the authors discussed that syntheses of such heterocyclic compounds were accomplished previously using various catalysts.[165–176] Besides, this present protocol offers other significant advantages, such as operational simplicity, mild reaction conditions, enhanced rates, ease of isolation of products, cleaner reaction profiles, and water as solvent. Hence, this could be useful and an attractive alternative method for the synthesis of γ-iminolactone derivatives of significant interest.

2.2.50 Entry-50: Synthesis of Functionalized bis(4*H*-Chromene) and 4*H*-Benzo[*g*]chromene derivatives

Type of reaction: C–C and C–O bond formation
Reaction conditions: Catalyst-free, acetonitrile, room temperature
Synthetic strategy: One-pot multicomponent reaction
Keywords: Isocyanides, dialkyl acetylenedicarboxylates, 2,5-dihydroxycyclohexa-2,5-diene-1,4-dione, 2-hydroxynaphthalene-1,4-dione, acetonitrile, catalyst-free, room temperature, one-pot multicomponent synthesis, functionalized bis(4*H*-chromene) and 4*H*-benzo[*g*]chromene derivatives

2.2.50.1 *General Reaction Scheme*

Shaabani and his group developed an efficient one-pot synthetic approach for the synthesis of highly functionalized bis(4*H*-chromene)-3,4-dicarboxylates (**5**) and 4*H*-benzo[*g*]chromene-3,4-dicarboxylate derivatives (**6**) in fairly good yields from the reactive intermediates generated by the addition of alkyl, aryl, and alicyclic isocyanides (**1**) to dialkyl acetylenedicarboxylates (**2**), followed by trapping them, respectively by 2,5-dihydroxycyclohexa-2,5-diene-1,4-dione (**3**) or 2-hydroxynaphthalene-1,4-dione (**4**) in acetonitrile at room temperature (Scheme 2.68).

2.2.50.2 *Proposed Mechanism*

The investigators proposed the following mechanistic course for the transformation (Scheme 2.69).[177]

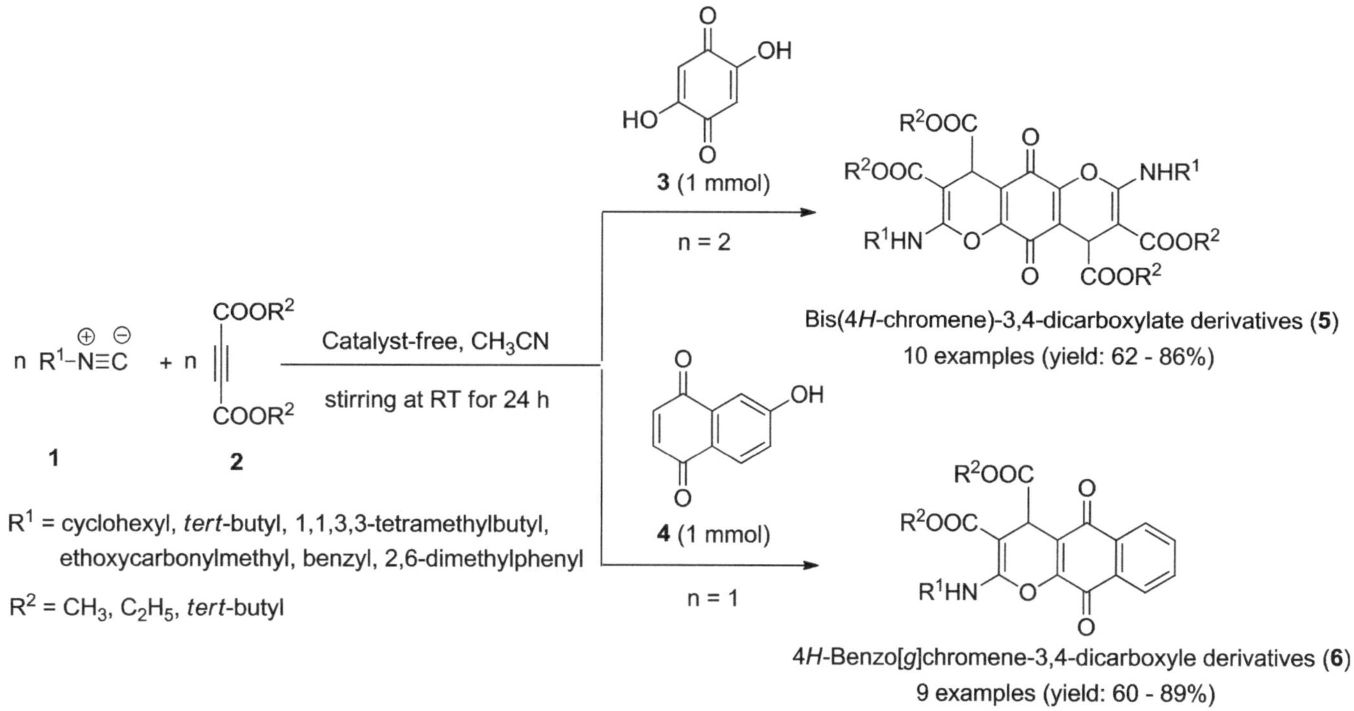

Scheme 2.68 Synthesis of functionalized bis(4*H*-chromene) and 4*H*-benzo[*g*]chromene derivatives (**5**/**6**).

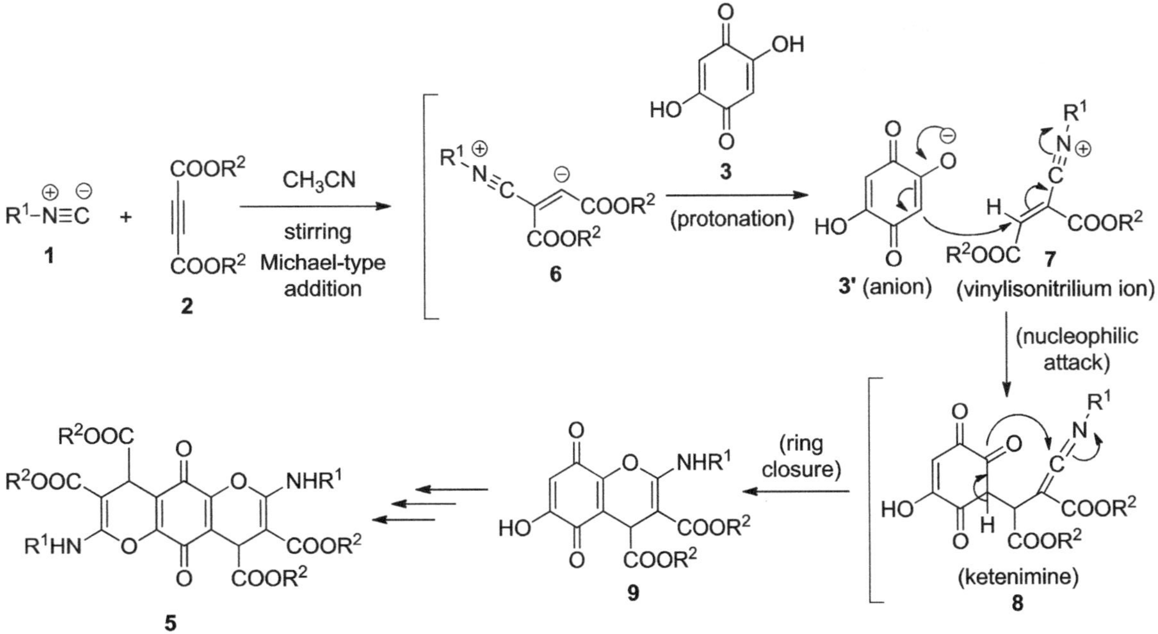

Scheme 2.69 Suggested mechanism for the generation of functionalized chromene derivatives (**5**/**6**).

2.2.50.3 Representative Examples

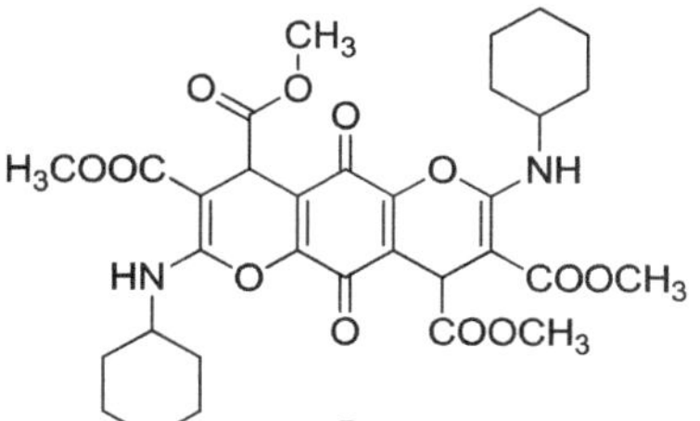

5a

Tetramethyl 2,7-*bis*(cyclohexylamino)-5,10-dioxo-4,5,9,10-tetrahydropyrano[2,3-*g*]chromene-3,4,8,9-tetracarboxylate (**5a**): brownish red powder, mp 257–258°C; reaction time: 24 h; yield: 86%

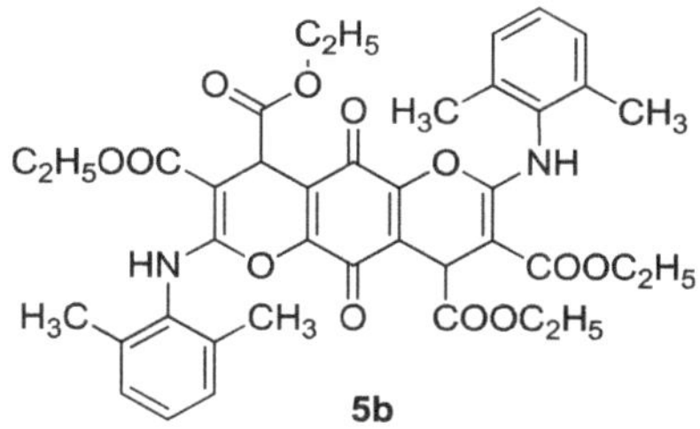

5b

Tetraethyl 2,7-*bis*((2,6-dimethylphenyl)amino)-5,10-dioxo-4,5,9,10-tetrahydropyrano[2,3-*g*]chromene-3,4,8,9-tetracarboxylate (**5b**): dark red powder, mp 245–248 °C; reaction time: 24 h; yield: 65%

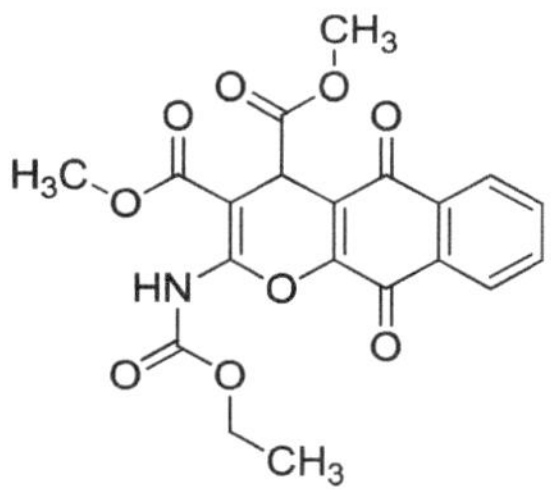

6a

Dimethyl 2-((ethoxycarbonyl)amino)-5,10-dioxo-5,10-dihydro-4*H*-benzo[*g*]chromene-3,4-dicarboxylate (**6a**): yellow powder, mp 167–169 °C; reaction time: 24 h; yield: 60%

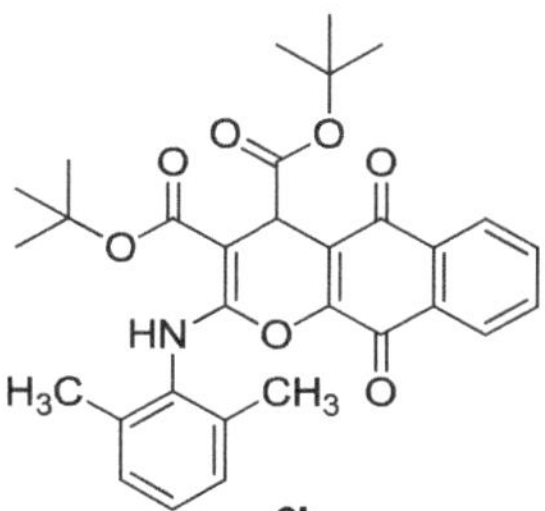

6b

Di-*tert*-butyl 2-((2,6-dimethylphenyl)amino)-5,10-dioxo-5,10-dihydro-4*H*-benzo[*g*]chromene-3,4-dicarboxylate (**6b**): dark yellow powder, mp 171–173 °C; reaction time: 24 h; yield: 64%

2.2.50.4 Experimental Procedure

Functionalized bis*(4H-chromeme) derivative* (**5**). To a magnetically stirred solution of 2,5-dihydroxycyclohexa-2,5-diene-1,4-dione (**3**; 1 mmol) and dialkyl acetylenedicarboxylate (**2**; 2 mmol) in acetonitrile (5 mL), a solution of isocyanide (**1**; 2 mmol) in acetonitrile (2 mL) was added at room temperature over 5 min. The mixture was then stirred for 24 h. After completion of the reaction, the precipitated product was separated out from the reaction mixture by filtration and was washed with 5 mL of *n*-hexane to afford bis-(4*H*-chromene)-3,4-dicarboxylate derivative (**5**) with reasonable good yield (62–86%). Each of the products was characterized by analytical analyses, as well as detailed spectral studies including IR, ^{1}H NMR, ^{13}C NMR, MS.

4H-Benzo[g]chromene derivative (**6**). To a magnetically stirred solution of 2-hydroxynaphthalene-1,4-dione (**4**, 1 mmol) and dialkyl acetylenedicarboxylate (**2**; 1 mmol) in acetonitrile (5 mL), a solution of isocyanide (**1**; 1 mmol) in acetonitrile (2 mL) was added at room temperature over 5 min. The mixture was then stirred for 24 h. After completion of the reaction, the precipitated product was separated out from the reaction mixture by filtration and was washed with 5 mL of *n*-hexane to afford 4*H*-benzo[*g*]chromene derivative (**6**) with reasonable good yield (60–89%). Each of the products was characterized by analytical analyses, as well as detailed spectral studies including IR, ^{1}H NMR, ^{13}C NMR, MS.

2.2.50.5 *Characterization Data of Two Representative Compounds*

Tetramethyl 2,7- bis(cyclohexylamino)-5,10-dioxo-4,5,9,10-tetrahydropyrano[2,3-*g*]chromene-3,4,8,9-tetracarboxylate (**5a**): brownish red powder, mp 257–258 °C; IR (KBr): 2933, 2859, 1751, 1673, 1602, 1446 cm^{-1}; ^{1}H NMR ($CDCl_3$, 300 MHz): δ 1.35–1.96 (20H, m, 10CH_2 of 2 cyclohexyl), 3.66 (6H, s, 2×O–CH_3), 3.71 (6H, s, 2×O–CH_3), 3.79 (2H, br s, CHNH), 4.64 (2H, s, 2×CH–CO_2Me), 8.57 (2H, br s, 2×CHNH); ^{13}C NMR ($CDCl_3$, 75 MHz): δ 24.3, 25.4, 33.5, (*C*-cyclohexyl), 35.2 (CH–CO_2Me), 50.3 (CHNH), 51.2, 52.8 (2×O–CH_3), 70.7, 117.7, 147.8, 158.3 (*C*-alkene), 169.1, 172.2, 176.9 (3C=O); MS at *m/z* (rel. %): 643 ([M + 1]$^{+}$, 2), 611 (3), 585 (5), 553 (3), 525 (4), 486 (5), 402 (5), 327 (5), 297 (5), 269 (5), 170 (5), 96 (20), 83 (30), 55 (80), 41 (80), 31 (100). Anal. calc. for $C_{32}H_{38}N_2O_{12}$: C, 59.81; H, 5.96; N, 4.36; found: C, 59.76; H, 5.84; N, 4.28.

Di-*tert*-butyl 2-((2,6-dimethylphenyl)amino)-5,10-dioxo-5,10-dihydro-4*H*-benzo[*g*] chromene-3,4-dicarboxylate (**6b**): dark yellow powder, mp 171–173 °C; IR (KBr): 2972, 2957, 1722, 1681, 1578 cm^{-1}; ^{1}H NMR ($CDCl_3$, 300 MHz): δ 1.44 (9H, s, C(CH_3)$_3$), 1.74 (9H, s, C(CH_3)$_3$), 2.17 (6H, s, 2×CH_3), 4.76 (1H, s, CH–CO_2*t*-Bu), 7.13 (3H, br s, Ar–H), 7.72–7.76 (2H, m, Ar–H), 8.02–8.12 (2H, m, Ar–H), 9.92 (1H, br s, N*H*); ^{13}C NMR ($CDCl_3$, 75 MHz): δ 18.4 (CH_3), 27.8, 28.5 (2×OC(CH_3)$_3$), 37.5 (CH–CO_2*t*-Bu), 74.0 (C-alkene), 80.5, 81.5 (2×OC(CH_3)$_3$), 121.4, 122.5, 127.5, 128.0, 130.6, 131.5, 133.7, 133.9, 134.0, 134.3, 135.6, 149.5, 157.7 (*C*-alkene, *C*-arom), 168.6, 170.7, 176.8, 182.7 (4C=O); MS at *m/z* (rel. %): 430 (M^{+}–101, 8), 374 (100), 105 (20), 77 (20), 57 (30), 41 (25). Anal. calc. for $C_{31}H_{33}NO_7$: C, 70.04; H, 6.26; N, 2.63; found: C, 70.11; H, 6.28; N, 2.71.

2.2.50.6 *Critical Views*

The key advantages of this method are the use of simple starting materials, neutral and mild reaction conditions at ambient temperature, easy workup procedure, and good functional group tolerance. Hence, this catalyst-free one-pot protocol may be of use to synthesize combinatorial libraries of such heterocycles with great diversity of substitution patterns for potential medicinal applications.

2.2.51 Entry-51: Synthesis of Substituted Cyclohepta[*b*]pyran-3,4-dicarboxylate Derivatives

Type of reaction: C–C and C–O bond formation
Reaction conditions: Catalyst-free, dichloromethane, room temperature
Synthetic strategy: One-pot multicomponent reaction
Keywords: Isocyanides, dialkyl acetylenedicarboxylates, α-tropolone, dichloromethane, catalyst-free, room temperature, one-pot multicomponent reaction, dialkyl 2-(alkylamino)-4,9-dihydro-9-oxocyclohepta[*b*]pyran-3,4-dicarboxylate derivatives.

2.2.51.1 General Reaction Scheme

Azizian *et al.*[178] synthesized dialkyl 2-(alkylamino)-4,9-dihydro-9-oxocyclohepta[*b*]pyran-3,4-dicarboxylate derivatives (**4**) *via* an efficient, catalyst-free, one-pot three-component reaction between alkyl isocyanides (**1**), dialkyl acetylenedicarboxylates (**2**) and α-tropolone (**3**) in dichloromethane under neutral conditions at room temperature (Scheme 2.70).

2.2.51.2 Representative Examples

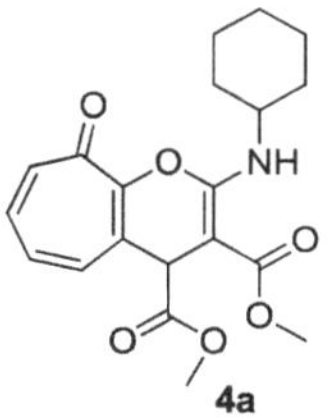

4a

Dimethyl 2-(cyclohexylamino)-4,9-dihydro-9-oxocyclohepta[*b*]pyran-3,4-dicarboxylate (**4a**): brown semisolid; yield: 86%

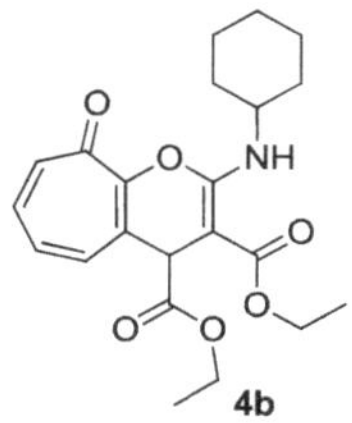

4b

Diethyl 2-(cyclohexylamino)-9-oxo-4,9-dihydrocyclohepta[*b*]pyran-3,4-dicarboxylate (**4b**): brown semisolid; yield: 83%

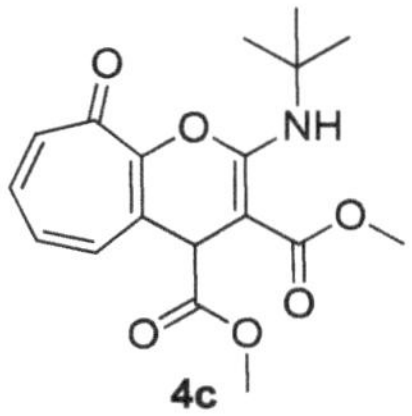

4c

Dimethyl 2-(*tert*-butylamino)-9-oxo-4,9-dihydrocyclohepta[*b*]pyran-3,4-dicarboxylate (**4c**): yellow powder, mp 115–117 °C; yield: 81%

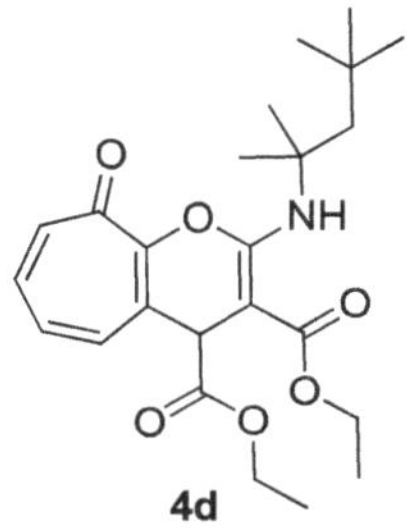

4d

Diethyl 4,9-dihydro-9-oxo-2-[(1,1,3,3-tetramethylbutyl)amino]cyclohepta[*b*]pyran-3,4-dicarboxylate (**4d**): yellow powder, mp 154–156 °C; yield: 79%

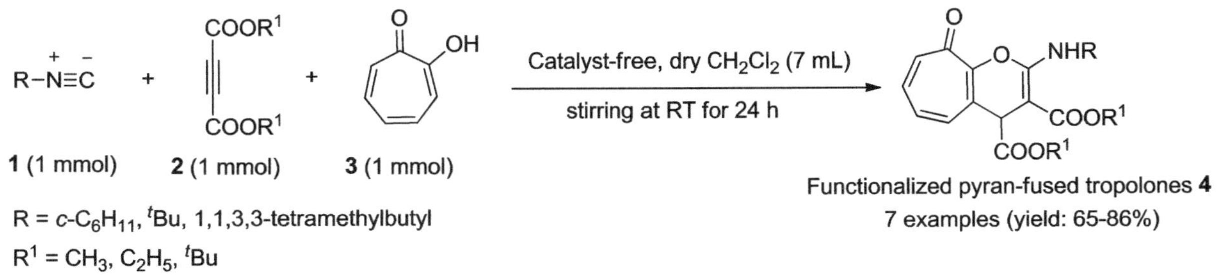

Scheme 2.70 Synthesis of substituted cyclohepta[*b*]pyran-3,4-dicarboxylates (**4**).

2.2.51.3 Experimental Procedure

To a magnetically stirred solution of tropolone (**3**; 1 mmol) and dialkyl acetylenedicarboxylate (**2**; 1 mmol) in dry dichloromethane (5 mL), a solution of alkyl isocyanides (**1**; 1 mmol) in dichloromethane (2 mL) was added at room temperature. The mixture was then stirred for 24 h. On completion of the reaction, the solvent was removed under reduced pressure and the residue was subjected to column chromatographic resolution using petroleum ether-ethyl acetate mixtures as eluents to obtain the pure product of dialkyl 2-(alkylamino)-4,9-dihydro-9-oxocyclohepta[*b*]pyran-3,4-dicarboxylates (**4**). The products were identified based on analytical and spectral studies.

2.2.51.4 Characterization Data of Two Representative Compounds

Dimethyl 2-(cyclohexylamino)-4,9-dihydro-9-oxocyclohepta[*b*]pyran-3,4-dicarboxylate (**4a**): brown semisolid; yield: 86%; IR (KBr): 3292 (NH), 2924, 1744, 1684, 1620, 1441, 1269, 1084 cm^{-1}; ^1H NMR ($CDCl_3$, 250 MHz): δ 1.16–2.07 (10H, m, 5×CH_2 of cyclohexyl), 3.63 (3H, s, OCH_3), 3.68 (3H, s, OCH_3), 3.90–4.11 (1H, m, N–CH), 4.55 (1H, s, CH); 6.83–6.91 (1H, m, 1H of tropone), 7.06 (1H, d, $J = 11.2$ Hz, 1H of tropone), 7.12–7.20 (2H, m, 2H of tropone), 8.26 (1H, br s, NH); ^{13}C NMR ($CDCl_3$, 62.9 MHz): δ 24.40, 24.45, 25.47, 33.44, 33.62 (5×CH_2 of cyclohexyl); 45.52 (CH); 50.20 (CHNH); 50.94, 52.73 (2 MeO); 70.22, 125.66, 130.34, 135.37, 135.97, 139.57, 154.67, 159.46 (C of alkene and tropone); 168.53, 172.32, 177.87 (3 C=O); EIMS (70 ev; rel.%): 373 (2, M^+), 314 (100, $[M-CO_2Me]^+$), 232 (48, $[314-C_6H_{10}]^+$), 200 (14, $[M+H-CO_2Me-C_6H_{10}-MeO]^+$), 172 (19, $[200-CO]^+$), 157 (9, $[172-NH_2]^+$). Anal. calc. for $C_{20}H_{23}NO_6$: C, 64.33; H, 6.21; N, 3.75; found: C, 64.26; H, 6.13; N, 3.68.

Diethyl 4,9-dihydro-9-oxo-2-[(1,1,3,3-tetramethylbutyl)amino]cyclohepta[*b*]-pyran-3,4-dicarboxylate (**4d**): yellow powder, mp 154–156 °C; yield: 79%; IR (KBr): 3269 (NH), 2952, 1727, 1669, 1615, 1434, 1177, 1082 cm^{-1}; ^1H NMR ($CDCl_3$, 250 MHz): δ 0.96 (9H, s, Me_3C), 1.160 (3H, t, $J = 7.0$ Hz, Me), 1.23 (3H, t, $J = 7.0$ Hz, Me), 1.53 (6H, s, Me_2C), 1.92 (2H, d, $J = 14.2$ Hz, CH_2); 4.01–4.17 (4H, m, 2×CH_2O), 4.52 (1H, s, CH), 6.80–6.92 (1H, m, 1 H of tropone), 7.08 (1H, d, $J = 11.6$ Hz, 1 H of tropone), 7.11–7.19 (2H, m, 2H of tropone), 8.58 (1H, br s, NH); ^{13}C NMR ($CDCl_3$, 62.9 MHz): δ 14.07, 14.55, 30.71, 30.96 (4 Me); 31.42 (Me_3C); 31.60 (C); 45.60 (CH); 52.11 (CH_2); 56.48 (C–NH); 59.35, 61.43 (2 CH2O); 70.53, 125.30, 130.13, 135.41, 135.84, 139.57, 154.56, 160.13 (C of alkene and tro pone); 168.41, 172.08, 177.77 (3 C=O); EIMS (70 ev, rel.%): 431 (1, M^+), 358 (29, $[M-CO_2Et]^+$), 328 (4, $[358-2\ Me]^+$), 246 (100, $[358-CMe_2CH_2CMe_3]^+$), 218 (9, $[246-Et]^+$), 200 (15, $[246-EtO]^+$), 172 (19, $[200-CO]^+$), 157 (8, $[172-NH_2]^+$). Anal. calc. for $C_{24}H_{33}NO_6$: C, 66.80; H, 7.71; N, 3.24; found: C, 66.92; H, 7.64; N, 3.31.

2.2.51.5 Critical Views

The investigators demonstrated a new and efficient one-pot method for accessing diverse heterocycle-fused tropolone derivatives, namely dialkyl 2-(alkylamino)-4,9-dihydro-9-oxocyclohepta[*b*]pyrans of potential synthetic and pharmacological interest with good yields from the readily available starting materials under neutral conditions at room temperature.[178]

2.2.52 Entry-52: Synthesis of 2-(Alkylimino)-7-oxo-1-oxa-6-azaspiro[4.4]nona-3,8-diene-3,4-dicarboxylates

Type of reaction: C–C and C–O bond formation
Reaction conditions: Catalyst-free, dichloromethane, room temperature
Synthetic strategy: One-pot multicomponent reaction
Keywords: Isocyanides, dialkyl acetylenedicarboxylates, *N*-arylmaleimides or -phthalimides, dichloromethane, catalyst-free, room temperature, one-pot multicomponent reaction, 2-(alkylimino)-7-oxo-1-oxa-6-azaspiro[4.4]nona-3,8-diene-3,4-dicarboxylates

2.2.52.1 General Reaction Scheme

Adib *et al.*[179] demonstrated a high-yielding, one-pot three-component protocol for the synthesis of pharmacologically interesting 2-(alkylimino)-7-oxo-1-oxa-6-azaspiro[4.4]nona-3,8-diene-3,4-dicarboxylates (**4**) from the reaction of isocyanides (**1**), dialkyl acetylenedicarboxylates (**2**), and *N*-arylmaleimides or -phthalimides (**3**) in dry dichloromethane in the absence of catalyst at room temperature (Scheme 2.71).

2.2.52.2 Representative Examples

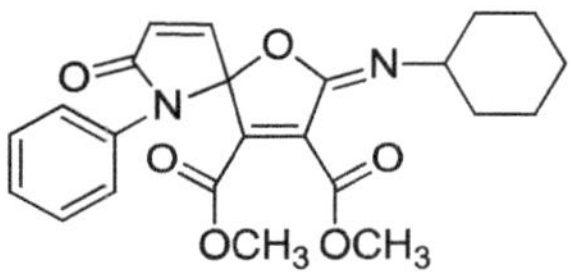

4a

Dimethyl 2-(cyclohexylimino)-7-oxo-6-phenyl-1-oxa-6-azaspiro[4.4]nona-3,8-diene-3,4-dicarboxylate (**4a**): colorless crystals, mp 187–188 °C; yield: 93%

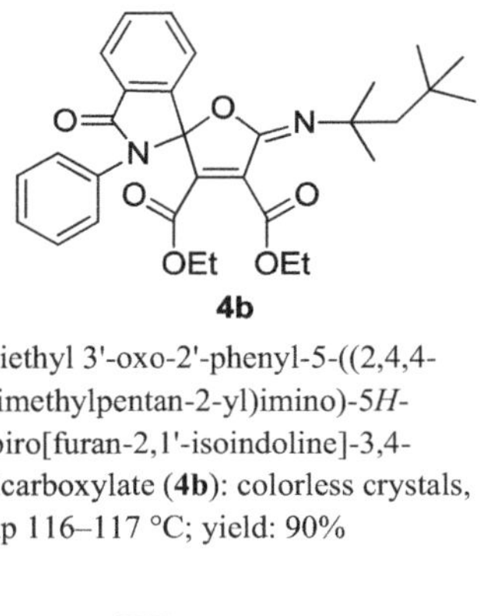

4b

Diethyl 3'-oxo-2'-phenyl-5-((2,4,4-trimethylpentan-2-yl)imino)-5*H*-spiro[furan-2,1'-isoindoline]-3,4-dicarboxylate (**4b**): colorless crystals, mp 116–117 °C; yield: 90%

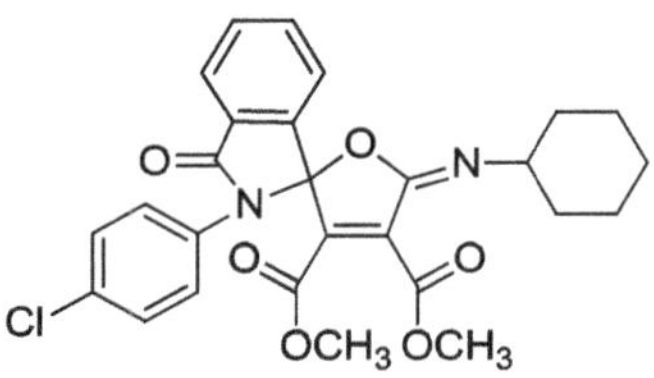

4c

Dimethyl 2'-(4-chlorophenyl)-5-(cyclohexylimino)-3'-oxo-5*H*-spiro[furan-2,1'-isoindoline]-3,4-dicarboxylate (**4c**): colorless crystals, mp 209–210 °C; yield: 92%

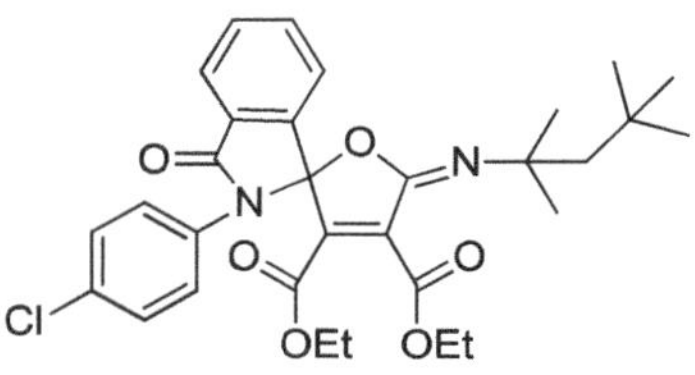

4d

Diethyl 2'-(4-chlorophenyl)-3'-oxo-5-((2,4,4-trimethylpentan-2-yl)imino)-5*H*-spiro[furan-2,1'-isoindoline]-3,4-dicarboxylate (**4d**): colorless crystals, mp 118–119 °C; yield: 97%

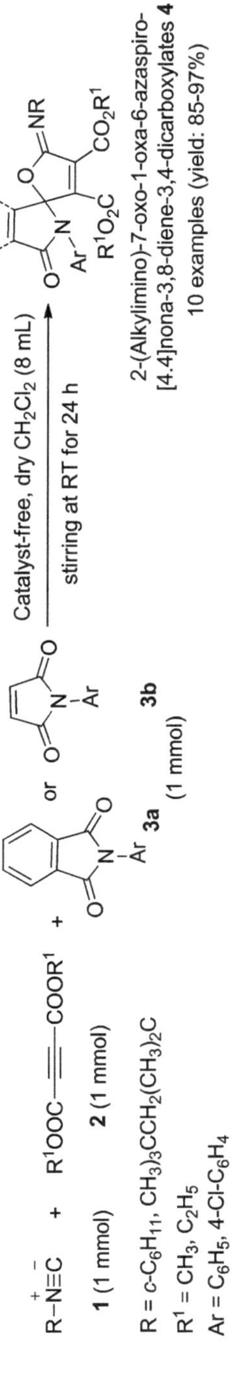

Scheme 2.71 Synthesis of 2-(alkylimino)-7-oxo-1-oxa-6-azaspiro[4.4]nona-3,8-diene-3,4-dicarboxylates (**4**).

2.2.52.3 Experimental Procedure

To a magnetically stirred solution of isocyanides (**1**; 1 mmol), dialkyl acetylenedicarboxylates (**2**; 1 mmol) in dry dichloromethane (6 mL), a solution of *N*-arylmaleimides or -phthalimides (**3**;1 mmol) in dichloromethane (2 mL) was added drop-wise at room temperature over 10 min. The mixture was then stirred for 24 h. On completion of the reaction, the solvent was removed under reduced pressure, and the residue was subjected to column chromatographic resolution using hexane–ethyl acetate (3 : 1) mixture as an eluent to obtain the pure product of 2-(alkylimino)-7-oxo-1-oxa-6-azaspiro[4.4]-nona-3,8-diene-3,4-dicarboxylates (**4**) after further recrystallization from hexane–ethyl acetate (1 : 1). The products were identified based on analytical and spectral studies.

2.2.52.4 Characterization Data of Two Representative Compounds

Dimethyl 2-(cyclohexylimino)-7-oxo-6-phenyl-1-oxa-6-azaspiro[4.4]nona-3,8-diene-3,4-dicarboxylate (**4a**): colorless crystals, mp 187–188 °C; yield: 93%; IR (KBr): 1788, 1753, 1732, 1701, 1659 (C=O), 1466, 1420, 1373, 1344, 1317, 1275, 1215, 1090, 1011, 945, 918, 878, 752 cm^{-1}; ^1H NMR ($CDCl_3$, 500.1 MHz): δ 1.20–1.74 (10H, m), 3.65–3.71 (1H, m), 3.73 (3H, s), 3.80 (3H, s), 6.47 (1H, d, J = 5.8 Hz), 6.91 (1H, d, J = 5.8 Hz), 7.15 (2H, d, J = 8.0 Hz), 7.30 (1H, t, J = 7.0 Hz), 7.33–7.37 (2H, m); ^{13}C NMR ($CDCl_3$, 125.8 MHz): δ 24.5, 24.6, 25.6, 33.3, 33.4, 52.8, 53.1, 57.1, 101.2, 126.8, 128.1, 129.3, 129.7, 133.8, 137.0, 140.6, 142.7, 151.0, 159.7, 161.2, 168.7; EIMS (rel.%): 424 (5, M^+), 373 (100), 289 (45), 169 (75), 92 (24), 77 (20). Anal. calc. for $C_{23}H_{24}N_2O_6$: C, 65.08; H, 5.70; N 6.60; found: C, 64.89; H, 5.79; N, 6.51.

Diethyl 2′-(4-chlorophenyl)-3′-oxo-5-((2,4,4-trimethylpentan-2-yl)imino)-5*H*-spiro[furan-2,1′-isoindoline]-3,4-dicarboxylate (**4d**): colorless crystals, mp 118–119 °C; yield: 97%; IR (KBr): 1733, 1682, 1657 (C=O), 1607, 1493, 1466, 1396, 1362, 1336, 1257, 1213, 1178, 1086, 1036, 903, 854, 820, 746, 688 cm^{-1}; ^1H NMR ($CDCl_3$, 250.1 MHz): δ 0.96 (9H, s), 1.03 (3H, t, J = 7.3), 1.30 (3H, t, J = 7.3 Hz), 1.31 (3H, s), 1.35 (3H, s), 1.52 (1H, d, ^2J = 14.5 Hz), 1.65 (1H, d, ^2J = 14.5 Hz), 3.94–4.13 (2H, 2 dq, ^2J = 11.0, ^3J = 7.3 Hz), 4.30 (2H, q, J = 7.3 Hz), 7.24 (2H, d, J = 8.8 Hz), 7.38 (2H, d, J = 8.8 Hz), 7.38–7.42 (1H, m), 7.58–7.70 (2H, m), 7.95–7.99 (1H, m); ^{13}C NMR ($CDCl_3$, 62.9 MHz): 13.6, 14.0, 29.8, 29.9, 31.7, 31.9, 55.5, 58.9, 61.9, 62.3, 100.5, 121.8, 124.3, 128.7, 129.6, 130.9, 130.9, 132.9, 133.4, 134.0, 135.8, 141.4, 142.4, 148.0, 159.2, 161.1, 166.9; EIMS (rel.%): 569 (<1, M^+ (^{37}Cl)), 567 (2, M^+ (^{35}Cl)), 495 (32), 439 (72), 376 (57), 338 (36), 296 (38), 194 (53), 57 (100). Anal. calc. for $C_{31}H_{35}ClN_2O_6$: C, 65.66; H, 6.22; N, 4.94; found: C, 65.48; H, 6.36; N, 4.88.

2.2.52.5 Critical Views

The investigators demonstrated a one-pot three-component reaction for the synthesis of 2-(alkylimino)-7-oxo-1-oxa-6-azaspiro[4.4]nona-3,8-diene-3,4-dicarboxylates of considerable synthetic and pharmacological interest with

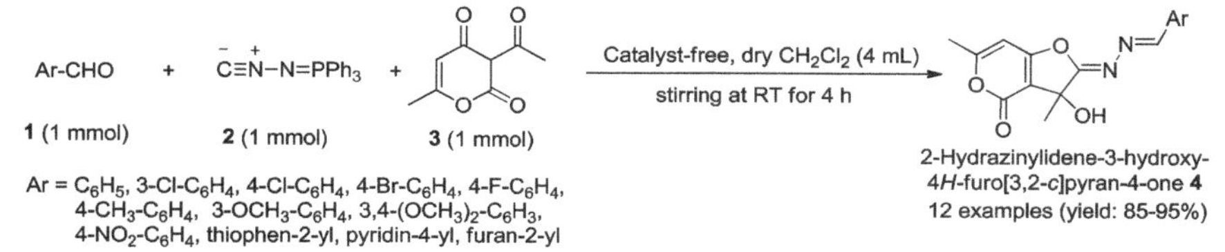

Scheme 2.72 Synthesis of 2-hydrazinylidene-3-hydroxy-4*H*-furo[3,2-*c*]pyran-4-ones (**4**).

good yields from the readily available starting materials under mild reaction conditions without any activation or modification of the starting materials.

2.2.53 Entry-53: Synthesis of 2-Hydrazinylidene-3-hydroxy-4*H*-furo[3,2-*c*]pyran-4-ones

Type of reaction: C–C and C–O bond formation
Reaction conditions: Catalyst-free, dichloromethane, room temperature
Synthetic strategy: One-pot multicomponent reaction
Keywords: Aromatic aldehydes, (*N*-isocyanimino) triphenylphosphorane, dehydroacetic acid (*i.e.* 3-acetyl-2-hydroxy-6-methyl-4*H*-pyran-4-one), dichloromethane, catalyst-free, room temperature, one-pot multicomponent reaction, tandem Passerini/intramolecular aza-Wittig reaction, 2-hydrazinylidene-3-hydroxy-4*H*-furo[3,2-*c*]pyran-4-ones

2.2.53.1 General Reaction Scheme

Adib *et al.*[180] designed a simple multicomponent reaction between aromatic aldehydes (**1**), (*N*-isocyanoimino) triphenylphosphorane (**2**) and dehydroacetic acid (*i.e.* 3-acetyl-2-hydroxy-6-methyl-4*H*-pyran-4-one; **3**) for the efficient one-pot synthesis of pharmacologically interesting 2-hydrazinylidene-3-hydroxy-4*H*-furo[3,2-*c*]pyran-4-ones (**4**) in dichloromethane under catalyst-free conditions at room temperature (Scheme 2.72).

2.2.53.2 Representative Examples

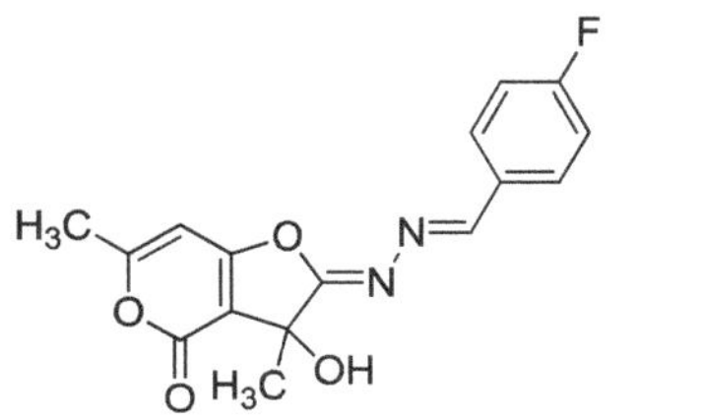

4a

(2*Z*)-2-[(2*E*)-(4-Fluorobenzylidene)hydrazinylidene]-2,3-dihydro-3-hydroxy-3,6-dimethyl-4*H*-furo[3,2-*c*]pyran-4-one (**4a**): cream powder, mp 166 °C; yield: 91%

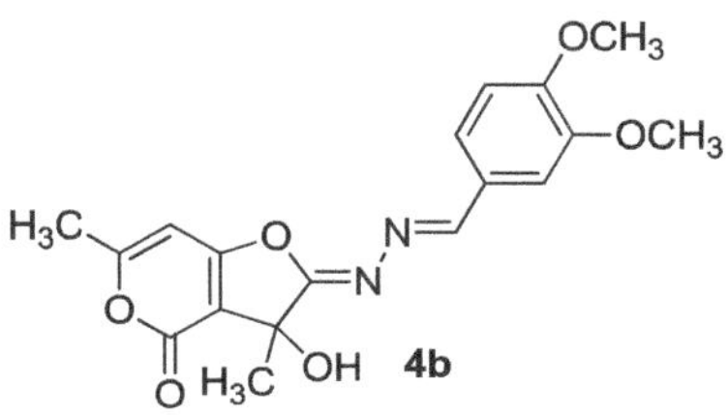

4b

(2*Z*)-2-[(2*E*)-(3,4-Dimethoxybenzylidene) hydrazinylidene]-2,3-dihydro-3-hydroxy-3,6-dimethyl-4*H*-furo[3,2-*c*]pyran-4-one (**4b**): cream powder, mp 163 °C; yield: 86%

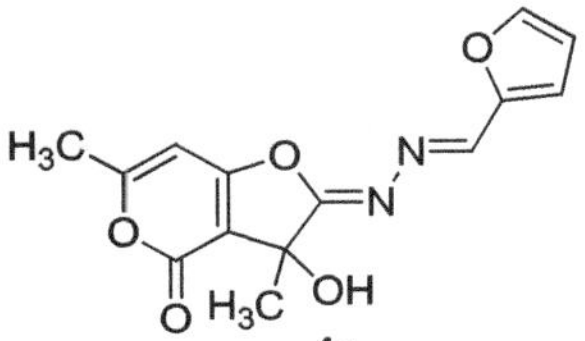

4c

(2Z)-2-{[(2E)-(Furan-2-yl)methylidene] hydrazinylidene}-2,3-dihydro-3-hydroxy-3,6-dimethyl-4H-furo[3,2-c]pyran-4-one (**4c**): cream powder, mp 218 °C; yield: 88%

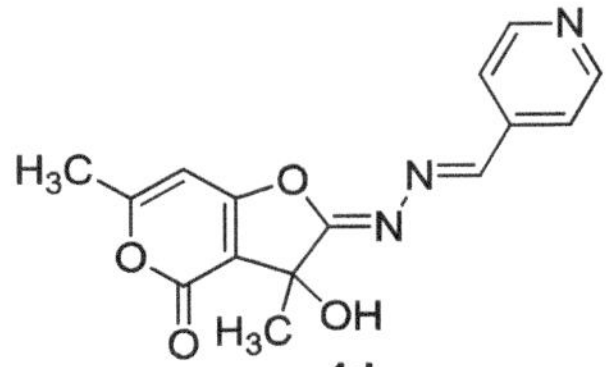

4d

(2Z)-2,3-Dihydro-3-hydroxy-3,6-dimethyl-2-{[(2E)-(pyridin-4-yl)methylidene] hydrazinylidene}-4H-furo[3,2-c]pyran-4-one (**4d**): pale-brown powder, mp 185 °C; yield: 89%

2.2.53.3 Experimental Procedure

A solution of aromatic aldehydes (**1**; 1 mmol), (*N*-isocyanimino)triphenylphosphorane (**2**; 0.302 g, 1 mmol) and 3-acetyl-6-methyl-2*H*-pyran-2,4(3*H*)-dione (dehydroacetic acid, **3**; 0.168 g, 1 mmol) in dichloromethane (4 mL) was stirred at room temperature for 4 h. On completion of the reaction, as monitored by TLC, the precipitated crude product was filtered and washed with dichloromethane (2×2 mL), followed by purification by column chromatography (SiO_2; hexane-ethyl acetate 4:1) to obtain pure products of 2-hydrazinylidene-3-hydroxy-4*H*-furo[3,2-*c*]pyran-4-ones (**4**) in good yields. All the compounds were characterized based on elemental analyses and spectral studies.

2.2.53.4 Characterization Data of Two Representative Compounds

(2*Z*)-2-[(2*E*)-(3,4-Dimethoxybenzylidene)hydrazinylidene]-2,3-dihydro-3-hydroxy-3,6-dimethyl-4*H*-furo[3,2-*c*]pyran-4-one (**4b**): cream powder, mp 163 °C; yield: 86%; IR (KBr): 3202 (OH), 1755 (C=O), 1697, 1649, 1599, 1576, 1512, 1461, 1463, 1425, 1381, 1273, 1236, 1213, 1150, 1167, 1065, 1034, 969, 916, 862, 808, 748 cm^{-1}; ^1H NMR (DMSO-d_6, 250.1 MHz): δ 1.66 (3H, s), 2.31 (3H, s), 3.82 (3H, s), 3.83 (s, 3H), 6.25 (1H, s), 6.67 (1H, s), 7.07 (1H, d, J=8.5 Hz), 7.38 (1H, d, J=8.5 Hz), 7.42 (1H, s), 8.36 (1H, s); ^{13}C NMR (DMSO-d_6, 62.9 MHz): δ 20.7, 24.2, 56.0, 56.1, 73.0, 95.5, 105.8, 110.0, 112.0, 123.8, 126.7, 149.4, 152.2, 158.7, 160.4, 161.9, 166.7, 168.7; EIMS (20 eV, rel.%): 358 (100, M^+), 343 (49), 328 (61), 301 (10), 206 (13), 191 (27), 179 (20), 163 (73), 150 (55), 137 (23), 120 (32), 107 (20), 92 (40), 77 (51), 65 (31), 43 (66). Anal. calc. for $C_{18}H_{18}N_2O_6$: C, 60.33; H, 5.06; N, 7.82; found: C, 60.38; H, 5.21; N, 7.57.

(2*Z*)-2,3-Dihydro-3-hydroxy-3,6-dimethyl-2-{[(2*E*)-(pyridin-4-yl)methylidene]hydrazinylidene}-4*H*-furo[3,2-*c*]pyran-4-one (**4d**): pale-brown powder, mp 185 °C; yield: 89%; IR (KBr): 3297 (OH), 1735 (C=O), 1680, 1641, 1605, 1584, 1452, 1406, 1249, 1142, 1130, 1067, 1042, 976, 914 cm^{-1}; ^1H NMR

(DMSO-d_6, 250.1 MHz): δ 1.69 (3H, s), 2.31 (3H, s), 6.36 (1H, s), 6.68 (1H, s), 7.76 (2H, d, J = 5.6 Hz), 8.46 (1H, s), 8.73 (2H, d, J = 5.6 Hz); ^{13}C–NMR (62.9 MHz, DMSO-d_6): δ 20.2, 23.4, 72.7, 94.8, 105.3, 121.7, 140.3, 150.4, 157.6, 158.0, 162.0, 166.0, 168.4; EIMS (20 eV, rel.%): 299 (3, M^+), 284 (51), 256 (7), 194 (7), 180 (17), 168 (15), 153 (74), 132 (20), 105 (25), 85 (24), 78 (29), 69 (24), 43 (100). Anal. calc. for $C_{15}H_{13}N_3O_4$: C, 60.20; H, 4.38; N, 14.04; found: C, 60.14; H, 4.42; N, 13.88.

2.2.53.5 Critical Views

A series of synthetically and pharmacologically potent fused *O*-heterocyclic 2-hydrazinylidene-3-hydroxy-4*H*-furo[3,2-*c*]pyran-4-one derivatives were prepared with good to excellent yields *via* one-pot multicomponent reaction under neutral conditions at ambient conditions without any activation or modification. Mild reaction conditions, good yields and operational simplicity render the present protocol interesting.

2.2.54 Entry-54: Synthesis of Polyfunctionalized Iminospiro-γ-lactones

Type of reaction: C–C and C–O bond formation
Reaction conditions: Catalyst-free, dichloromethane, room temperature
Synthetic strategy: One-pot multicomponent reaction
Keywords: Alkyl isocyanides, acetylenic esters, maleic anhydride, citraconic anhydride, dichloromethane, catalyst-free, room temperature, one-pot multicomponent reaction, polyfunctionalized iminospiro-γ-lactones

2.2.54.1 General Reaction Scheme

Bayat *et al.*[181] developed a simple, mild, one-pot three-component protocol for the synthesis of a novel series of biologically relevant polyfunctionalized iminospiro-γ-lactones (**4**) from the reaction of alkyl isocyanides (**1**) with electron-deficient acetylenic esters (**2**), in the presence of maleic anhydride (**3a**) or citraconic anhydride (**3b**), in good yields using dichloromethane, as a solvent, at room temperature (Scheme 2.73).

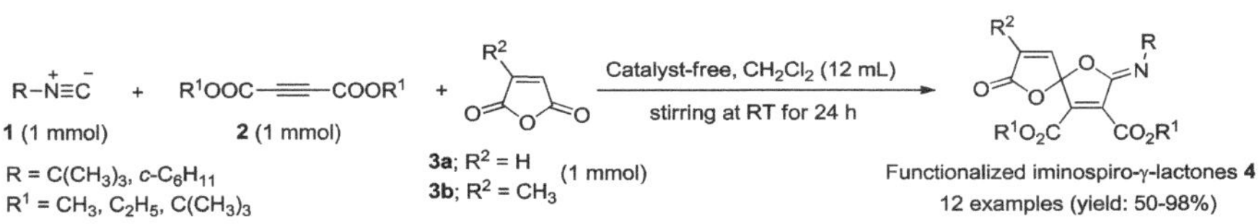

Scheme 2.73 Synthesis of polyfunctionalized iminospiro-γ-lactones (**4**).

2.2.54.2 Representative Examples

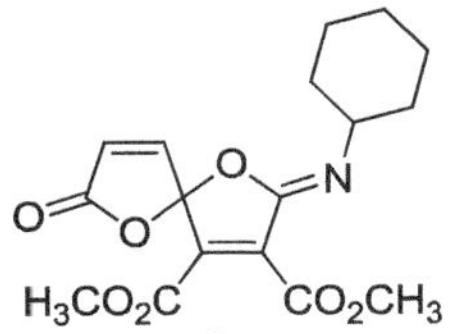

4a

(*Z*)-Dimethyl 2-(cyclohexylimino)-7-oxo-1,6-dioxaspiro[4.4]nona-3,8-diene-3,4-dicarboxylate (**4a**): yellow powder, mp 146–148 °C; yield: 98%

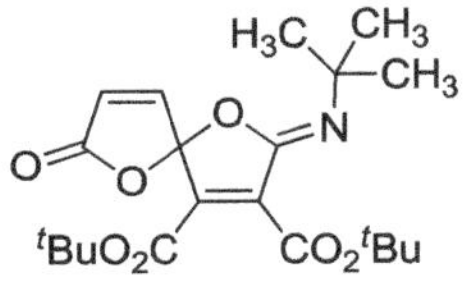

4b

(*Z*)-Di-*tert*-butyl 2-(tert-butylimino)-7-oxo-1,6-dioxaspiro[4.4]nona-3,8-diene-3,4-dicarboxylate (**4b**): pale yellow oil; yield: 65%

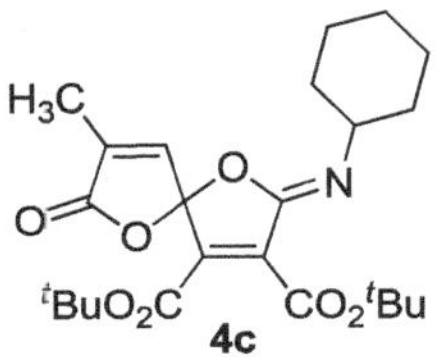

4c

(*Z*)-Di-*tert*-butyl 2-(cyclohexylimino)-8-methyl-7-oxo-1,6-dioxaspiro[4.4]nona-3,8-diene-3,4-dicarboxylate (**4c**):yellow powder,mp 158–160°C; yield: 60%

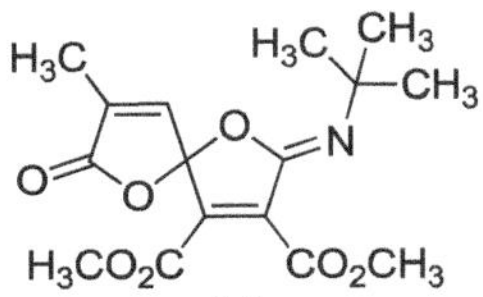

4d

(*Z*)-Dimethyl 2-(*tert*-butylimino)-8-methyl-7-oxo-1,6-dioxaspiro[4.4]nona-3,8-diene-3,4-dicarboxylate (**4d**): yellow powder, mp 135–137°C; yield: 94%

2.2.54.3 Experimental Procedure

To a magnetically stirred solution of acetylenic esters (**2**; 1 mmol) and maleic anhydride (**3a**; 1 mmol) or citraconic anhydride (**3b**; 1 mmol) in dichloromethane (10 mL), alkyl isocyanides (**1**; 1 mmol) in dichloromethane (2 mL) was added at room temperature over 10 min *via* a syringe. The mixture was then stirred at room temperature for 24 h. Upon completion of the reaction, the solvent was evaporated and the residue was purified by column chromatography (silica gel 70–230 mesh, hexane–ethyl acetate 4 : 1) to afford pure products of functionalized iminospiro-γ-lactones (**4**) in good yields. All the products were characterized by means of elemental analyses and detailed spectral studies.

2.2.54.4 Characterization Data of Two Representative Compounds

(*Z*)-Di-*tert*-butyl 2-(*tert*-butylimino)-7-oxo-1,6-dioxaspiro[4.4]nona-3,8-diene-3,4-dicarboxylate (**4b**): pale yellow oil; yield: 65%; IR (KBr): 1806, 1731, 1694 (C=O), 1643 (C=N) cm^{-1}; ^{1}H NMR ($CDCl_3$, 300 MHz): δ 1.35, 1.42, 1.47 (27H, 3s, 3×*t*-Bu); 6.38 (1H, d, $^3J = 5.5$ Hz, C_8-H); 7.20 (1H, d, $^3J = 5.5$ Hz, C_9-H); ^{13}C NMR ($CDCl_3$, 75 MHz): δ 27.95, 28.02, 28.65 (3×Me_3C); 56.74 (Me_3CN); 84.05, 85.92 (2×Me_3CO); 111.41 (C-5); 123.21 (C-8); 125.82 (C-3); 148.17 (C-4);

149.28 (C-2); 150.62 (C-9); 159.12, 160.81, 167.67 (3×C=O); EIMS (70 ev, rel.%): 407 (2, M^+), 240 (8), 198 (27), 166 (23), 155 (16), 135 (52), 109 (12), 86 (21), 84 (100), 58 (98), 41 (83). Anal. calc. for $C_{21}H_{29}NO_7$: C, 61.90; H, 7.17; N, 3.44; found: C, 61.8; H, 7.3; N, 3.3.

(*Z*)-Di-*tert*-butyl 2-(cyclohexylimino)-8-methyl-7-oxo-1,6-dioxaspiro[4.4]nona-3,8-diene-3,4-dicarboxylate (**4c**): yellow powder, mp 158–160 °C; yield: 60%; IR (KBr): 1820, 1736, 1697 (C=O), 1640 (C=N) cm^{-1}; ^1H NMR ($CDCl_3$, 300 MHz): δ 1.39, 1.51 (18H, 2s, 2×*t*-Bu); 1.46–1.81 (10H, m, 5×CH_2); 1.93 (3H, d, ^4J = 1.8 Hz, C_8-Me)); 3.35 (1H, m, CHN); 5.98 (1H, q, ^4J = 1.8 Hz, C_9-H); ^{13}C NMR ($CDCl_3$, 75 MHz): δ 13.32 (Me); 27.30, 27.82 (2×Me_3C); 27.72, 27.79, 28.01, 32.75, 32.84 (5×CH_2); 56.42 (CHN); 83.76, 84.02 (2×Me_3CO); 111.07 (C(5)); 119.41, 162.12 (C-8, C-9); 135.85 (C-3); 140.20 (C-4); 150.61 (C-2); 158.22, 159.78, 168.61 (3×C=O); EIMS (70 eV, rel.%): 336 (45), 335 (100), 318 (60), 290 (37), 237 (24), 148 (26), 98 (21), 58 (76). Anal. calc. for $C_{24}H_{33}NO_7$: C, 64.41; H, 7.43; N, 3.13; found: C, 64.3; H, 7.3; N, 3.1.

2.2.54.5 Critical Views

A catalyst-free and efficient method was developed under ambient conditions for the one-pot synthesis of a novel series of polyfunctionalized iminospiro-γ-lactones of potential synthetic interest. The present procedure has the advantage that the reaction runs under neutral conditions and, moreover, the substances can be mixed without any further activation or modification.

2.2.55 Entry-55: Synthesis of Functionalized 2*H*-Indeno[2,1-*b*]furans

Type of reaction: C–C and C–O bond formation
Reaction conditions: Catalyst-free, dichloromethane, room temperature
Synthetic strategy: One-pot multicomponent reaction
Keywords: Alcohols, dialkyl acetylenedicarboxylate, triphenylphosphine, ninhydrin, dichloromethane, catalyst-free, room temperature, one-pot multicomponent, diastereoselectivity, densely functionalized 2*H*-indeno[2,1-*b*]furans

2.2.55.1 General Reaction Scheme

Ramazani and his group[182] reported for the first time a one-pot straightforward three-component protocol for the diastereoselective synthesis of densely functionalized 2*H*-indeno[2,1-*b*]furans (**5**). This was achieved by the reaction of various alcohols (**2**), ninhydrin (**4**) and a highly reactive 1 : 1 intermediate, produced by the reaction of Ph_3P (**1**) and dialkyl acetylenedicarboxylates (**2**) in dichloromethane at room temperature (Scheme 2.74) to produce good yields.

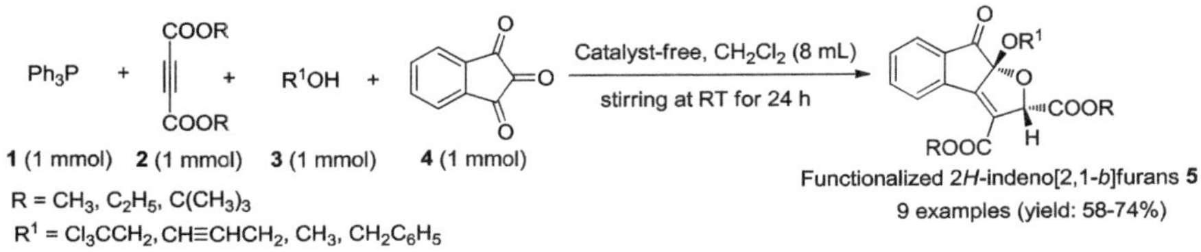

Scheme 2.74 Synthesis of functionalized 2*H*-indeno[2,1-*b*]furans (5).

2.2.55.2 Representative Examples

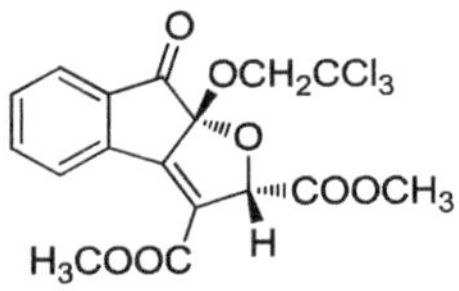

5a

(2*S*,8a*R*)-Dimethyl 8-oxo-8a-(2,2,2-trichloroethoxy)-8,8a-dihydro-2*H*-indeno[2,1-*b*]furan-2,3-dicarboxylate (**5a**): white solid, mp 134–138°C; yield: 60%

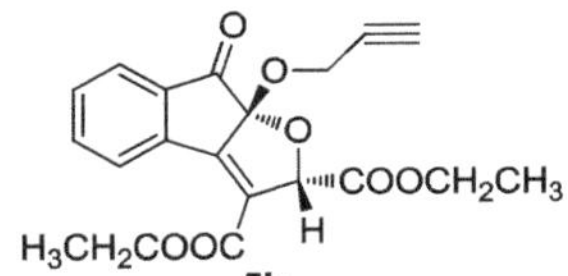

5b

(2*S*,8a*R*)-Diethyl 8-oxo-8a-(prop-2-yn-1-yloxy)-8,8a-dihydro-2*H*-indeno[2,1-*b*]furan-2,3-dicarboxylate (**5b**): white solid, mp 114–116 °C; yield: 58%

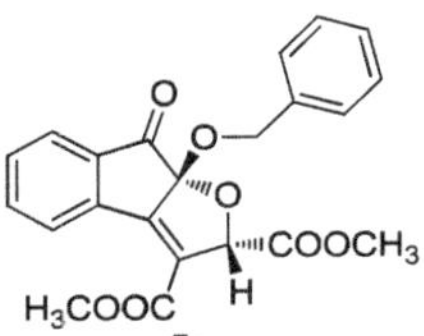

5c

(2*S*,8a*R*)-Dimethyl 8a-(benzyloxy)-8-oxo-8,8a-dihydro-2*H*-indeno[2,1-*b*]furan-2,3-dicarboxylate (**5c**): colorless oil; yield: 66%

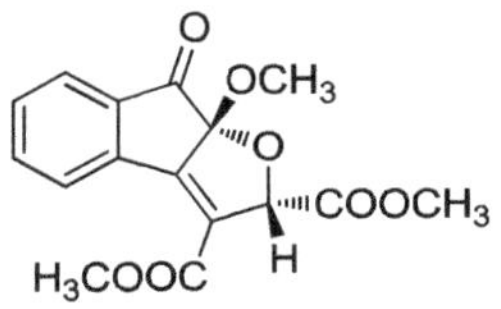

5d

(2*S*,8a*R*)-Dimethyl 8a-methoxy-8-oxo-8,8a-dihydro-2*H*-indeno[2,1-*b*]furan-2,3-dicarboxylate (**5d**): white solid, mp 115–117 °C; yield: 68%

2.2.55.3 Experimental Procedure

To a magnetically stirred solution of triphenylphosphine (**1**; 0.262 g, 1 mmol) and alcohol (**3**; 1 mmol) in dichloromethane (4 mL), diester (**2**; 1 mmol) in dichloromethane (4 mL) was added drop-wise at −10 °C over 15 minutes. The mixture was then allowed to become warmed up to room temperature, and powdered ninhydrin (**4**; 0.18 g, 1 mmol) was added to this resulting mixture. The combined mixture continued to be stirred for 24 h to complete the reaction. The solvent was then evaporated and the viscous residue as obtained was purified by flash column chromatography (silica gel, petroleum ether–ethyl acetate) to yield pure products of functionalized 2*H*-indeno[2,1-*b*]furans (**5**). All the products were characterized based on spectral and elemental analyses.

2.2.55.4 *Characterization Data of Two Representative Compounds*

(2*S*,8a*R*)-Diethyl 8-oxo-8a-(prop-2-yn-1-yloxy)-8,8a-dihydro-2*H*-indeno[2,1-*b*]furan-2,3-dicarboxylate (**5b**): white solid, mp 114–116 °C; yield: 58%; UV (95% EtOH): λ_{max} (log ε) 208 (4.16), 246 (4.44), 280 (4.34); IR (KBr): 3000 (m), 2151 (m), 1756 (s), 1729 (s), 1601 (m), 1402 (m), 1128 (s), 779 (w) cm^{-1}; ^1H NMR ($CDCl_3$, 250 MHz): δ 1.26, 1.34 (3H each, 2 t, $^3J=7.1$ Hz, 2×$MeCH_2$); 2.43 (1H, t, $^4J=-2.3$ Hz, $HCCCH_2$); 4.10–4.38 (4H, m, 2×$MeCH_2$); 4.36, 4.55 (2H, 2 dd, $^2J=-15.3$ Hz, $^4J=-2.3$ Hz, $HCCCH_2$); 5.99 (1H, s, H–C(2)); 7.62 (1H, t, $^3J=7.6$ Hz, Ar–H); 7.77 (1H, t, $^3J=7.6$ Hz, Ar–H); 7.85 (1H, d, $^3J=7.5$ Hz, Ar–H); 8.28 (1H, d, $^3J=7.7$ Hz, Ar–H); ^{13}C NMR ($CDCl_3$, 62.5 MHz): δ 13.94, 14.14 (2×$MeCH_2$); 53.16 (HCC*C*H_2); 61.55, 62.02 (2×$MeCH_2$); 74.95, 78.19 ($HCCCH_2$); 90.16 (C(2)); 113.24 (C(8a)); 125.42, 128.40, 132.33, 136.06 (4 arom. CH); 123.41, 138.30, 139.95, 148.54 (4 C); 161.49, 167.55 (2×C=O, 2 esters); 189.42 (C=O, ketone). Anal. calc. for $C_{20}H_{18}O_7$: C, 64.86, H, 4.90; found C, 64.53, H, 4.74.

(2*S*,8a*R*)-Dimethyl 8a-(benzyloxy)-8-oxo-8,8a-dihydro-2*H*-indeno[2,1-*b*]furan-2,3-dicarboxylate (**5c**): colorless oil; yield: 66%; IR (neat): 2980 (m), 1757 (s), 1720 (s), 1603 (s), 1449 (m), 1294 (m), 1248 (s), 1040 (m), 770 (w) cm^{-1}; ^1H NMR ($CDCl_3$, 250 MHz): δ 3.75, 3.82 (3H each, 2 s, 2×MeO); 4.72, 4.98 (1H each, 2 d, $^2J=-9.02$ Hz, $PhCH_2O$); 5.98 (1H, s, H–C(2)); 7.00–7.98 (8H, m, Ar–H); 8.22 (1H, d, $^3J=7.7$ Hz, Ar–H); ^{13}C NMR ($CDCl_3$, 62.5 MHz): δ 52.33, 52.90 (2×MeO); 67.35 ($PhCH_2O$); 89.67 (C(2)); 113.73 (C(8a)); 122.57, 125.38, 126.97, 127.54, 128.03, 128.47, 132.42, 136.01, 136.94, 138.25, 140.24, 149.32 (5 C, 9 CH); 162.07, 168.22 (2 C=O, 2 esters); 189.88 (C=O, ketone). Anal. calc. for $C_{22}H_{18}O_7$: C, 67.00, H, 4.60; found: C, 66.81, H, 4.62.

2.2.55.5 *Critical Views*

The present protocol offers, for the first time, a one-pot catalyst-free strategy for the efficient and completely diastereoselective synthesis of densely functionalized 2*H*-indeno[2,1-*b*]furans of potential synthetic and biological interests. The ease of workup, good yields, and fairly mild reaction conditions are noteworthy.

2.2.56 Entry-56: Synthesis of 1,4-Benzoxazinones

Type of reaction: C–N and C–O bond formation
Reaction conditions: Catalyst-free, methanol, room temperature
Synthetic strategy: One-step condensation
Keywords: 2-Aminophenols, dimethyl-2-oxo-glutarate, methanol, room temperature, condensation, 1,4-benzoxazinones

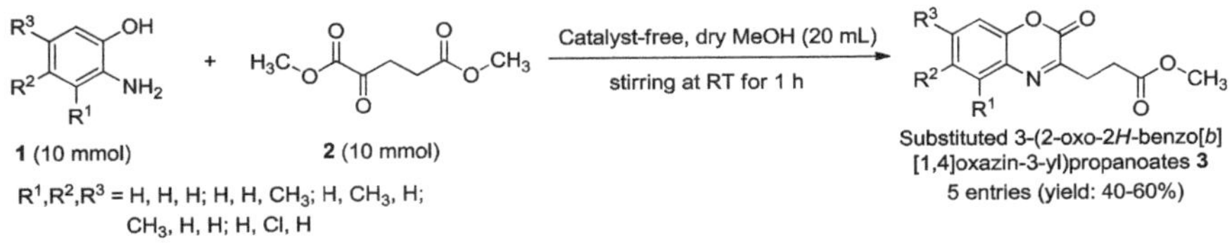

Scheme 2.75 Synthesis of 1,4-benzoxazinones (**3**).

2.2.56.1 General Reaction Scheme

Hachama *et al.*[183] synthesized a few novel antimicrobial 3-(2-oxo-2*H*-benzo[*b*][1,4]oxazin-3-yl)propanoates from the reaction of 2-aminophenols (**1**) with dimethyl-2-oxoglutarate (**2**) under mild conditions at room temperature (Scheme 2.75).

2.2.56.2 Representative Examples

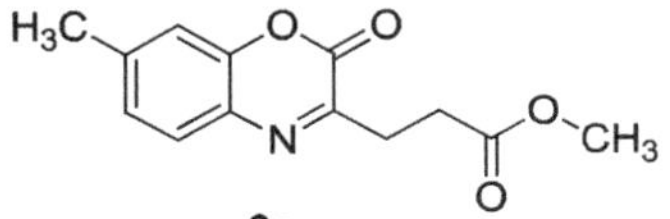

3a

Methyl 3-(7-methyl-2-oxo-2*H*-benzo[*b*][1,4]oxazin-3-yl)propanoate (**3a**): colorless crystals, mp 99–100 °C; yield: 40%

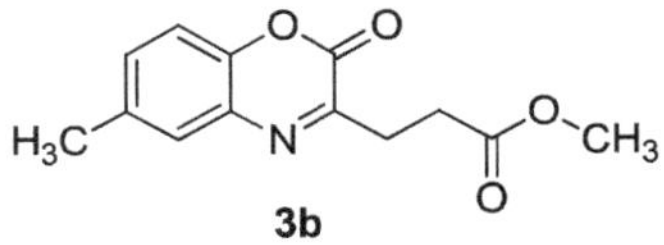

3b

Methyl 3-(6-methyl-2-oxo-2*H*-benzo[*b*][1,4]oxazin-3-yl)propanoate (**3b**): colorless crystals, mp 100–101 °C; yield: 50%

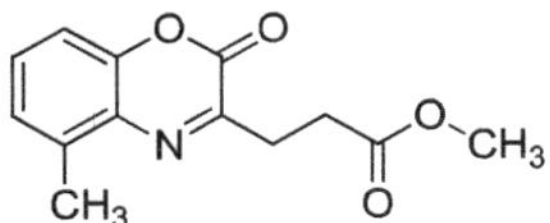

3c

Methyl 3-(5-methyl-2-oxo-2*H*-benzo[*b*][1,4]oxazin-3-yl)propanoate (**3c**): colorless crystals, mp 132–134 °C; yield: 50%

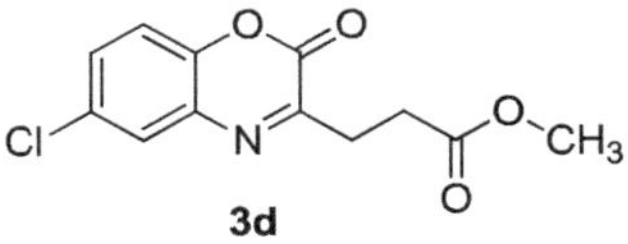

3d

Methyl 3-(6-chloro-2-oxo-2*H*-benzo[*b*][1,4]oxazin-3-yl)propanoate (**3d**): colorless crystals, mp 114–116 °C; yield: 50%

2.2.56.3 Experimental Procedure

A solution of dimethyl-2-oxoglutarate (**2**; 10 mmol, 1.74 g) in 10 mL of dry methanol was added drop-wise while stirring to a solution of the corresponding substituted 2-aminophenol (**1**; 10 mmol) in 20 mL of dry methanol at room temperature (20 °C). The reaction was almost completed after stirring for 1 h at room temperature; the crystals formed were filtered off and recrystallized from methanol to obtain pure products of substituted 3-(2-oxo-2*H*-benzo[*b*][1,4]oxazin-3-yl)propanoates (**3**) with 40–60% yield. All the products were characterized by elemental analyses, as well as detailed spectral studies.

2.2.56.4 Characterization Data of Two Representative Compounds

Methyl 3-(7-methyl-2-oxo-2H-benzo[*b*][1,4]oxazin-3-yl)propanoate (**3a**): colorless crystals, mp 99–100 °C; IR(KBr): 1740 (C=O), 1616 (N=C) cm^{-1}; ^{1}H NMR (DMSO-d_6, 300/400 MHz): δ 2.38 (3H, s, CH_3), 2.74 (2H, t, $J=7.1$ Hz, CH_2,), 3.01 (2H, t, $J=7.1$ Hz, CH_2,), 3.59 (3H, s, OCH_3), 7.18 (1H, d, $J=8.4$ Hz, arom-H), 7.20 (1H, s, arom-H), 7.54 (1H, d, $J=8.4$ Hz, arom-H); ^{13}C NMR (DMSO-d_6, 75/100 MHz): δ 20.98 (CH_3), 28.02 (CH_2), 29.12 (CH_2), 51.44 (OCH_3), 116.18 (C-8), 126.20 (C-6), 127.85 (C-5), 128.55 (C-7), 141.17 (C-4a), 145.93 (C-8a), 152.69 (C=O),155.21 (N=C), 172.66 (COO); EIMS (70 eV): *m/z* (%rel.) 247 (M^+, 88%), 216 (M^+–OCH_3, 51%), 188 (M^+–CO_2CH_3, 36%), 160 (M^+–CH_2–CH_2–CO_2CH_3, 100%). Anal. calc. for $C_{13}H_{13}NO_4$: C, 63.15; H, 5.30; N, 5.67; found: C, 63.08; H, 5.11; N, 5.61%.

Methyl 3-(6-chloro-2-oxo-2*H*-benzo[*b*][1,4]oxazin-3-yl)propanoate (**3b**): Colorless crystals, mp 114–116 °C; IR(KBr): 1740 (C=O), 1615 (N=C) cm^{-1}; ^{1}H NMR (DMSO-d_6, 300/400 MHz): δ 2.77 (2H, t, $J=7.2$ Hz, CH_2), 3.06 (2H, t, $J=7.2$ Hz, CH_2), 3.62 (3H, s, OCH_3), 7.45 (1H, d, $J=6.5$ Hz, arom-H), 7.59 (1H, dd, $J=6.5$ & 2.1 Hz, arom-H,), 7.74 (1H, d, $J=2.1$ Hz, arom-H); ^{13}C NMR (DMSO-d_6, 75/100 MHz): δ 28.19 (CH_2), 28.96 (CH_2), 51.46 (OCH_3), 117.94 (C-8), 127.27 (C-5), 128.73 (C-7), 129.97 (C-6), 131.32 (C-4a), 145.02 (C-8a), 152.12 (C=O), 158.19 (N=C), 172.49 (COO); EIMS (70 eV): *m/z* (%rel.) 267 (M^+, 16%), 236 (M^+–OCH_3, 15%), 208 (M^+–CO_2CH_3, 5%), 180 (M^+–CH_2–CH_2–CO_2CH_3, 100%). Anal. calc. for $C_{12}H_{10}ClNO_4$: C, 53.85; H, 3.77; N, 5.23; found: C, 53.83; H, 3.73; N, 5.16%.

2.2.56.5 Critical Views

This methods offers a catalyst-free way to synthesize biologically relevant 3-(2-oxo-2*H*-benzo[*b*][1,4]oxazin-3-yl)propanoates under mild conditions.

2.2.57 Entry-57: Synthesis of 1,3,4-Oxadiazoles

Type of reaction: C–N and C–O bond formation
Reaction conditions: Catalyst-free, dichloromethane, room temperature
Synthetic strategy: One-pot multicomponent reaction
Keywords: *N*-Isocyaniminotriphenylphosphorane, biacetyl, (*E*)-cinnamic acids, dichloromethane, catalyst-free, room temperature, one-pot multicomponent reaction, *aza*-Wittig reaction, 1,3,4-oxadiazoles

2.2.57.1 General Reaction Scheme

Ramazani *et al.*[184] synthesized sterically-congested 1,3,4-oxadiazole derivatives (**4**) from the one-pot multicomponent reaction of *N*-isocyaniminotriphenylphosphorane (**3**) with biacetyl (**1**) in the presence of

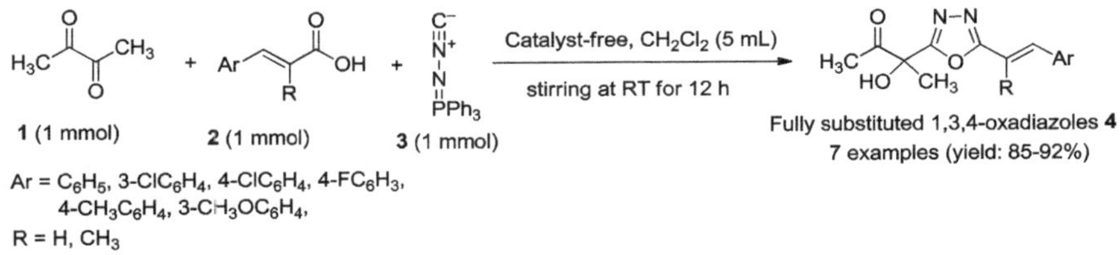

Scheme 2.76 Synthesis of 1,3,4-oxadiazoles (**4**).

(*E*)-cinnamic acids (**2**) in dichloromethane at room temperature and under neutral conditions (Scheme 2.76).

2.2.57.2 *Plausible Mechanism*

The investigators proposed a tentative mechanism for this transformation (Scheme 2.77).[184] In the first step, *N*-isocyaniminotriphenylphosphorane (**3**) undergoes nucleophilic addition to biacetyl (**1**), followed by protonation of the adduct with acid (**2**) leading to a nitrilium intermediate **6**. This intermediate is then attacked by the conjugate base (**5**) to form the 1 : 1 : 1 adduct (**7**). This adduct undergoes an intramolecular aza-Wittig reaction of iminophosphorane moiety with the ester carbonyl group to afford the isolated sterically-congested 1,3,4-oxadiazole derivative (**4**) by the removal of triphenylphosphine oxide (Ph_3PO) from the aza-Wittig intermediate (**8**).

2.2.57.3 *Representative Examples*

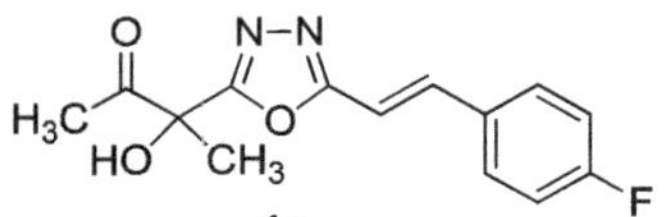

4a

(*E*)-3-(5-(4-Fluorostyryl)-1,3,4-oxadiazol-2-yl)-3-hydroxybutan-2-one (**4a**): white powder, mp 84–86 °C; yield: 90%

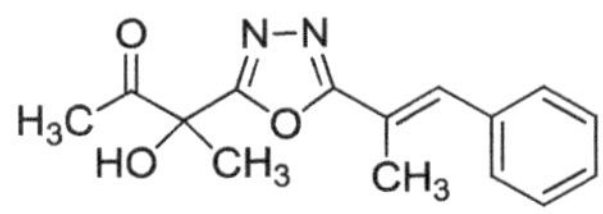

4b

(*E*)-3-Hydroxy-3-(5-(1-phenylprop-1-en-2-yl)-1,3,4-oxadiazol-2-yl)butan-2-one (**4b**): yellow viscous oil; yield: 86%

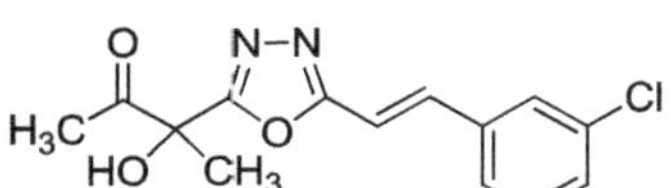

4c

(*E*)-3-(5-(3-Chlorostyryl)-1,3,4-oxadiazol-2-yl)-3-hydroxybutan-2-one (**4c**): white powder, mp 62–64°C; yield: 86%

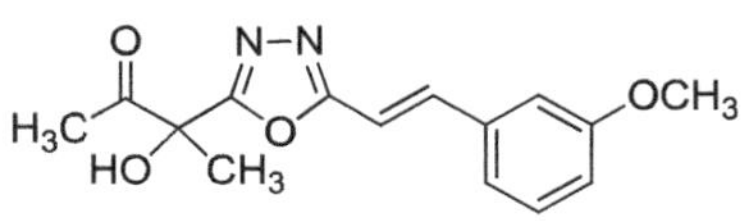

4d

(*E*)-3-Hydroxy-3-(5-(3-methoxystyryl)-1,3,4-oxadiazol-2-yl)butan-2-one (**4d**): white powder, mp 70–72 °C; yield: 87%

2.2.57.4 *Experimental Procedure*

A mixture of biacetyl (**1**; 1.0 mmol, 0.10 mL), (*E*)-cinnamic acid (**2**; 1.0 mmol), and *N*-isocyaniminotriphenylphosphorane (**3**; 1.0 mmol, 0.300 g) in

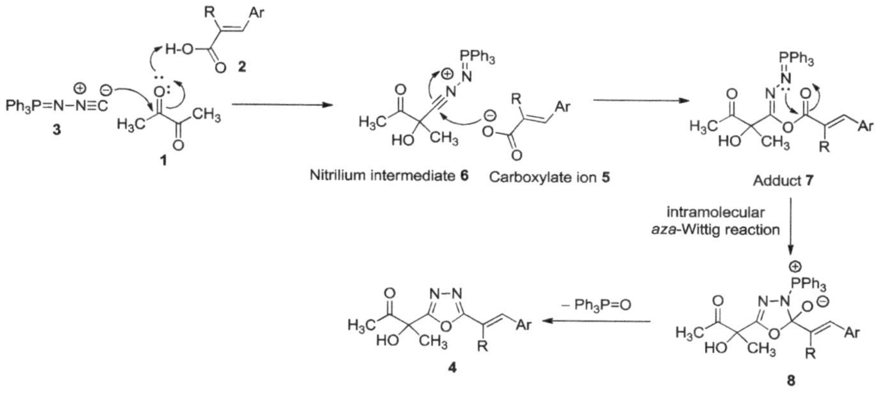

Scheme 2.77 Suggested mechanism for the formation of 1,3,4-oxadiazoles (**4**).

dichloromethane (5 mL) was stirred at room temperature for 12 h to complete the reaction. The solvent was removed under reduced pressure and the viscous residue was purified by HPLC (silica gel (F_{254}) powder; petroleum ether–ethyl acetate = 4 : 1) to obtain pre-fully substituted 1,3,4-oxadiazoles (**4**). All the products were characterized from spectral and analytical measurements.

2.2.57.5 Characterization Data of Two Representative Compounds

(*E*)-3-(5-(4-Fluorostyryl)-1,3,4-oxadiazol-2-yl)-3-hydroxybutan-2-one (**4a**): white powder, mp 84–86 °C; IR (neat): 3372, 2923, 1731, 1670, 1529, 1505, 1163, 818 cm^{-1}; ^{1}H NMR ($CDCl_3$, 250 MHz): δ 1.87 (3H, s, CH_3), 2.35 (3H, s, CH_3), 4.70 (1H, s, OH), 6.92 (1H, d, $^{3}J_{HH}$ = 16.5 Hz, =CH), 7.01–7.55 (5H, m, Ar–H and =CH); ^{13}C NMR ($CDCl_3$, 62.5 MHz): δ = 23.3 (CH_3), 23.7 (CH_3), 75.7 (C–OH), 109.0 (CH =), 138.9 (CH =), 116.2 (d, $^{2}J_{CF}$ = 28.3 Hz, Ar–CH), 129.4 (d, $^{3}J_{CF}$ = 8.2 Hz, Ar–CH), 130.7 (C), 162.5 (d, $^{1}J_{CF}$ = 440.3 Hz, C), 164.9 (C=N), 165.2 (C=N), 204.6 (C=O); MS: *m*/*z* (%) = 276 (M^{+}, 12), 234 (100), 191 (56), 164 (32), 147 (60), 121 (20), 101 (24), 43 (68). Anal. calc. for $C_{14}H_{13}FN_2O_3$ (276.09): C, 60.87; H, 4.74; N, 10.14; found: C, 60.81; H, 4.80; N, 10.08%.

(*E*)-3-Hydroxy-3-(5-(3-methoxystyryl)-1,3,4-oxadiazol-2-yl)butan-2-one (**4d**): white powder, mp 70–72 °C; IR (neat): 3236, 3004, 2922, 1730, 1650, 1580, 1466, 1108, 876, 795, 687 cm^{-1}; ^{1}H NMR ($CDCl_3$, 250 MHz): 1.93 (3H, s, CH_3), 2.35 (3H, s, CH_3), 3.83 (3H, s, OCH_3), 4.73 (1H, s, OH), 6.91–7.35 (5H, m, Ar–H and CH), 7.52 (1H, d, $^{3}J_{HH}$ = 16.5 Hz, CH); ^{13}C NMR ($CDCl_3$, 62.5 MHz): δ 23.3 (CH_3), 23.8 (CH_3), 55.3 (s, 3H, OCH_3), 75.8 (C–OH), 109.6 (CH , 140.1 (CH), 112.5 (CH), 116.1 (CH), 120.3 (CH), 130.0 (CH), 135.8 (C), 160.0 (C), 164.9 (C=N), 165.3 (C=N), 204.6 (C=O). Anal. calc. for $C_{15}H_{16}N_2O_4$ (288.11): C, 62.49; H, 5.59; N, 9.72; found: C, 62.44; H, 5.64; N, 9.67%.

2.2.57.6 Critical Views

1,3,4-Oxadiazoles are an important class of heterocyclic compounds with a wide range of pharmaceutical and biological activities. They also find huge applications as synthones in organic and medicinal chemistry for many useful compounds. Several methods have been reported in the literature for the synthesis of 1,3,4-oxadiazoles, which are multistep in nature and involve the use of catalysts and solvents. In contrast, the present protocol offers a mild, simple, and efficient route for the preparation of such sterically-congested 1,3,4-oxadiazole derivatives (**4**) *via* a one-pot multicomponent reaction with easy workup and high yields at ambient conditions.

2.2.58 Entry-58: Synthesis of 1,3-Thiazole-4(3*H*)-carboxylates

Types of reaction: C–N and C–S bond formation
Reaction conditions: Catalyst-free, acetone, room temperature
Synthetic strategy: One-pot multicomponent reaction
Keywords: Ammonium thiocyanate, benzylamine, ethyl bromopyruvate, benzoyl chloride, acetone, catalyst-free, room temperature, multicomponent reaction, 1,3-thiazole-4(3*H*)-carboxylates

2.2.58.1 General Reaction Scheme

A convenient and efficient synthetic route for accessing biologically promising 1,3-thiazole-4(3*H*)-carboxylates (**5**) *via* a four-component one-pot reaction between ammonium thiocyanate (**1**), acid chlorides (**2**), ethyl bromopyruvate (**3**), and primary amines (**4**) in acetone at room temperature was reported by Yavari and coworkers[185] (Scheme 2.78).

2.2.58.2 Plausible Mechanism

The authors proposed a tentative mechanism for this transformation (Scheme 2.79).[185] Initially, an isothiocyanate (**6**) is formed out of the reaction of ammonium thiocyanate with acyl chloride, followed by addition of amine (**4**) to generate **7**. Subsequent nucleophilic attack of **7** on ethyl bromopyruvate (**3**) yields the 1 : 1 adduct **8**. Intermediate (**8**) undergoes a cyclization

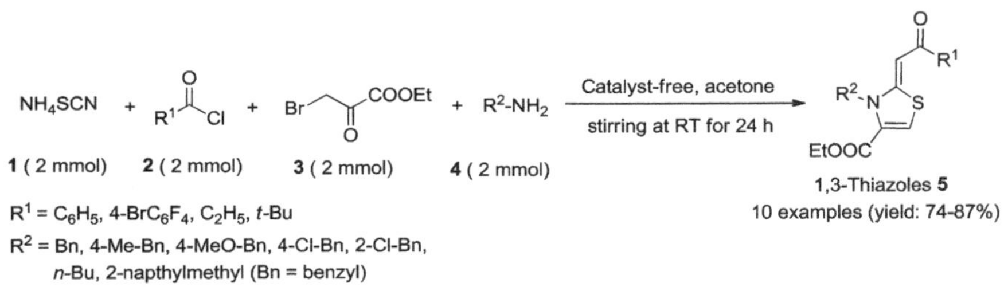

Scheme 2.78 Synthesis of 1,3-thiazole-4(3*H*)-carboxylates (**5**).

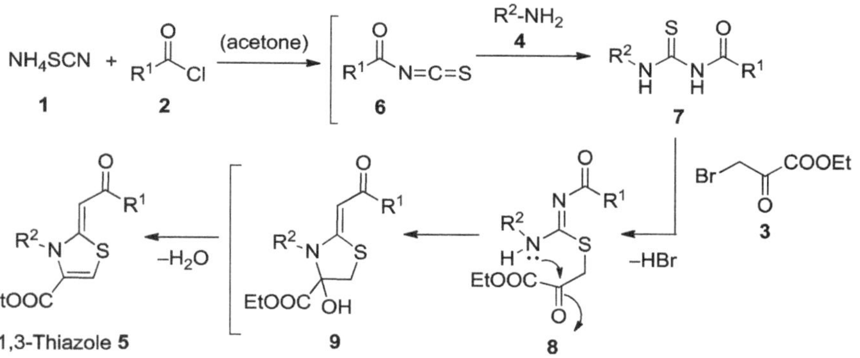

Scheme 2.79 Suggested mechanism for the formation of 1,3-thiazole-4(3*H*)-carboxylates (**5**).

reaction to give **9**, which is converted into the desired product **5** by the elimination of water.

2.2.58.3 Representative Examples

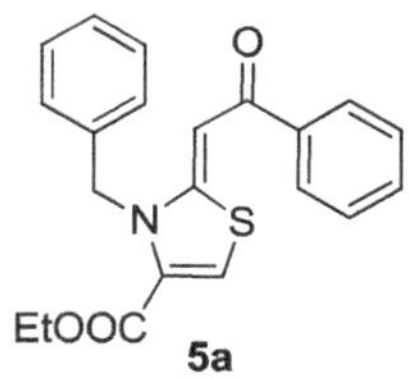

5a

(*Z*)-Ethyl 3-benzyl-2-(2-oxo-2-phenylethylidene)-2,3-dihydrothiazole-4-carboxylate (**5a**): colorless crystals, mp 119–121 °C; yield: 85%

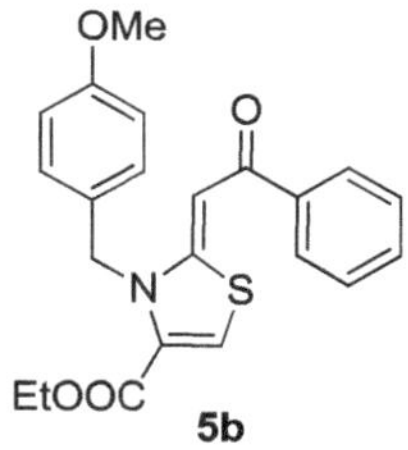

5b

(*Z*)-Ethyl 3-(4-methoxybenzyl)-2-(2-oxo-2-phenylethylidene)-2,3-dihydrothiazole-4-carboxylate (**5b**): yield: 78%

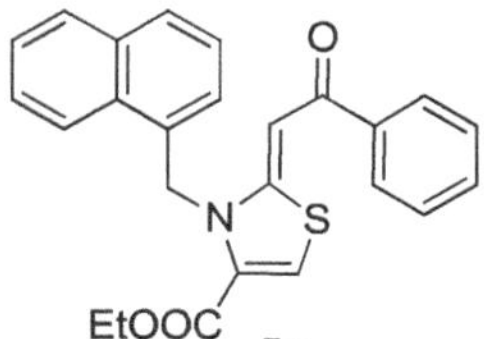

5c

(*Z*)-Ethyl 3-(naphthalen-1-ylmethyl)-2-(2-oxo-2-phenylethylidene)-2,3-dihydrothiazole-4-carboxylate (**5c**): yield: 75%

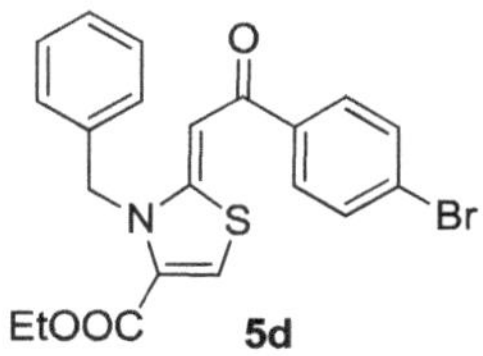

5d

(*Z*)-Ethyl 3-benzyl-2-(2-(4-bromophenyl)-2-oxoethylidene)-2,3-dihydrothiazole-4-carboxylate (**5d**): yield: 78%

2.2.58.4 Experimental Procedure

To a stirred solution of ammonium thiocyanate (**1**; 0.15 g, 2 mmol) in acetone (15 mL), benzoyl chloride (**2**; 0.28 g, 2 mmol) was added and the mixture was refluxed for 5 min; then a solution of ethyl bromopyruvate (**3**; 0.39 g, 2 mmol) in acetone (10 mL) was added drop-wise. After addition of the primary amine (**4**; 2 mmol) at room temperature, the resulting reaction mixture was stirred for 24 h, the solvent was removed under reduced pressure, and the residue was purified by column chromatography (SiO_2; hexane–EtOAc) to give 1,3-thiazole derivatives (**5**) with 74–97% of yield. All the products were characterized on the basis of detailed spectral and elemental analyses.

2.2.58.5 Characterization Data of One Representative Compound

(*Z*)-Ethyl 3-benzyl-2-(2-oxo-2-phenylethylidene)-2,3-dihydrothiazole-4-carboxylate (**5a**): colorless crystals, mp 119–121 °C; IR (KBr): 1716, 1658, 1587, 1477, 1429, 1369, 1369, 1304, 1231, 1105, 1054 cm^{-1}; ^1H NMR ($CDCl_3$, 500.1 MHz): δ 1.33 (3H, t, $^3J=7.2$ Hz, Me), 4.31 (2H, q, $^3J=7.2$ Hz, OCH_2), 6.07 (2H, s, NCH_2), 7.24–7.32 (3H, m, 3×CH), 7.38 (2H, d, $^3J=7.4$ Hz, 2×CH), 7.45 (2H, t, $^3J=7.2$ Hz, CH), 7.51 (1H, d, $^3J=7.2$ Hz, CH), 7.63 (1H, s, CH), 8.34 (2H, d, $^3J=7.2$ Hz, 2×CH); ^{13}C NMR ($CDCl_3$, 125 MHz): δ 14.0 (Me), 50.4 (CH_2), 61.9 (OCH_2), 120.8 (CH), 127.7 (2×CH), 128.1 (3×CH), 128.6 (2×CH), 129.5 (3×CH), 131.8 (C), 136.5 (C), 136.8 (C), 158.3 (C=O), 169.3 (C=N), 174.8 (C=O); MS: *m/z* (%rel) 366 (10) [M^+], 293 (75), 261 (62), 105 (100), 91 (82), 45 (84). Anal. calc. (%) for $C_{20}H_{18}N_2O_3S$ (366.43): C 65.56, H 4.95, N 7.64; found: C 65.63, H 4.99, N, 7.68.

2.2.58.6 Critical Views

Mild reaction conditions, good yields, operational simplicity and inexpensive starting materials are the main advantages of this one-pot protocol for the preparation of functionalized 1,3-thiazole-4(3*H*)-carboxylates.

2.2.59 Entry-59: Synthesis of Substituted 2-Aminothiazoles

Type of reaction: C–N and C–S bond formation
Reaction conditions: Catalyst-free, water, room temperature
Synthetic strategy: One-pot direct synthesis of 2-aminothiazoles
Keywords: Phenacyl bromides, thioureas, catalyst-free, water, room temperature, one-pot reaction, eco-friendly and clean, substituted 2-aminothiazoles

2.2.59.1 General Reaction Scheme

An efficient one-pot direct synthesis of substituted 2-aminothiazoles (**3**), from the reaction of phenacyl bromides (**1**) with thioureas (**2**) in water at

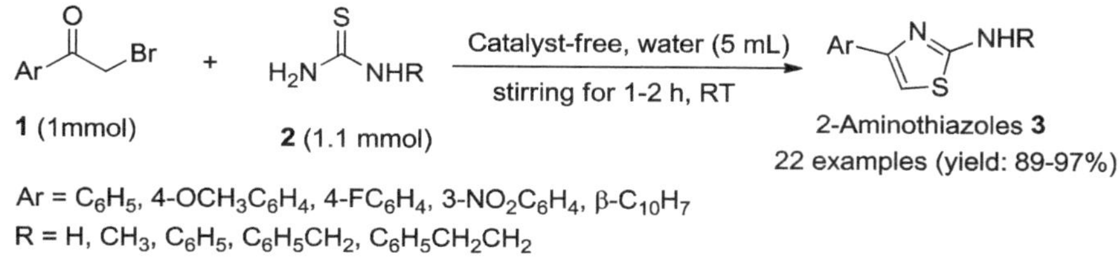

Scheme 2.80 Synthesis of substituted 2-aminothiazoles (**3**).

room temperature without any added catalyst or co-organic solvent, was reported by Srinivasan and his group[186] (Scheme 2.80).

2.2.59.2 Representative Examples

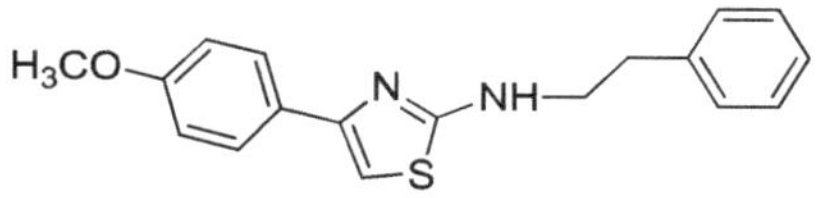

3a

4-(4-Methoxyphenyl)-*N*-phenethylthiazol-2-amine (**3a**): white powder, mp 121–122 °C, reaction time: 1.5 h, yield: 97%

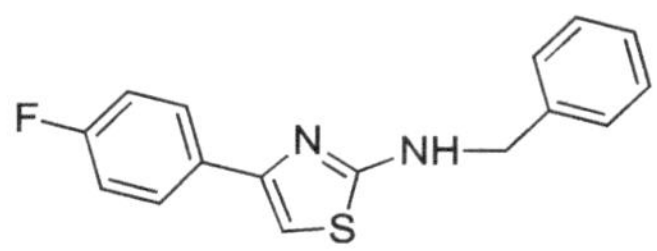

3b

N-Benzyl-4-(4-fluorophenyl)thiazol-2-amine (**3b**): white powder, mp 109–110 °C, reaction time: 1 h, yield: 96%

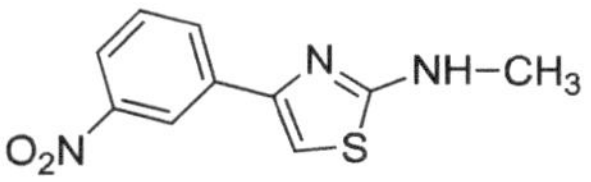

3c

N-Methyl-4-(3-nitrophenyl)thiazol-2-amine (**3c**): orange needles, mp 156–157 °C, reaction time: 2 h, yield: 92%

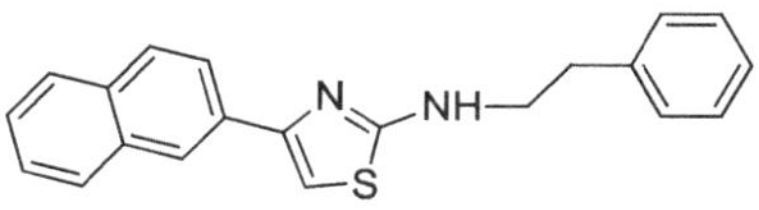

3d

4-(Naphthalen-2-yl)-*N*-phenethylthiazol-2-amine (**3d**): yellow solid, mp 125–126 °C, reaction time: 2 h, yield: 92%

2.2.59.3 Experimental Procedure

A mixture of phenacyl bromide (**1**; 1 mmol) and thiourea (**2**; 1.1 mmol) in water (5 mL) was stirred vigorously at room temperature under for the specified time (1–2 h) and the progress of the reaction was monitored by TLC. After completion of the reaction, the product was extracted using ethyl acetate (2×15 mL). The organic layer was separated from aqueous layer, and the combined organic layers were dried over anhydrous magnesium sulfate and evaporated under reduced pressure to obtain the crude solid product. The crude product was further purified by column chromatography using ethyl acetate–petroleum ether as the eluent to afford the pure product of 2-aminothiazole (**3**) in good to excellent yield. All the compounds were characterized on the basis of analytical and spectral studies.

2.2.59.4 Characterization Data of Two Representative Compounds

4-(4-Methoxyphenyl)-*N*-phenethylthiazol-2-amine (**3a**): white powder; yield: 97%; mp 121–122 °C; IR (KBr): 3228, 3019, 2958, 1549, 1492, 1333, 758 cm^{-1}; ^{1}H NMR ($CDCl_3$, 200 MHz): δ 2.93–3.0 (2H, t, $J = 6.9$ Hz, CH_2), 3.51–3.60 (2H, q, $J = 6.9$ Hz, CH_2N), 3.82 (3H, s, OCH_3), 5.32 (1H, br s, NH), 6.56 (1H, s, thiazole-H), 6.87–6.92 (2H, d, $J = 8.9$ Hz, ArH), 7.21–7.37 (5H, m, ArH), 7.69–7.74 (2H, d, $J = 8.9$ Hz, ArH); ^{13}C NMR ($CDCl_3$, 50 MHz): δ 35.3, 47.1, 55.2, 98.8, 113.8, 126.5, 127.2, 127.8, 128.6, 128.7, 138.4, 151.1, 159.1, 169.5. Anal. calc. for $C_{18}H_{18}N_2OS$: C, 69.65; H, 5.84; N, 9.02%; found: C, 69.42; H, 5.95; N, 9.11%.

N-Methyl-4-(3-nitrophenyl)thiazol-2-amine (**3c**): orange needles; yield: 92%; mp 156–157 °C; IR (KBr): 3431, 3019, 2923, 1591, 1565, 1534, 1517, 1353, 761 cm^{-1}; ^{1}H NMR ($CDCl_3$, 200 MHz): δ 3.02–3.04 (3H, d, $J = 5.12$ Hz, CH_3), 5.39 (1H, br s, NH), 6.86 (1H, s, thiazole-H), 7.48–7.56 (1H, t, $J = 7.9$ Hz, ArH), 8.09–8.13 (2H, dd, $J = 1.9$ and 7.9 Hz, ArH), 8.63–8.65 (1H, t, $J = 1.9$ Hz, ArH); ^{13}C NMR ($CDCl_3$, 50 MHz): δ 32.1, 103.0, 120.8, 122.1, 129.4, 131.7, 136.5, 148.5, 149.2, 170.8. Anal. calc. for $C_{10}H_9N_3O_2S$: C, 51.05; H, 3.86; N, 17.86%; found: C, 51.13; H, 3.77; N, 17.92%.

2.2.59.5 Critical Views

The investigators demonstrated a simple and highly efficient protocol for the synthesis of the biologically relevant heterocycles, 2-aminothiazole derivatives, in water at ambient conditions without any added catalyst or co-organic solvent. This method is superior to the earlier methods[187–193] in terms of avoidance of toxic volatile organic solvents and numerous catalysts employed in organic solvents to synthesize this class of compounds. In addition, the present protocol also offers several advantages including improved yields, simple experimental procedure, cleaner reactions, and low cost, thus making it a useful and attractive strategy in view of economic and environmental concerns. The investigators applied their protocol successfully in preparing fanetizole, a commercial anti-inflammatory drug. They assumed that the unique abilities of water such as hydrogen bonding, high dielectric constant, and polarity are primarily responsible for behaving as a more efficient medium for this reaction.

2.2.60 Entry-60: Synthesis of 5-Hydrazinoethylidene-2-iminothiazolidinones

Type of reaction: C–N and C–S bond formation
Reaction conditions: Catalyst-free, chloroform, room temperature
Synthetic strategy: Sequential three-component one-pot reaction
Keywords: 1,2-Diaza-1,3-dienes, amines, isothiocyanates, chloroform, catalyst-free, room temperature, sequential three-component reaction, one-pot synthesis, 5-hydrazinoethylidene-2-iminothiazolidinones

2.2.60.1 General Reaction Scheme

One-pot and catalyst-free efficient synthesis of functionalized 5-hydrazinoethylidene-2-iminothiazolidinones (**4**) from the sequential thia-Michael addition/condensation reactions of isothiocyanates (**2**) with primary amines (**1**) and 1,2-diaza-1,3-dienes (**3**) under ambient conditions was reported for the first time by Attanasi *et al.*[194] (Scheme 2.81).

2.2.60.2 Plausible Mechanism

A plausible mechanism is suggested, for such a convertion, in Scheme 2.82.[194] Initially, isothiocyanate (**2**) couples with the primary amine (**1**) to produce the corresponding thiourea (**5**), which then undergoes regioselective thia-Michael reaction with 1,2-diaza-1,3-diene (**3**) yielding the adduct (**6**). Intramolecular attack of the NH of the resulting isothiourea derivative (**6**) at the ester function on C-4 of the hydrazone chain, with subsequent loss of an alcohol molecule from the resulting heterocyclic intermediate (**7**), affords the 2-iminothiazolidinone compound (**4**).

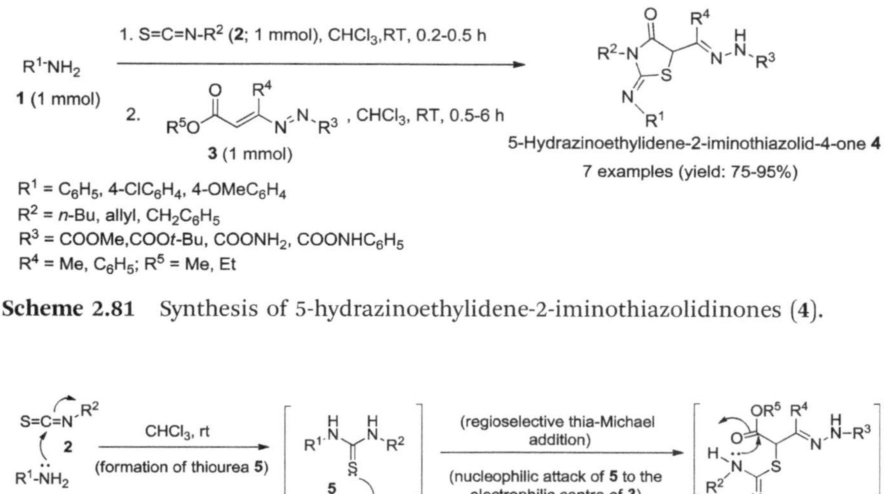

Scheme 2.81 Synthesis of 5-hydrazinoethylidene-2-iminothiazolidinones (**4**).

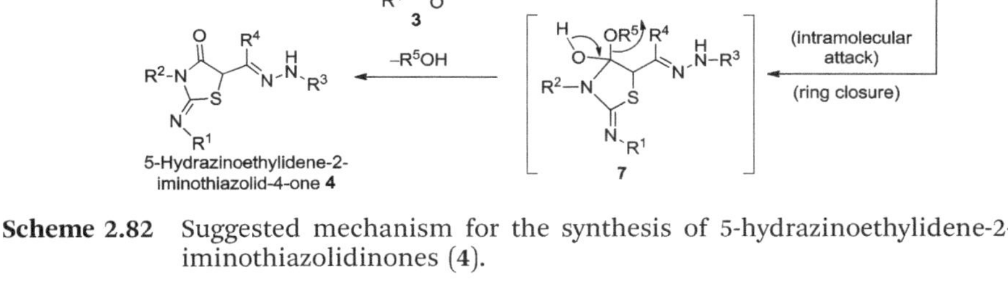

Scheme 2.82 Suggested mechanism for the synthesis of 5-hydrazinoethylidene-2-iminothiazolidinones (**4**).

2.2.60.3 Representative Examples

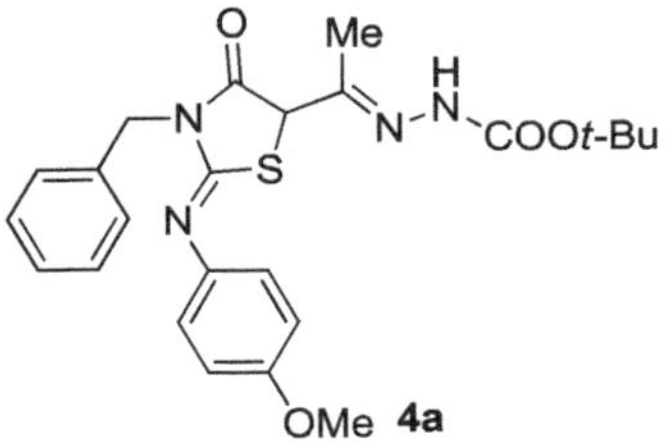
4a

tert-Butyl 2-(1-(3-benzyl-2-((4-methoxyphenyl)imino)-4-oxothiazolidin-5-yl)ethylidene)hydrazinecarboxylate (**4a**): light yellow foam; yield 95%

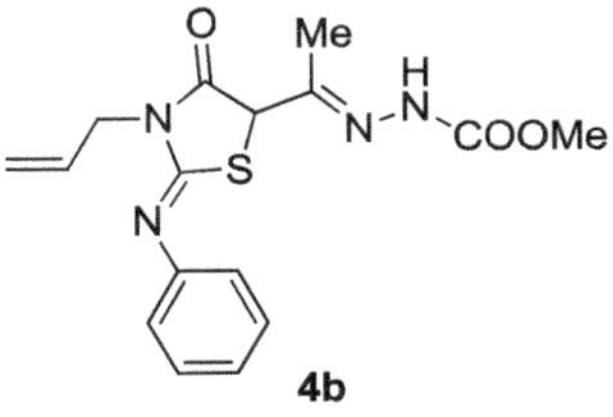
4b

Methyl 2-(1-(3-allyl-4-oxo-2-(phenylimino)thiazolidin-5-yl)ethylidene)hydrazinecarboxylate (**4b**): white powder; yield: 90%

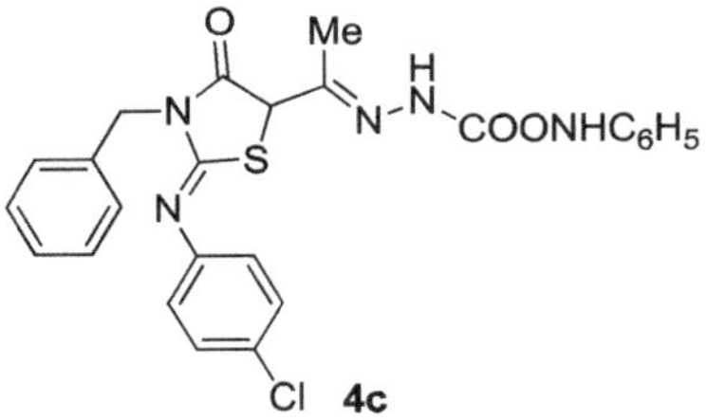
4c

3-Benzyl-2-((4-chlorophenyl)imino)-5-(1-(2-(((phenylamino)oxy)carbonyl)hydrazono)ethyl)thiazolidin-4-one (**4c**): white crystals, yield: 84%

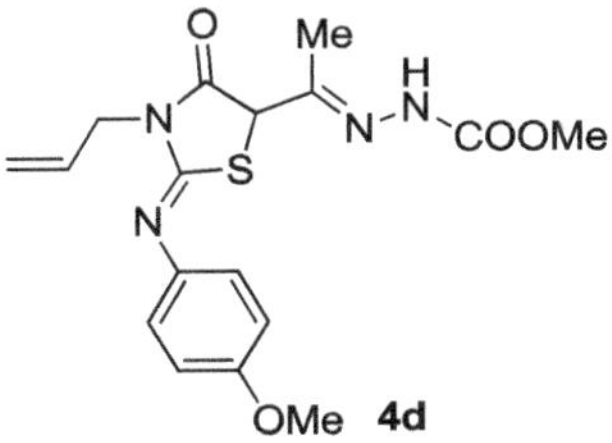
4d

Methyl 2-(1-(3-allyl-2-((4-methoxyphenyl)imino)-4-oxothiazolidin-5-yl)ethylidene)hydrazinecarboxylate (**4d**): light gray powder; yield: 85%

2.2.60.4 Experimental Procedure

Alkylthiocyanate (**2**; 1 mmol) was added to a stirred solution of a primary aromatic amine (**1**: 1 mmol) in chloroform (4 mL) and the mixture was stirred at room temperature. On the disappearance of the reagents forming the thiourea intermediate (within 0.2–0.5 h, as checked by TLC), 1,2-diaza-1,3-diene (**3**) was added, either solid or dissolved in $CHCl_3$ (1 mL), to the reaction vessel and the reaction mixture was allowed to stand at room temperature until the completion of the reaction (0.5–6 h). The solvent was then removed under reduced pressure and the crude reaction mixture was purified by column chromatography (cyclohexane–ethyl acetate mixtures) to obtain 5-hydrazinoethylidene-2-thiazolidin-4-one (**4**) as a mixture of C_5 *E*/*Z* isomers (approximately ratio 3 : 1). In the case of solid products, they were obtained by precipitation from the reaction medium and subsequent recrystallization from the appropriate solvents. All the products were characterized on the basis of elemental analyses and detailed spectral studies.

2.2.60.5 Characterization Data of Two Representative Compounds

3-Benzyl-2-((4-chlorophenyl)imino)-5-(1-(2-(((phenylamino)oxy)carbonyl)hydrazono) ethyl)thiazolidin-4-one (**4c**): white crystals; IR (nujol): 3276, 3201, 3099, 1711, 1698, 1652, 1629, 1603, 1573, 1547 cm^{-1}; ^1H NMR (DMSO-d_6, 400 MHz): δ 1.82 and 2.41 (3H, 2s), 4.94 and 4.99 (2H, 2s), 6.96 – 7.01 (2H, m), 7.25 – 7.55 (12H, m), 8.48 (1H, br s), 8.85 (1H, s), 8.94 and 9.99 (1H each, br s and s); ^{13}C NMR (DMSO-d_6, 100 MHz): δ 12.5 (q), 16.9 (q), 44.3 (t), 44.6 (t), 84.0 (s), 85.9 (s), 118.7 (d), 119.2 (d), 122.3 (d), 122.7 (d), 123.0 (d), 123.1 (d), 127.3 (d), 127.4 (d), 127.5 (s), 127.6 (d), 127.9 (s), 128.3 (d), 128.4 (d), 128.5 (d), 128.7 (d), 129.0 (d), 129.3 (d), 136.6 (s), 137.1 (s), 139.1 (s), 139.3 (s), 147.6 (s), 147.9 (s), 152.9 (s), 153.4 (s), 155.1 (s), 155.4 (s), 165.3 (s), 166.0 (s); MS (EI): *m/z* at 493 [$M^+ + 2$ (5)], 491 (M^+, 13), 398 (52), 274 (100), 241 (57), 179 (79), 119 (83). Anal. calc. for $C_{25}H_{22}ClN_5O_2S$ (491.99): C, 61.03; H, 4.51; N, 14.23; found: C, 61.12; H, 4.39; N, 14.32.

Methyl 2-(1-(3-allyl-2-((4-methoxyphenyl)imino)-4-oxothiazolidin-5-yl)ethylidene) hydrazinecarboxylate (**4d**): light gray powder; IR (nujol): 3317, 3290, 1742, 1653, 1616, 1585 cm^{-1}; ^1H NMR (DMSO-d_6, 400 MHz): δ 1.86 and 2.34 (3H, 2s), 3.58 and 3.62 (3H, 2s), 3.73 (3H, s), 4.33–4.38 (2H, m), 5.09–5.17 (2H, m), 5.84–5.94 (1H, m), 6.83–6.92 (4H, m), 8.75 and 9.82 (1H each, br and s), 9.52 (1H, br s); ^{13}C NMR (DMSO-d_6, 100 MHz): δ 16.2 (q), 16.7 (q), 43.4 (t), 43.5 (t), 52.4 (q), 55.2 (q), 84.7 (s), 86.7 (s), 114.4 (d), 114.6 (d), 116.5 (t), 116.7 (t), 122.1 (d), 122.3 (d), 132.0 (d), 132.4 (d), 141.9 (s), 142.1 (s), 150.8 (s), 155.7 (s), 155.9 (s), 156.7 (s), 156.8 (s), 165.4 (s), 165.8 (s); MS (EI): *m/z* at 376 (M^+, 46), 344 (8), 287 (2), 189 (36), 147 (100), 101 (53). Anal. calc. for $C_{17}H_{20}N_4O_4S$ (376.43): C, 54.24; H, 5.36; N, 14.88; found: C, 54.16; H, 5.49; N, 14.79.

2.2.60.6 Critical Views

The present work offers a synthetic strategy for regioselective functionalized 5-hydrazinoethylidene-2-iminothiazolidinones out of sequential addition of primary amine, isothiocyanate, and 1,2-diaza-1,3-diene, respectively under mild reaction conditions in the absence of any catalyst for the first time. The hydrazone side chain introduced by the conjugated heterodiene system at the 5-position of the heterocycle represents a valuable functionality for accessing novel 5-acyl derivatives difficult to obtain by other methods. The mild and simple reaction conditions make the protocol suitable for the generation of a library of functionalized 2-iminothiazolidin-4-ones.

2.2.61 Entry-61: Synthesis of Functionalized 2-Aminothiophenes

Types of reaction: C–C and C–S bond formation
Reaction conditions: Catalyst-free, dichloromethane, room temperature
Synthetic strategy: One-pot multicompoment reaction

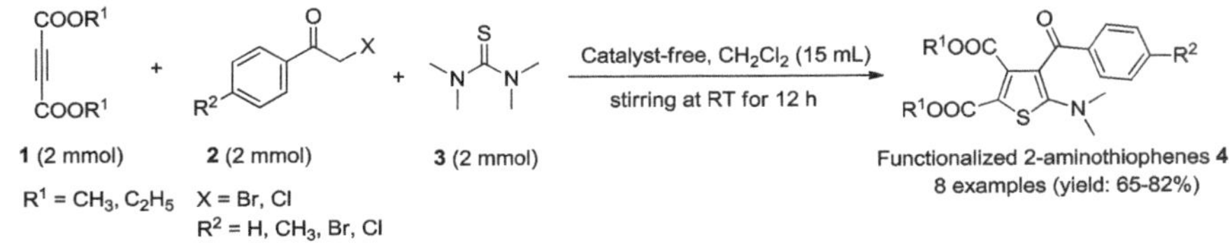

Scheme 2.83 Synthesis of functionalized 2-aminothiophenes (**4**).

Keywords: Diethyl/dimethyl acetylenedicarboxylate, phenacyl halides, tetramethylthiourea, dichloromethane, catalyst-free, room temperature, one-pot multicomponent reaction, functionalized 2-aminothiophenes

2.2.61.1 General Reaction Scheme

Yavari and Malekafzali[195] synthesized a series of functionalized 2-dimethylaminothiophenes (**4**) in good yields from the one-pot three-component reaction of dialkyl acetylenedicarboxylates (**1**), α-haloketones (**2**) and tetramethylthiourea (**3**) at room temperature (Scheme 2.83).

2.2.61.2 Plausible Mechanism

Mechanistically, it was assumed that the reaction involves the initial formation of a 1,5-dipolar intermediate (**5**) between tetramethylthiourea (**3**) and the acetylenic compound (**1**), which reacts with the α-haloketone (**2**) to produce adduct (**6**). Intermediate **6** is then converted to the 1,6-diionic compound (**7**) *via* elimination of HX. Cyclization of this intermediate leads to **8**, which is converted to the product **4** by elimination of dimethylamine (Scheme 2.84).

2.2.61.3 Representative Examples

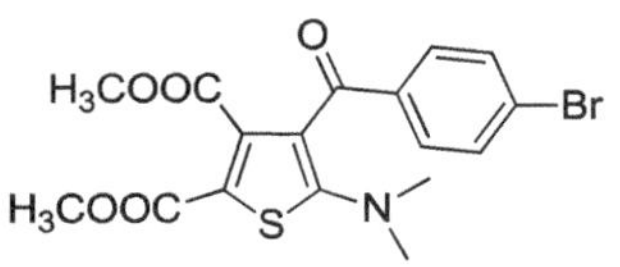

4a

Dimethyl 4-(4-bromobenzoyl)-5-(dimethylamino)thiophene-2,3-dicarboxylate (**4a**): yellow oil; yield: 70%

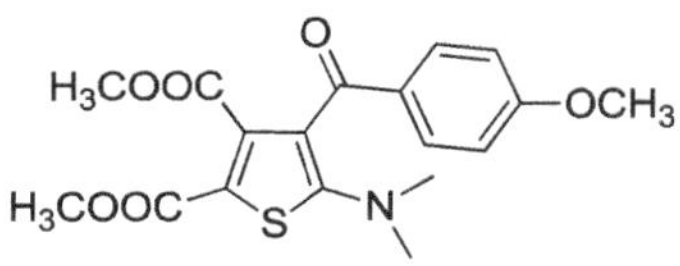

4b

Dimethyl 5-(dimethylamino)-4-(4-methoxybenzoyl)thiophene-2,3-dicarboxylate (**4b**): yellow oil; yield: 73%

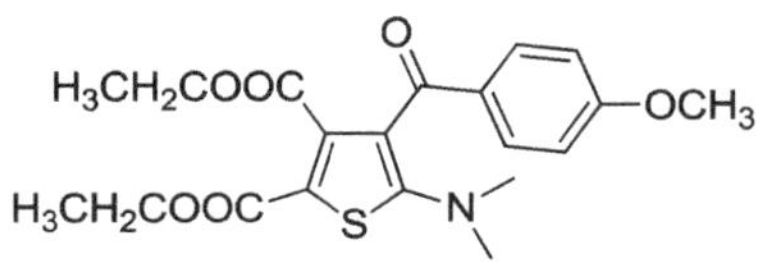

4c

Diethyl 5-(dimethylamino)-4-(4-methoxybenzoyl)thiophene-2,3-dicarboxylate (**4c**): yellow oil; yield: 70%

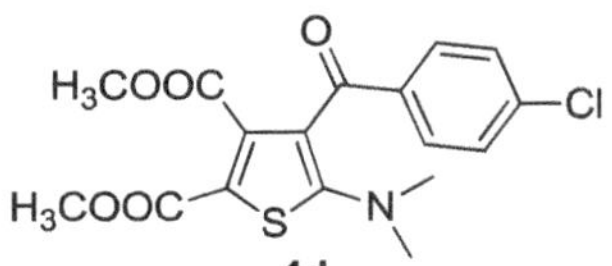

4d

Dimethyl 4-(4-chlorobenzoyl)-5-(dimethylamino)thiophene-2,3-dicarboxylate (**4d**): yellow oil; yield: 82%

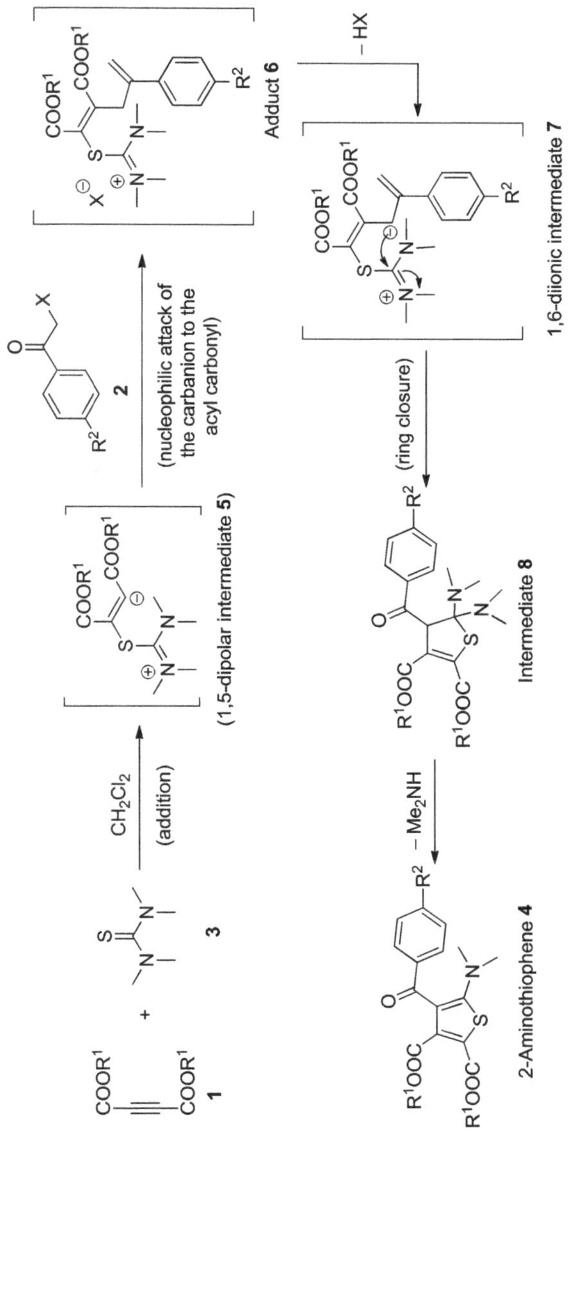

Scheme 2.84 Plausible mechanism for the synthesis of functionalized 2-aminothiophenes (4).

2.2.61.4 Experimental Procedure

To a stirred solution of acetylenic dicarboxylate (**1**; 2 mmol) and phenacyl halide (**2**; 2 mmol) in dichloromethane (15 mL), tetramethylthiourea (**3**; 0.26 g; 2 mmol) was added at room temperature, and the reaction mixture was stirred for 12 h. On completion of the reaction, the solvent was evaporated off under reduced pressure to produce a crude residue that was purified by column chromatography (SiO_2; *n*-hexane–EtOAc 6 : 1) to afford pure products of functionalized 2-aminothiophenes (**4**). All the products were characterized by IR, ^{1}H NMR, ^{13}C NMR, and EIMS, as well as elemental analyses.

2.2.61.5 Characterization Data of Two Representative Entries:

Dimethyl 4-(4-bromobenzoyl)-5-(dimethylamino)thiophene-2,3-dicarboxylate (**4a**): yellow oil; IR (KBr): 1734, 1630, 1542, 1436, 1260 cm^{-1}; ^{1}H NMR ($CDCl_3$, 500 MHz): δ 2.81 (6H, s, $(CH_3)_2N$), 3.86 (3H, s, CH_3O), 3.97 (3H, s, CH_3O), 7.62 (2H, d, ^3J = 8.6 Hz, Ar–H), 7.70 (2H, d, ^3J = 8.6 Hz, Ar–H); ^{13}C NMR ($CDCl_3$, 125 MHz): δ 43.3 ($(CH_3)_2N$), 52.8 (CH_3O), 53.1 (CH_3O), 124.9 (C), 127.9 (C), 130.6 (CH), 131.7 (CH), 133.3 (C), 134.4 (C), 138.1 (C), 153.2 (C), 162.2 (C=O), 164.2 (C=O), 186.0 (C=O); EIMS: *m/z* (rel. int.%) at 426 (M^+, 7), 184 (100), 170 (54), 156 (60), 59 (47), 44 (23). Anal. calc. for $C_{17}H_{16}BrNO_5S$ (426.28): C, 47.90; H, 3.78; N, 3.29; found: C, 48.1; H, 3.9; N, 3.4%.

Diethyl 5-(dimethylamino)-4-(4-methoxybenzoyl)thiophene-2,3-dicarboxylate (**4c**): yellow oil; IR (KBr): 1732, 1598, 1541, 1418, 1260 cm^{-1}; ^{1}H NMR ($CDCl_3$, 500 MHz):δ 1.34 (3H, t, ^3J = 7.1 Hz, CH_3), 1.41 (3H, t, ^3J = 7.1 Hz, CH_3), 2.88 (6H, s, $(CH_3)_2N$), 3.89 (3H, s, CH_3O), 4.32 (2H, q, ^3J = 7.1 Hz, CH_2O), 4.43 (2H, q, ^3J = 7.1 Hz, CH_2O), 6.96 (2H, d, ^3J = 8.8 Hz, CH), 7.86 (2H, d, ^3J = 8.8 Hz, CH); ^{13}C NMR ($CDCl_3$, 125 MHz): δ 13.9 (CH_3), 14.1 (CH_3), 43.3 ($(CH_3)_2N$), 55.5 (CH_3O), 61.9 (CH_2O), 62.2 (CH_2O), 113.7 (CH), 125.3 (C), 131.4 (C), 131.7 (CH), 133.7 (C), 134.7 (C), 153.0 (C), 160.7 (C), 163.6 (C=O), 165.7 (C=O), 186.0 (C=O); EIMS: *m/z* (rel. int.%) at 405 (M^+, 8), 135 (100), 121 (47), 107 (55), 73 (38), 45 (41), 44 (19); Anal. calc. for $C_{20}H_{23}NO_6S$ (405.46): C, 59.24; H, 5.72; N, 3.45; found: C, 59.0; H, 5.9; N, 3.3%.

2.2.61.6 Critical Views

This catalyst-free, one-pot synthetic method for such *S*-containing heterocyclic compounds of potential interest seems facile under neutral conditions at room temperature, and the starting material can be used without any activation or modification.

2.2.62 Entry-62: Synthesis of 1,3,4-Selenadiazines

Type of reaction: C–N and C–Se bond formation
Reaction conditions: Catalyst-free, dichloromethane, room temperature
Synthetic strategy: One-pot multicomponent reaction

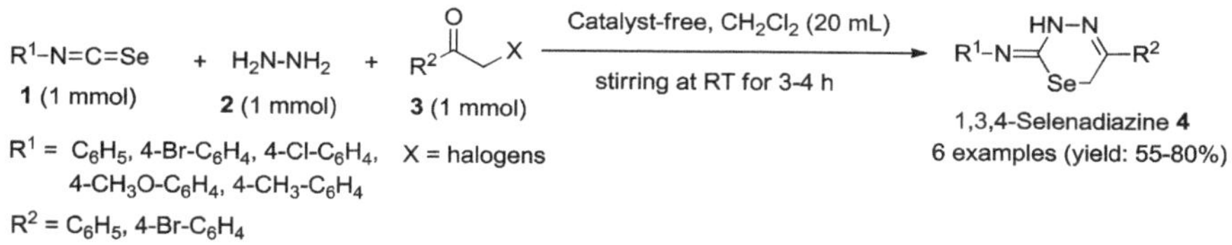

Scheme 2.85 Synthesis of 1,3,4-selenadiazines (**4**).

Keywords: Hydrazine hydrate, phenacyl halides, aryl isoselenocyanates, dichloromethane, catalyst-free, room temperature, one-pot multi-component, 1,3,4-selenadiazines

2.2.62.1 General Reaction Scheme

Sommen *et al.*[196] developed a convenient and useful protocol for the efficient synthesis of 1,3,4-selenadiazines (**4**) *via* a one-pot three-component reaction between various aryl isoselenocyanates (**1**) and different phenacyl halides (**3**) in the presence of hydrazine hydrate (**2**) in dichloromethane at room temperature (Scheme 2.85).

2.2.62.2 Representative Examples

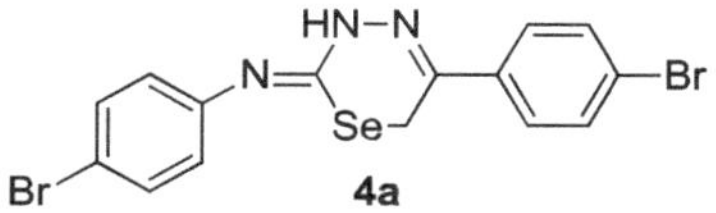

4-Bromo-*N*-(5-(4-bromophenyl)-3,6-dihydro-2*H*-1,3,4-selenadiazin-2-ylidene)aniline (**4a**): yellowish crystals, mp 176–178 °C; yield: 80%

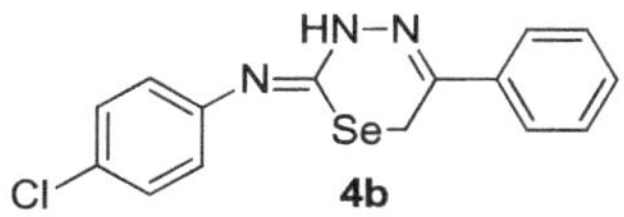

4-Chloro-*N*-(5-phenyl-3,6-dihydro-2*H*-1,3,4-selenadiazin-2-ylidene)aniline (**4b**): yellowish crystals, mp 178–180 °C; yield: 55%

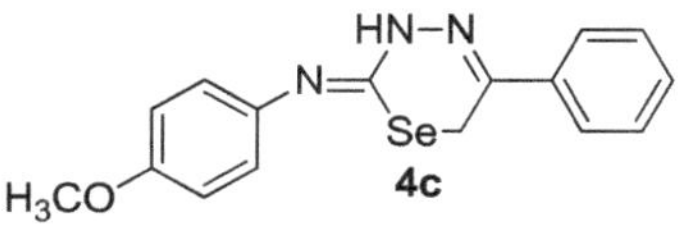

4-Methoxy-*N*-(5-phenyl-3,6-dihydro-2*H*-1,3,4-selenadiazin-2-ylidene)aniline (**4c**): yellowish crystals, mp 132–134 °C; yield: 67%

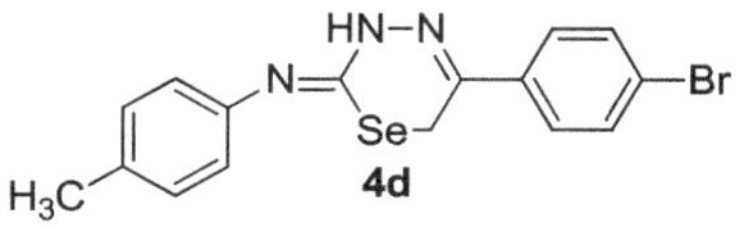

N-(5-(4-Bromophenyl)-3,6-dihydro-2*H*-1,3,4-selenadiazin-2-ylidene)-4-methylaniline (**4c**): yellowish crystals, mp 202–204 °C; yield: 78%

2.2.62.3 Experimental Procedure

A 25 mL round bottom flask equipped with a magnetic stirrer and condenser was charged with a mixture of isoselenocyanates (**4**; 1.0 mmol) and phenacyl halides (**3**; 1.0 mmol) in dichloromethane (20 mL). Then, hydrazine hydrate (**2**; 0.05 mL, 1.0 mmol) was added in one portion, and the mixture was stirred for 3 to 4 h at room temperature. Upon completion of the reaction, the solvent was removed under reduced pressure, and the crude product was purified by column chromatography (silica gel; hexane–ethyl acetate 100 : 0 to 50 : 50) to obtain pure 1,3,4-selenadiazines (**4**). All the products were characterized based on spectral and elemental analyses.

2.2.62.4 Characterization Data of Two Representative Compounds

4-Bromo-*N*-(5-(4-bromophenyl)-3,6-dihydro-2*H*-1,3,4-selenadiazin-2-ylidene)-aniline (**4a**): yellowish crystals, mp 176–178 °C; yield: 80%; IR: 3442 (m), 3155 (w), 3051(w), 2920 (m), 1626 (m), 1590 (vs), 1576 (s), 1554 (m), 1485 (m), 1407 (w), 1299 (w), 1271 (w), 1209 (m), 1172 (m), 1146 (w), 1101 (w), 1070 (m), 1000 (m), 890 (w), 828 (m), 725 (w), 707 (w), 653 (w), 604 (w) cm^{-1}; ^{1}H NMR (DMSO-d_6, 300 MHz): δ 3.88 (2H, s, CH_2), 6.85–7.10 (2H, br. m, Ar–H), 7.68 (2H, d-like, J = 8.2 Hz, Ar–H), 7.80 (2H, d-like, J = 8.2 Hz, Ar–H), 7.90–8.00 (2H, m, Ar–H), 11.79 (1H, br s, NH); ^{13}C NMR (DMSO-d_6, 75.5 MHz): δ 15.0 (t, CH_2), 122.8 (s, 2 arom. C), 124.2 (d, 2 arom. CH), 128.1 (d, 2 arom. CH), 131.3 (d, 2 arom. CH), 131.5 (d, 2 arom. CH), 134.5 (s, arom. C), 147.2 (s, arom. C), 155.5 (s, C(5)), 162.9 (s, C(2)); CIMS (rel.%): *m/z* 478 (7), 477 (9), 476 (52), 475 (19), 474 (100, $[M(^{80}Se,^{81}Br,^{79}Br) + 1]^+$), 473 (23), 472 (85, $[M(^{80}Se,^{79}Br,^{79}Br) + 1]^+$), 471 (19), 470 (35), 469 (8), 468 (9). Anal. calc. for $C_{15}H_{11}Br_2N_3Se$: C, 38.17; H, 2.35; N, 8.90; found: C, 38.01; H, 2.54; N 8.60.

4-Methoxy-*N*-(5-phenyl-3,6-dihydro-2*H*-1,3,4-selenadiazin-2-ylidene)aniline (**4c**): yellowish crystals, mp 132–134 °C; yield: 67%; IR: 3439 (m), 3346 (m), 2912 (m), 2836 (w), 1654 (s), 1638 (m), 1580 (vs), 1544 (s), 1509 (vs), 1447 (w), 1282 (m), 1249 (s), 1211 (w), 1178 (w), 1109 (w), 1077 (w), 1033 (w), 1011 (w), 892 (w), 826 (m), 800 (w), 757 (w), 713 (m), 692 (w) cm^{-1}; ^{1}H NMR (DMSO-d_6, 300 MHz): δ 3.82 (3H, s, *Me*O), 3.90 (2H, s, CH_2), 6.90–7.20 (4H, br m, d-like at 6.92, J = 8.2 Hz, Ar–H), 7.30–7.55 (2H, m, Ar–H), 7.75–8.00 (2H, m, Ar–H), 11.82 (1H, br s, NH); ^{13}C NMR (DMSO-d_6, 75.5 MHz): δ 15.1 (t, CH_2), 55.4 (q, MeO), 115.6 (d, 2 arom. CH), 124.2 (d, 2 arom. CH), 128.3 (d, 2 arom. CH), 129.2 (d, 2 arom. CH), 131.1 (d, 1 arom. CH), 133.9 (s, arom. C), 147.8 (s, arom. C), 153.7 (s, arom. C), 158.4 (s, C(5)), 166.1 (s, C(2)); CIMS (rel.%): *m/z* 350 (8), 349 (12), 348 (65), 347 (21), 346 (100, $[M(^{80}Se) + 1]^+$), 345 (19), 344 (52), 343 (15), 342 (14). Anal. calc. for $C_{16}H_{15}N_3OSe$: C, 55.82; H, 4.39; N, 12.21; found: C, 55.95; H, 4.67; N, 12.23.

2.2.62.5 Critical Views

Selenium-containing heterocycles create considerable interest because of their antitumor, antibacterial, and other biological and pharmaceutical activities. The present catalyst-free room temperature protocol offers a simple and efficient method for the synthesis of such selenium compounds

(1,3,4-selenadiazines) starting from isoselenocyanates for the first time with ease of workup and good yields.

2.2.63 Entry-63: Synthesis of Pyrazolyl 4*H*-Chromene Derivatives

Type of reaction: C–C, C–N and C–O bond formation
Reaction conditions: Catalyst-free, water, room temperature
Synthetic strategy: One-pot multicomponent reaction
Keywords: Salicylaldehydes, malononitrile, hydrazine hydrate, ethyl acetoacetate, aqueous medium, catalyst-free, room temperature, tandem Knoevenagel condensation and Michael addition, 6-*exo-dig* cyclization, one-pot four-component reaction, 2-amino-4(5-hydroxy-3-methyl, 1*H*-pyrazol-4-yl)-4*H*-chromene-3-carbonitriles, chromatography-free

2.2.63.1 General Reaction Scheme

A catalyst-free synthesis of diverse 2-amino-4-(5-hydroxy-3-methyl-1*H*-pyrazol-4-yl)-4*H*-chromene-3-carbonitrile derivatives (**5**) through a one-pot, four-component reaction between hydrazine hydrate (**1**), ethyl acetoacetate (**2**), 2-hydroxybenzaldehydes (**3**) and malononitrile (**4**) in water at room temperature was outlined by Kumaravel and Vasuki[197] (Scheme 2.86).

2.2.63.2 Plausible Mechanism

The authors proposed a plausible mechanism (Scheme 2.87) for the synthesis, which involves a series of the tandem reactions: (i) reaction between hydrazine hydrate (**1**) and ethyl acetoacetate (**2**) resulting in instantaneous formation of pyrazolone (**6**) that tautomerizes to **7**, (ii) Knoevenagel condensation between 2-hydroxybenzaldehyde (**3**) and malononitrile (**4**) forming 2-imino-2*H*-chromene-3-carbonitrile intermediate (**8**) by 6-*exo-dig* cyclization, and (iii) Michael addition of **7** to **8**, followed by subsequent rearrangement. The overall reaction is shown step-wise:

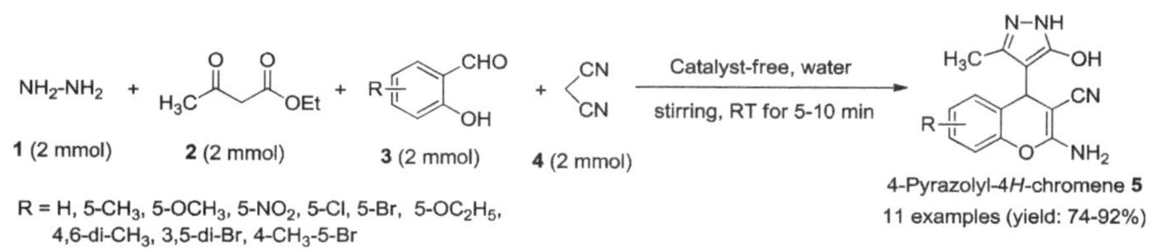

Scheme 2.86 Synthesis of pyrazolyl 4*H*-chromene derivatives (**5**).

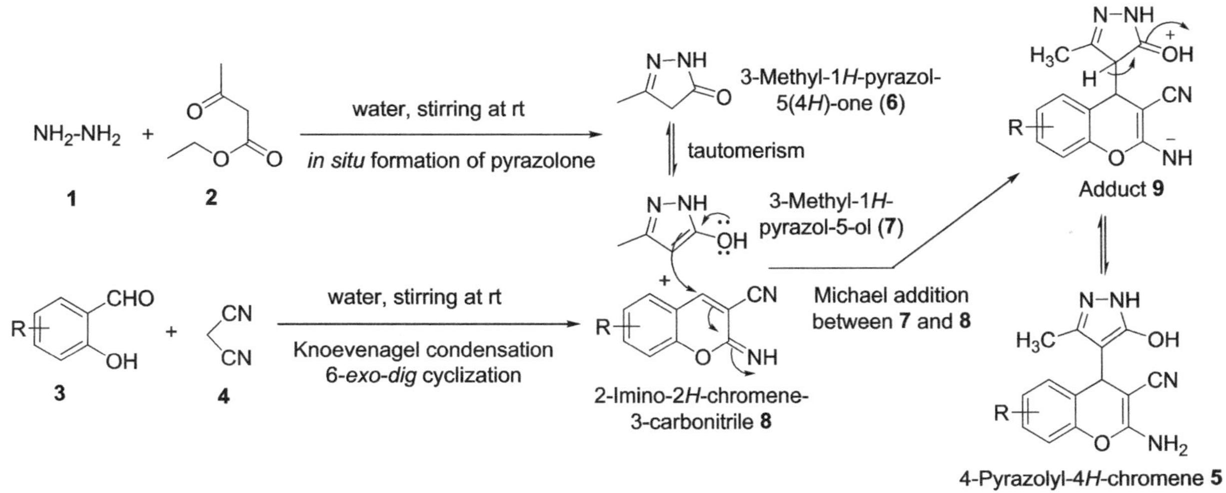

Scheme 2.87 Plausible mechanism for the synthesis of pyrazolyl 4*H*-chromene derivatives (5).

2.2.63.3 Representative Examples

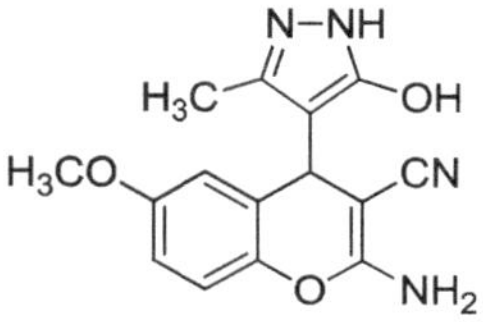

5a

2-Amino-4-(5-hydroxy-3-methyl-1*H*-pyrazol-4-yl)-6-methoxy-4*H*-chromene-3-carbonitrile (**5a**): white solid; yield: 78%

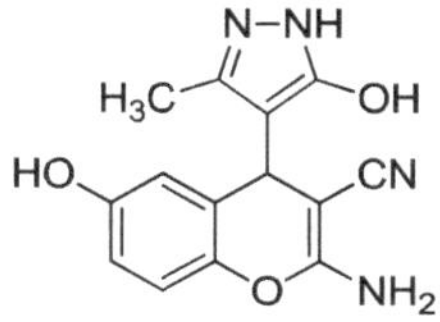

5b

2-Amino-6-hydroxy-4-(5-hydroxy-3-methyl-1*H*-pyrazol-4-yl)-4*H*-chromene-3-carbonitrile (**5b**): white solid; yield: 79%

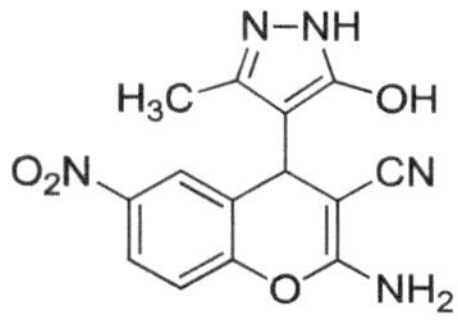

5c

2-Amino-4-(5-hydroxy-3-methyl-1*H*-pyrazol-4-yl)-6-nitro-4*H*-chromene-3-carbonitrile (**5c**): white solid; yield: 86%

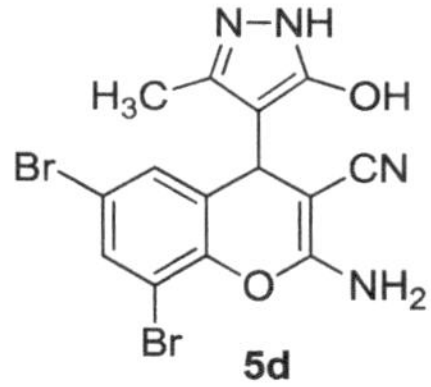

5d

2-Amino-6,8-dibromo-4-(5-hydroxy-3-methyl-1*H*-pyrazol-4-yl)-4*H*-chromene-3-carbonitrile (**5d**): white solid; yield: 74%

2.2.63.4 Experimental Procedure

To a stirred aqueous mixture of hydrazine hydrate (98%) (**1**; 0.107 g, 2 mmol) and ethyl acetoacetate (**2**; 0.260 g, 2 mmol), salicylaldehyde/its derivative (**3**; 2 mmol) and malononitrile (**4**; 0.132 g, 2 mmol) were added successively at room temperature under an open atmosphere with vigorous stirring for 5–10 min (TLC monitored). On completion of the reaction, the solid precipitate was filtered, washed with water and then with 5 mL of ethyl acetate–hexane mixture (1 : 1) to obtain pure product of 2-amino-4(5-hydroxy-3-methyl, 1*H*-pyrazol-4-yl)-4*H*-chromene-3-carbonitrile (**5**). Each of the products was characterized by IR, ^{1}H NMR, ^{13}C NMR and LCMS.

2.2.63.5 Characterization Data of Two Representative Compounds

2-Amino-6-hydroxy-4-(5-hydroxy-3-methyl-1*H*-pyrazol-4-yl)-4*H*-chromene-3-carbonitrile (**5b**): white solid; yield: 79%; IR (KBr): 3448, 3344, 2191, 1652, 1596, 1508, 1462, 1409 cm^{-1}; ^{1}H NMR (DMSO-d_6, 400 MHz): δ 1.93 (3H, s, –CH_3), 4.56 (1H, s, 4-CH), 6.39 (1H, d, Ar–H), 6.52 (2H, s, –NH_2), 6.54–6.57 (1H, m, Ar–H), 6.77 (1H, d, Ar–H), 9.17 (1H, s, –OH), 10.43 (2H, br s, –OH & –NH); ^{13}C NMR ($CDCl_3$, 100 MHz): δ 9.9, 28.8, 54.6, 104.8, 114.2, 114.5, 116.2, 121.0, 124.3, 136.6, 141.3, 153.6, 159.0 and 160.4; LCMS: m/z 285.4 $(M+1)^+$.

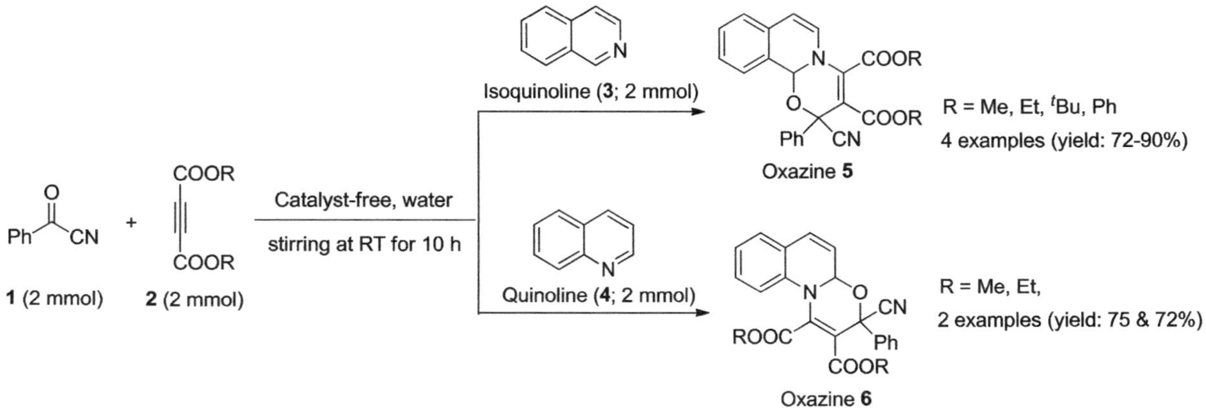

Scheme 2.88 Synthesis of oxazines (5/6).

2-Amino-6,8-dibromo-4-(5-hydroxy-3-methyl-1*H*-pyrazol-4-yl)-4*H*-chromene-3-carbonitrile (**5d**): white solid; yield: 74%; IR (KBr): 3448, 3344, 2191, 1652, 1596, 1508, 1462, 1408 cm^{-1}; ^{1}H NMR (DMSO-d_6, 400 MHz): δ 2.03 (3H, s, $-CH_3$), 4.6 (1H, s, 4-CH), 6.82 (2H, s, $-NH_2$), 7.26 (1H, s, Ar–H), 7.94 (1H, s, Ar–H), 9.75 (1H, br s, -*OH*) and 11. 16 (1H, br s, –NH); ^{13}C NMR ($CDCl_3$, 100 MHz): δ 9.6, 29.0, 55.2, 104.2, 110.2, 115.7, 120.1, 127.2, 130.6, 132.9, 136.5, 144.9, 158.9 and 159.4; LCMS: *m*/*z* 427.0 $(M+3)^+$.

2.2.63.6 Critical Views

The present method offers, for the first time, a catalyst-free multicomponent reaction for a combinatorial synthesis of novel highly functionalized 4-pyrazolyl-4*H*-chromene frameworks in water at ambient temperature in good to excellent yields, satisfying several green aspects such as high atom-economy, one-pot synthesis in water thus avoiding the use of conventional volatile organic solvents, ethanol and water as by-products, avoidance of catalyst and exhaustive column chromatographic purification of products. A variety of 2-hydroxybenzaldehyde were also found to undergo the reaction smoothly.

2.2.64 Entry-64: Synthesis of Oxazines

Type of reaction: C–C, C–N and C–O bond formation
Reaction conditions: Catalyst-free, water, room temperature
Synthetic strategy: One-pot multicomponent reaction
Keywords: Acetylenic esters, benzoyl cyanide, quinoline, isoquinoline, water, room temperature, one-pot multicomponent reaction, oxazines

2.2.64.1 General Reaction Scheme

A one-pot multicomponent protocol for direct synthesis of oxazine derivatives (**4**) in water was reported by Rostami-Charati and co-workers[198] from the reaction of benzoyl cyanide (**1**), acetylenic ester (**2**) and an *N*-nucleophile (such as isoquinoline **3** and quinoline **4**) in the absence of any catalyst at room temperature (Scheme 2.88).

2.2.64.2 Representative Examples

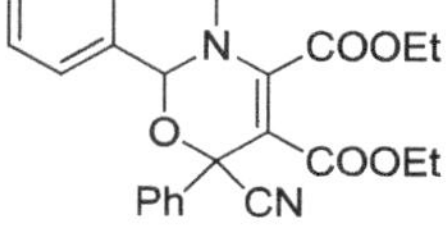

5a

Diethyl 2-cyano-2-phenyl-2,11b-dihydro-[1,3]oxazino[2,3-*a*]isoquinoline-3,4-dicarboxylate (**5a**): white powder, mp 148–150 °C; yield: 72%

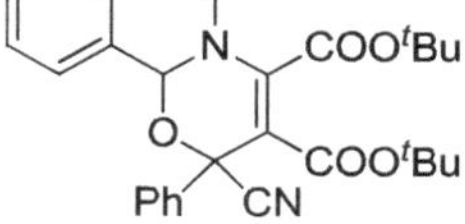

5b

Di-*tert*-butyl 2-cyano-2-phenyl-2,11b-dihydro-[1,3]oxazino[2,3-*a*]isoquinoline-3,4-dicarboxylate (**5b**): yellow powder, mp 69–71 °C; yield: 75%

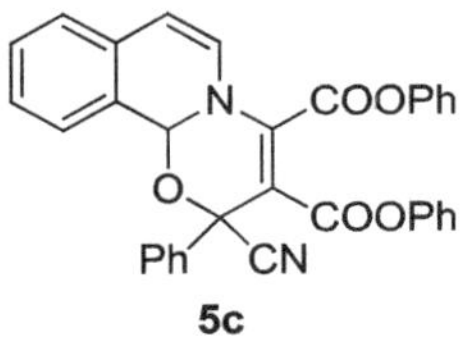

5c

Diphenyl 2-cyano-2-phenyl-2,11b-dihydro-[1,3]oxazino[2,3-*a*]isoquinoline-3,4-dicarboxylate (**5c**): yellow powder, mp 100–102 °C; yield: 82%

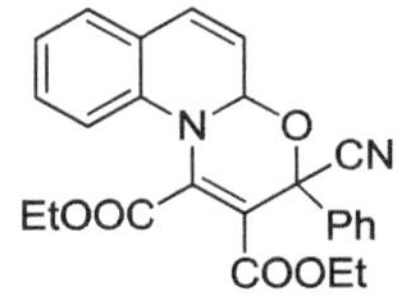

6a

Diethyl 3-cyano-3-phenyl-3,4a-dihydro-[1,3]oxazino[3,2-*a*]quinoline-1,2-dicarboxylate (**6a**): yellow powder, mp 150–152 °C; yield: 72%

2.2.64.3 Experimental Procedure

To a stirred solution of benzoyl cyanide (**1**; 2 mmol) and dialkyl acetylenedicarboxylate (**2**; 2 mmol) in water (10 mL), isoquinoline (**3**; 2 mmol) or quinoline (**4**; 2 mmol)was added and the reaction continued to be stirred for 10 h at room temperature. The completion of the reaction was confirmed by TLC (EtOAc–hexane 6 : 1). The resulting precipitate was separated by filtration and recrystallized from ethanol to afford pure oxazine derivatives (**5/6**). All the compounds were characterized on the basis of analytical and spectral studies.

2.2.64.4 Characterization Data of Two Representative Compounds

Di-*tert*-butyl 2-cyano-2-phenyl-2,11b-dihydro-[1,3]oxazino[2,3-*a*]isoquinoline-3,4-dicarboxylate (**5b**): yellow powder; yield: 75%; mp 69–71 °C; IR (KBr): 1726, 1696 (2 C=O), 1280, 1230, 1138, 763 cm^{-1}; ^{1}H NMR ($CDCl_3$, 500 MHz): δ 1.29 (9H, s, $C(CH_3)_3$), 1.67 (9H, s, $C(CH_3)_3$), 5.78 (1H, d, $^3J = 7.4$, CH), 5.80 (1H, s, CH), 6.43 (1H, d, $^3J = 7.0$, CH), 6.82 (1H, d, $^3J = 6.7$, CH), 7.05 (1H, d, $^3J = 6.7$, CH), 7.12 (1H, t, $^3J = 6.7$, CH), 7.25–7.26 (1H, m, CH), 7.35–7.36 (1H, m, CH), 7.48–7.50 (3H, m, 3×CH), 7.71–7.72 (1H, m, CH); ^{13}C NMR ($CDCl_3$,125 MHz): δ 27.7 (C(C $H_3)_3$), 27.8 ($C(CH_3)_3$), 75.1 (C), 78.0 (CH), 82.5 ($C(CH_3)_3$), 84.9 ($C(CH_3)_3$), 103.3 (C), 105.0 (CH), 119.0 (CN), 122.8 (CH), 125.1 (C), 125.2 (CH), 126.6 (C), 127.1 (CH), 127.4 (CH), 128.2 (2 CH), 128.7 (2 CH), 129.6 (CH), 129.7 (CH), 139.4 (C), 146.2 (C), 161.5 (C=O), 162.2 (C=O). Anal. calc. for $C_{29}H_{30}N_2O_5$ (486.56): C, 71.59; H, 6.21; N, 5.76%; found: C, 71.68; H, 6.24; N, 5.80%.

Diethyl 3-cyano-3-phenyl-3,4a-dihydro-[1,3]oxazino[3,2-*a*]quinoline-1,2-dicarboxylate (**6a**): yellow powder; yield: 72%; mp 150–152 °C; IR (KBr): 1738, 1725, 1715, 1658, 1458, 1325, 1245 cm^{-1}; ^{1}H NMR ($CDCl_3$, 500 MHz): δ 1.25 (3H, t, $^3J = 7.3$ Hz, CH_3), 1.32 (3H, t, $^3J = 7.3$ Hz, CH_3), 4.20 (2H, q, $^3J = 7.3$ Hz, CH_2O), 4.40 (2H, q, $^3J = 7.3$ Hz, OCH_2), 5.83 (1H, d, $^3J_{HH} = 7.2$ Hz, CH), 7.52 (2H, d, $^3J_{HH} = 7.4$ Hz, 2×CH), 7.62 (2H, t, $^3J_{HH} = 7.2$ Hz, 2×CH), 7.64 (2H, t, $^3J_{HH} = 7.3$ Hz, 2×CH), 7.75 (2H, d, $^3J_{HH} = 7.5$ Hz, 2×CH), 8.22 (2H, d, $^3J_{HH} = 7.2$ Hz, 2×CH), 8.25 (1H, d, $^3J_{HH} = 7.7$ Hz, CH); ^{13}C NMR ($CDCl_3$, 125 MHz): δ 14.1 (CH_3), 14.2 (CH_3), 62.5 (OCH_2), 62.7 (OCH_2), 76.2 (C), 83.5 (CH), 117.2 (C), 118.0 (2 CH), 118.6 (CN), 120.1 (2×CH), 124.5 (C), 125.6

(2×CH), 126.7 (C), 129.2 (2×CH), 129.5 (2×CH), 130.4 (CH), 132.5 (C), 133.4 (C), 164.8 (C=O), 172.7 (C=O). Anal. calc. for $C_{25}H_{22}N_2O_5$ (430.46): C, 69.76; H, 5.15; N, 6.51%; found: C, 69.68; H, 5.07; N, 6.47%.

2.2.64.5 Critical Views

The present method offers a catalyst-free facile route for the direct synthesis of oxazine derivatives, an important class of bridgehead nitrogen heterocycles, from the commercially available starting materials in aqueous medium under ambient conditions.

2.2.65 Entry-65: Synthesis of N^2-Alkyl-N^3-[2-(1,3,4-oxadiazol-2-yl)aryl]benzofuran-2,3-diamines

Type of reaction: C–C, C–N and C–O bond formation
Reaction conditions: Catalyst-free, ethanol, room temperature
Synthetic strategy: One-pot multicomponent reaction
Keywords: *N*-(Isocyanoimino)triphenylphosphorane, 2-aminobenzoic acid, 2-hydroxybenzaldehydes, isocyanides, ethanol, catalyst-free, room temperature, cyclocondensation, one-pot multicomponent, intramolecular aza-Wittig reaction, N^2-alkyl-N^3-[2-(1,3,4-oxadiazol-2-yl)aryl]benzofuran-2,3-diamines

2.2.65.1 General Reaction Scheme

Adib *et al.*[199] demonstrated a catalyst-free smooth synthesis of N^2-alkyl-N^3-[2-(1,3,4-oxadiazol-2-yl)aryl]benzofuran-2,3-diamines (**5**) *via* a one-pot four-component reaction of *N*-(isocyanoimino)triphenylphosphorane (**1**), 2-aminobenzoic acids (**2**), 2-hydroxybenzaldehydes (**3**) and isocyanides (**4**) in absolute ethanol at room temperature (Scheme 2.89).

2.2.65.2 Proposed Mechanism

The investigators outlined a plausible mechanism for this transformation (Scheme 2.90). Initially, generation of the imino intermediate (**6**) out of a condensation reaction between a 2-aminobenzoic acid (**2**) and an aldehyde **3** is assumed, which could undergo nucleophilic addition of the isocyanide (**4**), followed by protonation of the adduct by the adjacent carboxylic acid function, leading to a nitrilium carboxylate intermediate (**7**). This nitrilium intermediate **7** then undergoes intramolecular attack by the adjacent phenol OH group to form 3-aminobenzofuran-2(3*H*)-imine derivative (**8**), which could tautomerize to give benzofuran-2,3-diamine derivative (**9**). This COOH-substituted intermediate protonates (**1**), and the nitrilium intermediate (**11**) is then attacked by the conjugate base (**10**) to form the adduct (**12**). This adduct undergoes an intramolecular aza-Wittig reaction of the iminophosphorane moiety with the ester carbonyl to form the 1,3,4-oxadiazole ring by removal of Ph_3PO from zwitterion (**13**) and to afford the isolated product (**5**).

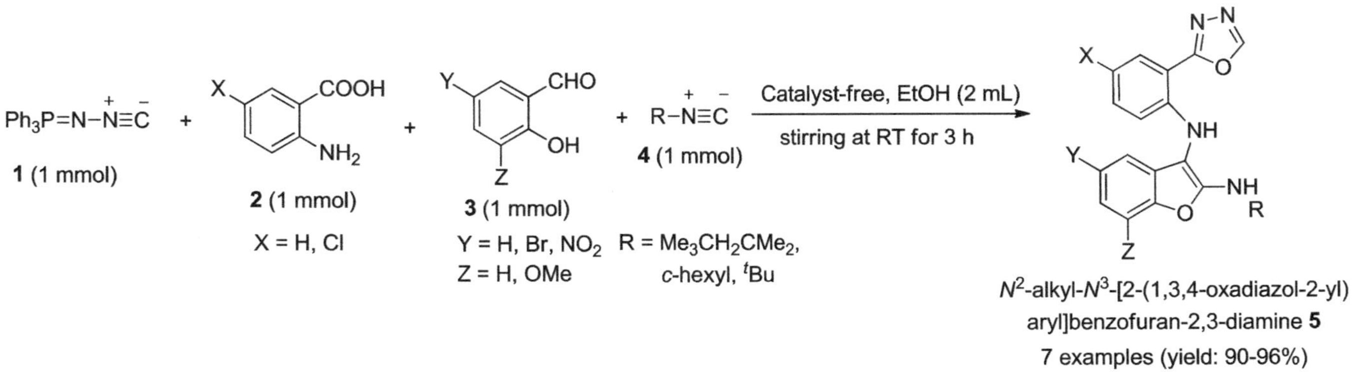

Scheme 2.89 Synthesis of N^2-alkyl-N^3-[2-(1,3,4-oxadiazol-2-yl)aryl]benzofuran-2,3-diamines (5).

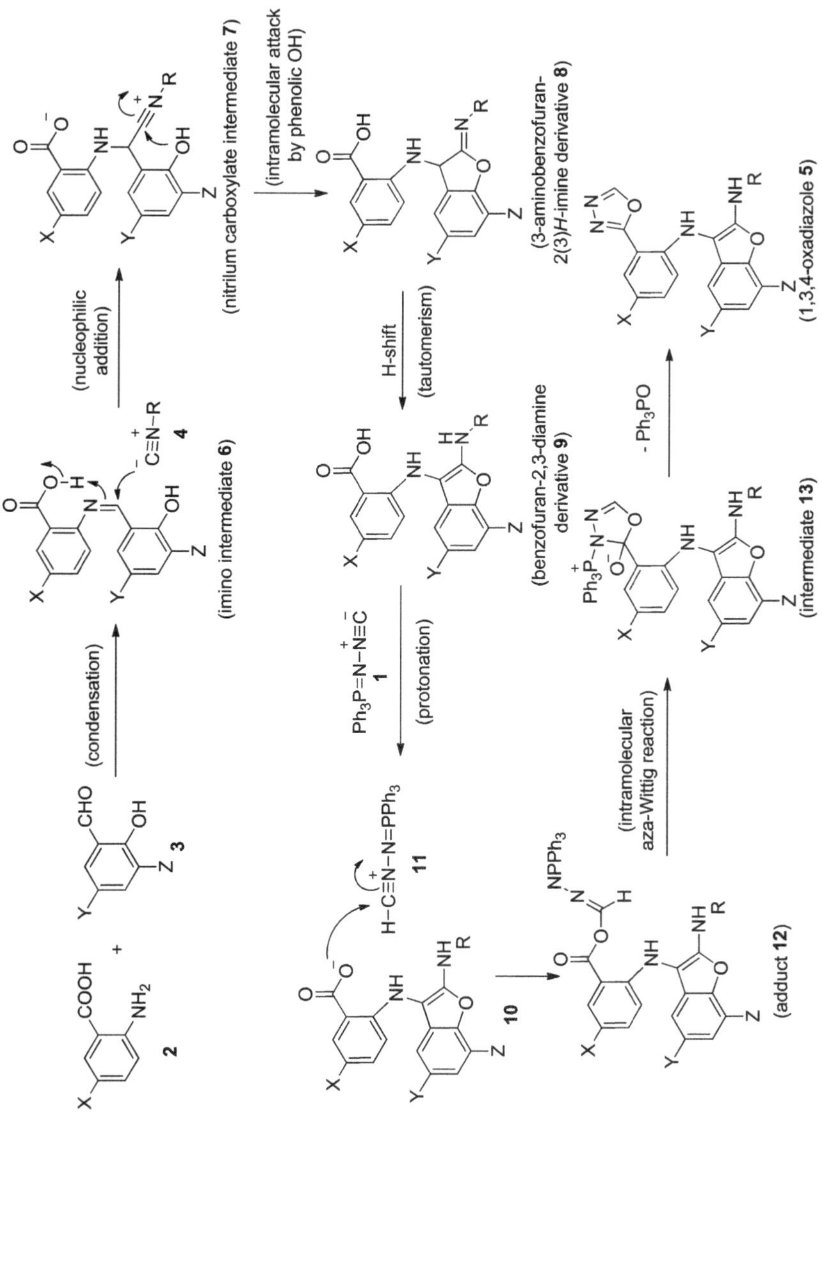

Scheme 2.90 Proposed mechanism for the formation of N^2-alkyl-N^3-[2-(1,3,4-oxadiazol-2-yl)aryl]benzofuran-2,3-diamines (**5**).

2.2.65.3 Representative Examples

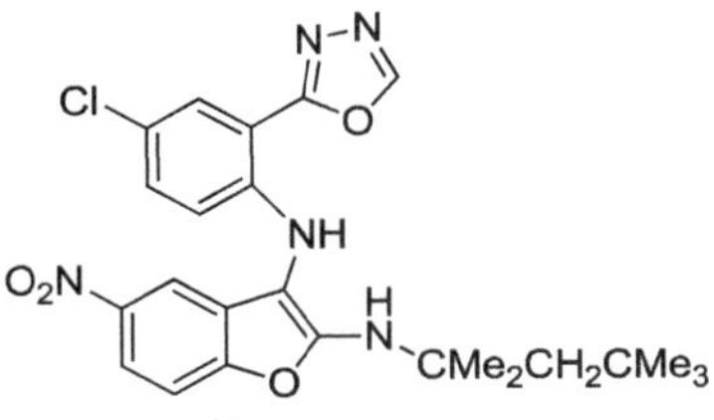

5a

N^3-(4-Chloro-2-(1,3,4-oxadiazol-2-yl)phenyl)-5-nitro-N^2-(2,4,4-trimethylpentan-2-yl)benzofuran-2,3-diamine (**5a**): orange crystals, mp 195–197 °C; yield: 90%

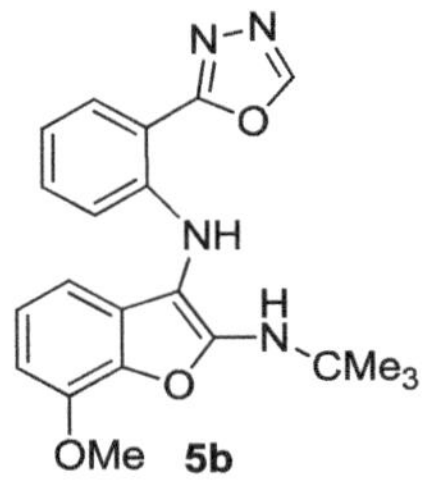

5b

N^3-(2-(1,3,4-oxadiazol-2-yl)phenyl)-N^2-(*tert*-butyl)-7-methoxybenzofuran-2,3-diamine (**5b**): yellow crystals, mp 223°C; yield: 91%

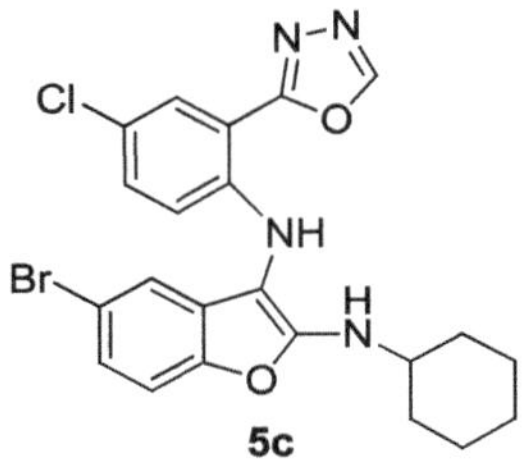

5c

5-Bromo-N^3-(4-chloro-2-(1,3,4-oxadiazol-2-yl)phenyl)-N^2-cyclohexylbenzofuran-2,3-diamine (**5c**): yellow crystals, mp 194–196 °C; yield: 93%

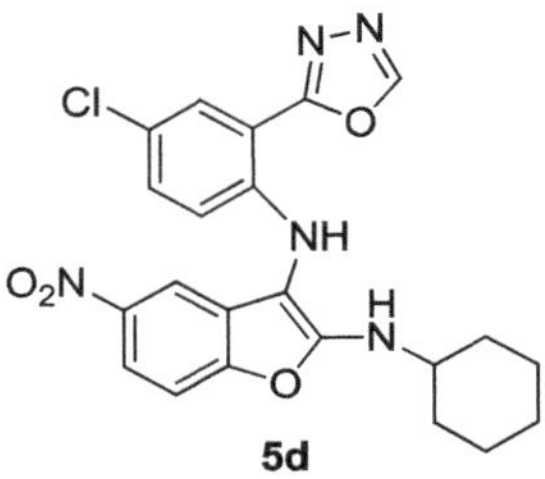

5d

N^3-(4-Chloro-2-(1,3,4-oxadiazol-2-yl)phenyl)-N^2-cyclohexyl-5-nitrobenzofuran-2,3-diamine (**5d**): orange crystals, mp 174–176 °C; yield: 96%

2.2.65.4 Experimental Procedure

A mixture of a 2-aminobenzoic acid derivative (**2**; 1 mmol), a 2-hydroxy-5-nitrobenzaldehyde derivative (**3**; 1 mmol), and isocyanide (**4**; 1 mmol) in EtOH (2 mL) was stirred at room temperature for 1 h. Next, *N*-(isocyanoimino)triphenylphosphorane (**1**; 0.302 g, 1 mmol) was added to the mixture, and the resulting mixture continued to be stirred at room temperature for a further 2 h. Then, the solvent was evaporated, and the residue purified using column chromatography (SiO_2, hexane–AcOEt 3 : 1) and recrystallization (hexane–AcOEt 1 : 1) to afford the desired product **5**. Each of the products was characterized by spectral and analytical analyses.

2.2.65.5 Characterization Data for Two Representative Compounds

N^3-(2-(1,3,4-oxadiazol-2-yl)phenyl)-N^2-(*tert*-butyl)-7-methoxybenzofuran-2,3-diamine (**5b**): yellow crystals, mp 223 °C; IR (KBr): 3315 and 3167 (NH), 1629, 1574, 1507, 1445, 1379, 1333, 1273, 1168, 1068, 950, 813, 717 cm^{-1};

^{1}H NMR (CDCl$_3$, 500.1 MHz): δ 1.40 (9H, s), 3.88 (1H, s), 4.04 (3H, s), 6.67 (1H, d, J = 8.1 Hz), 6.71 (1H, d, J = 7.7 Hz), 6.75 (1H, d, J = 8.5 Hz), 6.83 (1H, t, J = 7.7 Hz), 7.02 (1H, t, J = 7.9 Hz), 7.27 (1H, t, J = 7.7 Hz), 7.89 (1H, dd, J = 1.3 and 7.9 Hz), 8.47 (1H, s), 8.58 (1H, br s); ^{13}C NMR (CDCl$_3$, 125.8 MHz): δ 30.5, 53.8, 56.4, 99.5, 105.3, 105.9, 109.7, 113.9, 117.1, 123.4, 128.2, 129.6, 133.2, 138.4, 144.6, 147.1, 151.0, 154.5, 164.8; EIMS : *m/z* (%rel. intensity) 378 (<1,M$^+$), 277 (100), 236 (8), 199 (15), 183 (10), 151 (17), 109 (12), 91 (8), 77 (14), 65 (5), 51 (11). Anal. calc. for $C_{21}H_{22}N_4O_3$ (378.43): C, 66.65; H, 5.86; N, 14.81; found: C, 66.60; H, 5.78; N, 14.74.

N^3-(4-Chloro-2-(1,3,4-oxadiazol-2-yl)phenyl)-*N*^{2}-cyclohexyl-5-nitrobenzofuran-2,3-diamine (**5d**): orange crystals, mp 174–176 °C; IR (KBr): 3368, 3136, 1651, 1590, 1512, 1474, 1390, 1336, 1232, 1164, 1097, 1044, 1000, 952, 876, 808, 742 cm^{-1}; ^{1}H NMR (CDCl$_3$, 500.1 MHz): δ 1.14–2.05 (10H, m), 3.62–3.67 (1H, m), 3.98 (1H, br s), 6.61 (1H, d, J = 9.0 Hz), 7.23 (dd, J = 2.0, 8.9 Hz, 1H), 7.33 (d, J = 8.8 Hz, 1H), 7.82 (1H, s), 7.87 (1H, s), 7.94 (1H, dd, J = 2.0, 8.8 Hz), 8.50 (1H, s), 8.52 (1H, br s); ^{13}C NMR (125.8 MHz, CDCl$_3$): δ 24.9, 25.4, 34.4, 52.6, 91.9, 107.1, 110.0, 111.1, 114.9, 116.1, 122.4, 127.7, 129.9, 133.3, 144.7, 145.4, 151.3, 151.8, 156.9, 164.7; EIMS: *m/z* (%rel intensity) 455 (5, M$^+$(^{37}Cl)), 453 (11, M$^+$(^{35}Cl)), 370 (5), 354 (7), 337 (6), 279 (8), 167 (19), 149 (57), 97 (17), 81 (33), 69 (74), 57 (61), 43 (100). Anal. calc. for $C_{22}H_{20}ClN_5O_4$ (453.88): C, 58.22; H, 4.44; N, 15.43; found: C, 58.31; H, 4.51; N, 15.37.

2.2.65.6 Critical Views

The present method is advantageous over the traditional multistep synthetic routes[200–219] for the preparation of benzofuran derivatives. In addition, this technique offers a way to synthesize highly functionalized benzofurans of potential synthetic and pharmacological interest with excellent yields under catalyst-free one-pot multicomponent operation at ambient conditions.

2.2.66 Entry-66: Synthesis of Functionalized 1,3,4-Oxadiazoles

Type of reaction: C–C, C–N and C–O bond formation
Reaction conditions: Catalyst-free, acetonitrile, room temperature
Synthetic strategy: One-pot multicomponent reaction
Keywords: *N*-(Isocyanoimino)triphenylphosphorane, aromatic bis-aldehydes, aromatic carboxylic acids, acetonitrile, catalyst-free, room temperature, cyclocondensation, one-pot multicomponent, intramoleculer aza-Witting reaction, 1,3,4-oxadiazoles

2.2.66.1 General Reaction Scheme

Ramazani and his group[220] developed a one-pot synthesis for fully functionalized 1,3,4-oxadiazoles (**4**) in good yields from the reaction of (*N*-isocyanimino)triphenylphosphorane (**1**) with aromatic bis-aldehydes

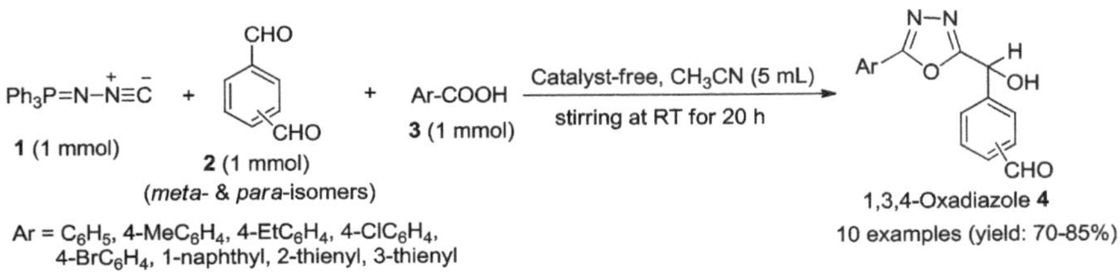

Scheme 2.91 Synthesis of functionalized 1,3,4-oxadiazoles (**4**).

(isophthalaldehyde and terphthalaldehyde) (**2**) in the presence of aromatic (or heteroaromatic) carboxylic acids (**3**) at room temperature and in neutral conditions without any side reactions (Scheme 2.91).

2.2.66.2 Proposed Mechanism

As per the proposition of the investigators (Scheme 2.92), the first step of this transformation involves a nucleophilic addition of the (*N*-isocyanimino) triphenylphosphorane (**1**) to bis-aldehyde (**2**), which facilitates its protonation with the acid (**3**), leading to nitrilium intermediate (**5**). This intermediate is then attacked by conjugate base of the acid (**3**) to form a 1 : 1 : 1 adduct (**6**) that smoothly undergoes an intramolecular aza-Wittig reaction with the iminophosphorane moiety and the ester carbonyl, which produces the isolated sterically-congested 1,3,4-oxadiazole derivative (**4**) by removal of triphenylphosphine oxide (Ph_3PO) from intermediate (**7**).

2.2.66.3 Representative Examples

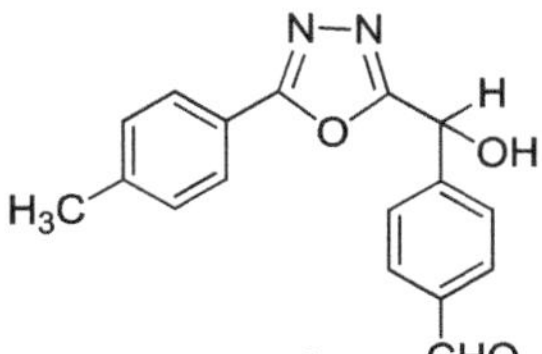

4a

4-(Hydroxy(5-(*p*-tolyl)-1,3,4-oxadiazol-2-yl)methyl)benzaldehyde (**4a**): yellow crystals, mp 91–93 °C; yield: 85%

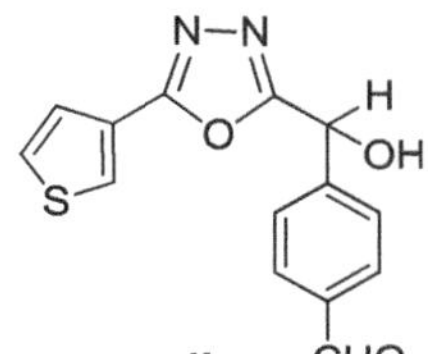

4b

4-(Hydroxy(5-(thiophen-3-yl)-1,3,4-oxadiazol-2-yl)methyl)benzaldehyde (**4b**): yellow oil; yield: 72%

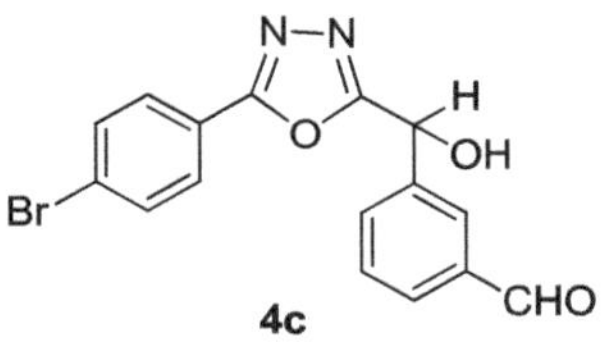

4c

3-((5-(4-Bromophenyl)-1,3,4-oxadiazol-2-yl)(hydroxy)methyl)benzaldehyde (**4c**): yellow crystals, mp 133–135°C; yield: 80%

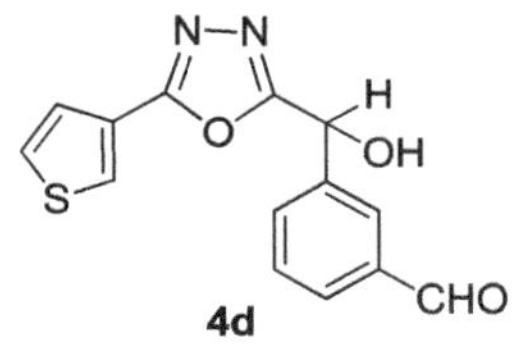

4d

3-(Hydroxy(5-(thiophen-3-yl)-1,3,4-oxadiazol-2-yl)methyl)benzaldehyde (**4d**): yellow oil; yield: 72%

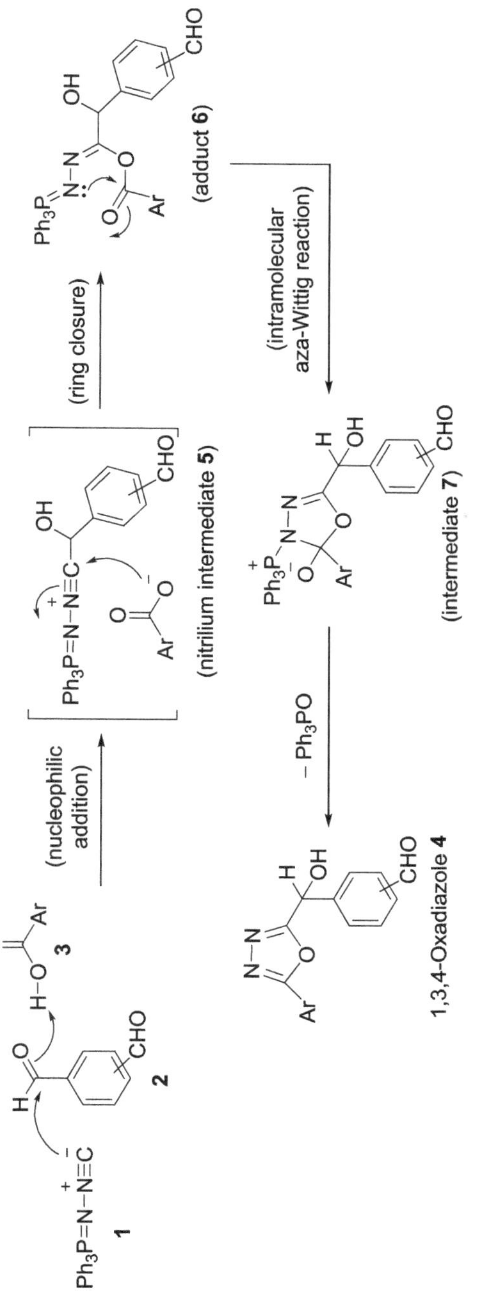

Scheme 2.92 Plausible mechanism for the synthesis of functionalized 1,3,4-oxadiazoles (4).

2.2.66.4 Experimental Procedure

A mixture of (*N*-isocyanimino)triphenylphosphorane (**1**; 1.0 mmol), aromatic bis-aldehyde (**2**; 1.0 mmol), and aromatic carboxylic acid (**3**; 1.0 mmol) in acetonitrile (5 mL) was stirred at room temperature for 20 h. Then, the solvent was removed under reduced pressure, and the viscous residue was purified by flash column chromatography using silica gel (petroleum ether–ethyl acetate (4 : 1)). The solvent was removed under reduced pressure to give the sterically-congested 1,3,4-oxadiazole (**4**). Each of the products was characterized by spectral and analytical analyses.

2.2.66.5 Characterization Data for Two Representative Compounds

4-(Hydroxy(5-(*p*-tolyl)-1,3,4-oxadiazol-2-yl)methyl)benzaldehyde (**4a**): yellow crystals, mp 91–93 °C; IR (KBr): 3389, 2922, 1700, 1607, 1480, 1202, 1068 cm^{-1}; ^{1}H NMR (250.13 MHz, $CDCl_3$): δ 2.39 (3H, s, CH_3), 4.25 (1H, br s, OH), 6.23 (1H, s, CH), 7.25–7.92 (8H, m, H–Ar), 10.02 (1H, s, CHO); ^{13}C NMR (62.53 MHz, $CDCl_3$): δ 21.66, 67.79, 126.99, 127.15, 129.78, 130.21, 129.18, 136.61, 142.86, 144.03, 164.85, 166.20; 191.72; MS: *m/z* (%rel. int.) 294 (M^+, 60), 159 (100), 133 (27), 119 (54), 105 (15), 91 (67), 77 (47), 57 (20), 43 (24). Anal. calc. for $C_{17}H_{14}N_2O_3$: C, 69.38; H, 4.79; N, 9.52%; found: C, 69.45; H, 4.74; N, 9.45.

3-(Hydroxy(5-(thiophen-3-yl)-1,3,4-oxadiazol-2-yl)methyl)benzaldehyde (**4d**): yellow oil; IR (KBr): 3399, 2920, 1698, 1605, 1437, 1057 cm^{-1}; ^{1}H NMR (250.13 MHz, $CDCl_3$): δ 2.70 (1H, br s, OH), 6.20 (1H, s, CH), 7.25–8.35 (7H, m, H–Ar), 9.95 (1H, s, CHO); ^{13}C NMR (62.53 MHz, $CDCl_3$): δ 66.96,125.87, 127.13, 12.66, 130.19, 129.18, 136.61, 144.03, 160.01, 169.80, 192.05. Anal. calc. for $C_{14}H_{10}N_2O_3S$: C, 58.73; H, 3.52; N, 9.78%; found: C, 58.67; H, 3.45; N, 9.75.

2.2.66.6 Critical Views

The present method offers a mild, simple, and efficient route for the preparation of sterically-congested 1,3,4-oxadiazole derivatives of potential pharmacological and biological activities. Ease of workup, good yields, and fairly mild reaction conditions make this protocol a useful addition to synthetic organic chemistry.

2.2.67 Entry-67: Synthesis of 2-(1,3,4-Oxadiazol-2-yl)-substituted 2-Hydroxy-1(2*H*)-acenaphthylenones

Type of reaction: C–C, C–N and C–O bond formation
Reaction conditions: Catalyst-free, acetonitrile, room temperature
Synthetic strategy: One-pot multicomponent reaction
Keywords: *N*-(Isocyanoimino)triphenylphosphorane, aromatic carboxylic acids, acenaphthaquinone, acetonitrile, catalyst-free, room temperature, cyclocondensation, one-pot multicomponent, intramolecular aza-Wittig reaction, 2-(1,3,4-oxadiazol-2-yl)-substituted 2-hydroxy-1(2*H*)-acenaphthylenones

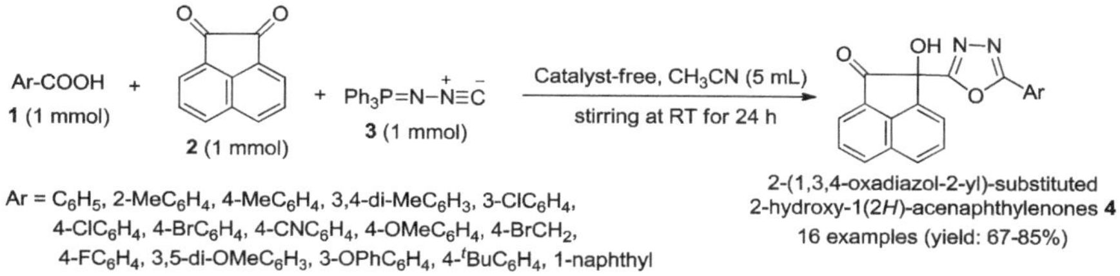

Scheme 2.93 Synthesis of 2-(1,3,4-oxadiazol-2-yl)-substituted 2-hydroxy-1(2*H*)-acenaphthylenones (**4**).

2.2.67.1 General Reaction Scheme

Ramazani and his group[221] extended their method to synthesize a new series of sterically-congested 1,3,4-oxadiazoles (**4**) in good yields from the reaction of (*N*-isocyanimino)triphenylphosphorane (**3**) with acenaphthaquinone (**2**) in the presence of aromatic carboxylic acids (**1**) at room temperature and in neutral conditions without any side reactions (Scheme 2.93).

2.2.67.2 Representative Examples

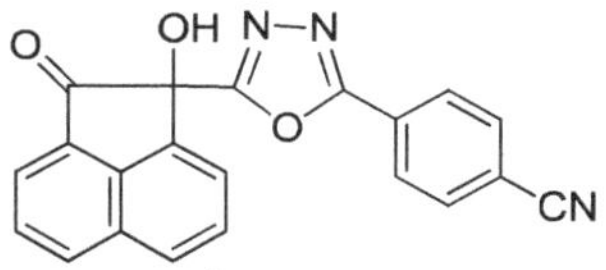

4a

4-(5-(1-Hydroxy-2-oxo-1,2-dihydroacenaphthylen-1-yl)-1,3,4-oxadiazol-2-yl)benzonitrile (**4a**): yellow powder, mp 161–163 °C; yield: 73%

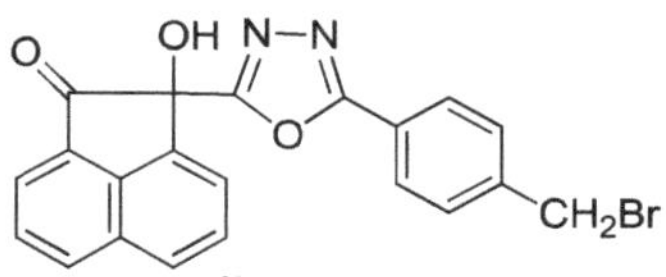

4b

2-(5-(4-(Bromomethyl)phenyl)-1,3,4-oxadiazol-2-yl)-2-hydroxyacenaphthylen-1(2*H*)-one (**4b**): yellow powder, mp 163–165 °C; yield: 82%

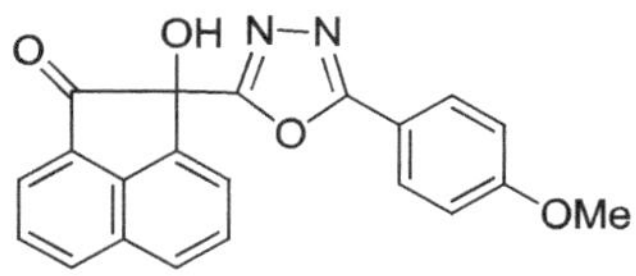

4c

2-Hydroxy-2-(5-(4-methoxyphenyl)-1,3,4-oxadiazol-2-yl)acenaphthylen-1(2*H*)-one (**4c**): yellow powder, mp 195–197 °C; yield: 68%

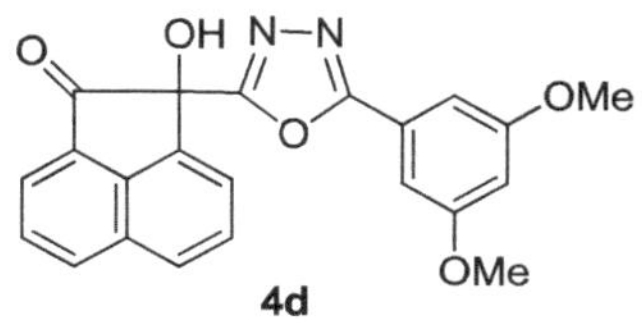

4d

2-(5-(3,5-Dimethoxyphenyl)-1,3,4-oxadiazol-2-yl)-2-hydroxyacenaphthylen-1(2*H*)-one (**4d**): yellow powder,mp 168–170 °C; yield: 67%

2.2.67.3 Experimental Procedure

A mixture of (*N*-isocyanimino)triphenylphosphorane (**3**; 1.0 mmol), acenaphthoquinone (**2**; 1.0 mmol), and aromatic carboxylic acid (**1**; 1.0 mmol) in 5 mL acetonitrile was stirred at room temperature for 24 h. The solvent was removed under reduced pressure, and the viscous residue was purified using a chromatographic technique (silica gel (F_{254}) powder; petroleum ether–ethyl acetate 4:1) to furnish the 1,3,4-oxadiazole (**4**). Each of the products was characterized by spectral and analytical analyses.

2.2.67.4 Characterization Data for Two Representative Compounds

2-(5-(4-(Bromomethyl)phenyl)-1,3,4-oxadiazol-2-yl)-2-hydroxyacenaphthylen-1(2*H*)-one (**4b**): yellow powder, mp 163–165 °C; IR (KBr): 3343, 3072, 1717, 1603, 1 486, 1012 cm^{-1}; ^{1}H NMR ($CDCl_3$, 250.13 MHz): δ 4.46 (2H, s, CH_2), 4.98 (1H, br s, OH), 7.43 (2H, d, $^3J_{HH} = 8.3$ Hz, Ar–H), 7.90 (2H, d, $^3J_{HH} = 8.3$ Hz, Ar–H), 7.70–7.86 (3H, m, Ar–H), 7.99–8.29 (3H, m, Ar–H); ^{13}C NMR ($CDCl_3$, 62.53 MHz): δ δ 32.09, 76.77, 122.56, 123.89, 126.95, 127.55, 129.07, 129.61, 132.62, 132.87, 1 23.10, 128.44, 128.79, 130.92, 135.24, 141.80, 165.12, 166.22, 198.15.

2-(5-(3,5-Dimethoxyphenyl)-1,3,4-oxadiazol-2-yl)-2-hydroxyacenaphthylen-1(2*H*)-one (**4d**): yellow powder, mp 168–170 °C; IR (KBr): 3380, 3094, 1725, 1604, 1463, 1162 cm^{-1}; ^{1}H NMR ($CDCl_3$, 250.13 MHz): δ 3.80 (6H, s, 2×OCH_3), 4.43 (1H, s, OH), 6.56–8.30 (9H, m, Ar–H); ^{13}C NMR ($CDCl_3$, 62.53 MHz): δ 55.64 (2×OCH_3), 87.00 (C–*O*H), 105.00, 107.05, 122.04, 123.94, 128.43, 132.59, 132.84, 124.75, 127.00, 131.04, 135.03, 142.10, 162.15, 164.02, 166.43, 189.51.

2.2.67.5 Critical Views

The key advantages of this present method are mild reaction conditions at ambient temperature and pressure, easy workup, high yields, and a clean reaction profile with no side reactions.

2.2.68 Entry-68: Synthesis of Sterically-congested 1,3,4-Oxadiazoles

Type of reaction: C–C, C–N and C–O bond formation
Reaction conditions: Catalyst-free, dichloromethane, room temperature
Synthetic strategy: One-pot multicomponent reaction
Keywords: *N*-(Isocyanoimino)triphenylphosphorane, primary amine, chloroacetone, aromatic carboxylic acids, dichloromethane, catalyst-free, room temperature, cyclocondensation, one-pot multicomponent, intramolecular aza-Wittig reaction, 1,3,4-oxadiazoles

2.2.68.1 General Reaction Scheme

Ramazani and his group[222] further extended their method to synthesize a new series of sterically-congested 1,3,4-oxadiazoles (**5**) in good yields from the reaction of chloroacetone (**1**), primary amine (**2**) and (*N*-isocyanimino)triphenylphosphorane (**4**) in the presence of aromatic carboxylic acids (**3**) at room temperature and in neutral conditions without any side reactions (Scheme 2.94).

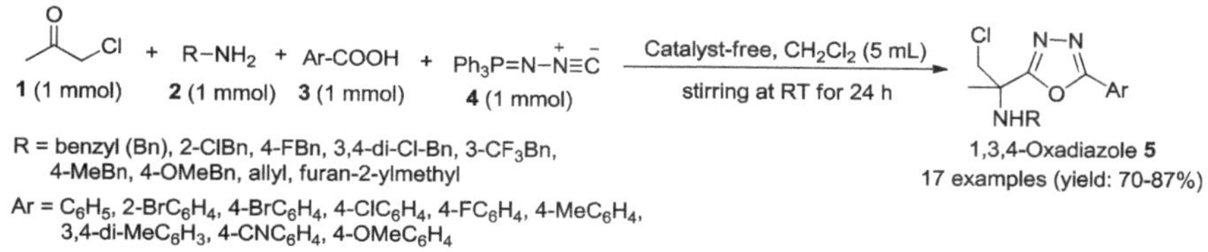

Scheme 2.94 Synthesis of sterically-congested 1,3,4-oxadiazoles (**5**).

2.2.68.2 Representative Examples

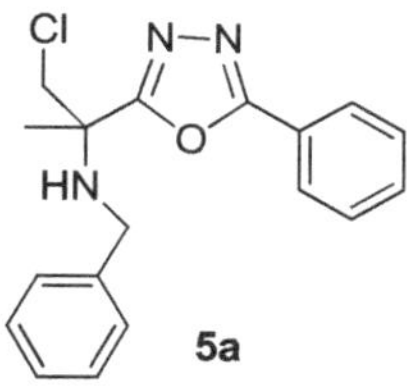

5a

N-Benzyl-1-chloro-2-(5-phenyl-1,3,4-oxadiazol-2-yl)propan-2-amine (**5a**): yellow powder, mp 82–84 °C; yield: 85%

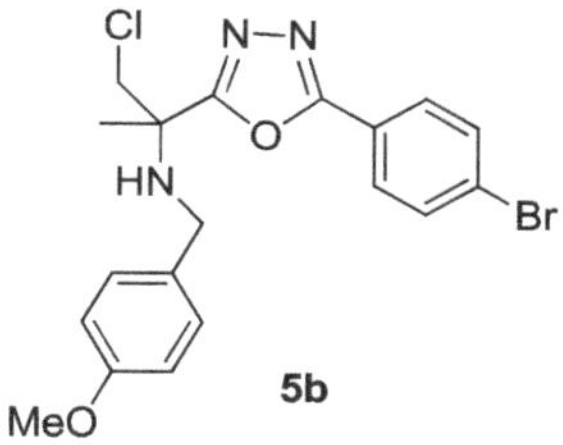

5b

2-(5-(4-Bromophenyl)-1,3,4-oxadiazol-2-yl)-1-chloro-N-(4-methoxybenzyl)propan-2-amine (**5b**): yellow powder, mp 95–97 °C; yield: 75%

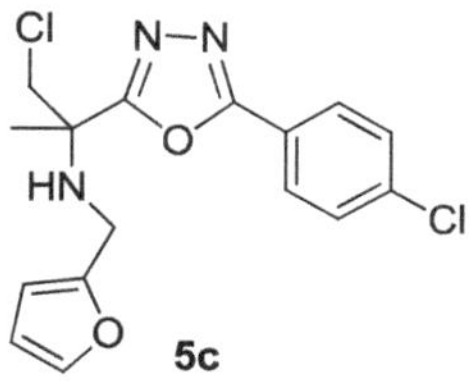

5c

1-Chloro-2-(5-(4-chlorophenyl)-1,3,4-oxadiazol-2-yl)-*N*-(furan-2-ylmethyl)propan-2-amine (**5c**): yellow powder, mp 79–81 °C; yield: 80%

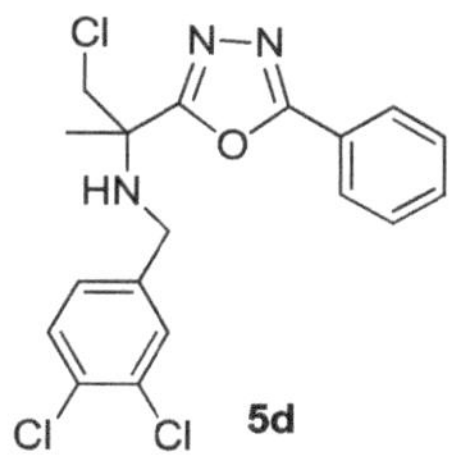

5d

1-Chloro-*N*-(3,4-dichlorobenzyl)-2-(5-phenyl-1,3,4-oxadiazol-2-yl)propan-2-amine (**5d**): yellow powder, mp 73–75 °C; yield: 85%

2.2.68.3 Experimental Procedure

To a magnetically stirred solution of chloroacetone (**1**; 1 mmol), primary amine (**2**; 1 mmol), and (*N*-isocyanoimino)triphenylphosphorane (**4**; 1 mmol) in dichloromethane (5 mL), a solution of benzoic acid derivative (**3**; 1 mmol) in dichloromethane (5 mL) was added drop-wise at room temperature over 15 min. The mixture was then well stirred for 12 h. The solvent was evaporated and the viscous residue purified by preparative TLC (silica gel F_{254}, petroleum ether–AcOEt 4 : 1) to obtain sterically-congested 1,3,4-oxadiazoles (**5**). Each of the products was characterized by spectral and analytical analyses.

2.2.68.4 Characterization Data for One Representative Compound

N-Benzyl-1-chloro-2-(5-phenyl-1,3,4-oxadiazol-2-yl)propan-2-amine (**5a**): yellow powder, mp 82–84 °C; IR (KBr): 3456, 3269, 2927, 1608, 1449, 1069 cm^{-1}; ^{1}H NMR ($CDCl_3$, 250 MHz): δ 1.77 (3H, s, Me); 2.23 (1H, s, NH), 3.69 (1H, d, J = 12.3 Hz), 3.76 (1H, d, J = 12.3 Hz), 3.96 (1H, d, J = 11.5 Hz), 4.07 (1H, d, J = 11.5 Hz), 7.24–8.07 (10H, m, Ar–H); ^{13}C NMR (62.5 MHz, $CDCl_3$): δ 22.53, 47.82, 50.24, 57.58, 127.00, 127.31, 128.27, 128.51, 129.07, 131.8, 123.70, 139.33, 165.24, 167.77; EIMS: *m*/*z* (%rel. int.) 327 (M^+), 187 (16), 149 (33), 105 (70), 91 (100), 76 (29), 43 (54). Anal. calc. for $C_{18}H_{18}ClN_3O$ (327.81): C, 65.95; H, 5.53; N, 12.82; found: C, 65.83; H, 5.49; N, 12.77.

2.2.68.5 Critical Views

The reported method offers a mild and simple route for the preparation of fully substituted 1,3,4-oxadiazole scaffolds with a handful of advantages such as ease of workup, high yields, and fairly mild reaction conditions.

2.2.69 Entry-69: Synthesis of 3-(5-Aryl-1,3,4-oxadiazol-2-yl)-3-hydroxybutan-2-ones

Type of reaction: C–C, C–N and C–O bond formation
Reaction conditions: Catalyst-free, dichloromethane, room temperature
Synthetic strategy: One-pot multicomponent reaction
Keywords: Biacetyl, (*N*-isocyanimino) triphenylphosphorane, aromatic carboxylic acids, dichloromethane, catalyst-free, room temperature, one-pot multicomponent reaction, tandem Passerini/intramolecular aza-Wittig reaction, 3-(5-aryl-1,3,4-oxadiazol-2-yl)-3-hydroxybutan-2-ones

2.2.69.1 General Reaction Scheme

The same Ramazani research group,[223] once again, reported a catalyst-free route for the preparation of 3-(5-aryl-1,3,4-oxadiazol-2-yl)-3-hydroxybutan-2-one derivatives (**4**) in high yields. This was achieved *via* a tandem Passerini/intramolecular aza-Wittig reaction of biacetyl (**3**) with (*N*-isocyanimino) triphenylphosphorane (**2**) in the presence of aromatic carboxylic acids (**1**) in dichloromethane under neutral conditions (Scheme 2.95).

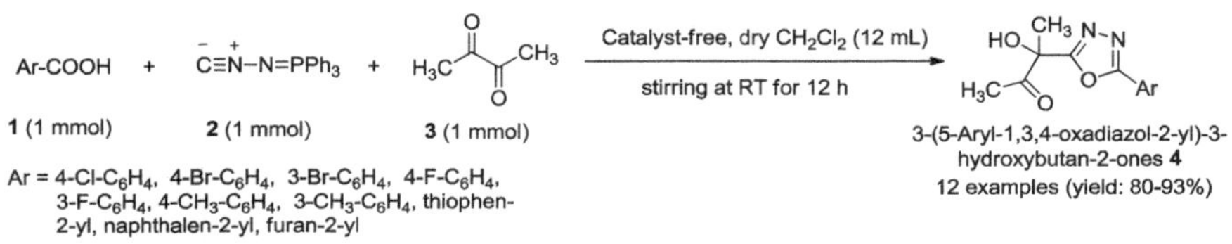

Scheme 2.95 Synthesis of 3-(5-aryl-1,3,4-oxadiazol-2-yl)-3-hydroxybutan-2-ones (**4**).

2.2.69.2 Plausible Mechanism

The investigators proposed a plausible mechanism for the transformation (Scheme 2.96). The first step involves nucleophilic addition of (**2**) to biacetyl (**3**), facilitated by its protonation with the acid (**1**), and leading to the formation of a nitrilium intermediate (**5**). This intermediate then may be attacked by the conjugate base of the acid (**6**) to form the 1 : 1 : 1 adduct (**6**) *via* the Passerini reaction. This adduct undergoes an intramolecular aza-Wittig reaction of the iminophosphorane moiety with the ester carbonyl group to afford the product (**4**) on removal of Ph_3PO from the intermediate (**8**/**8′**).

2.2.69.3 Representative Examples

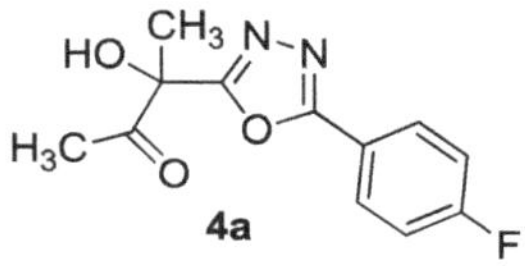

4a

3-(5-(4-Fluorophenyl)-1,3,4-oxadiazol-2-yl)-3-hydroxybutan-2-one (**4a**): white crystals, mp 93.2–95.0 °C; yield: 93%

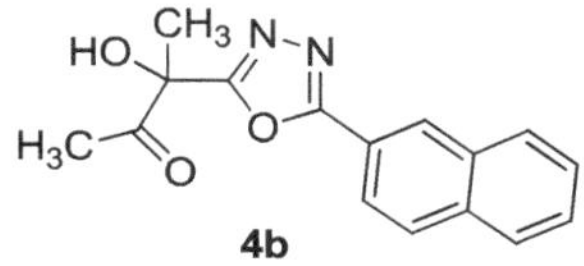

4b

3-Hydroxy-3-(5-(naphthalen-2-yl)-1,3,4-oxadiazol-2-yl)butan-2-one (**4b**): white crystals, mp 120.0–121.6 °C; yield: 86%

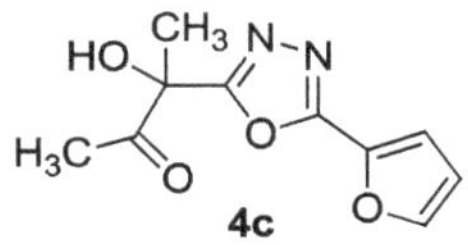

4c

3-(5-(Furan-2-yl)-1,3,4-oxadiazol-2-yl)-3-hydroxybutan-2-one (**4c**): white crystals, mp 100.6 °C (decomposed); yield: 80%

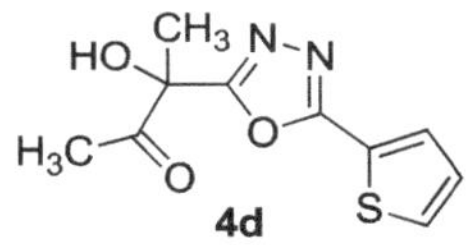

4d

3-Hydroxy-3-(5-(thiophen-2-yl)-1,3,4-oxadiazol-2-yl)butan-2-one (**4d**): white crystals, mp 70.0–71.5 °C; yield: 82%

2.2.69.4 Experimental Procedure

To a magnetically stirred solution of (*N*-isocyanimino)triphenylphosphorane (**2**; 1 mmol) and biacetyl (**3**; 1 mmol) in dichloromethane (7 mL), a solution of aromatic acid (**1**; 1 mmol) in dichloromethane (5 mL) was added dropwise at room temperature over 15 min. The mixture was stirred for 12 h. On completion of the reaction, the solvent was removed under reduced pressure and the viscous residue was purified using column chromatography (SiO_2; petroleum ether–ethyl acetate = 8 : 1) to yield the crystalline pure products of 3-(5-aryl-1,3,4-oxadiazol-2-yl)-3-hydroxybutan-2-ones. Each of the products was identified with the help of analytical and spectral studies.

2.2.69.5 Characterization Data of Two Representative Compounds

3-(5-(Furan-2-yl)-1,3,4-oxadiazol-2-yl)-3-hydroxybutan-2-one (**4c**): white crystals, mp 100.6 °C (decomposed); yield: 80%; IR (KBr): 3423 (OH), 2924, 1722

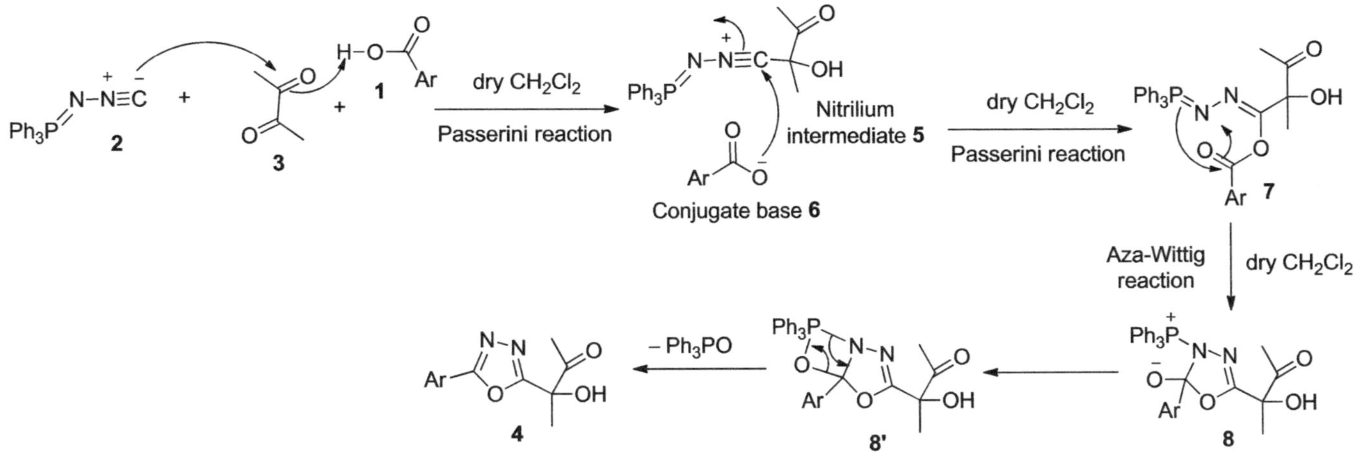

Scheme 2.96 Suggested mechanism for the generation of 3-(5-aryl-1,3,4-oxadiazol-2-yl)-3-hydroxybutan-2-ones (**4**).

(C=O), 1636, 1458, 1370, 752 cm^{-1}; ^1H NMR ($CDCl_3$, 250 MHz): δ 1.94, 2.35 (3H each, 2s, 2×Me), 4.66 (1H, s, OH, exchanged by D_2O addition), 6.56–6.63 (1H, m, 1 H of furan), 7.19 (1H, d, J = 3.2, 1 H of furan), 7.65 (1H, s, 1 H of furan); ^{13}C NMR ($CDCl_3$, 62.5 MHz): δ 23.64, 24.02 (2×Me); 76.02 (C–OH); 112.61, 115.28, 146.46 (3 CH of furan); 139.20 (C of furan); 163.01, 165.08 (2×C=N); 204.62 (C=O); EIMS (20 ev; %rel.): 222 (2, M^+), 196 (29), 180 (10), 153 (27), 127 (11), 111 (24), 95 (10), 83 (6), 43 (100), 41 (5). Anal. calc. for $C_{10}H_{10}N_2O_4$: C, 54.05; H, 4.54; N, 12.61; found: C, 54.00; H, 4.49; N, 12.58.

3-Hydroxy-3-(5-(thiophen-2-yl)-1,3,4-oxadiazol-2-yl)butan-2-one (**4d**): white crystals, mp 70.0–71.5 °C; yield: 82%; IR (KBr): 3246 (OH), 2924, 1724 (C=O), 1602, 1418, 1353, 727 cm^{-1}; ^1H NMR ($CDCl_3$, 250 MHz): δ 1.92 (3H, s, Me); 2.35 (3H, s, MeCO); 4.80 (1H, s, OH, exchanged by D_2O addition); 7.11–7.19 (1H, m, 1 H of thiophene); 7.56 (1H, d, J = 4.7, 1 H of thiophene); 7.75 (1H, d, J = 3.2, 1 H of thiophene); ^{13}C NMR ($CDCl_3$, 62.5 MHz): δ 23.56, 24.08 (2 Me); 76.05 (C–OH); 124.68 (C of thiophene); 128.54, 130.85, 131.12 (3 CH of thiophene); 162.30, 165.20 (2 C=N); 204.90 (C=O); EIMS (20 ev; %rel.): 238 (3, M^+), 196 (72), 194 (38), 168 (4), 153 (62), 127 (23), 111 (55), 97 (5), 83 (13), 69 (14). Anal. calc. for $C_{10}H_{10}N_2O_3S$: C, 50.41; H, 4.23; N, 11.76; found: C, 50.29; H, 4.19; N, 11.70.

2.2.69.6 *Critical Views*

The investigators offered a catalyst-free, mild and efficient one-pot route for the preparation of 3-(5-aryl-1,3,4-oxadiazol-2-yl)-3-hydroxybutan-2-one derivatives of biological interest *via* a tandem Passerini/intramolecular aza-Wittig reaction. High yields and fairly mild reaction conditions along with good scope of the reaction are noteworthy.

2.2.70 Entry-70: Synthesis of Hydrazino-substituted Chromeno[2,3-*c*]pyrroles

Type of reaction: C–C, C–N and C–O bond formation
Reaction conditions: Catalyst-free, acetonitrile, room temperature
Synthetic strategy: One-pot multicomponent domino addition-amination reaction
Keywords: 3-Formylchromones, cyclohexylisocyanide, azodicarboxylates, acetonitrile, catalyst-free, room temperature, one-pot multicomponent domino addition-amination reaction, hydrazino-substituted chromeno[2,3-*c*]pyrroles

2.2.70.1 *General Reaction Scheme*

An easy, one-pot three-component, convenient domino addition-amination reaction protocol involving 3-formylchromones (**1**), azodicarboxylates (**2**), and cyclohexylisocyanide (**3**), for the synthesis of a series of novel hydrazino-substituted chromeno[2,3-*c*]pyrroles (**4**) was described by Terzidis *et al.*[224]

Moderate to good yields were obtained in acetonitrile under room temperature conditions (Scheme 2.97).

2.2.70.2 Representative Examples

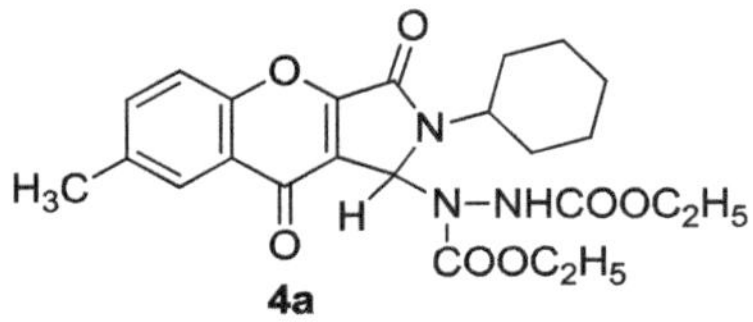

4a

Diethyl 1-(2-cyclohexyl-7-methyl-3,9-dioxo-1,2,3,9-tetrahydrochromeno[2,3-*c*]pyrrol-1-yl)hydrazine-1,2-dicarboxylate (**4a**): white solid, mp 194–196 °C; yield: 82%

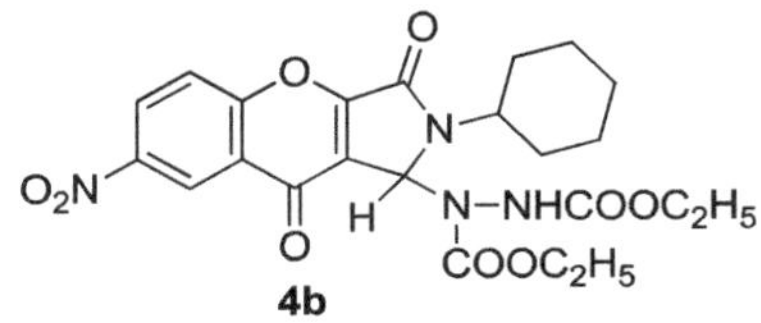

4b

Diethyl 1-(2-cyclohexyl-7-nitro-3,9-dioxo-1,2,3,9-tetrahydrochromeno[2,3-*c*]pyrrol-1-yl)hydrazine-1,2-dicarboxylate (**4b**): white solid, mp 208–210 °C; yield: 68%

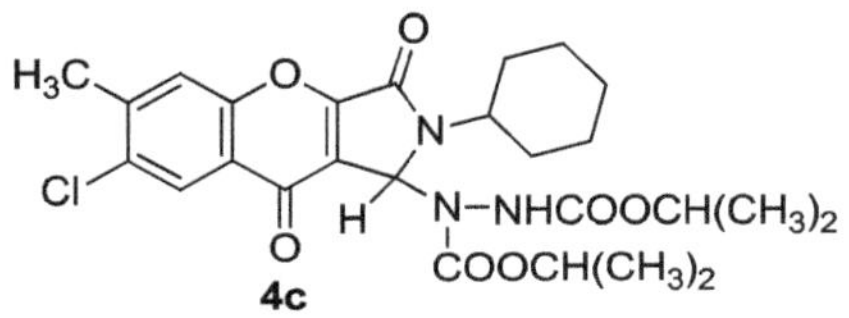

4c

Diisopropyl 1-(7-chloro-2-cyclohexyl-6-methyl-3,9-dioxo-1,2,3,9-tetrahydrochromeno[2,3-*c*]pyrrol-1-yl)hydrazine-1,2-dicarboxylate (**4c**): white solid, mp 268–266 °C; yield: 73%

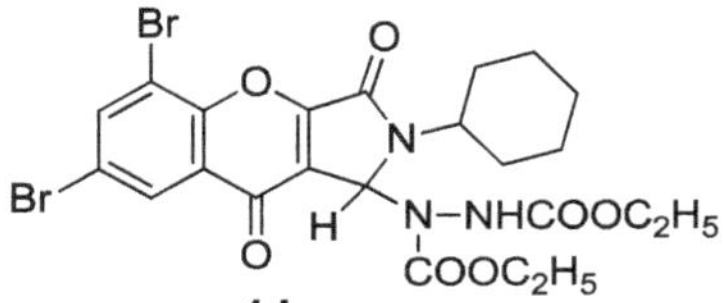

4d

Diethyl 1-(5,7-dibromo-2-cyclohexyl-3,9-dioxo-1,2,3,9-tetrahydrochromeno[2,3-*c*]pyrrol-1-yl)hydrazine-1,2-dicarboxylate (**4d**): green solid, mp 208–210 °C; yield: 73%

2.2.70.3 Experimental Procedure

To a magnetically stirred solution of 3-formylchromones (**2**; 1.0 mmol) and azodicarboxylates (**2**; 1.2 mmol) in acetonitrile (10 mL), cyclohexylisocyanide (1.2 mmol) was added *via* a syringe and the reaction mixture was stirred at room temperature for about 20–24 h until chromone (**1**) was consumed completely (monitored by TLC). On completion of the reaction the solvent was removed, and the product was crystallized out by addition of ether and was recrystallized from ethanol to furnish pure products of hydrazino-substituted chromeno[2,3-*c*]pyrroles (**4**). All the products were characterized based on spectral and elemental analyses.

2.2.70.4 Characterization Data of Two Representative Compounds

Diethyl 1-(2-cyclohexyl-7-nitro-3,9-dioxo-1,2,3,9-tetrahydrochromeno[2,3-*c*]pyrrol-1-yl)hydrazine-1,2-dicarboxylate (**4b**): white solid, mp 208–210 °C; yield: 68%; IR (KBr): 3273 (NH), 1749 (C=O), 1727 (C=O), 1663 (C=O) cm^{-1}; ^{1}H NMR ($CDCl_3$, 300 MHz): δ 1.00–2.10 (16H, m), 3.70–3.80 (1H, m, 1′-H), 3.85–4.05

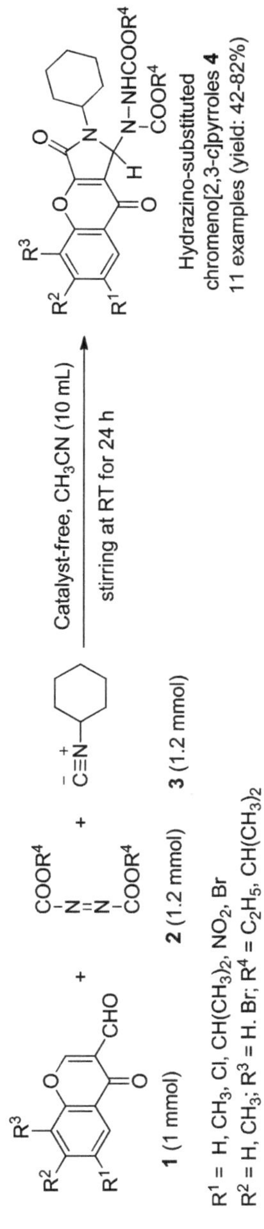

Scheme 2.97 Synthesis of hydrazino-substituted chromeno[2,3-*c*]pyrroles (**4**).

(2H, m, OCH_2), 4.30–4.45 (2H, m, OCH_2), 7.1 (1H, br s, 1-H), 7.70 (1H, br, NH), 7.80–7.95 (1H, br m, 5-H), 8.50–8.65 (2H, br m, 6,8-H); ^{13}C NMR ($CDCl_3$ + DMSO-d_6, 75 MHz): δ 14.3 (CH_2CH_3), 14.4 (CH_2CH_3), 24.7 and 24.8 (C-3′,5′), 25.7 (C-4′), 33.0 and 33.2 (C-2′,6′), 56.8 (C-1′), 62.5 (OCH_2), 63.9 (OCH_2), 86.6 (C-1), 119.2 (C-5), 121.2 (C-9a), 122.1 (C-8), 125.6 (C-8a), 130.3 (C-6), 145.1 (C-7), 149.4 (NHC=O), 154.4 (NC=O), 156.1 (C-3a), 156.3 (C-3), 159.8 (C-4a), 173.6 (C-9); LCMS (ESI, 1.65 eV): *m/z* (rel.%) 503 [70, $(M+H)^+$]. Anal. calc. for $C_{23}H_{26}N_4O_9$: C, 54.98; H, 5.22; N, 6.74; found: C, 55.16; H, 5.33; N 6.62.

Diethyl 1-(5,7-dibromo-2-cyclohexyl-3,9-dioxo-1,2,3,9-tetrahydrochromeno[2,3-*c*]pyrrol-1-yl)hydrazine-1,2-dicarboxylate (**4d**): green solid, mp 208–210 °C; yield: 73%; IR (KBr) ν_{max}: 3298 (NH), 1752 (C=O), 1712 (C=O), 1700 (C=O) cm^{-1}; ^1H NMR ($CDCl_3$ + DMSO-d_6, 300 MHz): δ 1.00–1.54 (11H, m, 2′, 3′, 4′, 5′, 6′-Hax, 2×CH_2CH_3), 1.75–1.92 (5H, m, 2′, 3′, 4′, 5′, 6′ -Heq), 3.70–3.85 (1H, m, 1′-H), 3.92–4.15 (2H, m, OCH_2), 4.20–4.40 (2H, m, OCH_2), 7.32 (1H, br s, 1-H), 8.19 (1H, s, 6-H), 8.21 (1H, s, 8-H), 9.1 (1H, br s, NH); ^{13}C NMR ($CDCl_3$ + DMSO-d_6, 75 MHz): δ 13.9 (2×CH_2CH_3), 23.9 and 24.0 (C-3′,5′), 25.6 (C-4′), 32.8 (C-2′,6′), 55.9 (C-1′), 60.7 (2×OCH2), 84.5 (C-1), 113.5 (C-5), 119.0 (C-7), 124.2 (C-9a), 127.3 (C-8a), 127.4 (C-8), 135.8 (C-6), 139.8 (C-6), 148.0 (NHC=O), 154.4 (NC=O), 155.8 (C-4a), 155.9 (C-3a), 161.5 (C-3), 172.2 (C-9); LCMS (ESI, 1.65 eV): *m/z* (rel.%) 614/616/618 [100, $(M+H)^+$]. Anal. calc. for $C_{23}H_{25}Br_2N_3O_7$: C, 44.90; H, 4.10; N, 6.83; found: C, 45.03; H, 4.05; N 6.74.

2.2.70.5 Critical Views

A new series of hydrazino-substituted chromeno[2,3-*c*]pyrroles were synthesized following a one-pot multicomponent domino addition-amination reaction involving 3-formylchromones, isocyanides, and azodicarboxylates for the first time. The facile and convenient reaction conditions and the formation of the otherwise inaccessible products in good yields from simple and easily available starting materials, without using any transition metal catalyst or extra organocatalyst, offer an extra flavor to the above reaction protocol.

2.2.71 Entry-71: Synthesis of Substituted 3,4-Dihydrocoumarins

Types of reaction: C–C and C–N/C–O bond formation
Reaction conditions: Catalyst-free, ethanol/methanol/dichloromethane, room temperature
Synthetic strategy: One-pot multicomponent
Keywords: 2-Hydroxybenzaldehydes, Meldrum's acid, isocyanides, aromatic/aliphatic alcohols, dichloromethane, room temperature, one-pot multicomponent reaction, Ugi reaction, substituted 3,4-dihydrocoumarins

2.2.71.1 General Reaction Scheme

Shaabani *et al.*[225] demonstrated facile and efficient synthesis of a series of densely functionalized 3,4-dihydrocoumarin derivatives (**5**) *via* a novel isocyanide-based four-component one-pot reaction between 2-hydroxybenzaldehydes (**1**), Meldrum's acid (**2**), isocyanides (**3**), and aromatic or aliphatic alcohols (**4**) in good to excellent yields under catalyst-free neutral conditions without any activation at ambient temperature (Scheme 2.98).

2.2.71.2 Proposed Mechanism

The investigators postulated that the transformation (Scheme 2.99) involves the initial formation of a conjugated electron-deficient heterodyne (**6**) by standard Knoevenogel condensation of 2-hydroxybenzaldehyde (**1**) and Meldrum's acid (**2**), followed by a [4 + 1] cycloaddition reaction or a Michael-type addition reaction with isocyanide (**3**) to afford an iminolactone intermediate (**7**). The hydroxy group of 2-hydroxybenzaldehyde moiety within this iminolactone intermediate (**7**) attacks the lactone carbonyl intramolecularly with the subsequent loss of acetone from **7** leading to the formation of **8**. Then, nucleophilic attack of alcohol **4** to the activated carbonyl moiety of **8**, affords substituted 3,4-dihydrocoumarin (**5**).

2.2.71.3 Representative Examples

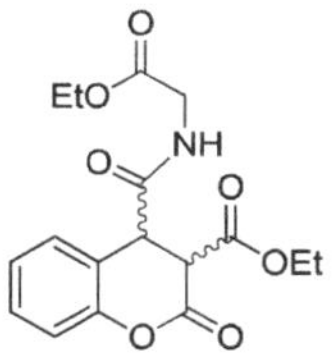

5a

Ethyl-4-((2-ethoxy-2-oxyethyl)carbamoyl)-3,4-dihydro-2-oxo-2*H*-chromene-3-carboxylate (**5a**): white powder, mp 126–129 °C; reaction time: 8h; yield: 75% (*cis*:*trans* = 72:28)

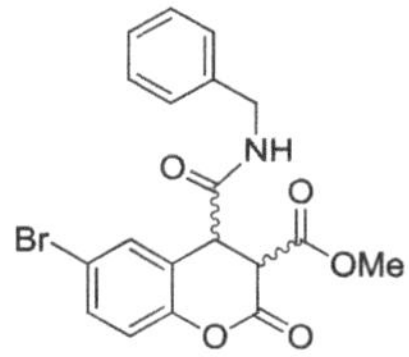

5b

Methyl 4-(benzylcarbamoyl)-6-bromo-3,4-dihydro-2-oxo-2*H*-chromene-3-carboxylate (**5b**): white powder; mp 177–180°C;reaction time: 8h; yield: 90% (*cis*:*trans* = 61:39)

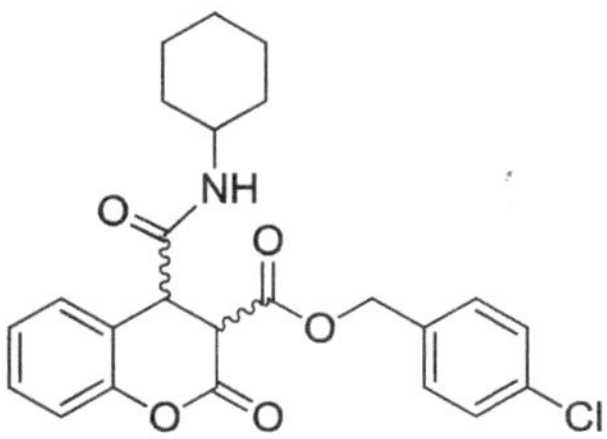

5c

4-Chlorobenzyl 4-(cyclohexylcarbamoyl)-3,4-dihydro-2-oxo-2*H*-chromene-3-carboxylate (**5c**): white powder; mp 148–152 °C; reaction time: 36h; yield: 78% (*cis*:*trans* = 63:37)

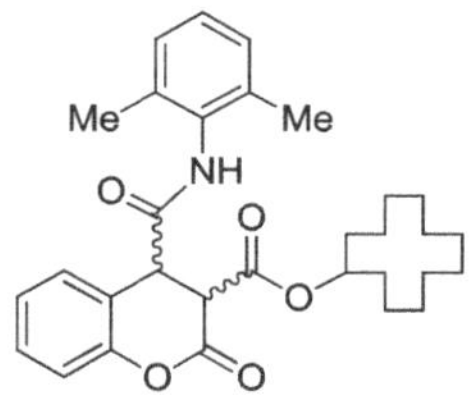

5d

Cyclododecyl 4-((2,6-dimethylphenyl) carbamoyl)-3,4-dihydro-2-oxo-2*H*-chromene-3-carboxylate (**5d**): white powder; mp 154–158 °C; reaction time: 36 h; yield: 75% (*cis*:*trans* = 100:0)

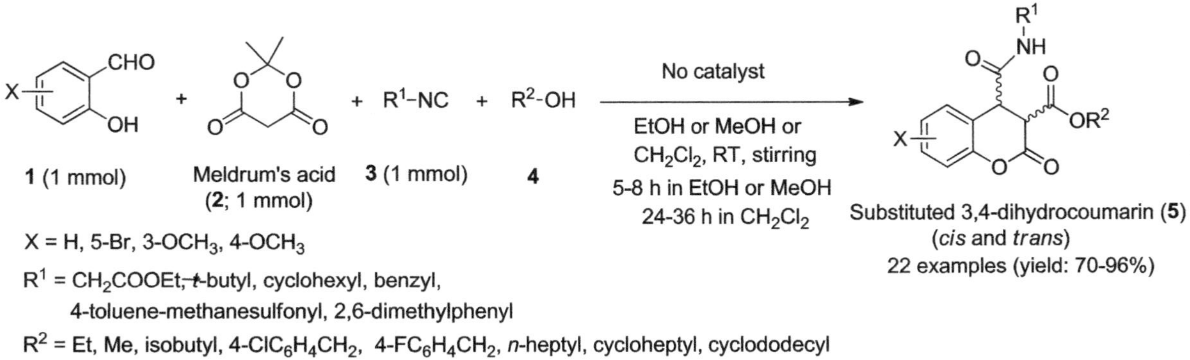

Scheme 2.98 Synthesis of substituted 3,4-dihydrocoumarins (5).

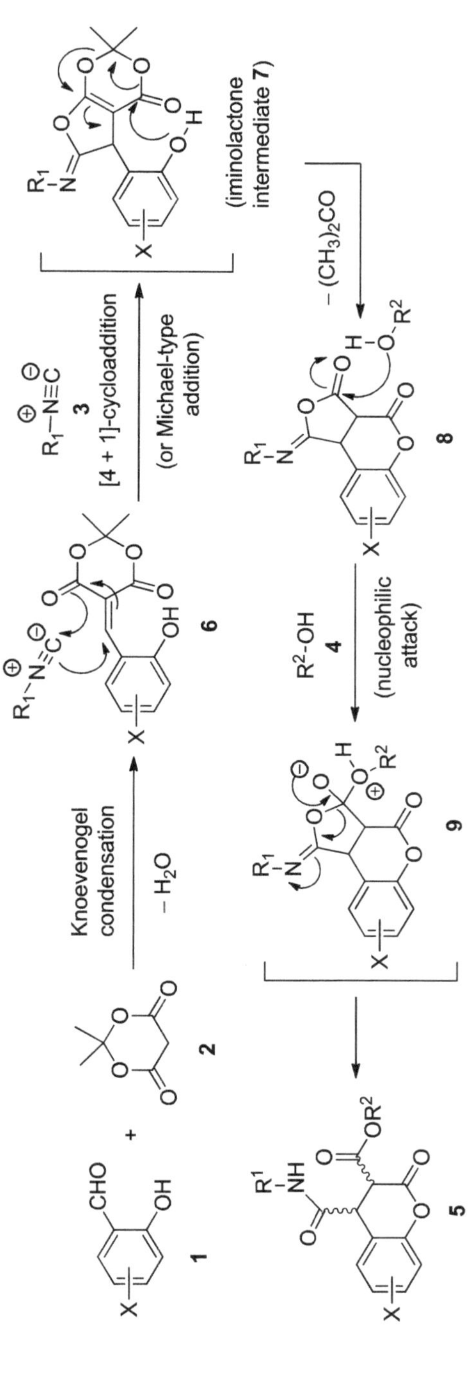

Scheme 2.99 Suggested mechanism for the formation of substituted 3,4-dihydrocoumarins (4).

2.2.71.4 Experimental Procedure

To a magnetically stirred solution of 2-hydroxybenzaldehyde (**1**, 1 mmol) and Meldrum's acid (**2**, 1 mmol) in ethanol or methanol (5 mL) (and alcohol **4** other than ethanol or methanol in dichloromethane) was added isocyanide (**3**, 1 mmol) and the reaction mixture was stirred at room temperature for the stipulated period (5–8 h in ethanol or methanol and 24–36 h in dichloromethane). After completion of the reaction, as indicated by TLC (ethyl acetate–*n*-hexane, 2 : 1), the precipitate was washed with ethanol or diethyl ether to afford substituted 3,4-dihydrocoumarin **5** as white powder in good to excellent yield (70–96%). (In the case of dichloromethane medium, the solvent was removed under vacuum to leave the crude mass before washing.) Each of the products was characterized by analytical, as well as detailed spectral studies including IR, ^{1}H NMR, ^{13}C NMR, and mass.

2.2.71.5 Characterization Data of Two Representative Compounds

Methyl 4-(benzylcarbamoyl)-6-bromo-3,4-dihydro-2-oxo-2*H*-chromene-3-carboxylate (**5b**: *cis* : *trans* = 61 : 39): white powder; mp 177–180 °C; IR (KBr): 3360, 2930, 2851, 1782, 1654, 1540 cm^{-1}; ^{1}H NMR (DMSO-d_6, 300 MHz) *cis* and *trans*: δ 1.08–1.78 (20H, m, 10CH$_2$ of two cyclohexyls, *cis* and *trans*), 3.43 (2H, br s, 2CH–N of cyclohexyls), 3.61 (OCH$_3$, *cis*), 3.70 (OCH$_3$, *trans*), 4.01 (1H, d, $^3J_{HH}$ = 4.1 Hz, CH-4, *cis*), 4.26 (1H, d, $^3J_{HH}$ = 3.45 Hz, CH-3, *trans*), 4.32 (1H, d, $^3J_{HH}$ = 3.4 Hz, CH-3, *trans*), 4.36 (1H, d, $^3J_{HH}$ = 4.1 Hz, CH-4 *cis*), 7.04–7.10 (2H, m, H–Ar, *cis* and *trans*), 7.48–7.52 (2H, m, H–Ar, *cis* and *trans*), 7.65 (2H, brs, H–Ar, *cis* and *trans*), 8.17 (1H, brs, NH, *cis*), 8.39 (1H, brs, NH, *trans*); ^{13}C NMR (DMSO-d_6, 75 MHz) *cis*: δ 24.69, 24.72, 25.52, 32.49, 32.55 (C-cyclohexyls), 44.84 (CH-4), 47.90 (CH–N of cyclohexyl), 48.22 (CH-3), 53.59 (OCH3), 116.68, 119.48, 123.42, 130.87, 132.56, 150.45 (C–Ar), 162.93, 167.41, 168.19 (3C=O); 13C NMR (75 MHz, DMSO-*d6*) *trans*: δ 24.60, 24.65, 25.60, 32.36, 32.73 (*C*-cyclohexyls), 44.37 (CH-4), 46.52 (CH–N of cyclohexyl), 48.07 (CH-3), 52.78 (OCH$_3$), 116.11, 119.24, 125.47, 130.69, 132.16, 150.94 (C–Ar), 163.57, 167.24, 169.17 (3C=O); MS at *m/z* (%): 419 (M$^+$, ^{81}Br, 5), 417 (M$^+$, ^{79}Br, 5), 387 (20), 385 (20), 353 (10), 351 (10), 227 (25), 226 (20), 225 (25), 224 (20), 198 (20), 196 (20), 91 (100), 65 (20), 31 (15). Anal. calc. for $C_{19}H_{16}BrNO_5$: C, 54.56; H, 3.86; N, 3.35; found: C, 54.34; H, 3.95; N, 3.47.

Cyclododecyl 4-((2,6-dimethylphenyl) carbamoyl)-3,4-dihydro-2-oxo-2*H*-chromene-3-carboxylate (**5d**; *cis* : *trans* = 100 : 00): White powder; mp 154–158 °C; IR (KBr): 3256, 2940, 2859, 1780, 1743, 1651, 1590 cm^{-1}; ^{1}H NMR (DMSO-d_6, 300 MHz) *cis* and *trans*: δ 1.31–1.72 (22H, m, 11CH$_2$ of cyclododecyl), 2.09 (3H, s, CH$_3$), 2.17 (3H, s, CH$_3$), 4.09 (1H, d, $^3J_{HH}$ = 6.0 Hz, CH-4), 4.54 (1H, d, $^3J_{HH}$ = 6.0 Hz, CH-3), 5.05 (1H, br s, OCH), 6.79–7.32 (7H, m, H–Ar), 10.28 (1H, br s, NH); ^{13}C NMR (DMSO-d_6, 75 MHz): δ 17.52, 17.82 (2CH$_3$), 20.75, 20.85, 23.16, 23.22, 23.39, 23.45, 23.74, 23.91, 24.08, 28.99, 29.05, (11CH$_2$ of cyclododecyl), 48.71 (CH-4), 54.14 (CH-3), 73.98 (OCH), 115.82, 119.66,

122.64, 128.67, 128.86, 129.76, 130.06, 130.92, 132.11, 139.03, 136.97, 155.39 (C-Ar), 168.06, 171.22, 175.65 (3C=O); MS at *m/z* (%): 505 ($[M+1]^+$, 1), 339 (10), 321 (50), 295 (10), 147 (100), 118 (40), 82 (30), 57 (30), 41 (50). Anal. calc. for $C_{31}H_{39}NO_5$: C, 73.63; H, 7.77; N, 2.77; found: C, 73.54; H, 7.87; N, 2.88.

2.2.71.6 Critical Views

The present method offers an easy access of fully substituted 3,4-dihydrocoumarins of medicinal interest starting from simple and readily available starting materials *via* a four-component one-pot reaction at room temperature under neutral conditions without using any catalyst or activation. The reaction shows good functional group tolerance and is high yielding and product isolation is very straightforward.

2.2.72 Entry-72: Synthesis of bis(2-Arylimino-1,3-thiazolidin-4-one) Derivatives

Type of reaction: C–C, C–N and C–S bond formation
Reaction conditions: Catalyst-free, dichloromethane, room temperature
Synthetic strategy: One-pot multicomponent reaction
Keywords: Dialkyl but-2-ynedioates, isothiocyanatobenzene, aliphatic diamines, benzene-1,2-diamines, dichloromethane, catalyst-free, room temperature, one-pot multicomponent reaction, bis(2-arylimino-1,3-thiazolidin-4-ones), 2-(arylimino)-*N*-(enaminoaryl)-1,3-thiazolidin-4-ones

2.2.72.1 General Reaction Scheme

Alzadeh and research group[226] developed an efficient chemoselective one-pot pseudo-five-component protocol for the synthesis of biologically interesting bis(2-arylimino-1,3-thiazolidin-4-ones) (**5**) in relatively good yields *via* the reaction of dialkyl but-2-ynedioates (**1**), isothiocyanatobenzene (**2**) and aliphatic diamines (**3**) in anhydrous dichloromethane under ambient conditions. Additionally, they also demonstrated that the use of benzene-1,2-diamines (**4**) instead of aliphatic diamines (**3**) leads to 2-(arylimino)-*N*-(enaminoaryl)-1,3-thiazolidin-4-ones (**6**) under the same reaction conditions (Scheme 2.100).

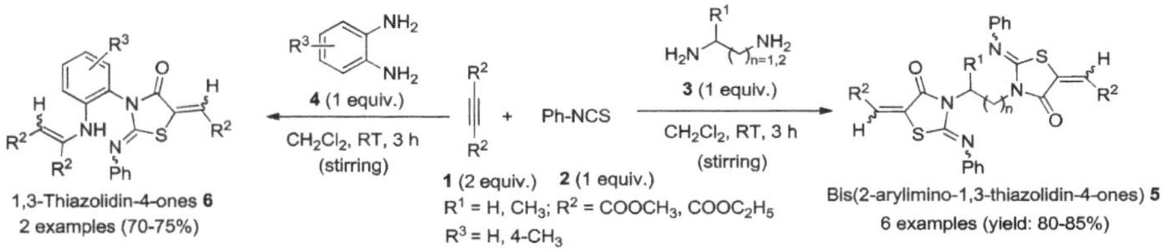

Scheme 2.100 Synthesis of bis(2-arylimino-1,3-thiazolidin-4-one) derivatives (**5**).

2.2.72.2 Representative Examples

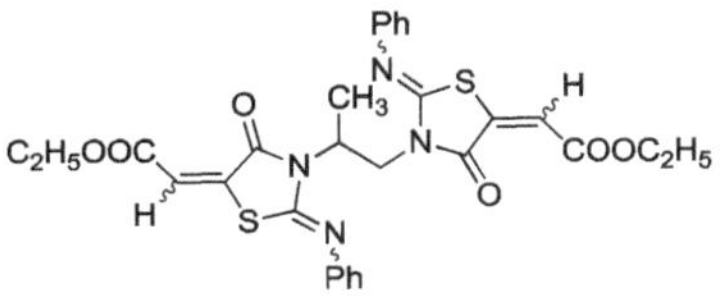

5a

Diethyl 2,2′-{propane-1,2-diyl*bis*[4-oxo-2-(phenylimino)-1,3-thiazolidin-3-yl-5-ylidene]}diethanoate (**5a**): yellow powder, mp 141–143°C; yield: 80%

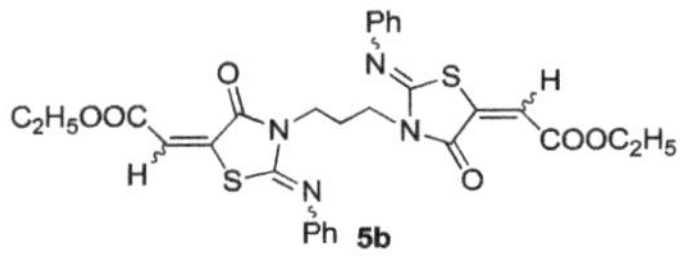

5b

Diethyl 2,2′-{propane-1,3-diyl*bis*[4-oxo-2-(phenylimino)-1,3-thiazolidin-3-yl-5-ylidene]}diethanoate (**5b**): yellow powder, mp 147–150 °C; yield: 80%

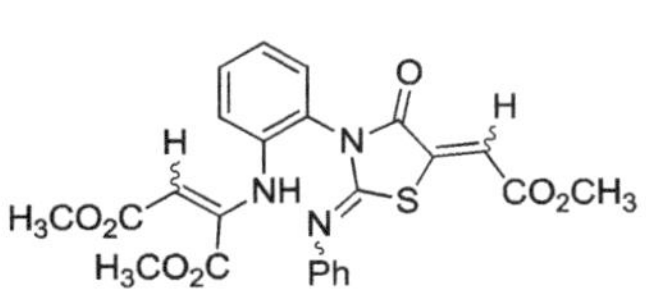

6a

Dimethyl 2-({2-[5-(2-Methoxy-2-oxoethylidene)-4-oxo-2-(phenylimino)-1,3-thiazolidin-3-yl]phenyl}-amino)but-2-enedioate (**6a**): yellow powder, mp 159–161°C; yield: 75%

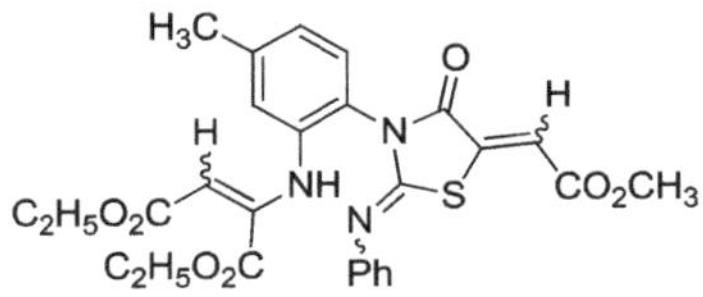

6b

Diethyl 2-({2-[5-(2-Ethoxy-2-oxoethylidene)-4-oxo-2-(phenylimino)-1,3-thiazolidin-3-yl]-4-methylphenyl} amino)but-2-enedioate (**6b**): red liquid; yield: 70%

2.2.72.3 Experimental Procedure

Dialkyl but-2-ynedioates (**1**; 2 mmol) were added slowly to a magnetically stirred solution of isothiocyanatobenzene (**2**; 1 mmol) and aliphatic diamines (**3**; 2 mmol) in dichloromethane (3 mL) at room temperature. The mixture was then stirred for 3 h under ambient conditions to complete the reaction. Upon completion (monitored by TLC), the solvent was evaporated, and the residue was subjected to column chromatography using hexane–ethyl acetate mixtures as eluent to furnish pure products of bis(2-arylimino-1,3-thiazolidin-4-ones) (**5**). Similarly, pure products of 2-(arylimino)-*N*-(enaminoaryl)-1,3-thiazolidin-4-ones (**6**) were prepared using benzene-1,2-diamines (**4**) instead of aliphatic diamines (**3**) under the same reaction conditions. All the products were characterized by means of elemental analyses and detailed spectral studies.

2.2.72.4 Characterization Data of Two Representative Compounds

Diethyl 2,2′-{propane-1,3-diylbis[4-oxo-2-(phenylimino)-1,3-thiazolidin-3-yl-5-ylidene]}diethanoate (**5b**): yellow powder, mp 147–150 °C; yield: 80%; IR (KBr): 1721 (2×CO_2Me), 1688 (2×NCO), 1642 (2×C=N), 1612 (2×C=C), 1193, 1152 (2×C–O) cm^{-1}; ^1H NMR ($CDCl_3$, 500.1 MHz): δ 1.30 (6H, t, J=7.1 Hz,

2×CH_3CH_2), 2.32–2.34 (2H, m, CH_2), 4.12 (4H, t, *J* = 7.1 Hz, 2×CH_2N), 4.26 (4H, q, *J* = 7.1 Hz, 2×$MeCH_2O$), 6.89 (2H, s, 2×=CH), 6.98 (4H, d, *J* = 7.4 Hz), 7.18 (2H, t, *J* = 7.4 Hz), 7.35 (4H, t, *J* = 7.9 Hz); ^{13}C NMR ($CDCl_3$, 125.7 MHz): δ 14.13, 25.78, 40.55, 61.64, 116.63, 121.08, 125.20, 129.32, 141.24, 147.17, 150.94, 164.92, 165.84. EIMS (20 eV, rel.%): *m/z* 592 (10, M^+), 546 (6), 519 (1), 501 (35), 316 (100), 301 (24), 277 (18), 117 (14), 104 (27), 91 (31), 77 (49), 59 (11), 41 (7). Anal. calc. for $C_{29}H_{28}N_4O_6S_2$: C, 58.77; H, 4.76; N, 9.45; found: C, 58.75; H, 4.79; N, 9.44.

Diethyl 2-({2-[5-(2-Ethoxy-2-oxoethylidene)-4-oxo-2-(phenylimino)-1,3-thiazolidin-3-yl]-4-methylphenyl}amino)but-2-enedioate (**6b**): red liquid; yield: 70%; IR (KBr): 1730, 1710 (3×CO_2Me), 1668 (NCO), 1642 (C=N), 1611 (NC=C), 1605 (SC=C), 1267, 1195 (2×C–O) cm^{-1}; ^{1}H NMR ($CDCl_3$, 500.1 MHz): δ 0.97 (3H, t, *J* = 7.0 Hz, CH_3CH_2O), 1.27 (3H, t, *J* = 7.0 Hz, CH_3CH_2O), 1.30 (3H, t, *J* = 7.0 Hz, CH_3CH_2O), 2.34 (3H, s, Me), 4.04 (2H, q, *J* = 7.0 Hz, $MeCH_2O$), 4.16 (2H, q, *J* = 7.0 Hz, $MeCH_2O$), 4.25 (2H, q, *J* = 7.0 Hz, $MeCH_2O$), 5.53 (1H, s, NC=CH), 6.89 (s, 1H), 6.98 (1H, s, =CH), 7.07–7.17 (2H, m), 7.24–7.32 (5H, m), 9.50 (1H, s, NH); ^{13}C NMR ($CDCl_3$, 125.7 MHz): δ 13.99, 14.12, 14.30, 21.17, 59.92, 61.66, 61.86, 95.78, 117.06, 120.99, 125.08, 126.62, 128.14, 128.95, 129.04, 130.83, 138.41, 140.46, 141.26, 147.47, 148.50, 151.18, 163.88, 164.08, 165.89, 168.90; EIMS (20 eV, rel.%): *m/z*: 551 (8, M^+), 504 (7), 478 (10), 459 (100), 393 (5), 365 (20), 308 (25), 275 (8), 254 (40), 185 (14), 135 (10), 108 (10), 77 (20). Anal. calc. for $C_{28}H_{29}N_3O_7S$: C, 60.97; H, 5.30; N, 7.62; found: C, 60.95; H, 5.32; N, 7.65.

2.2.72.5 Critical Views

The investigators offered a general way to prepare synthetically and pharmacologically promising heterocyclic scaffolds, both bis(2-arylimino-1,3-thiazolidin-4-one) and 2-(arylimino)-*N*-(enaminoaryl)-1,3-thiazolidin-4-one, from easily available starting materials *via* a catalyst-free one-pot multicomponent reaction in good yields within reasonable a reaction time. Operational simplicity, easily accessible starting materials, relatively good yields, use of ambient conditions and no need of prior activation or modification of the starting compounds are the notable advantages of this present method.

2.2.73 Entry-73: Synthesis of 2-(4-Oxo-1,3-thiazinan-5-yl)acetic acids

Type of reaction: C–C, C–N and C–S bond formation
Reaction conditions: Catalyst-free, dichloromethane, room temperature
Synthetic strategy: One-pot multicomponent reaction
Keywords: Primary alkylamines, phenyl isothiocyanate/benzoyl isothiocyanate, itaconic anhydride, dichloromethane, catalyst-free, room temperature, one-pot multicomponent reaction, 2-(4-oxo-1,3-thiazinan-5-yl)acetic acids

2.2.73.1 General Reaction Scheme

Yavari *et al.*[227] reported on a one-pot synthesis of 2-(4-oxo-1,3-thiazinan-5-yl)acetic acids (**4**) from a three-component reaction of primary alkylamines (**1**), phenyl isothiocyanate or benzoyl isothiocyanate (**2**), and itaconic anhydride (**3**) in dichloromethane under ambient conditions (Scheme 2.101).

2.2.73.2 Representative Examples

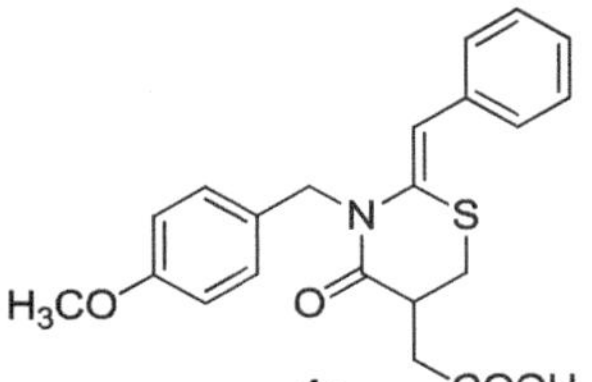

4a

(*Z*)-2-(2-Benzylidene-3-(4-methoxybenzyl)-4-oxo-1,3-thiazinan-5-yl)acetic acid (**4a**): yellow oil; yield: 84%

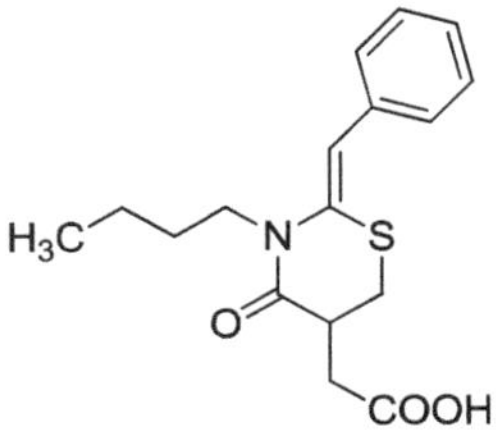

(*Z*)-2-(2-Benzylidene-3-butyl-4-oxo-1,3-thiazinan-5-yl)acetic acid (**4b**): white powder, mp 108–110 °C; yield: 84%

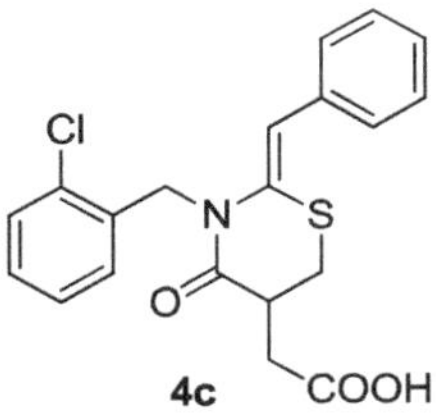

4c

(*Z*)-2-(2-Benzylidene-3-(2-chlorobenzyl)-4-oxo-1,3-thiazinan-5-yl)acetic acid (**4c**): yellow oil; yield: 72%

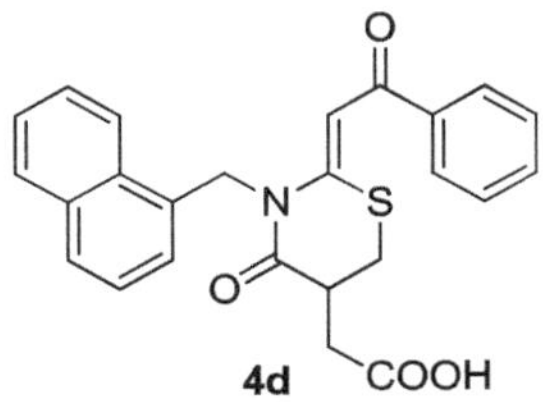

4d

(*Z*)-2-(3-(Naphthalen-1-ylmethyl)-4-oxo-2-(2-oxo-2-phenylethylidene)-1,3-thiazinan-5-yl)acetic acid (**4d**): brown powder, mp 130–132 °C; yield: 80%

2.2.73.3 Experimental Procedure

To a magnetically stirred solution of primary alkylamines (**1**; 2 mmol) and phenyl isothiocyanate or benzoyl isothiocyanate (**2**; 2 mmol) in dichloromethane (10 mL) was added itaconic anhydride (**3**; 2 mmol) at room temperature for 2–5 h. Upon completion of the reaction as indicated by TLC, the solvent was removed under reduced pressure to leave a residue that was purified by column chromatography (SiO_2; hexane–ethyl acetate 5 : 1) to afford the pure products of 2-(4-oxo-1,3-thiazinan-5-yl)acetic acids (**4**), fully characterized by means of elemental analyses and detailed spectral studies.

2.2.73.4 Characterization Data of Two Representative Compounds

(*Z*)-2-(2-Benzylidene-3-butyl-4-oxo-1,3-thiazinan-5-yl)acetic acid (**4b**): white powder, mp 108–110 °C; yield: 84%; IR (KBr): 3385 (br.), 1705, 1700, 1668,

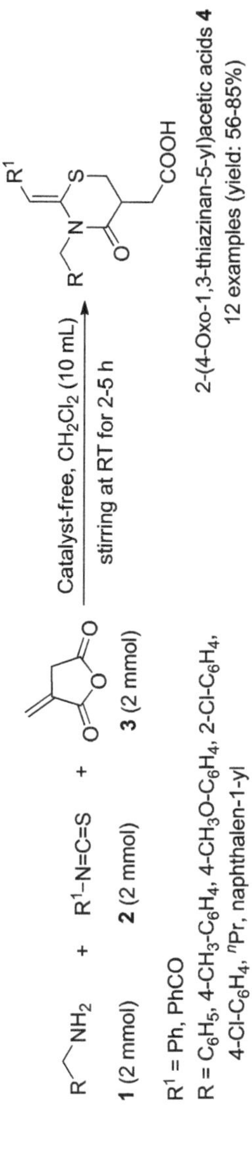

Scheme 2.101 Synthesis of 2-(4-oxo-1,3-thiazinan-5-yl)acetic acids (**4**).

1584, 1435, 1390, 1304, 1181, 1121, 953, 767 cm^{-1}; ^{1}H NMR ($CDCl_3$, 500.1 MHz): δ 0.95 (3H, t, $^{3}J = 7.5$ Hz, Me), 1.37–1.40 (2H, m, CH_2), 1.67–1.70 (2H, m, CH_2), 2.66 (1H, dd, $^{3}J = 6.7$ Hz, $^{2}J = 17.1$ Hz, CH), 2.84 (1H, dd, $^{3}J = 3.8$ Hz, $^{2}J = 12.8$ Hz, CH), 2.99 (1H, dd, $^{3}J = 6.0$ Hz, $^{2}J = 16.9$ Hz, CH), 3.10 (1H, dd, $^{3}J = 12.5$ Hz, $^{2}J = 12.8$ Hz, CH), 3.30–3.33 (1H, m, CH), 4.16–4.19 (2H, m, CH_2), 6.85 (2H, d, $^{3}J = 8.0$ Hz, CH), 7.12 (1H, t, $^{3}J = 7.4$ Hz, CH), 7.34 (2 H, t, $^{3}J = 7.8$ Hz, CH), 10.70 (1H, br. s, OH); ^{13}C NMR ($CDCl_3$, 125.7 MHz): δ 13.8 (Me), 20.1 (CH_2), 26.5 (CH_2), 29.8 (CH_2-S), 35.0 (CH_2), 41.0 (CH), 44.3 (CH_2–N), 120.6 (2×CH), 124.2 (CH), 129.1 (2×CH), 148.1 (C), 150.4 (C=N), 170.6 (C=O); 176.9 (CO_2); EIMS (70 eV; rel.%): *m/z* 153 (8), 134 (45), 77 (80), 59 (10), 45 (5). Anal. calc. for $C_{16}H_{20}N_2O_3S$: C, 59.98; H, 6.29; N, 8.74; found: C, 59.88; H, 6.20; N, 8.63.

(*Z*)-2-(3-(Naphthalen-1-ylmethyl)-4-oxo-2-(2-oxo-2-phenylethylidene)-1,3-thiazinan-5-yl)acetic acid (**4d**): brown powder, mp 130–132 °C; yield: 80%; IR (KBr): 3371 (br.), 1725, 1700, 1690, 1634, 1490, 1432, 1411, 1303, 1269, 1215, 1165, 1084 cm^{-1}; ^{1}H NMR ($CDCl_3$, 500.1 MHz): δ 2.89 (1H, dd, $^{3}J = 6.0$ Hz, $^{2}J = 17.5$ Hz, CH), 3.04 (1H, dd, $^{3}J = 5.2$ Hz, $^{2}J = 17.5$ Hz, CH), 3.25 (1H, dd, $^{3}J = 3.5$ Hz, $^{2}J = 12.5$ Hz, CH), 3.58 (1H, dd, $^{3}J = 10.7$ Hz, $^{2}J = 11.1$ Hz, CH), 3.60–3.62 (1H, m, CH), 5.94 (2H, AB, $\Delta\nu_{AB} = 82.3$, $^{2}J = 17.8$ Hz, CH_2), 7.17 (t, $^{3}J = 7.5$ Hz, 2H, CH), 7.33 (1H, d, $^{3}J = 7.0$ Hz, CH), 7.40–7.43 (3H, m, CH), 7.58–7.61 (2H, m, CH), 7.82 (1H, d, $^{3}J = 7.9$ Hz, CH), 7.95 (2H, d, $^{3}J = 7.5$ Hz, CH), 8.22 (1H, d, $^{3}J = 8.0$ Hz, CH), 10.49 (1H, br s, OH); ^{13}C NMR ($CDCl_3$, 125.7 MHz): δ 26.8 (CH_2-S), 35.0 (CH_2), 41.8 (CH), 45.9 (CH_2–N), 123.3 (CH), 123.7 (CH), 126.2 (CH), 126.5 (CH), 127.1 (CH), 128.0 (CH), 128.9 (2 CH), 129.6 (CH), 130.2 (2×CH), 132.4 (C), 133.4 (C), 133.5 (CH), 134.7 (C), 135.9 (C), 161.5 (C=N), 167.7 (C=O), 172.7 (CO_2), 176.2 (C=O); EIMS (70 eV; rel.%): *m/z* 153 (5), 136 (80), 77 (85), 59 (5), 45 (10). Anal. calc. for $C_{24}H_{20}N_2O_4S$: C, 66.65; H, 4.66; N, 6.48; found: C, 66.52; H, 4.51; N, 6.52.

2.2.73.5 Critical Views

The investigators synthesized a novel series of polyfunctionalized heterocycles, 2-(4-oxo-1,3-thiazinan-5-yl)acetic acids, *via* a simple, efficient and convenient one-pot method from inexpensive starting materials without any prior activation or modification.

2.2.74 Entry-74: Synthesis of Functionalized Pyrano[3,2-*c*]chromen-5(4*H*)-ones

Type of reaction: C–C and C–O bond formation
Reaction conditions: Catalyst-free, water-ethanol, room temperature
Synthetic strategy: One-pot multicomponent reaction
Keywords: Aldehydes, acetohenones, 4-hydroxycoumarin, aqueous ethanol, catalyst-free, room temperature, one-pot multicomponent reaction, 2-aryl-4-alkyl/aryl-pyrano[3,2-*c*]chromen-5(4*H*)-ones, 3-/4-(5-oxo-2-aryl-4,5-dihydropyrano-[3,2-*c*]chromen-4-yl)benzaldehydes

2.2.74.1 General Reaction Scheme

Recently, Brahmachari and Nurjamal[228] synthesized a series of pharmaceutically interesting diverse 2-aryl-4-alkyl/aryl-pyrano[3,2-*c*]chromen-5(4*H*)-ones (**4**) and 3-/4-(5-oxo-2-aryl-4,5-dihydropyrano[3,2-*c*]chromen-4-yl)benzaldehydes (**4′**) from the one-pot multicomponent reaction between alkyl/aryl aldehydes (**1**/**1′**), acetophenones (**2**) and 4-hydroxycoumarin (**3**) in aqueous ethanol at ambient conditions without the aid of any catalyst (Scheme 2.102). The key advantages of this newly developed protocol are the clean reaction profile, no use of catalyst and no toxic organic solvents, mild reaction conditions at room temperature, energy efficiency, use of commercially available low-cost starting materials, high atom-economy, and good to excellent yields.

2.2.74.2 Plausible Mechanism

The investigators suggested a plausible mechanism for this catalyst-free transformation based on pH monitoring throughout the progress of the reaction as well as isolation and characterization of intermediate as outlined in Scheme 2.103.[228]

2.2.74.3 Representative Examples

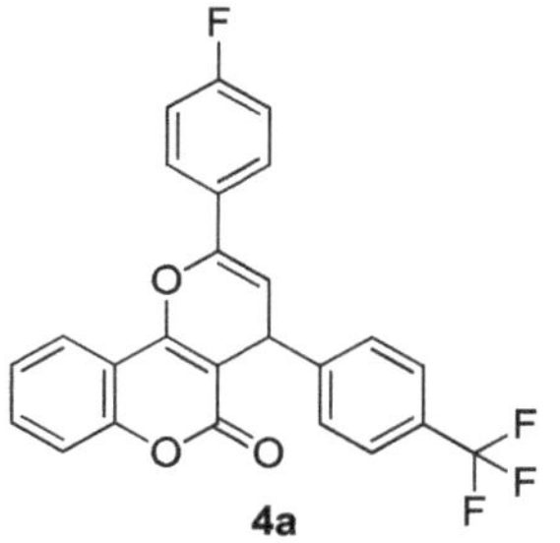

4a

4-(4-(Trifluoromethyl)phenyl)-2-(4-fluorophenyl)pyrano[3,2-*c*]chromen-5(4*H*)-one (**4a**): white powder, mp 254–256 °C, time: 12 h; yield: 84%

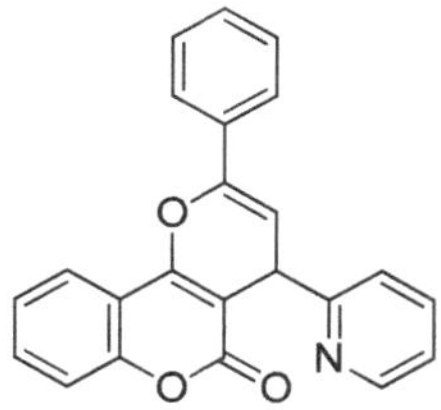

4b

2-Phenyl-4-(pyridin-2-yl)pyrano[3,2-*c*]chromen-5(4*H*)-one (**4b**): white powder, mp 238–240 °C, time: 12 h; yield: 91%

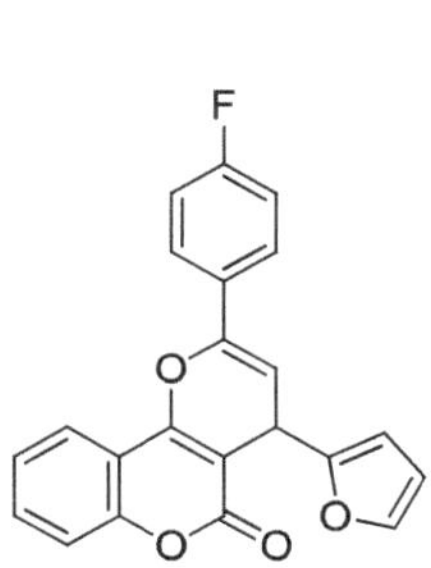

4c

2-(4-Fluorophenyl)-4-(furan-2-yl)pyrano[3,2-*c*]chromen-5(4*H*)-one (**4c**): dark gray powder, mp 194–195 °C, time: 12 h; yield: 82%

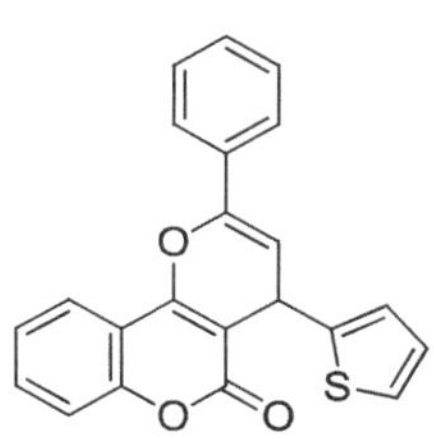

4d

2-Phenyl-4-(thiophen-2-yl)pyrano[3,2-*c*]chromen-5(4*H*)-one (**4d**): pale yellow powder, mp 187–189 °C; yield: 86%

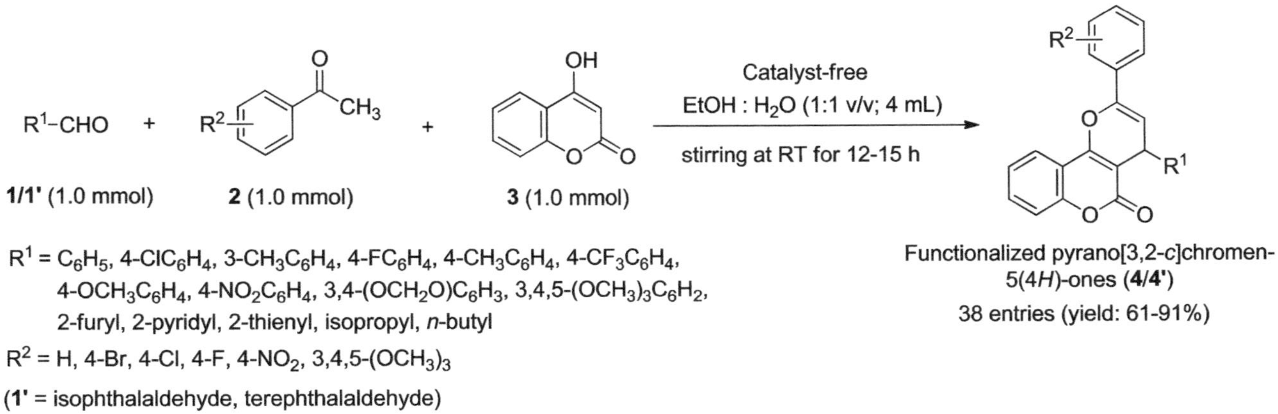

Scheme 2.102 Synthesis of functionalized pyrano[3,2-*c*]chromen-5(4*H*)-ones (**4**/**4′**).

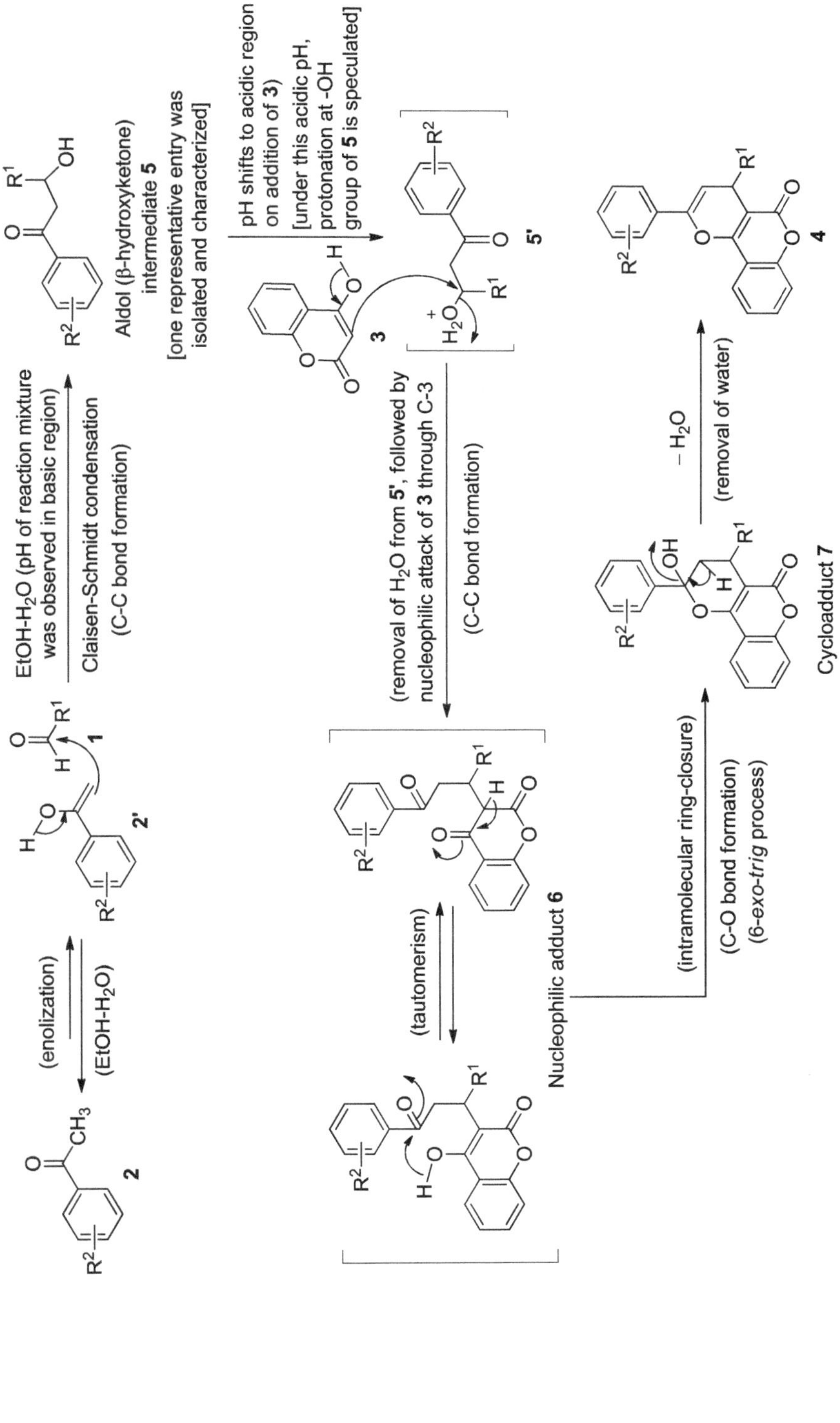

Scheme 2.103 Plausible mechanism for the gerneration of pyrano[3,2-*c*]chromen-5(4*H*)-ones (**4**/**4'**).

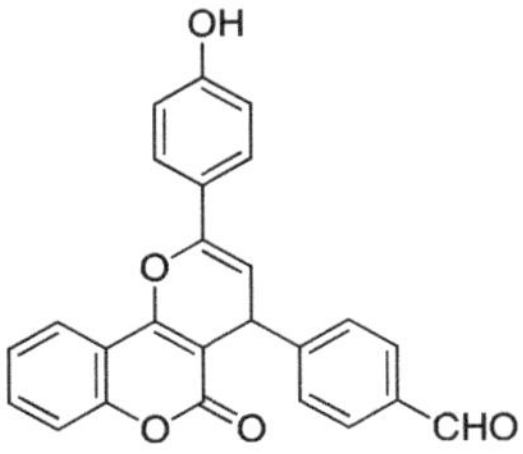

4'a

4-(2-(4-Hydroxyphenyl)-5-oxo-4,5-dihydropyrano[3,2-*c*]chromen-4-yl)benzaldehyde (**4'a**):white solid,mp 176–178°C, time: 14 h; yield: 69%

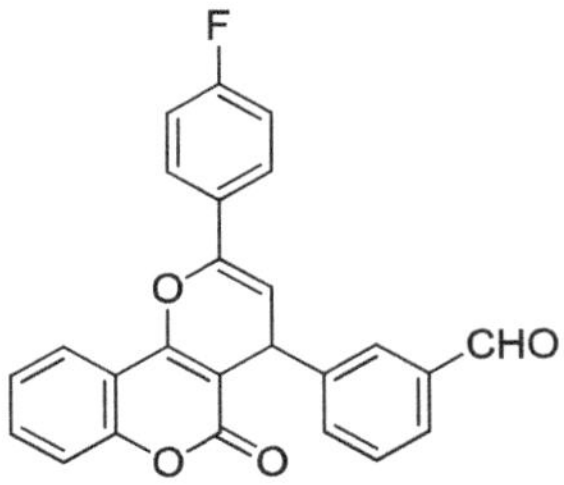

4'b

3-(2-(4-Fluorophenyl)-5-oxo-4,5-dihydropyrano[3,2-*c*]chromen-4-yl)benzaldehyde (**4'b**): white solid, mp 193–194 °C, time: 15 h; yield: 79%

2.2.74.4 Experimental Procedure

An oven-dried sealed tube was charged with a magnetic stir bar, aldehyde (**1**/**1**′ : 1 mmol), acetophenone (**2**; 1 mmol) and 4 mL aqueous ethanol (1 : 1 v/v) in a sequential manner at ambient conditions (25–30 °C), and the reaction mixture was then stirred vigorously for about 2–3 h. After that 4-hydroxycoumarin (**3**; 1 mmol) was added to the stirred reaction mixture and the stirring was continued for up to 10–12 h under room temperature conditions. The progress of the reaction was monitored by TLC. On completion of the reaction, a solid mass precipitated out which was filtered off, followed by purification of the crude product either by washing with cold aqueous ethanol and recrystallization from ethyl acetate or by using column chromatography (petroleum ether–ethyl acetate mixtures as eluents) to furnish pure product (**4**/**4**′). The structure of each purified compound was confirmed by analytical, as well as spectral, studies.

2.2.74.5 Characterization Data of Three Representative Compounds

4-(4-(Trifluoromethyl)phenyl)-2-(4-fluorophenyl)pyrano[3,2-*c*]chromen-5(4*H*)-one (**4a**): white powder, mp 254–256 °C, yield: 84%; IR (KBr): 3074, 3017, 2990, 2727, 1670, 1609, 1563, 1493, 1449, 1348, 1327, 1218, 1171, 1122, 1065, 1020, 957, 836, 768, 671, 545, 464 cm^{-1}. ^{1}H NMR (DMSO-d_6, 400 MHz): δ 7.88 (2H, d, $J = 7.6$ Hz, Ar–H), 7.59–7.56 (4H, m, Ar–H), 7.37–7.33 (4H, m, Ar–H), 7.32–7.28 (2H, m, Ar–H), 6.38 (1H, br s, vinylic H), 4.29 (1H, br s, –CH); ^{13}CNMR (DMSO-d_6, 100 MHz): δ 166.51 (CF), 164.57 (CO), 152.40 (2C), 146.38, 131.69 (2C), 127.51 (2C), 126.28, 125.97, 124.87, 124.05, 123.48 (2C), 123.22 (2C), 118.67 (2C), 115.83 (2C), 103.39 (2C), 36.34 (C-4), 22.26 (d, $J = 41$ Hz, –CF_3); TOF-MS: *m/z* calc. for $[C_{25}H_{14}F_4O_3Na]^+ = [M + Na]^+ = 461.0777$; found 461.0764. Elemental analysis: calc. (%) for $C_{25}H_{14}F_4O_3$: C, 68.50; H, 3.22; found: C, 68.62; H, 3.21.

2-(4-Fluorophenyl)-4-(furan-2-yl)pyrano[3,2-*c*]chromen-5(4*H*)-one (**4c**): Dark gray powder, mp 194–195 °C, yield: 82%; IR (KBr): 3108, 3074, 1660, 1608, 1567,

1500, 1455, 1348, 1312, 1265, 1211, 1184, 1101, 1096, 1014, 951, 905, 811, 770, 729, 674, 602, 552, 514, 462 cm^{-1}. ^1H NMR (DMSO-d_6, 400 MHz): δ 7.91 (2H, d, $J = 7.6$ Hz, Ar–*H*), 7.58 (2H, t, $J = 7.6$ Hz, Ar–H), 7.45 (1H, br s, Ar–H), 7.35–7.29 (4H, m, Ar–H), 6.31 (1H, br s, Ar–H), 6.27 (1H, br s, Ar–H), 6.05 (1H, br s, vinylic H), 4.85 (1H, br s, –CH); ^{13}CNMR (DMSO-d_6, 100 MHz):δ 165.71 (CO), 164.37, 153.27, 152.26 (2C), 141.40 (2C), 132.01 (2C), 124.02 (2C), 123.78 (2C), 118.09, 116.00 (2C), 110.28 (2C), 106.00, 102.61 (2C), 31.60 (C-4): TOF-MS: *m/z* calc. for $[C_{22}H_{13}FO_4Na]^+ = [M+Na]^+ = 383.0696$; found 383.0690. Elemental analysis: calc. (%) for $C_{22}H_{13}FO_4$: C, 73.33; H, 3.64; found: C, 73.41; H, 3.62.

4-(2-(4-Hydroxyphenyl)-5-oxo-4,5-dihydropyrano[3,2-*c*]chromen-4-yl)benzaldehyde (**4′a**): white solid, mp 176–178 °C, yield: 69%; IR (KBr): 3634, 3092, 3035, 1705, 1660, 1608, 1564, 1494, 1449, 1350, 1309, 1273, 1214, 1175, 1096, 1053, 1024, 958, 912, 816, 769, 663, 551, 496 cm^{-1}. ^1H NMR (DMSO-d_6, 400 MHz): δ 9.93 (1H, s, –CHO), 7.86 (2H, dd, $J = 7.6$, 1.6 & 1.2 Hz, Ar–H), 7.76 (2H, d, $J = 8.0$ Hz, Ar–H), 7.57 (2H, dt, $J = 8.0$, 7.6 & 1.6 Hz, Ar–H), 7.37–7.32 (4H, m, Ar–H), 7.29 (2H, dt, $J = 7.6$, 7.2 & 0.8 Hz, Ar–H), 6.38 (1H, br s, vinylic H), 4.16 (1H, br s, –CH) (phenolic OH did not appear up to 18 ppm); ^{13}CNMR (DMSO-d_6, 100 MHz):δ 193.01 (CHO), 167.21 (CO), 164.99 (2C), 152.87, 149.79, 134.41 (2C), 132.02 (2C), 129.87 (2C), 127.92 (2C), 124.52 (2C), 123.84 (2C), 119.33 (2C), 116.23 (2C), 103.80 (2C), 37.18 (C-4); TOF-MS: *m/z* calc. for $[C_{25}H_{16}O_5Na]^+ = [M+Na]^+ = 419.0895$; found 419.0902. Elemental analysis: calc. (%) for $C_{25}H_{16}O_5$: C, 75.75; H, 4.07; found: C, 75.79; H, 4.08.

2.2.74.6 Critical Views

A simple, straightforward and highly efficient catalyst-free synthesis of pharmaceutically-interesting,[229–234] diverse kind of functionalized 2-aryl-4-alkyl/aryl-pyrano[3,2-*c*]chromen-5(4*H*)-ones (**4a-4z″**) and 3-/4-(5-oxo-2-aryl-4,5-dihydropyrano[3,2-*c*]chromen-4-yl)benzaldehydes (**4′a–4′j**) was designed by the present investigators.[228] The synthesis was based on a one-pot multicomponent reaction in aqueous ethanol at ambient conditions, and this protocol is advantageous over the earlier methods,[235–247] particularly in terms of the mild reaction conditions, use of no catalyst, good to excellent yields, high atom-economy, eco-friendliness, easy isolation of products, and reusability of reaction media.

2.2.75 Entry-75: Synthesis of Functionalized Pyrano[3,2-*c*]chromen-5(4*H*)-ones

Type of reaction: C–C and C–O bond formation
Reaction conditions: Catalyst-free, water-ethanol, room temperature
Synthetic strategy: One-pot multicomponent reaction
Keywords: Barbituric/2-thiobarbituric acids, aldehydes, 2-hydroxy-1,4-naphthoquinone, aqueous ethanol, catalyst-free, room temperature, one-pot multicomponent reaction, 5-aryl-2-oxo-/thioxo-2,3-dihydro-1*H*-benzo[6,7]-chromeno[2,3-*d*]pyrimidine-4,6,11(5*H*)-triones

2.2.75.1 General Reaction Scheme

A novel series of functionalized 5-aryl-2-oxo-/thioxo-2,3-dihydro-1*H*-benzo[6,7]chromeno[2,3-*d*]pyrimidine-4,6,11(5*H*)-triones (**4**) has recently been designed by Brahmachari and Nayek.[248] The syntheses utilize a convenient, clean, facile and catalyst-free one-pot multicomponent reaction between barbituric/2-thiobarbituric acids (**1**), aromatic aldehydes (**2**) and 2-hydroxy-1,4-naphthoquinone (**3**) in aqueous ethanol at ambient conditions (Scheme 2.104).

2.2.75.2 Plausible Mechanism

The investigators suggested a plausible mechanism for this catalyst-free transformation based on pH monitoring throughout the progress of the reaction as well as isolation and characterization of the intermediate as outlined in Scheme 2.105.[248]

2.2.75.3 Representative Examples

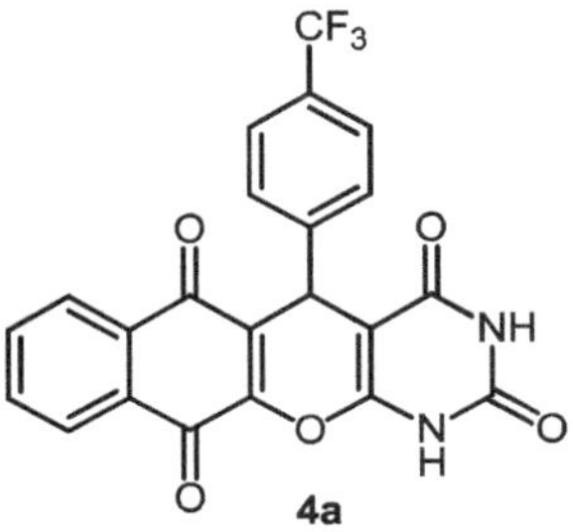

4a

5-(4-(Trifluoromethyl)phenyl)-1*H*-benzo[6,7]chromeno[2,3-*d*]pyrimidine-2,4,6,11(3*H*,5*H*)-tetraone (**4a**): reddish-yellow solid, mp 188–190 °C; yield: 96%

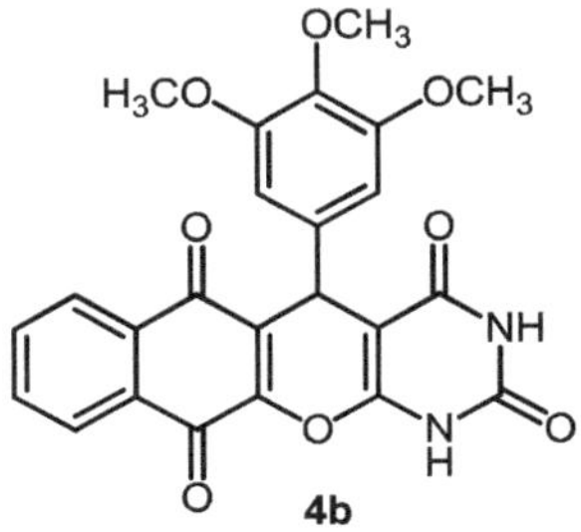

4b

5-(3,4,5-Trimethoxyphenyl)-1*H*-benzo[6,7]chromeno[2,3-*d*]pyrimidine-2,4,6,11(3*H*,5*H*)-tetraone (**4b**): pale yellow solid, mp 209–210 °C; yield: 98%

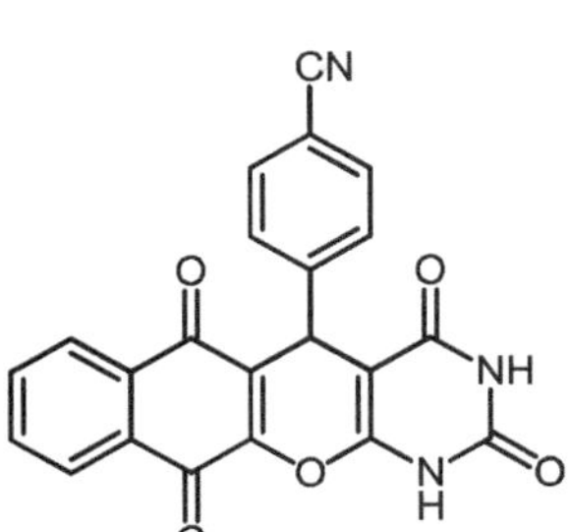

4c

4-(2,4,6,11-Tetraoxo-2,3,4,5,6,11-hexahydro-1*H*-benzo[6,7]chromeno[2,3-d]pyrimidin-5-yl)benzonitrile (**4c**): pale yellow solid, mp 188–190°C; yield: 98%

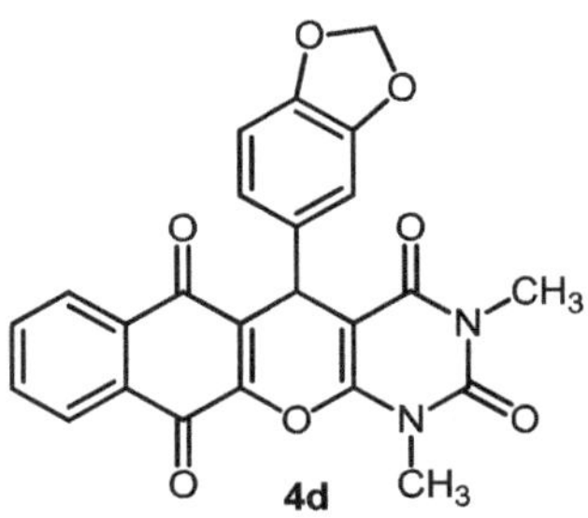

4d

5-(Benzo[*d*][1,3]dioxol-5-yl)-1,3-dimethyl-1*H*-benzo[6,7]chromeno[2,3-*d*]pyrimidine-2,4,6,11(3*H*,5*H*)-tetraone (**4d**): golden yellow solid, mp 178–179 °C; yield: 98%

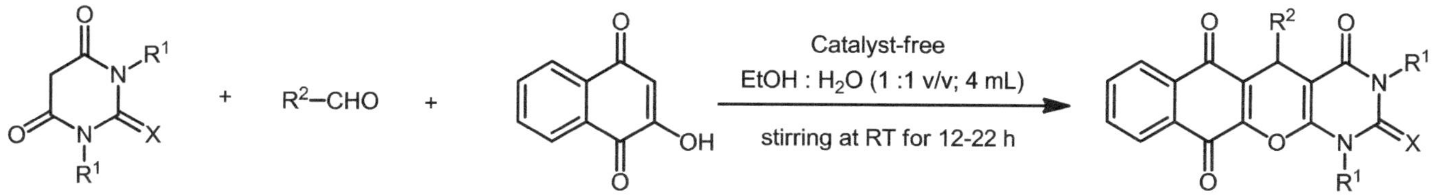

R^1 = H, CH_3; X = O, S

R^2 = C_6H_5, 3-BrC_6H_4, 4-BrC_6H_4, 4-CNC_6H_4, 4-$CF_3C_6H_4$, 4-$OCH_3C_6H_4$, 2,5-di-$OCH_3C_6H_3$, 3,4,5-tri-$OCH_3C_6H_2$, 3,4-$(OCH_2O)C_6H_3$, 4-$CH_3C_6H_4$, 3-$CHOC_6H_4$, 4-$CHOC_6H_4$, 3-$NO_2C_6H_4$, 4-$NO_2C_6H_4$, 2-furyl

Scheme 2.104 Synthesis of functionalized 1*H*-benzo[6,7]chromeno[2,3-*d*]pyrimidines (**4**).

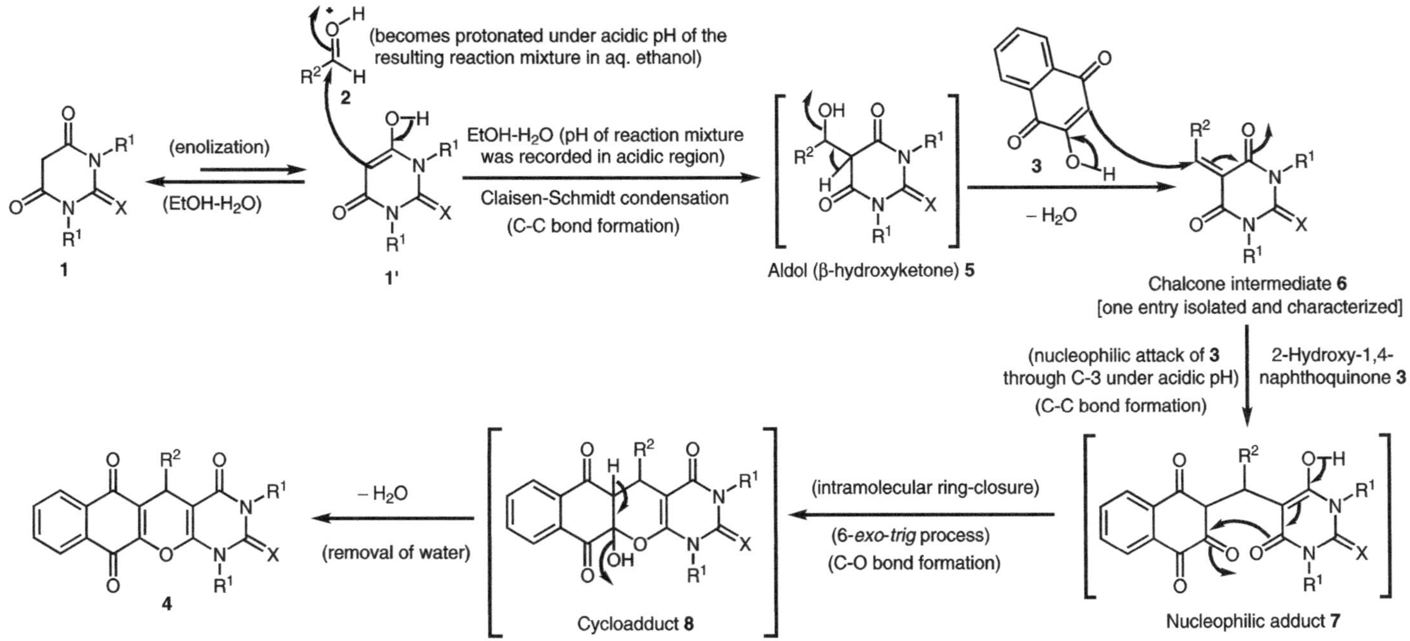

Scheme 2.105 Plausible mechanism for the gerneration of 1*H*-benzo[6,7]chromeno[2,3-*d*]pyrimidines (**4**).

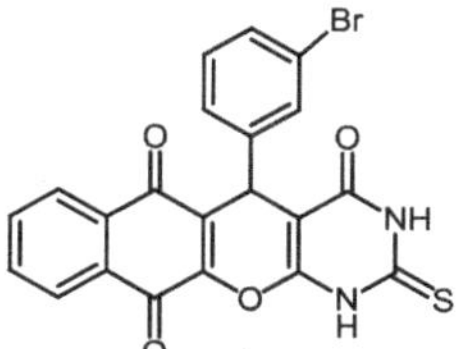

4e

5-(3-Bromophenyl)-2-thioxo-2,3-dihydro-1*H*-benzo[6,7]chromeno[2,3-*d*]pyrimidine-4,6,11(5*H*)-trione (**4e**): pale yellow solid, mp 193–195 °C; yield: 98%

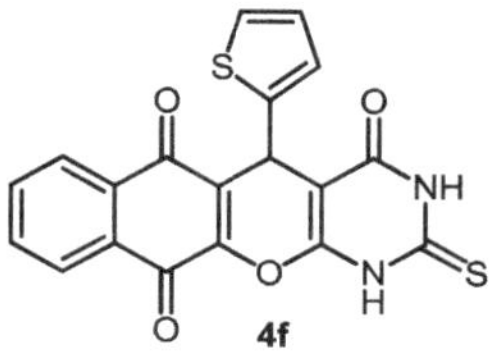

4f

5-(Furan-2-yl)-2-thioxo-2,3-dihydro-1*H*-benzo[6,7]chromeno[2,3-*d*]pyrimidine-4,6,11(5*H*)-trione (**4f**): dark yellow solid, mp 162–166 °C; yield: 96%

2.2.75.4 *Experimental Procedure*

An oven-dried sealed tube was charged with a magnetic stir bar, aldehyde (**1**/**1′**; 1 mmol), acetophenone (**2**; 1 mmol) and 4 mL aqueous ethanol (1 : 1 v/v) in a sequential manner at ambient conditions (25–30 °C), and the reaction mixture was then stirred vigorously for about 2–3 h. After that 4-hydroxycoumarin (**3**; 1 mmol) was added to the stirred reaction mixture and the stirring was continued for up to 10–12 h under room temperature conditions. The progress of the reaction was monitored by TLC. On completion of the reaction, a solid mass precipitated out, which was filtered off, followed by purification of the crude product either by washing with cold aqueous ethanol and recrystallization from ethyl acetate or by using column chromatography (petroleum ether–ethyl acetate mixtures as eluents) to furnish pure product (**4**/**4′**). The structure of each purified compound was confirmed by analytical, as well as spectral, studies.

2.2.75.5 *Characterization Data of Three Representative Compounds*

5-(4-(Trifluoromethyl)phenyl)-1*H*-benzo[6,7]chromeno[2,3-*d*]pyrimidine-2,4,6,11(3*H*,5*H*)-tetraone (**4a**): reddish-yellow solid, mp 188–190 °C; yield: 96%; IR (KBr): 3234 (NH), 3190, 2979, 2869, 2366, 1711(CO), 1691(CO), 1670 (CONH), 1605, 1595, 1565, 1465, 1370, 1324, 1282, 1164, 1120, 1068, 953, 790, 732, 601, 540, 457 cm^{-1}; ^{1}H NMR (DMSO-d_6, 400 MHz): δ 9.95 (2H, br s, NH), 7.97 (1H, d, $J = 7.2$ Hz, Ar–H), 7.90 (1H, dd, $J = 7.6$ & 1.2 Hz, Ar–H), 7.78 (1H, dt, $J = 7.6$ & 1.2 Hz, Ar–H), 7.73–7.69 (1H, m, Ar–H), 7.65 (1H, t, $J = 8.0$ & 6.0 Hz, Ar–H), 7.50 (2H, d, $J = 8.0$ Hz , Ar–H), 7.27 (1H, d, $J = 8.0$ Hz, Ar–H), 6.37 (1H, br s, –CH); ^{13}CNMR (DMSO-d_6, 100 MHz):δ 183.84 (2×CO), 172.59 (CONH), 165.81 (*C*ONH), 151.55, 151.05, 148.96, 134.51 (2 C), 132.91, 131.41, 130.76, 129.25, 128.04 (2 C), 126.41, 126.28, 125.78, 125.13 (*C*F_3), 125.10, 88.81, 49.04 (–CH). Elemental analysis: calc. (%) for $C_{22}H_{11}N_2O_5F_3$: C, 60.01; H, 2.52, N, 6.36; found: C, 60.03; H, 2.51, N, 6.34.

5-(Benzo[*d*][1,3]dioxol-5-yl)-1,3-dimethyl-1*H*-benzo[6,7]chromeno[2,3-*d*]pyrimidine-2,4,6,11(3*H*,5*H*)-tetraone (**4d**): golden yellow solid, mp 178–179 °C;

yield: 98%; IR (KBr): 3097, 2992, 2957, 2791, 2361, 1705 (CO), 1673 ($CONCH_3$), 1647, 1588, 1494, 1449, 1420, 1368, 1350, 1301, 1275, 1241, 1216, 1169, 1140, 1123, 1100, 1035, 987, 969, 931, 861, 814, 791, 745, 698, 658, 608, 576, 513, 477, 418 cm^{-1}; ^1H NMR (DMSO-d_6, 400 MHz): δ 7.99–7.95 (2H, m, Ar–H), 7.85–7.77 (2H, m, Ar–H), 7.08 (1H, d, $J=8.0$ Hz, Ar–H), 6.79–6.78 (1H, m, Ar–H), 6.74 (1H, d, $J=8.0$ Hz, Ar–H), 5.93 (2H, s, OCH_2O), 4.85 (1H, br s, –CH), 3.01 (6H, s, $-NCH_3$); ^{13}CNMR (DMSO-d_6, 100 MHz):δ 184.27 (2×CO), 168.05 (*C*ONCH$_3$), 165.37 ($CONCH_3$), 156.11, 152.24 (3 C), 147.12 (2 C), 135.29, 133.77 (2 C), 126.48 (2 C), 126.12, 107.97 (3 C), 102.77, 101.18 ($-OCH_2O-$), 43.62 (–CH), 29.07 ($-NCH_3$), 28.44 (NCH_3); elemental analysis: calc. (%) for $C_{24}H_{16}N_2O_7$: C, 64.87; H, 3.63; N, 6.30; found: C, 64.84; H, 3.61; N, 6.28.

5-(Furan-2-yl)-2-thioxo-2,3-dihydro-1*H*-benzo[6,7]chromeno[2,3-*d*]pyrimidine-4,6,11(5*H*)-trione (**4f**): dark yellow solid, mp 162–166 °C; yield: 96%; IR (KBr): 3139 (NH), 3071, 3014, 2360, 1701 (CO), 1645 (CONH), 1555, 1523, 1449, 1378, 1352 (C=S), 1285, 1228, 1174, 1144, 1108, 1031, 980, 961, 926, 875, 786, 747, 733, 676, 636, 586, 527, 462, 409 cm^{-1}; ^1H NMR ($CDCl_3$, 400 MHz): δ 8.11 (4H, d, $J=7.6$ Hz, Ar–H), 7.79 (2H, dt, $J=7.6$ & 0.8 Hz, Ar–H), 7.72 (1H, dt, $J=7.6$, 7.0 & 1.2 Hz, Ar–H), 7.45 (2H, br s, –NH), 6.36 (1H, s, –CH); ^{13}CNMR (DMSO-d_6, 100 MHz):δ 185.14 (C=S), 182.09 (2×CO), 167.49 (CO), 156.51, 152.87, 135.43, 133.29, 133.04, 129.57, 126.85 (2 C), 126.65, 120.50, 117.16, 110.87 (2 C), 107.34, 48.03 (–CH). Elemental analysis: calc. (%) for $C_{19}H_{10}N_2O_5S$: C, 60.31; H, 2.66; N, 7.40; found: C, 60.26; H, 2.65; N, 7.38.

2.2.75.6 Critical Views

A novel series of pharmaceutically interesting functionalized 5-aryl-2-oxo-/thioxo-2,3-dihydro-1*H*-benzo[6,7]chromeno[2,3-*d*]pyrimidine-4,6,11(5*H*)-trione derivatives has been synthesized for the first time in aqueous ethanol at ambient temperature. The key advantages of this newly developed protocol are the clean reaction profile, use of no catalyst and no toxic organic solvents, mild reaction conditions at room temperature, energy efficiency, use of commercially available low-cost starting materials, no need of column chromatographic purification, high atom-economy, low E-factors, and excellent yields.[248]

2.3 Concluding Remarks

A good deal of innovative synthetic protocols occurring at room temperature and pressure without the aid of any catalyst are presented in this chapter. Each reaction is described in detail, covering all possible aspects including reaction type, nature of bonds and compounds generated, reaction conditions, general reaction scheme, illustrative examples, experimental procedure, and characterization data of representative compounds with a motto that a researcher and/or an advanced chemistry student can easily synthesize a compound of choice following a particular synthetic protocol in laboratory, followed by its characterization. In addition, each reaction strategy is critically overviewed in respect of how it attains green credentials and how it is advantageous over previously reported method(s), if any. This chapter underlines the fact that the design of reaction protocols has the power to materialize catalyst-free syntheses of a huge array of diverse molecular scaffolds in a convenient manner, even without any activation or modification of starting meterials. It is believed that this trend of research will be explored more in the near future.

References

1. P. T. Anastas and J. C. Warner, *Green Chemistry: Theory and Practice*, Oxford University Press, Oxford, 2000.
2. G. Brahmachari, *Green Synthetic Approaches for Biologically Relevant Heterocycles*, Elsevier, Amsterdam, The Netherlands, 2015.
3. S. Abou-Shehada, P. Mampuys, B. U. W. Maes, J. H. Clark and L. Summerton, *Green Chem.*, 2017, **19**, 249.
4. N. J. Willis, C. A. Fisher, C. M. Alder, A. Harsanyi, L. Shukla, J. P. Adams and G. Sandford, *Green Chem.*, 2016, **18**, 1313.
5. F. Roschangar, R. A. Sheldon and C. H. Senanayake, *Green Chem.*, 2015, **17**, 752.
6. C. Jimenez-Gonzalez, D. J. C. Constable and C. S. Ponder, *Chem. Soc. Rev.*, 2012, **41**, 1485.
7. C. Jimenez-Gonzalez, C. S. Ponder, Q. B. Broxterman and J. B. Manley, *Org. Process Res. Dev.*, 2011, **15**, 912.
8. J. Auge, *Green Chem.*, 2008, **10**, 225.
9. D. J. C. Constable, A. D. Curzons and V. L. Cunningham, *Green Chem.*, 2002, **4**, 521.
10. C. Jimenez-Gonzalez, A. D. Curzons, D. J. C. Constable, M. R. Overcash and V. L. Cunningham, *Clean Prod. Processes*, 2001, **3**, 35.
11. G. Brahmachari, *Room Temperature Organic Synthesis*, Elsevier, Amsterdam, The Netherlands, 2015.
12. G. Brahmachari and B. Banerjee, *Curr. Green Chem.*, 2015, **2**, 274.
13. P. Galletti, M. Pori and D. Giacomini, *Eur. J. Org. Chem.*, 2011, 3896.
14. M. S. Sigman and E. N. Jacobsen, *J. Am. Chem. Soc.*, 1998, **120**, 4901.
15. S. Masumoto, H. Usuda, M. Suzuki, M. Kanai and M. Shibasaki, *J. Am. Chem. Soc.*, 2003, **125**, 5634.
16. B. F. Li, K. Yuan, M. J. Zhang, H. Wu, L. X. Dai, Q. R. Wang and X. L. Hou, *J. Org. Chem.*, 2003, **68**, 6264.
17. J. S. Yadav, B. V. S. Reddy, B. Eeshwaraiah and M. Srinivas, *Tetrahedron*, 2004, **60**, 1767.
18. L. Royer, S. K. De and R. A. Gibbs, *Tetrahedron Lett.*, 2005, **46**, 4595.
19. G. K. S. Prakash, T. Mathew, C. Panja, S. Alconcel, H. Vaghoo, C. Do and G. Olah, *Proc. Natl. Acad. Sci. U. S. A.*, 2007, **104**, 3703.
20. A. Majhi, S. S. Kim and H. S. Kim, *Appl. Organomet. Chem.*, 2008, **22**, 466.

21. Z. L. Shen, S. J. Ji and T. P. Loh, *Tetrahedron*, 2008, **64**, 8159.
22. P. Merino, E. Marqués-López, T. Tejero and R. P. Herrera, *Tetrahedron*, 2009, **65**, 1219.
23. B. Karmakar and J. Banerji, *Tetrahedron Lett.*, 2010, **51**, 2748.
24. J. Wang, Y. Masui and M. Onaka, *Eur. J. Org. Chem.*, 2010, 1763.
25. M. G. Dekamin and Z. Mokhtatari, *Tetrahedron*, 2012, **68**, 922.
26. G. Brahmachari and B. Banerjee, *Asian J. Org. Chem.*, 2012, **1**, 251.
27. J.-J. Yu, L.-M. Wang, J.-Q. Liu, F.-L. Guo, Y. Liu and N. Jiao, *Green Chem.*, 2010, **12**, 216.
28. B. A. Marco, T. A. Manuel, P. M. Itzia, M. M. Francisco, E. Georgina, M. Elies and E. Enrique, *J. Chem. Crystallogr.*, 1999, **29**, 759.
29. G. Cravotto, A. Demetri, G. M. Nano, G. Palmmisano, A. Penoni and S. Tagliapietra, *Eur. J. Org. Chem.*, 2003, 4438.
30. K. M. Khan, G. M. Maharvi, M. T. H. Khan, A. J. Shaikh, S. Perveen, S. Begun and M. I. Choudhary, *Bioorg. Med. Chem.*, 2006, **14**, 344.
31. D. B. Ramachary and M. Kishor, *J. Org. Chem.*, 2007, **72**, 5056.
32. A. Olyaei, E. C. Parashkuhi, S. Raoufmoghaddam and M. Sadeghpour, *Synth. Commun.*, 2010, **40**, 3609.
33. N. Azizi, E. Akbari, F. Ebrahimi and M. R. Saidi, *Monatsh. Chem.*, 2010, **141**, 323.
34. A. D. Innocenti, A. Capperucci, A. Cerreti, S. Pollicino, S. Scappecchi, I. Malesci and G. Castagnoli, *Synlett*, 2005, 3063.
35. A. Z. Halimehjani, A. Jalali, M. Khalesi, A. Ashouri and K. Marjani, *Synth. Commun.*, 2011, **41**, 1638.
36. K. Marjani, M. Khalesi, A. Ashouri, A. Jalali and A. Ziyaei-Halimehjani, *Synth. Commun.*, 2011, **41**, 451.
37. G. L. Khatik, R. Kumar and A. K. Chakraborti, *Org. Lett.*, 2006, **8**, 2433.
38. N. Azizi, F. Aryanasab and M. R. Saidi, *Org. Lett.*, 2006, **8**, 5275.
39. N. Azizi, F. Ebrahimi, E. Aakbari, F. Aryanasab and M. R. Saidi, *Synlett*, 2007, 2797.
40. Ch. Bhujanga Rao, B. Chinnababu and Y. Venkateswarlu, *J. Org. Chem.*, 2009, **74**, 8856.
41. M. Bayat, H. Imanieh and S. H. Hossieni, *Chin. Chem. Lett.*, 2009, **20**, 656.
42. L. B. Liu, T. S. Jin, L. S. Han, Li, M. Qina and T. S. Li, *E-J. Chem.*, 2006, **3**, 117.
43. S. Kantevari, R. Bantu and L. Nagarapu, *J. Mol. Catal. A Chem.*, 2007, **269**, 53.
44. M. Kidwai, V. Bansal, P. Mothsra, S. Saxena, R. K. Somvanshi, S. Dey and T. J. Singh, *J. Mol. Catal. A Chem.*, 2007, **268**, 76.
45. G. Kaupp, M. R. Naimi-Jarnal and J. Schmeyers, *Tetrahedron*, 2003, **59**, 3753.
46. G. Cravotto, G. M. Nano, G. Palmisano and S. Tagliapietra, *Synthesis*, 2003, 1286.
47. G. Kaupp and M. R. Naimi-Jamal, *Chem. – Eur. J.*, 2002, **8**, 594.
48. G. Kaupp, J. Schmeyers and J. Boy, *J. Prakt. Chem.*, 2000, **342**, 269.

49. S. Das and A. J. Thakur, *Eur. J. Org. Chem.*, 2011, 2301.
50. B. L. Lam and L. N. Pridgen, *J. Org. Chem.*, 1986, **51**, 2592.
51. J. Azizian, M. R. Mohammadizadeh, F. Teimouri, A. A. Mohammadi and A. R. Karimi, *Synth. Commun.*, 2006, **36**, 3631.
52. R. Bansal, R. S. Kumar, G. Kumar, S. Thota, S. Thamotharan, V. Parthasarathi and A. Linden, *J. Heterocycl. Chem.*, 2008, **45**, 1789.
53. L. Spacilova, P. Dzubak, M. Hajduch, S. Krupkova, P. Hradila and J. Hlavác, *Bioorg. Med. Chem. Lett*, 2007, **17**, 6647.
54. M. Dabiri, H. Arvin-Nezhad, H. R. Khavasi and A. Bazgir, *Tetrahedron*, 2007, **63**, 1770.
55. D. Shi, J. Shi and S. Rong, *Chin. J. Chem.*, 2010, **28**, 791.
56. J. Taran, A. Ramazani, S. W. Joo, K. Ślepokura and T. Lis, *Helv. Chim. Acta*, 2014, **97**, 1088.
57. U. Kazmaier and S. Ackermann, *Org. Biomol. Chem.*, 2005, **3**, 3184.
58. B. A. Trofimov, L. V. Andriyankova, L. P. Nikitina, K. V. Belyaeva, A. G. Mal'kina, A. V. Afonin and I. A. Ushakov, *Synlett*, 2012, **23**, 2069.
59. A. Alizadeh and N. Zohreh, *Synlett*, 2012, **23**, 428.
60. A. Dandia, R. Singh, M. Saha and A. Shivpuri, *Pharmazie*, 2002, **57**, 602.
61. A. D. Mishra, *J. Nepal Chem. Soc.*, 2009, **24**, 49.
62. A. V. Velikorodov, O. Y. Poddubnyi, A. K. Kuanchalieva and O. O. Krivosheev, *Russ. J. Org. Chem.*, 2010, **46**, 1826.
63. R. Abonia, J. Castillo, B. Insuasty, J. Quiroga, M. Nogueras and J. Cobo, *ACS Comb. Sci.*, 2013, **15**, 2.
64. Y. Wu, J. Wang, P. Li and F. Y. Kwong, *Synlett*, 2012, **23**, 788.
65. P. Zaderenko, M. C. López and P. Ballesteros, *J. Org. Chem.*, 1996, **61**, 6825.
66. M. I. Uddin, K. Nakano, Y. Ichikawa and H. Kotsuki, *Synlett*, 2008, 1402.
67. J. Moran, P. Dornan and A. M. Beauchemin, *Org. Lett.*, 2007, **9**, 3893.
68. I. Yavari, S. Souri and M. Sirouspour, *Synlett*, 2008, 1633.
69. A. Hasaninejad and S. Firoozi, *Mol. Diversity*, 2013, **17**, 459.
70. V. K. Aggarwal, J. Vicente and R. V. Bonnert, *J. Org. Chem.*, 2003, **68**, 5381.
71. G. Varvounis, Y. Fiamegos and G. Pilidis, *Tetrahedron Lett.*, 2004, **48**, 8819.
72. H. J. Cristau, P. P. Cellier, J. F. Spindler and M. Taillefer, *Eur. J. Org. Chem.*, 2004, 695.
73. R. Martin, M. R. Rivero and S. L. Buchwald, *Angew. Chem., Int. Ed.*, 2006, **45**, 7079.
74. S. T. Heller and S. R. Natarajan, *Org. Lett.*, 2006, **8**, 2675.
75. L. Zhu, P. Guo, G. Li, J. Lan, R. Xie and J. You, *J. Org. Chem.*, 2007, **72**, 8535.
76. Z. Xi, F. Liu, Y. Zhou and W. Chen, *Tetrahedron*, 2008, **64**, 4254.
77. R. Goikhman, T. L. Jacques and D. Sames, *J. Am. Chem. Soc.*, 2009, **131**, 3042.
78. J. J. Neumann, M. Suri and F. Glorius, *Angew. Chem., Int. Ed.*, 2010, **49**, 903.

79. J. Wang, P.-F. Li, S. H. Chan, A. S. C. Chan and F. Y. Kwong, *Tetrahedron Lett.*, 2012, **53**, 2887.
80. A. Alizadeh, *Helv. Chim. Acta*, 2005, **88**, 2777.
81. Y. Haghighatnia, S. Balalaie and H. R. Bijanzadeh, *Helv. Chim. Acta*, 2012, **95**, 818.
82. P. B. Thakur and H. M. Meshram, *RSC Adv.*, 2014, **4**, 6019.
83. J. Venkata Prasad, M. Prabhakar, K. Manjulatha, D. Rambabu, K. Anand Solomon, G. Gopi Krishna and K. Anil Kumar, *Tetrahedron Lett.*, 2010, **51**, 3609.
84. S. M. Ebrahimi, M. Mahdavi, S. Emami, M. Saeedi, M. Asadi, L. Firoozpour, M. Khoobi, K. Divsalar, A. Shafiee and A. Foroumadi, *Synth. Commun.*, 2014, **44**, 665.
85. A. Shaabani, F. Hajishaabanha, H. Mofakham, M. Mahyari and B. Lali, *Helv. Chim. Acta*, 2012, **95**, 246.
86. S. Kumar, P. Sharma, K. K. Kapoor and M. S. Hundal, *Tetrahedron*, 2008, **64**, 536.
87. S. Margarita, O. Estael, V. Yamila, P. Beatriz, M. Lourdes, M. Nazario, Q. Margarita, S. Carlos, L. S. Jose, N. Hector, B. Norbert and M. P. Oswald, *Tetrahedron*, 1999, **55**, 875.
88. L. Ohberg and J. Westman, *Synlett*, 2001, 1296.
89. A. Agarwal and P. M. S. Chauhan, *Tetrahedron Lett.*, 2005, **46**, 1345.
90. R. Sridhar and P. T. Perumal, *Tetrahedron*, 2005, **61**, 2465.
91. L.-M. Wang, J. Sheng, L. Zhang, J.-W. Han, Z. Fan, H. Tian and C.-T. Qian, *Tetrahedron*, 2005, **61**, 1539.
92. S. Ko and C.-F. Yao, *Tetrahedron*, 2006, **62**, 7293.
93. J. C. Legeay, J. Y. Goujon, J. J. V. Eynde, L. Toufet and J. P. Bazureau, *J. Comb. Chem.*, 2006, **8**, 829.
94. A. Kumar and R. A. Maurya, *Tetrahedron*, 2007, **63**, 1946.
95. O. A. Attanasi, L. De Crescentini, G. Favi, S. Nicolini, F. R. Perrulli and S. Santeusanio, *Org. Lett.*, 2011, **13**, 353.
96. D. J. Martin, *J. Prakt. Chem.*, 1991, **333**, 261.
97. E. Atay, I. B. Blogoeva, A. J. Jirby and I. G. Pojarieff, *J. Chem. Soc., Perkin Trans. 2*, 1998, 2289.
98. I. B. Blogoeva, M. M. Toteva, N. Ouarti and M.- F. Ruasse, *J. Org. Chem.*, 2001, **66**, 2123.
99. J. Matthews and R. A. Rivero, *J. Org. Chem.*, 1997, **62**, 6090.
100. S. W. Kim, S. Y. Ahn, J. S. Koh, J. H. Lee, S. Ro and H. Y. Cho, *Tetrahedron Lett.*, 1997, **38**, 4603.
101. A. Boeijen, J. A. W. Kruijtzer and R. M. J. Liskamp, *Bioorg. Med. Chem. Lett.*, 1995, **5**, 47.
102. M. Beller, M. Eckert, W. A. Moradi and H. Neumann, *Angew. Chem., Int. Ed.*, 1999, **38**, 1454.
103. J. M. Ignacio, S. Macho, S. Marcaccini, R. Pepino and T. Torroba, *Synlett*, 2005, 3051.
104. A. Volonterio, C. R. de Arellano and M. Zanda, *J. Org. Chem.*, 2005, **70**, 2161.

105. O. A. Attanasi, S. Bartoccini, G. Favi, G. Giorgi, F. R. Perrulli and S. Santeusanio, *J. Org. Chem.*, 2012, **77**, 1161.
106. R. Ottanà, R. Maccari, M. L. Berreca, G. Bruno, A. Rotondo, A. Rossi, G. Chiricosta, R. D. Paola, L. Sautebin, S. Cuzzocrea and M. G. Vigorita, *Bioorg. Med. Chem.*, 2005, **13**, 4243.
107. T. Kline, H. B. Felise, K. C. Barry, S. R. Jackson, H. V. Nguyen and S. I. Miller, *J. Med. Chem.*, 2008, **51**, 7065.
108. A. R. Kazemizadeh, N. Hajaliakbari, R. Hajian, N. Shajari and A. Ramazani, *Helv. Chim. Acta*, 2012, **95**, 594.
109. L. Moradi, C. Bahrami, A. Mirzaei, M. R. Zarehbin and I. Yavari, *Tetrahedron Lett.*, 2015, **56**, 1510.
110. X. Wang, S.-Y. Wang and S.-J. Ji, *Org. Lett.*, 2013, **15**, 1954.
111. H. Liu and A. Domling, *J. Org. Chem.*, 2009, **74**, 6895.
112. W. Wang, E. Herdtweck and A. Domling, *Chem. Commun.*, 2010, **46**, 770.
113. D. Zheng, S. Li, Y. Luo and J. Wu, *Org. Lett.*, 2011, **13**, 6402.
114. D. Zheng, S. Li and J. Wu, *Org. Lett.*, 2012, **14**, 2655.
115. F.-C. Yu, X.-P. Hao, X.-R. Lin, S.-J. Yan and J. Lin, *Tetrahedron*, 2015, **71**, 4084.
116. M. Borthakur, S. Gogoi, J. Gogoi and R. C. Boruah, *Tetrahedron Lett.*, 2010, **51**, 5160.
117. U. Kucklaender, R. Bollig, W. Frank, A. Gratz and J. Jose, *Bioorg. Med. Chem.*, 2011, **19**, 2666.
118. M. Shiri, *Chem. Rev.*, 2012, **112**, 3508.
119. A. A. Esmaeili and O. Ghareghani, *Helv. Chim. Acta*, 2007, **90**, 1712.
120. R. Akbarzadeh, T. Amanpour, A. Abolhasani Soorki and A. Bazgir, *Helv. Chim. Acta*, 2012, **95**, 483.
121. A. N. Nicholson, B. M. Stone and C. H. Clarke, *Br. J. Clin. Pharmacol.*, 1977, **4**, 567.
122. H. Kruse, *Drug Dev. Res.*, 1982, **2**(S1), 145.
123. T. F. Herpin, K. G. Van Kirk, J. M. Salvino, S. T. Yu and R. F. Labaudinière, *J. Comb. Chem.*, 2000, **2**, 513.
124. M. S. Balakrishna and B. Kaboudin, *Tetrahedron Lett.*, 2001, **42**, 1127.
125. H.-Y. Zhao and G. Liu, *J. Comb. Chem.*, 2007, **9**, 1164.
126. C.-W. Kuo, S. V. More and C.-F. Yao, *Tetrahedron Lett.*, 2006, **47**, 8523.
127. I. Yavari, S. Seyfi, M. Nematpour and Z. Hossaini, *Helv. Chim. Acta*, 2010, **93**, 1413.
128. A. Alizadeh, A. Rezvanian and L.-G. Zhu, *Helv. Chim. Acta*, 2007, **90**, 2414.
129. A. Alizadeh and N. Zohreh, *Helv. Chim. Acta*, 2008, **91**, 844.
130. A. Alizadeh and N. Zohreh, *Helv. Chim. Acta*, 2010, **93**, 1221.
131. W. Liu, Y. Xu, X. Sun, D. Lu and L. Guo, *Synlett*, 2014, **25**, 1093.
132. M. Li, W. Kong, L.-R. Wen and F.-H. Liu, *Tetrahedron*, 2012, **68**, 4838.
133. M. Adib, M. H. Sayahi, N. Mahmoodi and H. R. Bijanzadeh, *Helv. Chim. Acta*, 2006, **89**, 1176.

134. H.-B. Zhang, L. Liu, Y.-J. Chen, D. Wang and C.-J. Li, *Eur. J. Org. Chem.*, 2006, 869.
135. M. C. Pirrung, K. Park and Z. Li, *Org. Lett.*, 2001, **3**, 365.
136. M. C. Pirrung, L. Deng, Z. Li and K. Park, *J. Org. Chem.*, 2002, **67**, 8374.
137. A. K. Maiti and P. Bhattacharyya, *J. Chem. Res., Synop.*, 1997, 424–425.
138. J. S. Yadav, B. V. S. Reddy and T. Swamy, *Synthesis*, 2004, 106.
139. J. S. Yadav, B. V. S. Reddy and T. Swamy, *Tetrahedron Lett.*, 2003, **44**, 9121.
140. M. C. Pirrung, Y. Liu, L. Deng, D. K. Halstead, Z. Li, J. F. May, M. Wedel, D. A. Austin and N. J. G. Webster, *J. Am. Chem. Soc.*, 2005, **127**, 4609.
141. M. C. Pirrung, K. Fujita and K. Park, *J. Org. Chem.*, 2005, **70**, 2537.
142. P. B. Thakur and H. M. Meshram, *RSC Adv.*, 2014, **4**, 5343.
143. S. Rostamizadeh, M. Z. Kassaee, N. Shadjou and H. Zandi, *Monatsh. Chem.*, 2013, **144**, 703.
144. H. I. El-Subbagh, S. M. Abu-Zaid, M. A. Mahran, F. A. Badria and A. M. Al-Obaid, *J. Med. Chem.*, 2000, **43**, 2915.
145. A. G. Hammam, M. A. Sharaf and N. A. Abdel-Hafez, *Indian J. Chem.*, 2001, **40B**, 213.
146. M. A. Al-Omar, K. M. Youssef, M. A. El-Sherbeny, A. A. Awadalla and H. I. El-Subbagh, *Arch. Pharm. Chem.*, 2005, **338**, 175.
147. R. R. Kumar, S. Perumal, P. Senthilkumar, P. Yogeeswarib and D. Sriram, *Bioorg. Med. Chem. Lett.*, 2007, **17**, 6459.
148. S. A. F. Rostom, G. S. Hassan and H. I. El-Subbagh, *Arch. Pharm. Chem.*, 2009, **2009**(342), 584.
149. Z.-G. Han, S.-J. Tu, B. Jiang, S. Yan, X.-H. Zhang, S.-S. Wu, W.-J. Hao, X.-D. Cao, F. Shi, G. Zhang and N. Ma, *Synthesis*, 2009, 1639.
150. S.-L. Wang, Z.-G. Han, S.-J. Tu, X.-H. Zhang, S. Yan, W.-J. Hao, F. Shi, X.-D. Cao and S.-S. Wu, *J. Heterocycl. Chem.*, 2009, **46**, 828.
151. A. S. Girgis, N. S. M. Ismail and H. Farag, *Eur. J. Med. Chem.*, 2011, **46**, 2397.
152. A. Habibi and A. Rahmani, *Helv. Chim. Acta*, 2011, **94**, 1806.
153. A. E. Rubtsov, N. V. Kovylyaeva and V. V. Zalesov, *Pharm. Chem. J.*, 2005, **39**, 11.
154. Q.-H. Chen, P. N. P. Rao and E. E. Knaus, *Bioorg. Med. Chem.*, 2006, **14**, 7898.
155. D. Iannazzo, A. Piperno, G. Romeo, R. Romeo, U. Chiacchio, A. Rescifina, E. Balestrieri, B. Macchi, A. Mastino and R. Cortese, *Bioorg. Med. Chem.*, 2008, **16**, 9610.
156. D. Matiadis, K. C. Prousis and O. Igglessi-Markopoulou, *Molecules*, 2009, **14**, 3914.
157. M. Adib, M. H. Sayahi, S. A. Koloogani and P. Mirzaei, *Helv. Chim. Acta*, 2006, **89**, 299.
158. C. Jutz, R. M. Wagner, A. Kroatz and H. G. Lobering, *Justus Liebigs Ann. Chem.*, 1975, **5**, 874.
159. P. Ribereau and G. Queguiner, *Can. J. Chem.*, 1983, **61**, 334.
160. P. Ribereau and G. Queguiner, *Tetrahedron*, 1983, **39**, 3593.
161. T. Umemoto and G. Tomizawa, *Tetrahedron Lett.*, 1987, **28**, 2705.

162. J. L. Duffy, T. A. Rano, N. J. Kevin, K. T. Chapman, W. A. Schleif, D. B. Olsen, M. Stahlhut, C. A. Rutkowski, L. C. Kuo, L. Jin, J. H. Lin, E. A. Emini and J. R. Tata, *Bioorg. Med. Chem. Lett.*, 2003, **13**, 2569.
163. L.-X. Zhao, Y.-S. Moon, A. Basnet, E.-K. Kim, Y. Jahng, J. G. Park, T. C. Jeong, W.-J. Cho, S.-U. Choi, C. O. Lee, S.-Y. Lee, C.-S. Lee and E.-S. Lee, *Bioorg. Med. Chem. Lett.*, 2004, **14**, 1333.
164. A. Ramazani, A. Rezaei, A. T. Mahyari, M. Rouhani and M. Khoobi, *Helv. Chim. Acta*, 2010, **93**, 2033.
165. E. Winterfeldt, D. Schumann and H.-J. Dillinger, *Chem. Ber.*, 1969, **102**, 1656.
166. Y. Ito, H. Kato and T. Saegusa, *J. Org. Chem.*, 1982, **47**, 741.
167. V. Nair, A. U. Vinod, J. S. Nair, A. R. Sreekanth and N. P. Rath, *Tetrahedron Lett.*, 2000, **41**, 6675.
168. S. Ma and H. Xie, *J. Org. Chem.*, 2002, **67**, 6575.
169. V. Nair, A. U. Vinod, N. Abhilash, R. S. Menon, V. Santhi, R. L. Varma, S. Viji, S. Mathew and R. Srinivas, *Tetrahedron*, 2003, **59**, 10279.
170. N. Chatani, M. Oshita, M. Tobisu, Y. Ishii and S. Murai, *J. Am. Chem. Soc.*, 2003, **125**, 7812.
171. G. Bez and C.-G. Zhao, *Org. Lett.*, 2003, **5**, 4991.
172. V. Nair, R. S. Menon, A. Deepthi, B. Rema Devi and A. T. Biju, *Tetrahedron Lett.*, 2005, **46**, 1337.
173. S. Ma and H. Xie, *Tetrahedron*, 2005, **61**, 251.
174. V. Nair and A. Deepthi, *Tetrahedron Lett.*, 2006, **47**, 2037.
175. M. B. Teimouri, A. Shaabani and R. Bazhrang, *Tetrahedron*, 2006, **62**, 1845.
176. A. Shaabani, E. Soleimani and A. Sarvary, *Monatsh. Chem.*, 2008, **139**, 629.
177. A. Shaabani, R. Ghadari, A. Sarvary and A. H. Rezayan, *J. Org. Chem.*, 2009, **69**, 4372.
178. J. Azizian, A. Ramazani and M. Hajia, *Helv. Chim. Acta*, 2011, **94**, 371.
179. M. Adib, S. Moghimi, M. H. Sayahi and H. R. Bijanzadeh, *Helv. Chim. Acta*, 2009, **92**, 944.
180. M. Adib, S. Ansari, L.-G. Zhu and M. Rahimi-Nasrabadi, *Helv. Chim. Acta*, 2013, **96**, 675.
181. M. Bayat, H. Imanieh and H. Abbasi, *Helv. Chim. Acta*, 2010, **93**, 757.
182. A. Ramazani, N. Noshiranzadeh, A. Ghamkhari, K. Ślepokura and T. Lis, *Helv. Chim. Acta*, 2008, **91**, 2252.
183. K. Hachama, M. Khodja, S. Moulay, H. Boutoumi, L. Hennig and D. Sicker, *J. Heterocycl. Chem.*, 2013, **50**, 413.
184. A. Ramazani, B. Abdian, F. Z. Nasrabadi and M. Rouhani, *Phosphorus, Sulfur Silicon Relat. Elem.*, 2013, **188**, 642.
185. I. Yavari, Z. Hossaini, F. Shirgahi-Talari and S. Seyfi, *Synlett*, 2008, 1631.
186. T. M. Potewar, S. A. Ingale and K. V. Srinivasan, *Tetrahedron*, 2008, **64**, 5019.
187. N. Bailey, A. W. Dean, D. B. Judd, D. Middlemiss, R. Storer and P. W. Stephen, *Bioorg. Med. Chem. Lett.*, 1996, **6**, 1409.

188. P. C. Kearney, M. Fernandez and J. A. Flygare, *J. Org. Chem.*, 1998, **62**, 196.
189. S. E. I. Kazzouli, S. B. Raboin, A. Mouadbib and G. Guillaumet, *Tetrahedron Lett.*, 2002, **43**, 3193.
190. M. Narender, M. Somi Reddy, R. Sridhar, Y. V. D. Nageswar and K. Rama Rao, *Tetrahedron Lett.*, 2005, **46**, 5953.
191. B. Das, V. Saidi Reddy and R. Ramu, *J. Mol. Catal. A: Chem.*, 2006, **252**, 235.
192. H. L. Siddiqui, A. Iqbal, S. Ahmed and G. Weaver, *Molecules*, 2006, **11**, 206.
193. W. K. George and R. M. Arjun, *Tetrahedron Lett.*, 2006, **47**, 5171.
194. O. A. Attanasi, S. Bartoccini, G. Favi, G. Giorgi, F. R. Perrulli and S. Santeusanio, *J. Org. Chem.*, 2012, **77**, 1161.
195. I. Yavari and A. Malekafzali, *J. Heterocycl. Chem.*, 2013, **50**, 992.
196. G. L. Sommen, A. Linden and H. Heimgartner, *Helv. Chim. Acta*, 2006, **89**, 1322.
197. K. Kumaravel and G. Vasuki, *Green Chem.*, 2009, **11**, 1945.
198. F. Rostami-Charati, Z. Hossaini, E. Gharaee and M. A. Khalilzadeh, *J. Heterocycl. Chem.*, 2013, **50**, E174.
199. M. Adib, E. Sheikhi, M. Karimzadeh, H. R. Bijanzadeh and M. Amanlou, *Helv. Chim. Acta*, 2012, **95**, 788.
200. J. B. Wright, *J. Org. Chem.*, 1960, **25**, 1867.
201. R. Royer, E. Bisagni, C. Hudry, A. Cheutin and M. L. Desvoye, *Bull. Soc. Chim. Fr.*, 1963, 1003.
202. Y. Kawase, S. Takata and E. Hikishima, *Bull. Chem. Soc. Jpn.*, 1971, **44**, 749.
203. M. Kalyanasundaram, K. Rajagopalan and S. Swaminathan, *Tetrahedron Lett.*, 1980, **21**, 4391.
204. T. Horaguchi, H. Iwanami, T. Tanaka, E. Hasegawa and T. Shimizu, *J. Chem. Soc., Chem. Commun.*, 1991, 44.
205. T. Horaguchi, H. Kobayashi, K. Miyazawa, E. Hasegawa, T. Shimizu, T. Suzuki and K. Tanemura, *J. Heterocycl. Chem.*, 1990, **27**, 935.
206. S. Torii, L. H. Xu and H. Okumoto, *Synlett*, 1992, 515.
207. G. Dyker, *J. Org. Chem.*, 1993, **58**, 6426.
208. R. C. Larock, E. K. Yum, M. J. Doty and K. K. C. Sham, *J. Org. Chem.*, 1995, **60**, 3270.
209. T. L. Boehm and H. D. H. Showalter, *J. Org. Chem.*, 1996, **61**, 6498.
210. D. Fancelli, M. C. Fagnola, D. Severino and A. Bedeschi, *Tetrahedron Lett.*, 1997, **38**, 2311.
211. N. G. Kundu, M. Pal, J. S. Mahanty and M. De, *J. Chem. Soc., Perkin Trans. 1*, 1997, 2815.
212. D. Fancelli, M. C. Fagnola, D. Severino and A. Bedeschi, *Tetrahedron Lett.*, 1997, **38**, 2311.
213. N. G. Kundu, M. Pal, J. S. Mahanty and M. De, J. Chem. Soc. Perkin Trans. *1*, 1997, 2815.

214. K. C. Nicolaou, S. A. Snyder, A. Bigot and J. A. Pfefferkorn, *Angew. Chem., Int. Ed.*, 2000, **39**, 1093.
215. G. W. Kabalka, L. Wang and R. M. Pagni, *Tetrahedron*, 2001, **57**, 8017.
216. B. Gabriele, R. Mancuso, G. Salerno and M. Costa, *J. Org. Chem.*, 2007, **72**, 9278.
217. B. Gabriele, R. Mancuso and G. Salerno, *J. Org. Chem.*, 2008, **73**, 7336.
218. X. C. Huang, Y. L. Liu, Y. Liang, S. F. Pi, F. Wang and J. H. Li, *Org. Lett.*, 2008, **10**, 1525.
219. V. Fiandanese, D. Bottalico, G. Marchese and A. Punzi, *Tetrahedron*, 2008, **64**, 53.
220. A. Ramazani, Y. Ahmadi and A. Mahyar, *Synth. Commun.*, 2011, **41**, 2273.
221. A. Ramazani, F. Z. Nasrabadi, A. M. Malekzadeh and Y. Ahmadi, *Monatsh. Chem.*, 2011, **142**, 625.
222. A. Ramazani, F. Z. Nasrabadi and Y. Ahmadi, *Helv. Chim. Acta*, 2011, **94**, 1024.
223. A. Ramazani, M. Rouhani, A. Rezaei, N. Shajari and A. Souldozi, *Helv. Chim. Acta*, 2011, **94**, 282.
224. M. A. Terzidis, V. G. Tsiaras, N. M. Drosos, P. M. Kasapidou, J. Stephanidou-Stephanatou, C. A. Tsoleridis, G. Buth and G. E. Kostakis, *Tetrahedron Lett.*, 2014, **55**, 5601.
225. A. Shaabani, E. Soleimani, A. H. Rezayan, A. Sarvary and H. R. Khavasi, *Org. Lett.*, 2008, **10**, 2581.
226. A. Alizadeh, Z. Noaparast, H. Sabahno and N. Zohreh, *Helv. Chim. Acta*, 2010, **93**, 1401.
227. I. Yavari, M. J. Bayat, S. Souri and M. Sirouspour, *Helv. Chim. Acta*, 2009, **92**, 1903.
228. G. Brahmachari and K. Nurjamal, *ChemistrySelect*, 2017, **2**, 3695.
229. G. Brahmachari, *Handbook of Pharmaceutical Natural Products*, Wiley-VCH, Weinheim, 2010, vol. 1 and 2.
230. S. Omura, F. Kuno, K. Otoguro, T. Sunazuka, K. Shiomi, R. Masuma and Y. Iwai, *J. Antibiot.*, 1995, **48**, 745.
231. F. Kuno, K. Otoguro, K. Shiomi, R. Masuma, Y. Iwai and S. Omura, *J. Antibiot.*, 1996, **49**, 742.
232. F. Kuno, K. Shiomi, K. Otoguro, T. Sunazuka and S. Omura, *J. Antibiot.*, 1996, **49**, 748.
233. K. Otoguro, K. Shiomi, Y. Yamaguchi, N. Arai, T. Sunazuka, R. Masuma, Y. Iwai and S. Omura, *J. Antibiot.*, 2000, **53**, 50.
234. L. Lei, Y.-b. Xue, Z. Liu, S.-s. Peng, Y. He, Y. Zhang, R. Fang, J.-p. Wang, Z.-w. Luo, G.-m. Yao, J.-w. Zhang, G. Zhang, H.-p. Song and Y.-h. Zhang, *Sci. Rep.*, 2015, **5**, 13544, DOI: 10.1038/srep13544.
235. G. Appendino, G. Cravotto, S. Tagliapietra, G. M. Nano and G. Palmisano, *Helv. Chem. Acta*, 1990, **73**, 1865.
236. G. Cravotto, G. M. Nano and S. Tagliapietra, *Synthesis*, 2001, 49.

237. Y. Jacquot, B. Refouvelet, L. Bermont, G. L. Adessi, G. Leclercq and A. Xicluna, *Pharmazie*, 2002, **57**, 233.
238. X.-S. Wang, J.-X. Zhou, Z.-S. Zeng, Y.-L. Li, D.-Q. Shi and S.-J. Tu, *ARKIVOC*, 2006, 107.
239. K. C. Majumdar, P. Debnath and P. K. Maji, *Tetrahedron Lett.*, 2007, **48**, 5265.
240. X. Lin, X. Dai, Z. Mao and Y. Wang, *Tetrahedron*, 2009, **65**, 9233.
241. S. Berger and E. Haak, *Tetrahedron Lett.*, 2010, **51**, 6630.
242. Z. He, X. Lin, Y. Zhu and Y. Wang, *Heterocycles*, 2010, **81**, 965.
243. R. Sarma, M. M. Sarma, K. C. Lekhok and D. Prajapati, *Synlett*, 2010, 2847.
244. L. Yunkui, Z. Jie, Q. Jianqiang, B. Jiang and X. Zhenyuan, *J. Org. Chem.*, 2011, **76**, 9096.
245. A. Naseem and B. Venkata Babu, *Synth. Commun.*, 2013, **43**, 3044.
246. H. Mukut, H. Johannes, V. Tonder, C. Barend and B. Bezuidenhoudt, *Tetrahedron Lett.*, 2013, **54**, 3773.
247. S. Yaragorla, O. L. Saini and G. Singh, *Tetrahedron Lett.*, 2015, **56**, 1649.
248. G. Brahmachari and N. Nayek, *ACS Omega*, 2017, **2**, 5025.

CHAPTER 3

Catalyst-free Organic Reactions with Conventional Heating

3.1 Introduction

Conventional heating and/or heating under reflux is common practice in organic chemistry in carrying out so many organic reactions that are unreasonably slow and that can take an extended period of time to achieve any noticeable effect; so simply heating or heating under reflux is often used to increase the rate of reaction. Chemists have observed that certain useful organic transformations can also be achieved fruitfully with just conventional heating and without the aid of any catalyst/additive. With this view in mind, the present chapter summarizes a few comprehensively screened catalyst-free organic synthetic protocols which utilize heat for the generation of certain synthetically and pharmaceutically useful chemical compounds.

3.2 Organic Transformations with Conventional Heating

A good number of catalyst-free organic reactions occurring under conventional heating/refluxing conditions to furnish a variety of organic compounds have been reported so far. This section presents such useful organic transformations in an entry-based format highlighting the key aspects for each of them.

3.2.1 Entry-1: Synthesis of 1,2,4-Oxadiazoles

Type of reaction: C–N and C–O bond formation
Reaction conditions: Catalyst-free, water, refluxing

Green Chemistry Series No. 51
Catalyst-free Organic Synthesis
By Goutam Brahmachari

Published by the Royal Society of Chemistry, www.rsc.org

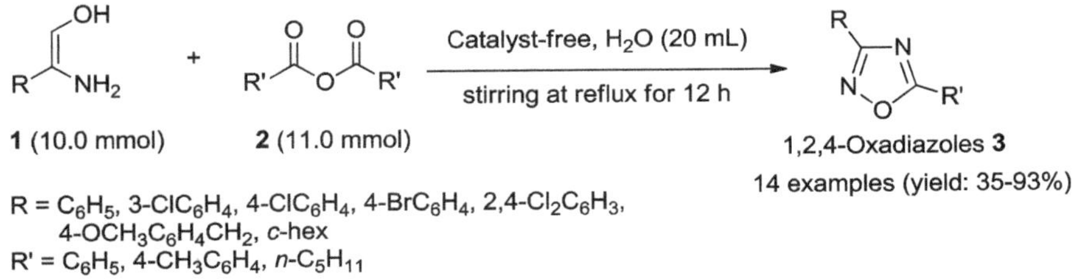

Scheme 3.1 Synthesis of 1,2,4-oxadiazoles (**3**).

Synthetic strategy: Cyclization
Keywords: Amidoximes, acid anhydrides, catalyst-free, water, 1,2,4-oxadiazoles

3.2.1.1 *General Reaction Scheme*

Kaboudin and Malekzadeh[1] developed a simple and efficient protocol for the synthesis of 1,2,4-oxadiazoles (**3**), in moderate to good yields, from the reaction of amidoximes (**1**) with anhydrides (**2**), in water under reflux conditions in the absence of any added catalyst (Scheme 3.1).

3.2.1.2 *Representative Examples*

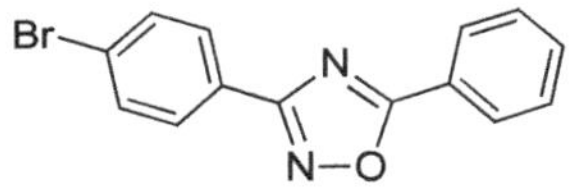

3a

3-(4-Bromophenyl)-5-phenyl-1,2,4-oxadiazole (**3a**): white solid, mp 110–112 °C; yield: 81%

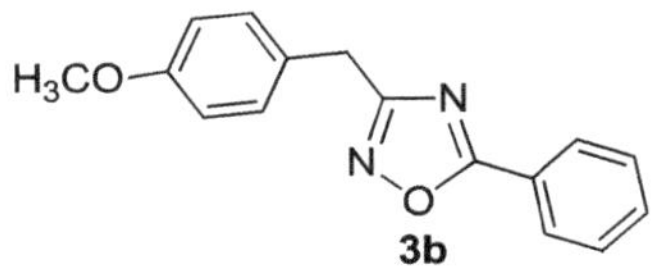

3b

3-(4-Methoxybenzyl)-5-phenyl-1,2,4-oxadiazole (**3b**): white solid, mp 96–98 °C; yield: 69%

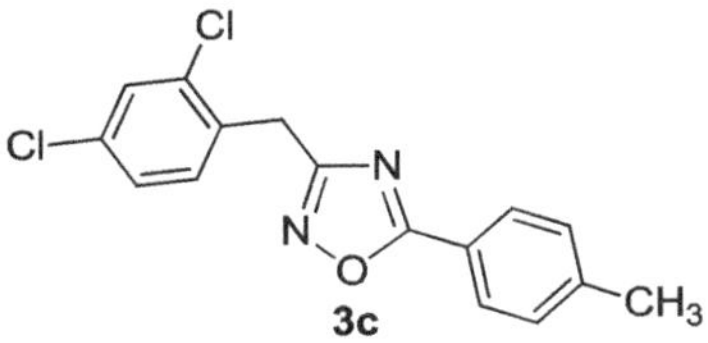

3c

3-(2,4-Dichlorobenzyl)-5-(*p*-tolyl)-1,2,4-oxadiazole (**3c**): white solid, mp 126–128 °C; yield: 72%

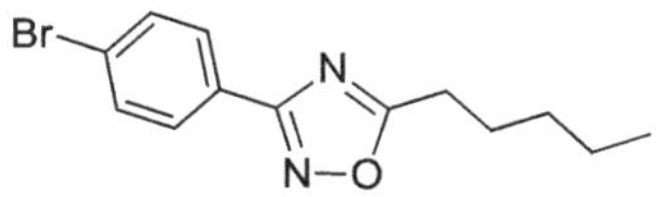

3d

3-(4-Bromophenyl)-5-pentyl-1,2,4-oxadiazole (**3d**): white solid, mp 96–98 °C; yield: 48%

3.2.1.3 *Experimental Procedure*

A mixture of amidoxime (**1;** 10 mmol) and acid anhydride (**2;** 11 mmol) in 20 mL of water solution was stirred for 12 h under reflux. After stirring, the reaction mixture was washed with diethyl ether (3×50 mL). The resulting mixture was subjected to column chromatography on silica gel with

EtOAc–*n*-hexane mixture, followed by evaporation of the solvent under reduced pressure to afford pure products of 1,2,4-oxadiazoles (**3**) in 35–93% yields. All the products were characterized based on analytical and spectral measurements.

3.2.1.4 Characterization Data of Two Representative Compounds

3-(2,4-Dichlorobenzyl)-5-(*p*-tolyl)-1,2,4-oxadiazole (**3c**): white solid, mp 126–128 °C; yield: 72%; ^{1}H NMR ($CDCl_3$, 400 MHz): δ 2.49 (3H, s), 7.39 (2H, d, *J* = 8.0 Hz), 7.43 (1H, dd, *J* = 2 and 8.4 Hz), 7.61 (1H, d, *J* = 2 Hz), 8.03 (1H, d, *J* = 8.4 Hz), 8.13 (2H, d, *J* = 8 Hz); ^{13}C NMR ($CDCl_3$, 100 MHz): δ 22.72, 121.22, 124.98, 127.35, 128.22, 129.91, 130.86, 132.59, 134.37, 137.21, 143.84, 167.00, 175.8. Anal. calc. for $C_{15}H_{10}Cl_2N_2O$: C, 59.01; H, 3.30; N, 9.21; found: C, 59.23; H, 3.15; N, 9.10.

3-(4-Bromophenyl)-5-pentyl-1,2,4-oxadiazole (**3d**): white solid, mp 96–98 °C; yield: 48%; ^{1}H NMR ($CDCl_3$, 400 MHz): δ 0.94 (3H, t, *J* = 6.8 Hz), 1.35–1.46 (4H, m), 1.89 (2H, quin., *J* = 7.6 Hz), 2.95 (2H, t, *J* = 7.6 Hz), 7.63 (2H, d, *J* = 8.4 Hz), 7.96 (2H, d, *J* = 8.4 Hz); ^{13}C NMR ($CDCl_3$, 100 MHz): δ 14.1, 22.2, 26.3, 26.6, 29.7, 31.2, 125.6, 126.0, 128.9, 132.1, 167.5, 180.3. Anal. calc. for $C_{13}H_{15}BrN_2O$: C, 53.05; H, 5.14; N, 9.52; found: C, 52.86; H, 5.15; N, 9.43.

3.2.1.5 Critical Views

This protocol for the catalyst-free preparation of 1,2,4-oxadiazoles using water as the solvent medium is operationally simple with moderate to good yields, and appeared to be advantageous over previously reported methods for this scaffold.[2–9]

3.2.2 Entry-2: Synthesis of Functionalized 5-Arylfuro[2,3-*d*]pyrimidin-4-ols

Type of reaction: C–C and C–O bond formation
Reaction conditions: Catalyst-free, water, heating
Synthetic strategy: [3 + 2]-Heterocyclization
Keywords: 2-Aminopyrimidine-4,6-diol/2-methylpyrimidine-4,6-diol, nitroolefins, catalyst-free, water, 5-arylfuro[2,3-*d*]pyrimidin-4-ols

3.2.2.1 General Reaction Scheme

Li and Zhang[10] prepared a series of densely functionalized 5-arylfuro[2,3-*d*]pyrimidin-4-ols (**3**) *via* the [3 + 2] cyclization of 2-amino/2-methyl-pyrimidine-4,6-diol (**1**) and a variety of nitroolefins (**2**), under catalyst-free conditions in water with conventional heating at 90 °C (Scheme 3.2).

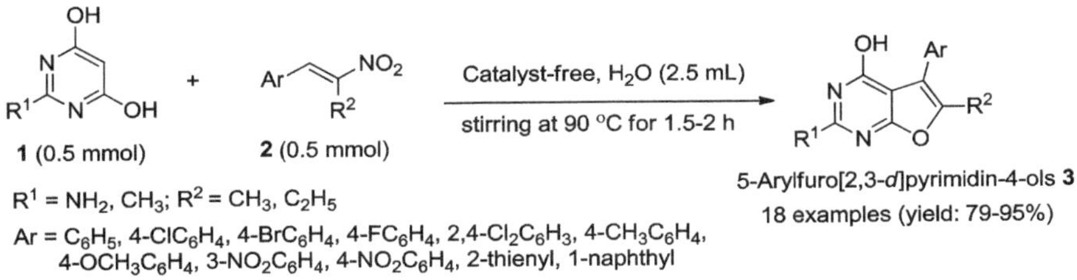

Scheme 3.2 Synthesis of functionalized 5-arylfuro[2,3-*d*]pyrimidin-4-ols (**3**).

3.2.2.2 Representative Examples

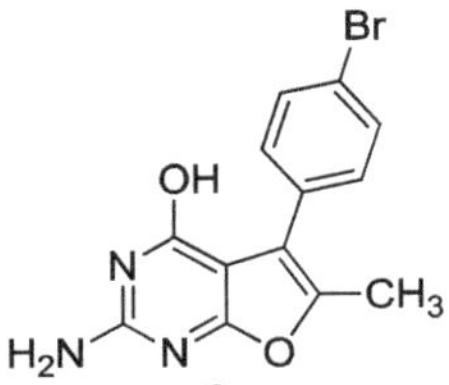

3a

2-Amino-5-(4-bromophenyl)-6-methylfuro[2,3-*d*]pyrimidin-4-ol (**3a**): brown solid, mp >300 °C; yield: 90%

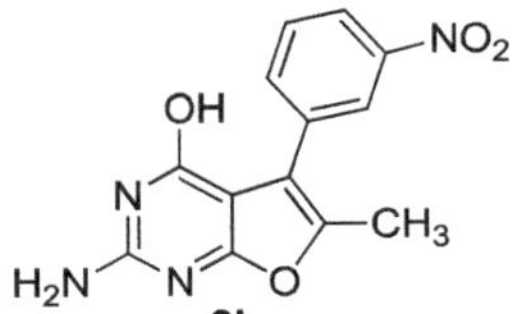

3b

2-Amino-6-methyl-5-(3-nitrophenyl)furo[2,3-*d*]pyrimidin-4-ol (**3b**): light brown solid, mp >300 °C; yield: 88%

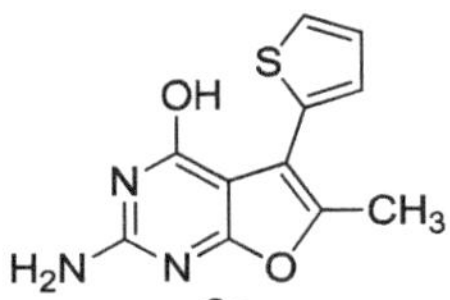

3c

2-Amino-6-methyl-5-(thiophen-2-yl)furo[2,3-*d*]pyrimidin-4-ol (**3c**): black solid, mp >300 °C; yield: 85%

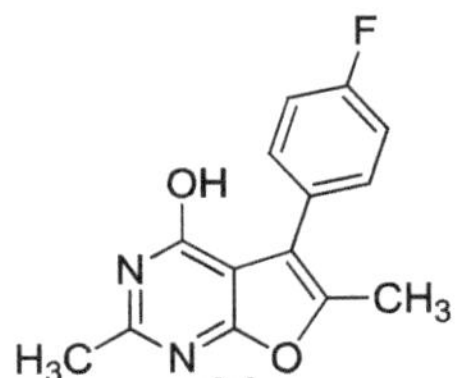

3d

5-(4-Fluorophenyl)-2,6-dimethylfuro[2,3-*d*]pyrimidin-4-ol (**3d**): light yellow solid, mp >300 °C; yield: 94%

3.2.2.3 Experimental Procedure

2-Amino-/2-methyl-pyrimidine-4,6-diol (**1**, 0.5 mmol), nitroolefins (**2**, 0.5 mmol, 1.0 equiv) and water (2.5 mL) were placed into a 10 mL reaction vial. Then, the reaction vial was closed and stirred at 90 °C for 1.5–2 h. Upon completion, the reaction mixture was cooled to room temperature and diluted with cold water (30 mL). The solid product was collected by filtration and was purified by recrystallization from hot 95% ethanol to afford 5-arylfuro[2,3-*d*]pyrimidin-4-ol (**3**) with excellent yields. All the products were characterized based on spectral studies.

3.2.2.4 Characterization Data of Two Representative Compounds

2-Amino-6-methyl-5-(thiophen-2-yl)furo[2,3-*d*]pyrimidin-4-ol (**3c**): black solid, mp >300 °C; yield: 85%; ^{1}H NMR (DMSO-d_6, 400 MHz): δ 10.91 (1H, br, s, –O*H*), 7.83–7.82 (1H, m, Ar–*H*), 7.53–7.51 (1H, m, Ar–*H*), 7.11–7.09 (1H, m, Ar–*H*), 6.70 (2H, br, s, –NH_2), 2.45 (3H, s, –CH_3); ^{13}C NMR (DMSO-d_6, 100 MHz): δ 158.6, 154.4, 143.1, 142.8, 134.1, 133.0, 128.8, 127.6, 125.6, 96.5, 13.5; HRMS (ESI): *m*/*z* $[M+H]^+$ calc. for $C_{11}H_{10}N_3O_2S^+$: 248.0488; found: 248.0493.

5-(4-Fluorophenyl)-2,6-dimethylfuro[2,3-*d*]pyrimidin-4-ol (**3d**): light yellow solid, mp >300 °C; yield: 94%; ^{1}H NMR (DMSO-d_6, 400 MHz): δ 12.45 (1H, br, s, –O*H*), 7.57–7.54 (2H, m, Ar–*H*), 7.26 (2H, t, $J = 8.8$ Hz, Ar–*H*), 2.37 (6H, d, $J = 10.4$ Hz, –CH_3); ^{13}C NMR (DMSO-d_6, 100 MHz): δ 164.5, 163.0, 160.6, 159.3, 156.3, 146.6, 132.2, 132.1, 127.6, 116.9, 115.4, 115.2, 103.9, 21.3, 12.6; HRMS (ESI): *m*/*z* $[M+H]^+$ calc. for $C_{14}H_{12}FN_2O_2^{\ +}$: 259.0877; found: 259.0880.

3.2.2.5 Critical Views

Synthetic methodology for such bicyclic furo[2,3-*d*]pyrimidines is still limited; an earlier report involved the cyclization between 2-aminopyrimidine-4,6-diol and aryl-substituted-(*Z*)-(2-chloro-2-nitrovinyl)benzenes in refluxing ethanol and butanone using DBU as the catalyst.[11] The present protocol for the synthesis of fully substituted 5-arylfuro[2,3-*d*]pyrimidin-4-ols *via* [3 + 2]-heterocyclization, thus, features many advantages such as water as green solvent, short reaction time, simple operation, and high yields.[10]

3.2.3 Entry-3: Synthesis of 2-Methyl-4-amino-1,2,3,4-tetrahydroquinolines

Type of reaction: C–C and C–N bond formation
Reaction conditions: Catalyst-free, water, refluxing
Synthetic strategy: Aza-Diels–Alder-type reaction
Keywords: 1-Vinylpyrrolidin-2-one, anilines, catalyst-free, water, aza-Diels–Alder-type reaction, diastereoselectivity, 2-methyl-4-amino-1,2,3,4-tetrahydroquinoline derivatives

3.2.3.1 General Reaction Scheme

Wu *et al.*[12] developed a catalyst-free method for the diastereoselective synthesis of 2-methyl-4-amino-1,2,3,4-tetrahydroquinolines (**3**) from the reaction between 1-vinylpyrrolidin-2-one (**1**) and anilines (**2**) *via* an aza-Diels–Alder-type reaction in refluxing water (Scheme 3.3).

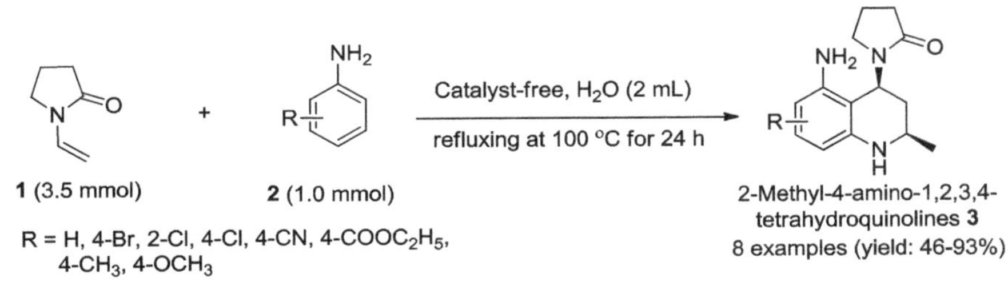

Scheme 3.3 Synthesis of 2-methyl-4-amino-1,2,3,4-tetrahydroquinolines (**3**).

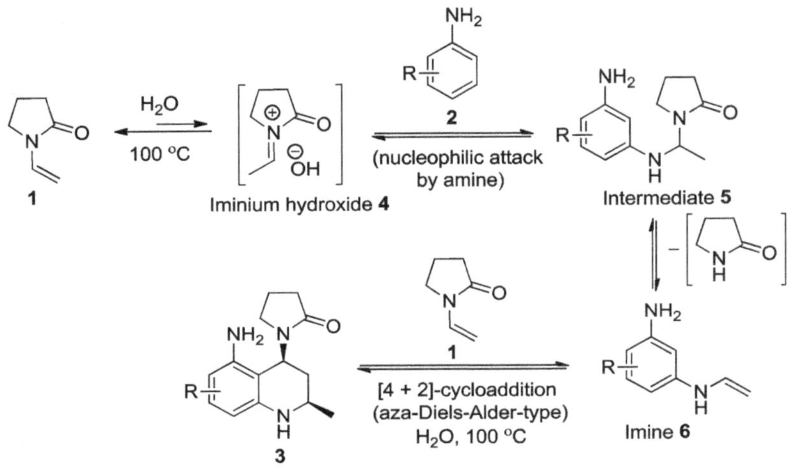

Scheme 3.4 Plausible mechanism for the formation of 2-methyl-4-amino-1,2,3,4-tetrahydroquinolines (**3**).

3.2.3.2 *Plausible Mechanism*

The authors proposed a plausible mechanism for this transformation as depicted in Scheme 3.4.

3.2.3.3 *Representative Examples*

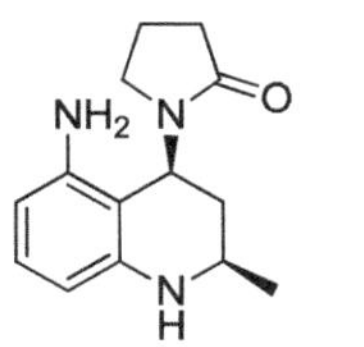

3a

1-((2*R*,4*S*)-5-Amino-2-methyl-1,2,3,4-tetrahydroquinolin-4-yl)pyrrolidin-2-one (**3a**): white solid, mp 72–74 °C; yield: 93%

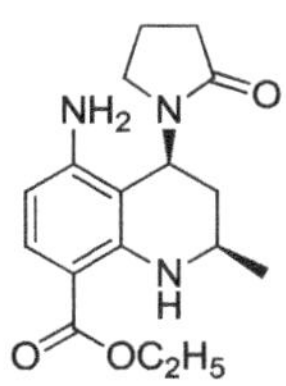

3b

(2*R*,4*S*)-Ethyl 5-amino-2-methyl-4-(2-oxopyrrolidin-1-yl)-1,2,3,4-tetrahydroquinoline-8-carboxylate (**3b**): pale yellow solid, mp 140–142 °C; yield: 48%

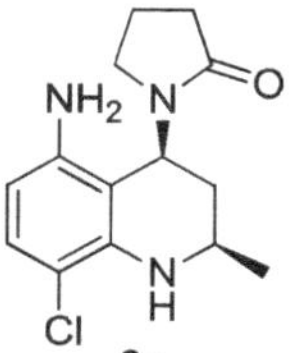

3c

1-((2*R*,4*S*)-5-Amino-8-chloro-2-methyl-1,2,3,4-tetrahydroquinolin-4-yl)pyrrolidin-2-one (**3c**): pale yellow solid, mp 147–149 °C; yield: 81%

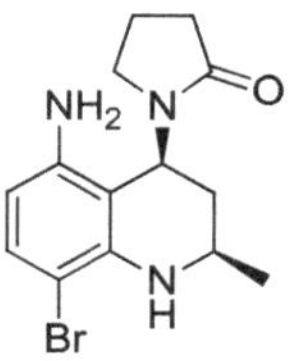

3d

1-((2*R*,4*S*)-5-Amino-8-bromo-2-methyl-1,2,3,4-tetrahydroquinolin-4-yl)pyrrolidin-2-one (**3d**): white solid, mp 212–213 °C; yield: 74%

3.2.3.4 Experimental Procedure

To a mixture of 1-vinylpyrrolidin-2-one (**1**; 0.389 g, 3.5 mmol) and anilines (**2**; 1.0 mmol), 2.0 mL water was added. The reaction mixture was stirred vigorously at 100 °C for up to 24 h, and on completion (as indicated by TLC), the reaction mixture was extracted with ethyl acetate (3×10 mL). The combined organic phase was washed with brine, dried with anhydrous sodium sulfate, and evaporated under reduced pressure. The resulting residue was purified by flash column chromatography, with ethyl acetate and petroleum ether as eluents, to afford the pure products of 2-methyl-4-amino-1,2,3,4-tetrahydroquinolines (**3**). All the products were characterized based on spectral studies.

3.2.3.5 Characterization Data of Two Representative Compounds

(2*R*,4*S*)-Ethyl 5-amino-2-methyl-4-(2-oxopyrrolidin-1-yl)-1,2,3,4-tetrahydroquinoline-8-carboxylate (**3b**): pale yellow solid, 140–142 °C; yield: 48%; IR (KBr): 3315, 2950, 1726, 1014 cm^{-1}; ^{1}H NMR ($CDCl_3$, 300 MHz): δ 1.26 (3H, d, $J=6.2$ Hz), 1.34 (3H, t, $J=7.1$ Hz), 1.65–1.77 (1H, m), 1.94–2.08 (3H, m), 2.45–2.61 (2H, m), 3.18–3.23 (2H, m), 3.65–3.70 (1H, m), 4.18 (1H, br s), 4.29 (2H, d, $J=7.1$ Hz), 5.53 (1H, dd, $J=12.2$, 5.6 Hz), 6.46 (1H, d, $J=8.4$ Hz), 7.50 (1H, s), 7.70 (1H, d, $J=8.5$ Hz); ^{13}C NMR ($CDCl_3$, 100 MHz): δ 176.0, 166.8, 149.7, 130.2, 128.6, 118.8, 117.6, 113.5, 60.33, 47.8, 46.8, 42.3, 33.7, 31.5, 22.1, 18.4, 14.5; HRMS: calc. for $C_{17}H_{22}N_2O_3$: 302.1630 $[M]^+$; found, 302.1601.

1-((2*R*,4*S*)-5-Amino-8-bromo-2-methyl-1,2,3,4-tetrahydroquinolin-4-yl)pyrrolidin-2-one (**3d**): white solid, mp 212–213 °C; yield: 74%; IR (KBr): 3278, 2921, 1671, 1490, 1272, 802 cm^{-1}; ^{1}H NMR ($CDCl_3$, 300 MHz): δ 1.23 (3H, d, $J=6.2$ Hz), 1.64–1.76 (1H, m), 1.90–2.05 (3H, m), 2.47–2.59 (2H, m), 3.15–3.26 (2H, m), 3.54–3.59 (1H, m), 3.87 (1H, br s), 5.51 (1H, dd, $J=11.7$, 5.7 Hz), 6.40 (1H, d, $J=8.5$ Hz,), 6.90 (1H, s), 7.09 (1H, d, $J=7.0$ Hz); ^{13}C NMR ($CDCl_3$, 100 MHz): δ 175.9, 145.0, 131.0, 129.2, 121.1, 116.3, 109.5, 47.9, 47.0, 42.3, 33.7, 31.5, 22.3, 18.4; HRMS: calc. for $C_{14}H_{17}BrN_2O$: 308.0524 $[M]^+$; found, 308.0524.

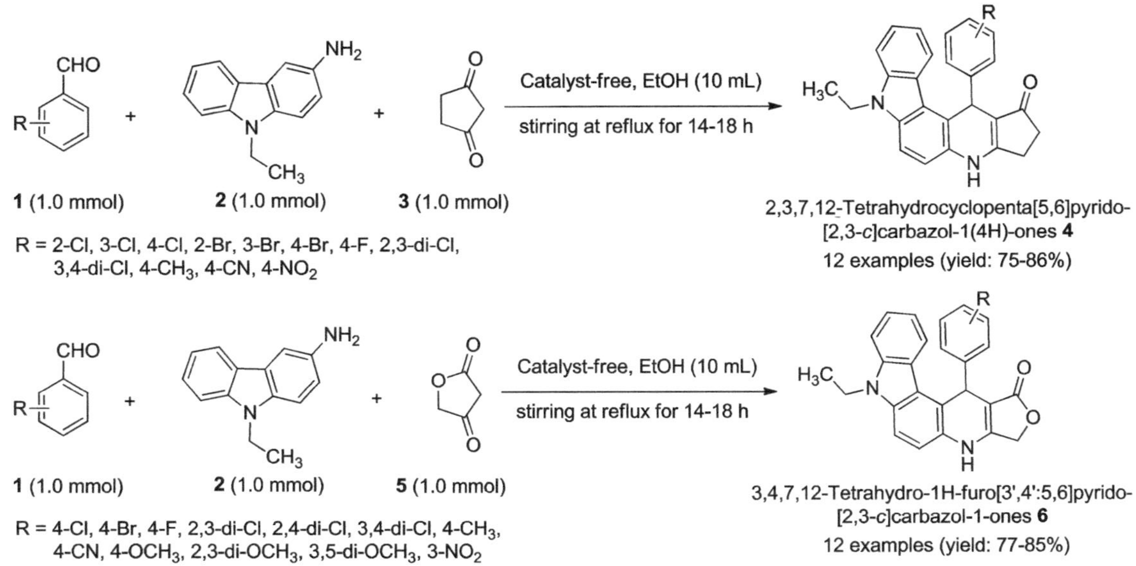

Scheme 3.5 Synthesis of substituted pyrido[2,3-*c*]carbazole derivatives (**5**/**6**).

3.2.3.6 Critical Views

The present protocol offers diastereoselective synthesis of 2-methyl-4-amino-1,2,3,4-tetrahydroquinoline derivatives *via* an aza-Diels–Alder-type reaction in water under catalyst-free conditions with moderate to good yields.[12] Previous endeavors for constructing this important organic scaffold report the use of strong acidic conditions at high temperature,[13–21] and several transition metal catalysts such as In,[22] Cu,[23] Ru,[24] and Pd.[25–27] Hence, the current reaction protocol renders an attractive alternative to the other existing methods.[28–34]

3.2.4 Entry-4: Synthesis of 2,3,7,12-Tetrahydrocyclopenta[5,6]pyrido[2,3-*c*]carbazol-1(4*H*)-ones and 3,4,7,12-Tetrahydro-1*H*-furo[3′,4′:5,6]pyrido[2,3-*c*]carbazol-1-ones

Type of reaction: C–C and C–N bond formation
Reaction conditions: Catalyst-free, ethanol, refluxing
Synthetic strategy: One-pot multicomponent reaction
Keywords: Aromatic aldehydes, 9-ethyl-9*H*-carbazol-3-amine, cyclopentane-1,3-dione/tetronic acid, catalyst-free, ethanol, 2,3,7,12-tetrahydrocyclopenta-[5,6]pyrido[2,3-*c*]carbazol-1(4*H*)-ones/3,4,7,12-tetrahydro-1*H*-furo[3′,4′:5,6]pyrido-[2,3-*c*]carbazol-1-ones

3.2.4.1 General Reaction Scheme

A novel series of pentacyclic heterocyles, such as 2,3,7,12-tetrahydrocyclopenta[5,6]pyrido[2,3-*c*]carbazol-1(4*H*)-ones (**4**) and 3,4,7,12-tetrahydro-1*H*-furo[3′,4′:5,6]pyrido[2,3-*c*]carbazol-1-ones (**6**), was synthesized by Wang and his group in a one-pot three-component reaction of aromatic aldehydes (**1**) and 9-ethyl-9*H*-carbazol-3-amine (**2**), with cyclopentane-1,3-dione (**3**) or tetronic acid (**5**), respectively, in ethanol under reflux without the aid of a catalyst (Scheme 3.5).[35]

3.2.4.2 Representative Examples

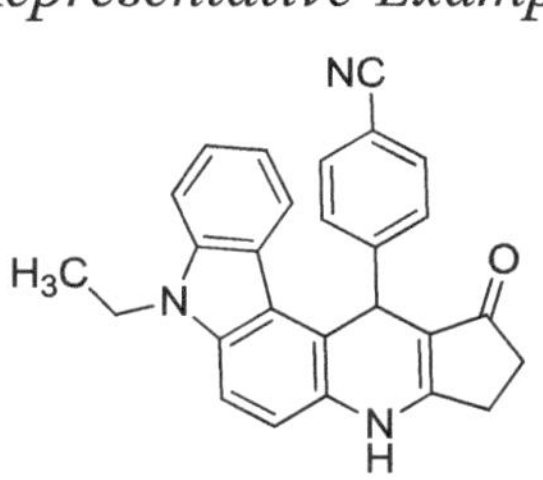

4a

4-(7-Ethyl-1-oxo-1,2,3,4,7,12-hexahydrocyclopenta[5,6]pyrido[2,3-*c*]carbazol-12-yl)benzonitrile (**4a**): solid, mp >300 °C; yield: 86%

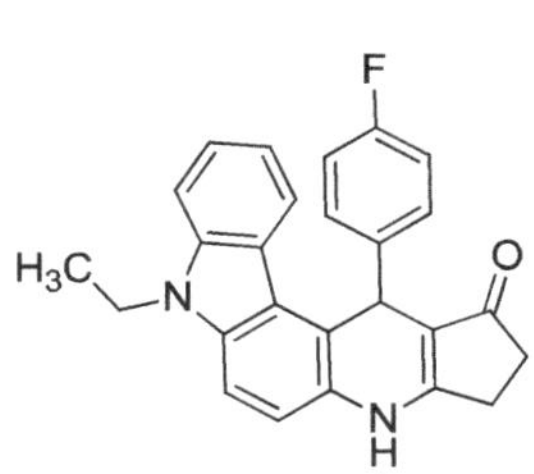

4b

7-Ethyl-12-(4-fluorophenyl)-2,3,7,12-tetrahydrocyclopenta[5,6]pyrido[2,3-*c*]carbazol-1(4*H*)-one (**4b**): solid, mp >300 °C; yield: 80%

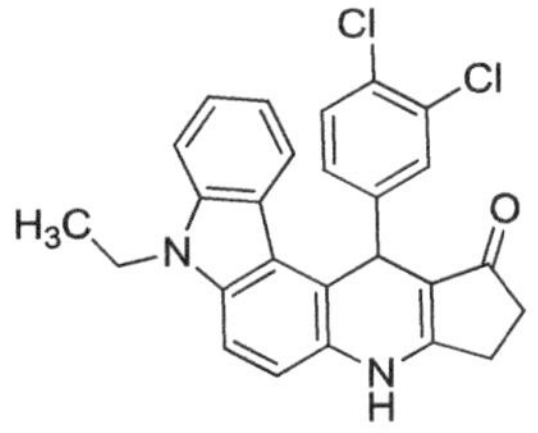

4c

12-(3,4-Dichlorophenyl)-7-ethyl-2,3,7,12-tetrahydrocyclopenta[5,6]pyrido[2,3-*c*]carbazol-1(4*H*)-one (**4c**): solid, mp >300 °C; yield: 79%

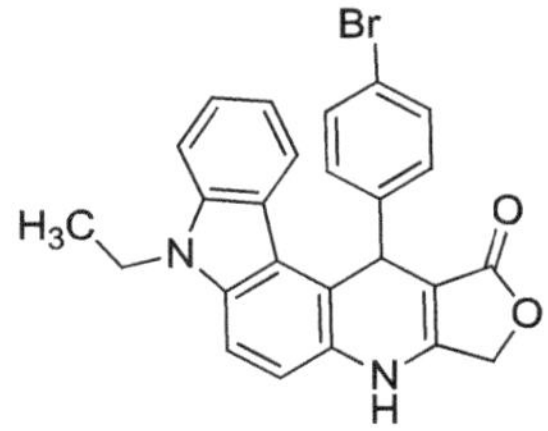

6a

12-(4-Bromophenyl)-7-ethyl-3,4,7,12-tetrahydro-1*H*-furo[3',4':5,6]pyrido[2,3-*c*]carbazol-1-one (**6a**): solid, mp >300 °C; yield: 83%

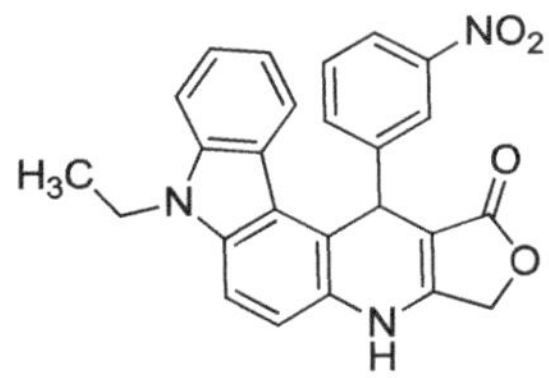

6b

7-Ethyl-12-(3-nitrophenyl)-3,4,7,12-tetrahydro-1*H*-furo[3',4':5,6]pyrido[2,3-*c*]carbazol-1-one (**6b**): solid, mp >300 °C; yield: 80%

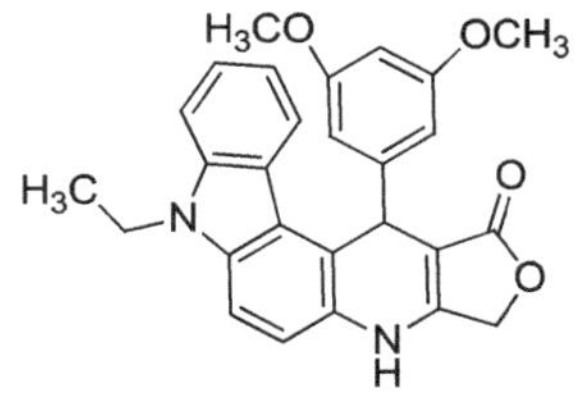

6c

12-(3,5-Dimethoxyphenyl)-7-ethyl-3,4,7,12-tetrahydro-1*H*-furo[3',4':5,6]pyrido[2,3-*c*]carbazol-1-one (**6c**): solid, mp >300 °C; yield: 78%

3.2.4.3 *Experimental Procedure*

10 mL of ethanol was added to a mixture of aromatic aldehydes (**1**; 1.0 mmol) and 9-ethyl-9*H*-carbazol-3-amine (**2**; 210 mg, 1.0 mmol) plus cyclopentane-1,3-dione (**3**; 98 mg, 1.0 mmol) or tetronic acid (**5**; 100 mg, 1.0 mmol) in a dry 50 mL flask; The reaction mixture was then stirred at reflux for 14–18 h. After the completion of the reaction, as indicated by TLC, the pure products of 2,3,7,12-tetrahydrocyclopenta[5,6]pyrido[2,3-*c*]carbazol-1(4*H*)-ones (**4**) or 3,4,7,12-tetrahydro-1*H*-furo[3′,4′:5,6]pyrido[2,3-*c*]carbazol-1-ones (**6**) were obtained by simple filtration when the reaction was cooled to room temperature. All the products were characterized based on spectral studies.

3.2.4.4 *Characterization Data of Two Representative Compounds*

12-(3,4-Dichlorophenyl)-7-ethyl-2,3,7,12-tetrahydrocyclopenta[5,6]pyrido[2,3-*c*]carbazol-1(4*H*)-one (**4c**): solid, mp >300 °C; yield: 79%; IR (KBr): 3235, 2971, 2926, 1637, 1549, 1458, 1276, 1247, 1224, 1155, 1053, 1030, 879, 798, 773 cm^{-1}; ^{1}H NMR (DMSO-d_6, 400 MHz): δ 1.30 (3H, t, $J = 7.2$ Hz, $-CH_3$), 2.27–2.34 (2H, m, $-CH_2$), 2.64–2.68 (2H, m, $-CH_2$), 4.42 (2H, q, $J = 7.2$ Hz,

$-CH_2$), 5.87 (1H, s, $-CH$), 7.02–7.06 (1H, m, Ar–H), 7.09 (1H, dd, $J=2.0$ and 8.4 Hz, Ar–H), 7.30 (1H, d, $J=8.8$ Hz, Ar–H), 7.34–7.38 (1H, m, Ar–H), 7.41–7.43 (2H, m, Ar–H), 7.55 (1H, d, $J=8.0$ Hz, Ar–H), 7.63 (1H, d, $J=8.8$ Hz, Ar–H), 7.85 (1H, d, $J=8.0$ Hz, Ar–H), 10.29 (1H, s, $-NH$); HRMS (ESI, m/z): calc. for $C_{26}H_{19}Cl_2N_2O$ $[M-H]^-$ 445.0875; found 445.0877.

12-(3,5-Dimethoxyphenyl)-7-ethyl-3,4,7,12-tetrahydro-1*H*-furo[3′,4′:5,6]-pyrido[2,3-*c*]carbazol-1-one (**6c**): solid, mp >300 °C; yield: 78%; IR (KBr): 3227, 3079, 2977, 1731, 1634, 1605, 1587, 1482, 1431, 1347, 1311, 1291, 1201, 1029, 1017, 812, 802, 739, 700 cm^{-1}; ^1H NMR (DMSO-d_6, 400 MHz): δ 1.30 (3H, t, $J=7.2$ Hz, $-CH_3$), 3.61 (6H, s, 2×$-OCH_3$), 4.42 (2H, q, $J=7.2$ Hz, $-CH_2$), 4.84 (1H, d, $J=15.6$ Hz, $-CH$), 4.92 (1H, d, $J=15.6$ Hz, $-CH$), 5.76 (1H, s, $-CH$), 6.22–6.23 (1H, m, Ar–H), 6.32 (2H, d, $J=2.4$ Hz, Ar–H), 7.02–7.06 (1H, m, Ar–H), 7.24 (1H, d, $J=8.4$ Hz, Ar–H), 7.34–7.37 (1H, m, Ar–H), 7.55 (1H, d, $J=8.0$ Hz, Ar–H), 7.61 (1H, d, $J=8.4$ Hz, Ar–H), 7.94 (1H, d, $J=9.2$ Hz, Ar–H), 10.12 (1H, s, NH); HRMS (ESI, m/z): calc. for $C_{27}H_{23}N_2O_4$ $[M-H]^-$ 439.1658, found 439.1658.

3.2.4.5 Critical Views

The investigators developed an efficient method for the synthesis of a new series of biologically interesting pyrido[2,3-*c*]carbazole derivatives under catalyst-free conditions from a three-component reaction of aldehydes, 9-ethyl-9*H*-carbazol-3-amine plus cyclopentane-1,3-dione or tetronic acid, under reflux in ethanol; the process possesses the advantages of mild reaction conditions, good yields, one-pot operational simplicity and the use of no added catalyst.

3.2.5 Entry-5: Synthesis of Substituted Pyrimidine Derivatives

Type of reaction: C–N bond formation
Reaction conditions: Catalyst-free, *N*,*N*-dimethylformamide (DMF), refluxing
Synthetic strategy: One-pot multicomponent reaction/domino reaction
Keywords: Diethyl but-2-ynedioate, aliphatic/aromatic amines, formaldehyde, catalyst-free, *N*,*N*-dimethylformamide (DMF), one-pot multicomponent reaction, domino reaction, multisubstituted tetrahydropyrimidines

3.2.5.1 General Reaction Scheme

Zhang *et al.*[36] reported a convenient one-pot synthesis of multisubstituted pyrimidine analogs (**5**) *via* multicomponent reactions between diethyl but-2-ynedioate (**1**), aliphatic/aromatic amines (**2**/**4**) and formaldehyde (**3**) at reflux in *N*,*N*-dimethylformamide (DMF) under catalyst-free conditions (Scheme 3.6); the reaction proceeds through three domino sequences such as two-component hydroamination, three-component Mannich-type reaction, and two-component aminealdehyde dehydration-cyclization process.

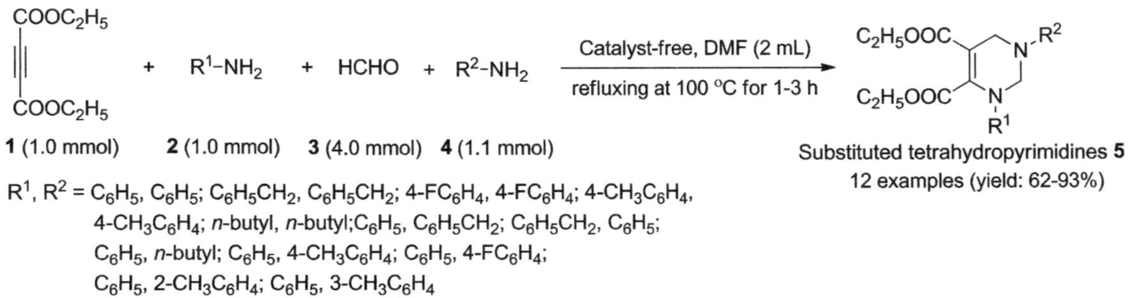

Scheme 3.6 Synthesis of substituted pyrimidine derivatives (5).

3.2.5.2 Representative Examples

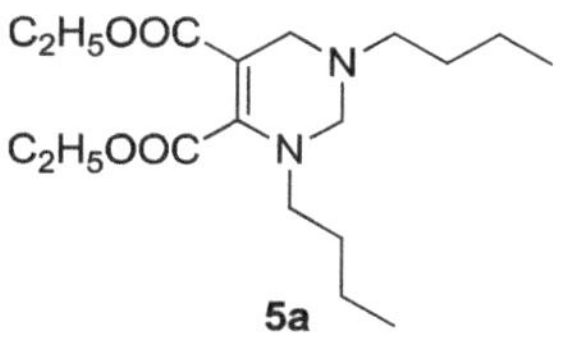

5a

Diethyl 1,3-dibutyl-1,2,3,6-tetrahydropyrimidine-4,5-dicarboxylate (**5a**): yellow oil; yield: 96%

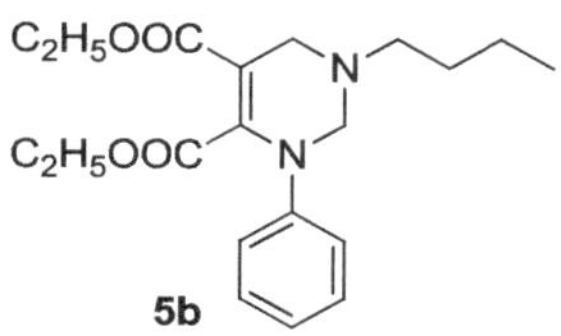

5b

Diethyl 1-butyl-3-phenyl-1,2,3,6-tetrahydropyrimidine-4,5-dicarboxylate (**5b**): yellow oil; yield: 90%

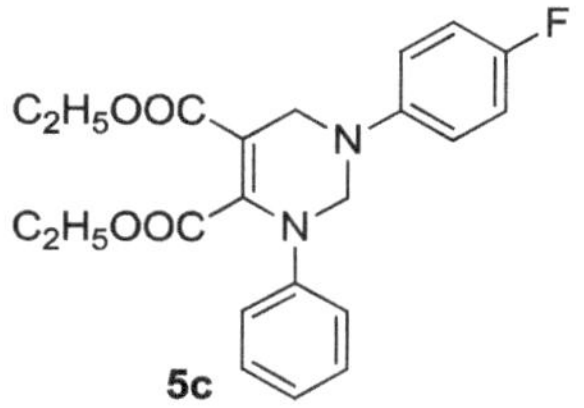

5c

Diethyl 1-(4-fluorophenyl)-3-phenyl-1,2,3,6-tetrahydropyrimidine-4,5-dicarboxylate (**5c**): yellow oil; yield: 80%

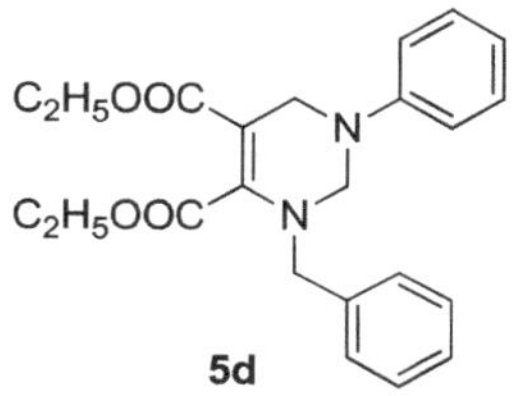

5d

Diethyl 3-benzyl-1-phenyl-1,2,3,6-tetrahydropyrimidine-4,5-dicarboxylate (**5d**): yellow oil; yield: 86%

3.2.5.3 Experimental Procedure

To a stirring mixture of diethyl but-2-ynedioate (**1**; 170 mg, 1.0 mmol) and aromatic amine (**2**; 1.0 mmol), 2 mL of *N*,*N*-dimethylformamide, aliphatic/aromatic amine (**4**; 1.1 mmol) and 35% formaldehyde (**3**; 4.0 mmol, 345 mg) were added successively. The mixture was stirred at 100 °C for 1–3 h. After completion of the reaction (monitored by TLC), the mixture was cooled to room temperature, diluted with water and extracted with diethyl ether (5×15 mL), and then the ether layer was washed with saturated brine (5×10 mL) and dried with anhydrous magnesium sulfate. After the solvent was removed *in vacuo*, the crude product was purified by preparative TLC with hexane–ethyl acetate (10:1) as the eluent to afford the desired products of substituted tetrahydropyrimidines (**5**) with good yields. The compounds were identified by means of analytical and spectral studies.

3.2.5.4 Characterization Data of Two Representative Compounds

Diethyl 1-butyl-3-phenyl-1,2,3,6-tetrahydropyrimidine-4,5-dicarboxylate (**5b**): yellow oil; yield: 90%; IR (KBr): 3446, 2960, 1740, 1685, 1581, 1460, 1399, 1283, 1245, 1145, 1039 cm^{-1}; ^1H NMR ($CDCl_3$, 400 MHz): δ 4.33 (2H, q, $J=7.2$ Hz,), 4.07 (2H, q, $J=7.2$ Hz), 3.93 (2H, s), 3.46 (2H, s), 2.99 (2H, t, $J=7.6$ Hz), 2.49 (2H, t, $J=7.6$ Hz), 1.50–1.35 (4H, m), 1.34–1.32 (4H, m), 1.26–1.17 (6H, m), 0.92–0.87 (6H, m); MS (EI): *m/z* (%rel): 340 (M^+, 22). Anal.

calc. for $C_{18}H_{32}N_2O_4$: C, 63.50; H, 9.47; N, 8.23; found: C, 63.71; H, 9.55; N, 7.92.

Diethyl 3-benzyl-1-phenyl-1,2,3,6-tetrahydropyrimidine-4,5-dicarboxylate (**5d**): yellow oil; yield: 86%; IR (KBr): 3446, 2984, 1742, 1681, 1579, 1456, 1392, 1269, 1208, 1106, 1038 cm^{-1}; ^1H NMR ($CDCl_3$, 400 MHz): δ 7.28–7.16 (7H, m), 6.80–6.78 (3H, m), 4.39 (2H, s), 4.29 (2H, q, $J = 7.2$ Hz), 4.21 (2H, s), 4.17 (2H, q, $J = 7.2$ Hz), 4.09 (2H, s), 1.30–1.19 (6H, m); MS (EI): *m/z* (%rel): 394 (M^+, 25), 303 (33).

3.2.5.5 *Critical Views*

A few methods are reported in the literature for constructing multiply-substituted tetrahydropyrimidine rings, which suffer from many drawbacks as low yields, long reaction time, and cumbersome experimental processes.[37–47] Compared with these methods, the present one is convenient and operationally simple for accessing a series of densely functionalized tetrahydropyrimidine derivatives; the key advantages of this process are the use of no catalyst, a short reaction time, good to excellent yields, and the generation of no toxic by-products.[36]

3.2.6 Entry-6: Synthesis of 2-Aryl-5-cyano-4-methylsulfanylpyrimidin-6-ones

Type of reaction: C–N bond formation
Reaction conditions: Catalyst-free, neat, heating
Synthetic strategy: Condensation reaction
Keywords: Ketene *N,S*-acetal, aryl aldehydes, catalyst- and solvent-free, condensation, 2-aryl-5-cyano-4-methylsulfanylpyrimidin-6-ones

3.2.6.1 *General Reaction Scheme*

Hagimori and co-workers[48] reported a facile one-pot procedure for the synthesis of 2-aryl-5-cyano-4-methylsulfanylpyrimidin-6-one derivatives (**3**) with fluorescent property by heating a mixture of ketene *N,S*-acetal [(*E*)-3-amino-2-cyano-3-(methylthio)acrylamide; **1**] and aryl aldehydes (**2**) under solvent and catalyst-free conditions with moderate yields (Scheme 3.7).

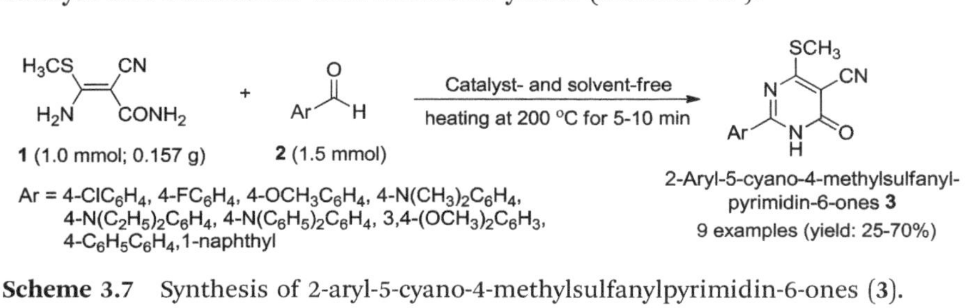

Scheme 3.7 Synthesis of 2-aryl-5-cyano-4-methylsulfanylpyrimidin-6-ones (**3**).

3.2.6.2 Representative Examples

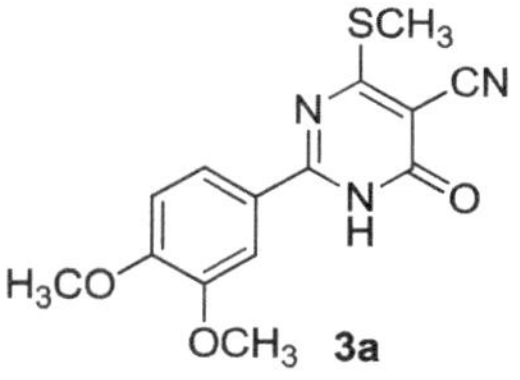

3a

2-(3,4-Dimethoxyphenyl)-4-(methylthio)-6-oxo-1,6-dihydropyrimidine-5-carbonitrile (**3a**): slight yellow needles, mp 345–348 °C; yield: 52%

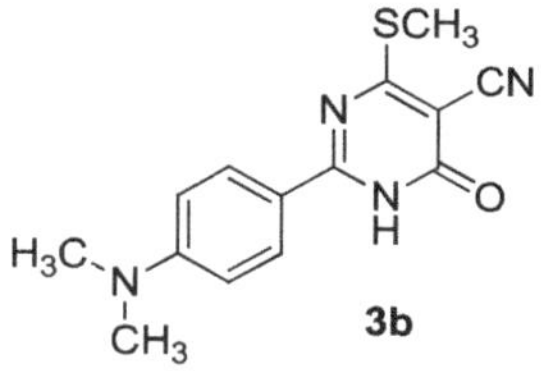

3b

2-(4-(Dimethylamino)phenyl)-4-(methylthio)-6-oxo-1,6-dihydropyrimidine-5-carbonitrile (**3b**): yellow needles, mp 333–334 °C; yield: 70%

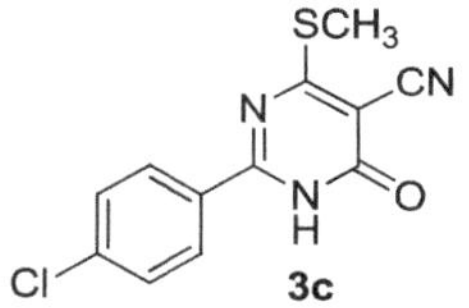

3c

2-(4-Chlorophenyl)-4-(methylthio)-6-oxo-1,6-dihydropyrimidine-5-carbonitrile (**3c**): slight yellow needles, mp 309–310 °C; yield: 47%

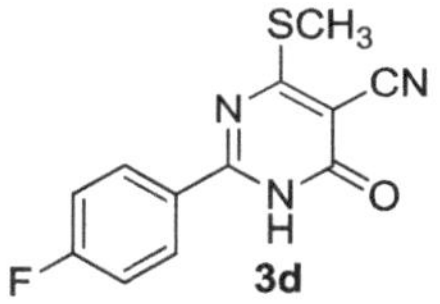

3d

2-(4-Fluorophenyl)-4-(methylthio)-6-oxo-1,6-dihydropyrimidine-5-carbonitrile (**3d**): colorless needles, mp 315–317 °C; yield: 35%

3.2.6.3 Experimental Procedure

A mixture of ketene *N,S*-acetal (**1**; 0.157 g, 1.0 mmol) and aryl aldehydes (**2**; 1.5 mmol) was stirred for 5–10 min at 200 °C to completion of the reaction with the formation of a solid. The reaction mixture was then cooled, washed with methanol, and recrystallized from methanol to furnish pure products of 2-aryl-5-cyano-4-methylsulfanylpyrimidin-6-ones (**3**) with moderate yields. All the products were characterized by means of elemental and spectral studies.

3.2.6.4 Characterization Data of Two Representative Compounds

2-(3,4-Dimethoxyphenyl)-4-(methylthio)-6-oxo-1,6-dihydropyrimidine-5-carbonitrile (**3a**): Slight yellow needles, mp 345–348 °C; yield: 52%; UV (EtOH): λ_{max} (log ε): 275 (4.34), 325 (4.38) nm; IR (KBr): 3450 (OH, br), 2225 (CN), 1670, 1660, 1600, 1550, 1525, 1503 cm^{-1}; ^1H NMR ($CDCl_3 + CF_3COOH$, 500 MHz): δ 2.81 (3H, s, S–CH_3), 4.03 (3H, s, OCH_3), 4.04 (3H, s, OCH_3), 7.11 (1H, d, $J = 8.8$ Hz, 5′-H), 7.72 (1H, d, $J = 2.3$ Hz, 2′-H), 8.03 (1H, dd, $J = 2.3$, 8.8 Hz, 6′-H); MS: *m/z* (%rel) 304 ($[M+1]^+$, 17), 303 (M^+, 100), 288 (11), 287 (44), 164 (28), 140 (89), 44 (12). Anal. calc. for $C_{14}H_{13}N_3O_3S$: C, 55.43; H, 4.32; N, 13.85; found: C, 55.64; H, 4.29; N, 13.99.

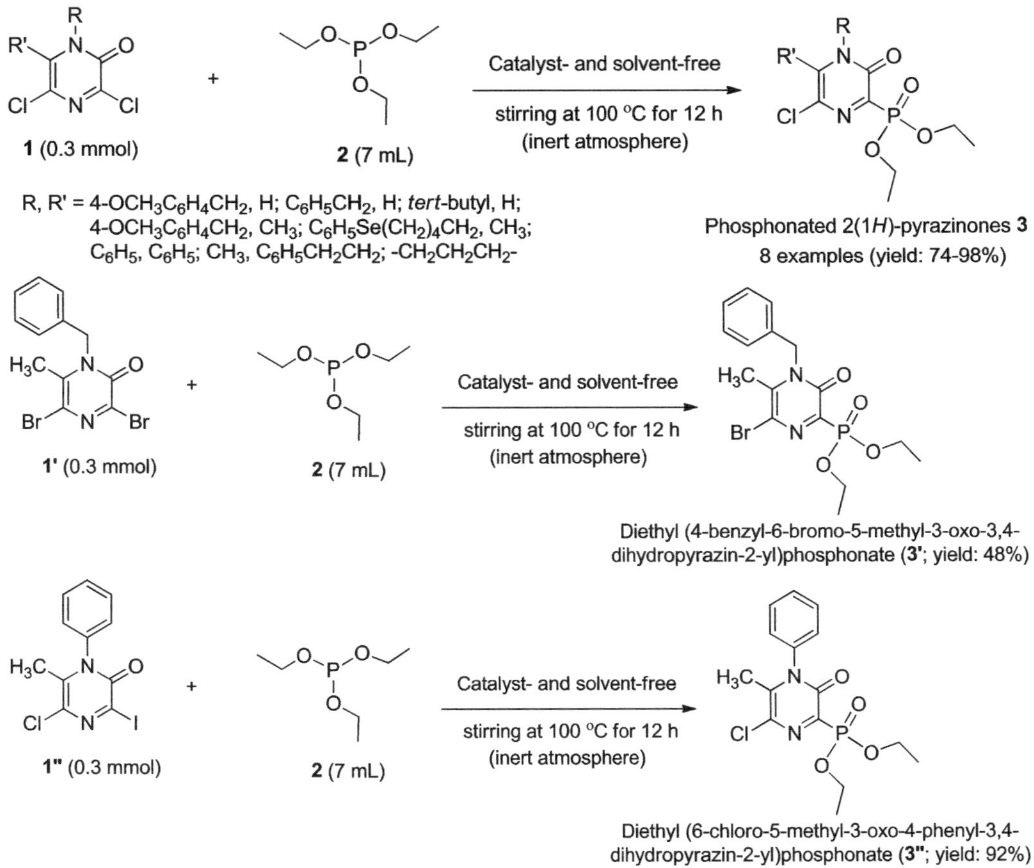

Scheme 3.8 Synthesis of phosphonated 2(1*H*)-pyrazinones (**3**).

2-(4-(Dimethylamino)phenyl)-4-(methylthio)-6-oxo-1,6-dihydropyrimidine-5-carbonitrile (**3b**): yellow needles, mp 333–334 °C; yield: 70%; UV (EtOH): λ_{max} (log ε): 359 (4.51), 406 (4.58) nm; IR (KBr): 3450 (OH, br), 2220 (CN), 1635, 1605, 1510, 1500 cm^{-1}; ^1H NMR (DMSO-d_6, 500 MHz): δ 2.66 (3H, s, S–CH_3), 3.05 (6H, s, $N(CH_3)_2$), 6.78 (2H, d, $J=9.4$ Hz, 3′,5′-H), 8.15 (2H, d, $J=9.4$ Hz, 2′,6′-H), 12.83 (1H, br s, –N*H*); MS: *m/z* (%rel) 287 ($[M+1]^+$, 11), 286 (M^+, 52), 147 (22), 145 (20), 140 (24), 54 (38), 46 (100). Anal. calc. for $C_{14}H_{14}N_4OS$: C, 58.72; H, 4.93; N, 19.57; found: C, 58.81; H, 4.90; N, 19.62.

3.2.6.5 Critical Views

This present method represents a facile preparation of highly-functionalized and fluorescence-active pyrimidine derivatives as useful building blocks under solvent and catalyst-free conditions; however, the yields obtained were moderate amounts.

3.2.7 Entry-7: Synthesis of Phosphonated 2(1*H*)-Pyrazinones

Type of reaction: C–P bond formation
Reaction conditions: Catalyst-free, neat, heating
Synthetic strategy: Arbuzov-type reaction
Keywords: 3,5-Dichloropyrazinones, triethyl phosphite, catalyst- and solvent-free, Arbuzov-type reaction, phosphonated 2(1*H*)-pyrazinones

3.2.7.1 General Reaction Scheme

Alen *et al.*[49] synthesized a series of phosphonated 2(1*H*)-pyrazinones (**3**) for the first time by developing a simple and catalyst-free method *via* an Arbuzov-type reaction between 3,5-dichloropyrazinones (**1**) and triethyl phosphite (**2**) with heating in the absence of any other added solvent (Scheme 3.8). The method was also found to proceed in the case of 3-bromo- and 3-iodopyrazinones.

3.2.7.2 Representative Examples

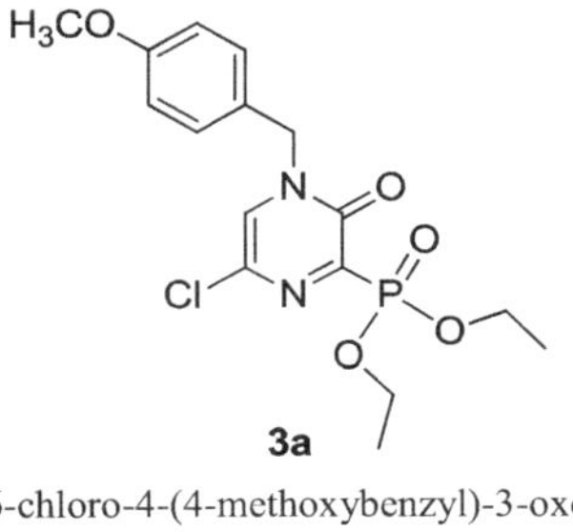

3a

Diethyl (6-chloro-4-(4-methoxybenzyl)-3-oxo-3,4-dihydropyrazin-2-yl)phosphonate (**3a**): oil; yield: 98%

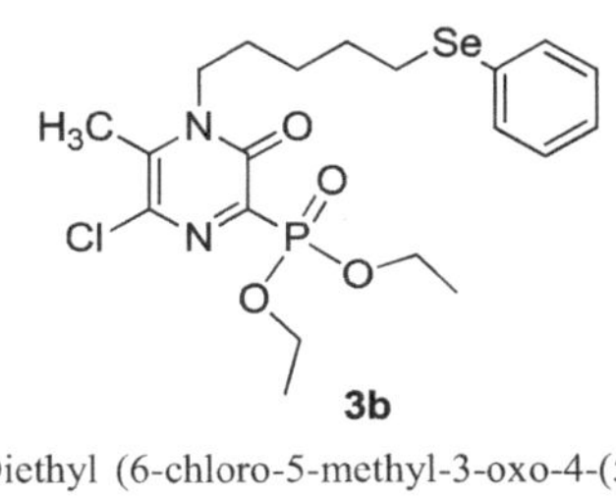

3b

Diethyl (6-chloro-5-methyl-3-oxo-4-(5-(phenylselanyl)pentyl)-3,4-dihydropyrazin-2-yl)phosphonate (**3b**): oil; yield: 92%

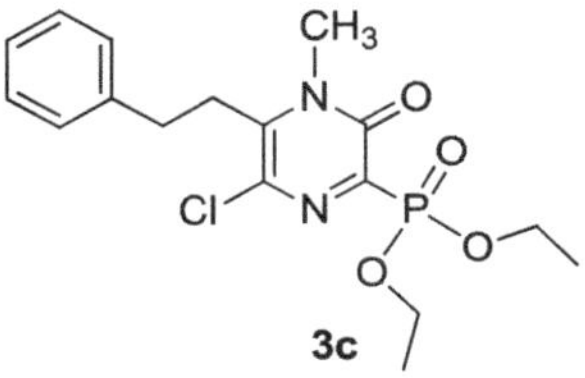

3c

Diethyl (6-chloro-4-methyl-3-oxo-5-phenethyl-3,4-dihydropyrazin-2-yl)phosphonate (**3c**): solid, mp 102 °C; yield: 94%

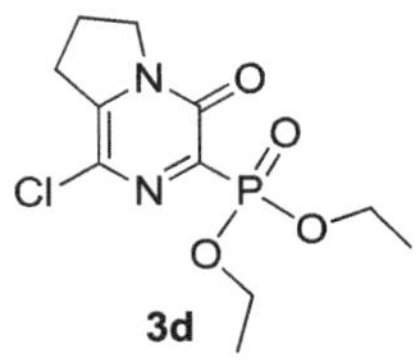

3d

Diethyl (1-chloro-4-oxo-4,6,7,8-tetrahydropyrrolo[1,2-*a*]pyrazin-3-yl)phosphonate (**3d**): oil; yield: 78%

3.2.7.3 Experimental Procedure

Pyrazinones (**1**; 0.3 mmol) were stirred in 7 mL of triethyl phosphite (**2**) for 12 h at 100 °C under an inert atmosphere. After completion of the reaction, excess triethyl phosphite was evaporated off. The crude product was purified by column chromatography over silica gel using dichloromethane–methanol (97:3) as the eluent, to give pure products of pyrazinone-phosphonates (**3**) with good to excellent yields. The compounds were identified by means of spectral analyses.

3.2.7.4 Characterization Data of Two Representative Compounds

Diethyl (6-chloro-5-methyl-3-oxo-4-(5-(phenylselanyl)pentyl)-3,4-dihydropyrazin-2-yl)phosphonate (**3b**): oil; yield: 92%; IR (NaCl): 1657, 1547 cm^{-1}; ^{1}H NMR ($CDCl_3$, 400 MHz): δ 7.47 (2H, d, *J* = 7.5 and 1.9 Hz), 7.27–7.21 (3H, m), 4.38–4.30 (4H, m), 4.02 (2H, t, *J* = 15.8 Hz), 2.90 (2H, t, *J* = 14.5 Hz), 2.52 (3H, d, *J* = 1.3 Hz) 1.79–1.72 (2H, m), 1.71–1.64 (2H, m), 1.55–1.47 (2H, m), 1.39 (6H, t, *J* = 7.1 Hz); ^{13}C NMR ($CDCl_3$, 100 MHz): δ 154.5 (d, *J* = 31 Hz), 144.4 (d, *J* = 233 Hz), 140.3 (d, *J* = 3 Hz), 132.82, 129.9, 128.8, 126.6 (d, *J* = 28 Hz), 126.5, 63.7 (d, *J* = 6 Hz), 45.9, 29.2, 27.1, 26.9, 26.6, 16.8, 16.1 (d, *J* = 6 Hz); EIMS: *m/z* (%rel) 506 (M^+, 100), 369 (22), 349 (81); HRMS: calc. for $C_{20}H_{28}ClN_2O_4PSe$: 506.0640; found: 506.0641.

Diethyl (1-chloro-4-oxo-4,6,7,8-tetrahydropyrrolo[1,2-*a*]pyrazin-3-yl)phosphonate (**3d**): oil; yield: 78%; IR (NaCl): 1658, 1568 cm^{-1}; ^{1}H NMR ($CDCl_3$, 400 MHz): δ 4.33 (4H, quin., *J* = 1.9 and 0.4 Hz), 4.26 (2H, t, *J* = 1.9 Hz), 3.29 (2H, t, *J* = 2.0 Hz), 2.37 (2H, quin., *J* = 1.9 Hz), 1.40 (6H, t, *J* = 1.8 Hz); ^{13}C NMR ($CDCl_3$, 75 MHz): δ 153.5 (d, *J* = 32 Hz), 145.6, 143.4 (d, *J* = 231 Hz), 121.4 (d, *J* = 28 Hz), 62.7 (d, *J* = 7 Hz), 50.1, 30.7, 19.2, 15.4 (d, *J* = 5 Hz); EIMS: *m/z* (rel%): 306 (M^+, 36), 170 (100); HRMS: calc. for $C_{11}H_{16}ClN_2O_4P$: 306.0536; found: 306.0526.

3.2.7.5 Critical Views

The present protocol offers a practical way to synthesize monochlorinated pyrazinones containing a phosphonate moiety as an important building

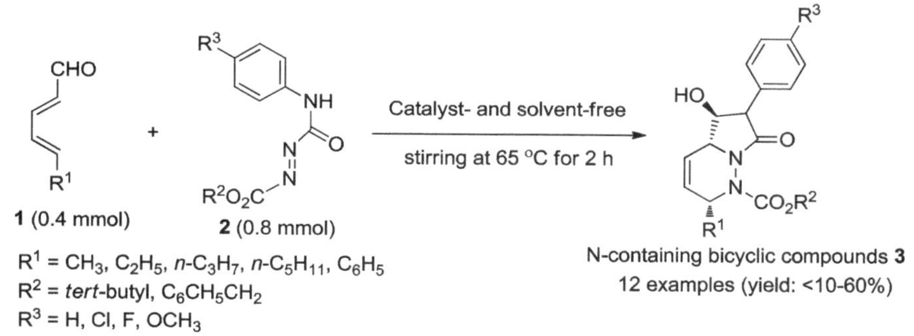

Scheme 3.9 Synthesis of nitrogen-containing bicyclic derivatives (**3**).

block under catalyst- and solvent-free conditions with good to excellent yields.[49]

3.2.8 Entry-8: Synthesis of Nitrogen-containing Bicyclic Derivatives

Type of reaction: C–N bond formation
Reaction conditions: Catalyst-free, neat, heating
Synthetic strategy: Hetero-Diels–Alder reaction
Keywords: Dienals, diazenes, catalyst- and solvent-free, hetero-Diels–Alder reaction, diastereoselectivity, nitrogen-containing bicyclic derivatives

3.2.8.1 General Reaction Scheme

Crouillebois *et al.*[50] developed a solvent- and catalyst-free synthesis of a series of nitrogen-containing bicyclic derivatives (**3**) from the reaction between dienals (**1**) and diazenes (**2**) involving the hemiaminal formation/hetero-Diels–Alder reaction in a highly diastereoselective manner with heat (Scheme 3.9).

3.2.8.2 Representative Examples

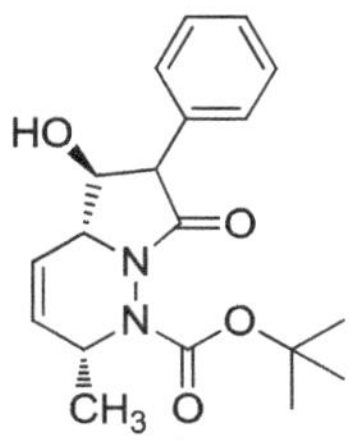

3a

(2*R*,4a*R*,5*S*)-*tert*-Butyl 5-hydroxy-2-methyl-7-oxo-6-phenyl-4a,5,6,7-tetrahydropyrrolo[1,2-*b*]pyridazine-1(2*H*)-carboxylate (**3a**): oil; yield: 60%

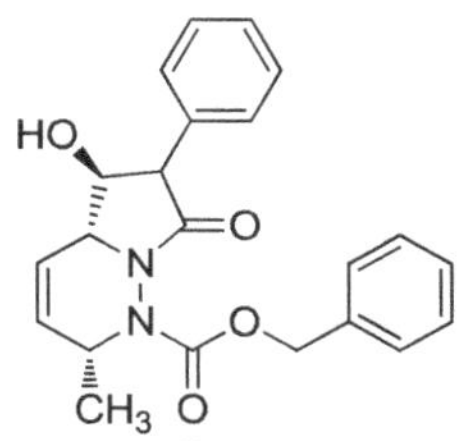

3b

(2*R*,4a*R*,5*S*)-Benzyl 5-hydroxy-2-methyl-7-oxo-6-phenyl-4a,5,6,7-tetrahydropyrrolo[1,2-b]pyridazine-1(2*H*)-carboxylate (**3b**): oil; yield: 67%

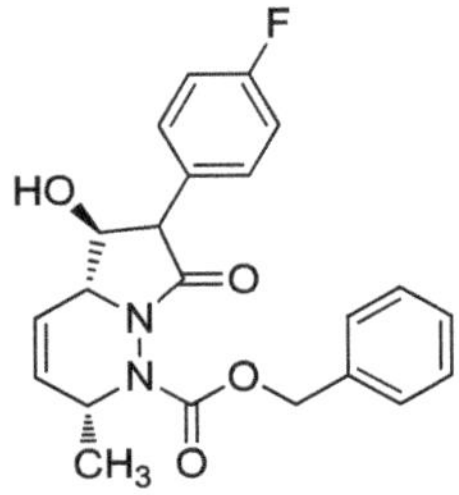

3c

(2*R*,4a*R*,5*S*)-Benzyl 6-(4-fluorophenyl)-5-hydroxy-2-methyl-7-oxo-4a,5,6,7-tetrahydropyrrolo[1,2-*b*]pyridazine-1(2*H*)-carboxylate (**3c**): oil; yield: 46%

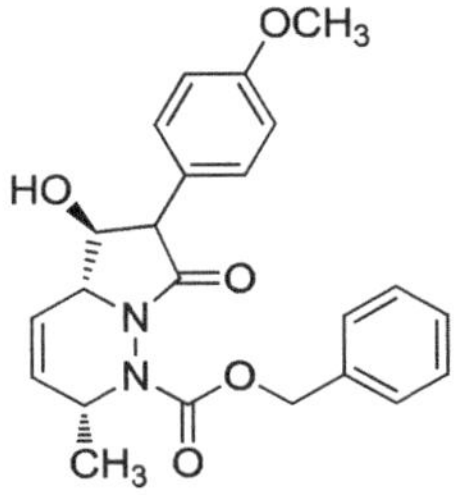

3d

(2*R*,4a*R*,5*S*)-Benzyl 5-hydroxy-6-(4-methoxyphenyl)-2-methyl-7-oxo-4a,5,6,7-tetrahydropyrrolo[1,2-*b*]pyridazine-1(2*H*)-carboxylate (**3d**): oil; yield: 47%

3.2.8.3 Experimental Procedure

Dienal (**1**; 0.4 mmol) was added to a diazene derivative (**2**; 0.8 mmol) and the mixture was stirred at 65 °C for 2 h, when a crude product was formed that was directly purified by flash chromatography (eluent pentane–ethyl acetate) to afford the desired pure product of bicyclic compounds (**3**). The compounds were identified by means of spectral analyses.

3.2.8.4 Characterization Data of Two Representative Compounds

(2*R*,4a*R*,5*S*)-*tert*-Butyl 5-hydroxy-2-methyl-7-oxo-6-phenyl-4a,5,6,7-tetrahydropyrrolo[1,2-*b*]pyridazine-1(2*H*)-carboxylate (**3a**): oil; yield: 60%; FTIR (neat): 3327, 2978, 2931, 1695 cm^{-1}; ^1H NMR ($CDCl_3$, 200 MHz): δ 1.35 (3H, d, $J=8.0$ Hz), 1.52 (9H, s), 4.06 (1H, br d, $J=4.0$ Hz), 4.47 (1H, br s), 5.28 (1H, br s), 5.72–5.85 (2H, m), 7.10 (1H, t, $J=8.0$ Hz), 7.26–7.34 (2H, m), 7.62–7.68 (2H, m); ^{13}C NMR ($CDCl_3$, 50 MHz): δ 20.1, 28.4 (3C), 51.4, 58.8, 81.6, 82.5, 120.5 (2C), 123.0, 124.5, 129.0 (2C), 131.7, 138.0, 155.2, 156.5; HRMS (ESI): calc. for $C_{18}H_{23}N_3O_4Na$ $[M+Na]^+$: 368.1586; found: 368.1582.

(2*R*,4a*R*,5*S*)-Benzyl 6-(4-fluorophenyl)-5-hydroxy-2-methyl-7-oxo-4a,5,6,7-tetrahydropyrrolo[1,2-*b*]pyridazine-1(2*H*)-carboxylate (**3c**): oil; yield: 46%; FTIR (neat): 3318, 2975, 1700 cm^{-1}; ^1H NMR ($CDCl_3$, 300 MHz): δ 1.37 (3H, d, $J=7.0$ Hz), 2.65 (1H, s), 4.13 (1H, s), 4.52–4.67 (1H, m), 5.17–5.26 (3H, m), 5.76–5.94 (2H, m), 7.00 (2H, t, $J=8.4$ Hz), 7.37 (5H, br s), 7.56 (2H, br s); ^{13}C NMR ($CDCl_3$, 75 MHz): δ 20.4, 51.4, 59.1, 68.7, 82.3, 115.8 (d, $J=21.8$ Hz, 2C), 122.7, 123.1 (d, $J=8.2$ Hz, 2C), 128.1 (2C), 128.6, 128.7 (2C), 131.3, 133.7, 135.6, 156.3, 156.7, 160.0 (d, $J=242.2$ Hz); HRMS (ESI): calc. for $C_{21}H_{21}FN_3O_4$ $[M+H]^+$: 398.1516; found: 398.1511.

3.2.8.5 Critical Views

A new series of nitrogen-containing bicyclic compounds were synthesized with the help of the present protocol.[50] The key benefits of this method are the use of no solvent or catalyst and high diasteroselectivity, but the method suffers from low yields in most of the cases.

3.2.9 Entry-9: Synthesis of Trifluoromethyl-substituted Bicyclic Pyridines

Type of reaction: C–C and C–N bond formation
Reaction conditions: Catalyst-free, neat, refluxing
Synthetic strategy: One-pot multicomponent reaction
Keywords: Heterocyclic ketene aminals (HKAs), triethoxymethane, aryl-substituted 4,4,4-trifluoro-1,3-butanediones, catalyst- and solvent-free, one-pot multicomponent reaction, trifluoromethyl-substituted bicyclic pyridines

3.2.9.1 General Reaction Scheme

A facile and catalyst-free method for the synthesis of functionalized trifluoromethyl-substituted bicyclic pyridines (**4**) was developed by Zhu *et al.*[51] from a one-pot three-component reaction, between heterocyclic ketene aminals (**1**), triethoxymethane (**2**), and aryl-substituted 4,4,4-trifluoro-1,3-butanediones (**3**), heated at reflux under solvent-free conditions with good to excellent yields (Scheme 3.10).

3.2.9.2 Representative Examples

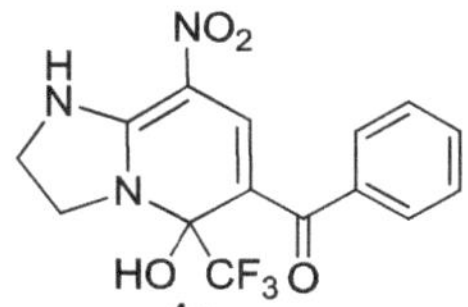

4a

(5-Hydroxy-8-nitro-5-(trifluoromethyl)-1,2,3,5-tetrahydroimidazo[1,2-*a*]pyridin-6-yl)(phenyl)methanone (**4a**): white solid, mp 173–175 °C; yield: 93%

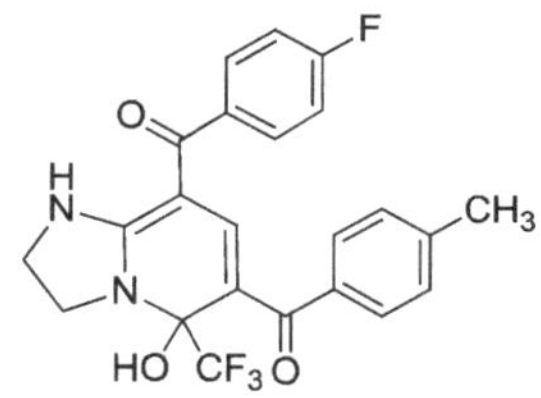

4b

(8-(4-Fluorobenzoyl)-5-hydroxy-5-(trifluoromethyl)-1,2,3,5-tetrahydroimidazo[1,2-*a*]pyridin-6-yl)(*p*-tolyl)methanone (**4b**): white solid, 210–213 °C; yield: 89%

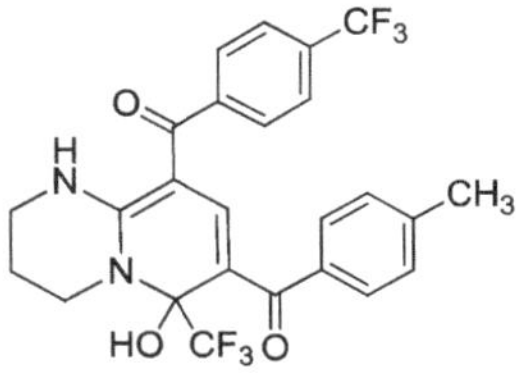

4c

(6-Hydroxy-6-(trifluoromethyl)-9-(4-(trifluoromethyl)benzoyl)-2,3,4,6-tetrahydro-1*H*-pyrido[1,2-*a*]pyrimidin-7-yl)(*p*-tolyl)methanone (**4c**): white solid, 186–189 °C; yield: 90%

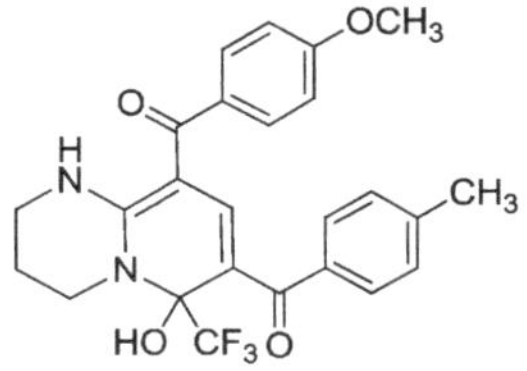

4d

(6-Hydroxy-7-(4-methylbenzoyl)-6-(trifluoromethyl)-2,3,4,6-tetrahydro-1*H*-pyrido[1,2-*a*]pyrimidin-9-yl)(4-methoxyphenyl)methanone (**4d**): white solid, mp 213–216 °C; yield: 91%

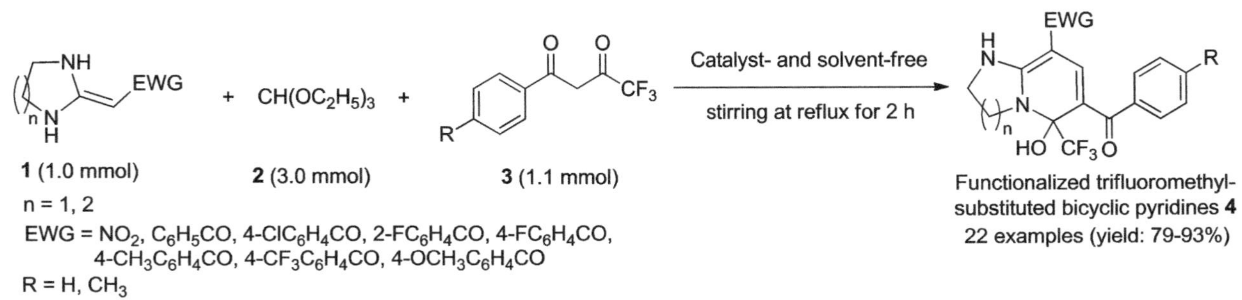

Scheme 3.10 Synthesis of trifluoromethyl-substituted bicyclic pyridines (**4**).

3.2.9.3 Experimental Procedure

A mixture of heterocyclic ketene aminals (HKAs, **1**; 1.0 mmol), triethoxymethane (**2**; 3 mmol), aryl-substituted 4,4,4-trifluoro-1,3-butanediones (**3**; 1.1 mmol) placed into a 25 mL round-bottom flask was stirred under refluxing conditions for 2 h until the HKAs **1** were completely consumed. The mixture was diluted with ethyl acetate (2×50 mL) and quenched with water (50 mL). The organic layer was dried with sodium sulfate, concentrated, and purified by flash column chromatography (petroleum ether–EtOAc, 3:1) to afford trifluoromethyl-substituted bicyclic pyridines (**4**). The products were identified by spectral studies including FTIR, ^{1}H and ^{13}C NMR, and HRMS.

3.2.9.4 Characterization Data of Two Representative Compounds

(5-Hydroxy-8-nitro-5-(trifluoromethyl)-1,2,3,5-tetrahydroimidazo[1,2-*a*]pyridin-6-yl)(phenyl)methanone (**4a**): white solid, mp 173–175 °C; yield: 93%; IR (KBr): 3338, 2915, 1587, 1377, 1183, 1211, 984 cm^{-1}; ^{1}H (CDCl$_3$, 500 MHz): δ 8.79 (1H, br s, –N*H*), 8.40 (1H, br s, OH), 7.98 (1H, s, C=C*H*), 7.68–7.73 (2H, m, Ar–*H*), 7.60–7.64 (1H, m, Ar–*H*), 7.47–7.54 (2H, m, Ar–*H*), 4.16–4.22 (1H, m, –NC*H*$_2$), 4.05–4.11 (1H, m, Ar–*H*), 3.95–4.02 (2H, m, –NC*H*$_2$); ^{13}C NMR (CDCl$_3$, 125 MHz): δ 198.7, 154.6, 139.0, 137.3, 133.5, 129.8, 129.2, 124.4 (d, J = 292.5 Hz), 109.1, 85.3 (q, J = 33.8 Hz), 44.7, 43.8; HRMS (ESI-TOF): *m/z* calc. for $C_{15}H_{13}F_3N_3O_4$ ([M + H]$^+$), 356.0853; found, 356.0850

(6-Hydroxy-6-(trifluoromethyl)-9-(4-(trifluoromethyl)benzoyl)-2,3,4,6-tetrahydro-1*H*-pyrido[1,2-*a*]pyrimidin-7-yl)(*p*-tolyl)methanone (**4c**): white solid, 186–189 °C; yield: 90%; IR (KBr): 3429, 3068, 1592, 1328, 1256, 1168, 743 cm^{-1}; ^{1}H (CDCl$_3$, 400 MHz): δ 10.06 (1H, br s, –N*H*), 7.63 (2H, d, J = 8.0 Hz, Ar–*H*), 7.50 (2H, d, J = 8.0 Hz, Ar–*H*), 7.36 (2H, d, J = 8.0 Hz, Ar–*H*), 7.21 (1H, s, C=C*H*), 7.09 (2H, d, J = 8.0 Hz, Ar–*H*), 3.97–4.01 (1H, m, –NC*H*$_2$), 3.57–3.66 (2H, m, –NC*H*$_2$), 3.43–3.48 (1H, m, –NC*H*$_2$), 2.30 (3H, s, –C*H*$_3$), 2.02–2.14 (m, 2H, –C*H*$_2$); ^{13}C NMR (CDCl$_3$, 100 MHz): δ 197.6, 190.4, 155.3, 148.7, 143.3, 142.6, 134.7, 131.7 (q, J = 32.4 Hz), 128.8, 128.8, 128.5, 125.1, 125.1, 123.0 (d, J = 106.5 Hz), 104.5, 94.7, 84.9 (q, J = 32.5 Hz), 39.4, 39.0, 21.3, 19.7; HRMS (ESI-TOF): *m/z* calc. for $C_{25}H_{21}F_6N_2O_3$ ([M + H]$^+$), 511.1451; found, 511.1454.

3.2.9.5 Critical Views

The present method offers an easy and practical route for the synthesis of a novel series of highly functional bicyclic pyridines under solvent- and catalyst-free conditions at reflux with good to excellent yields. This made it to some extent advantageous over previously reported methods involving the use of expensive metal reagents, tedious workup procedures, and a lack of diversity of the products.[52–55]

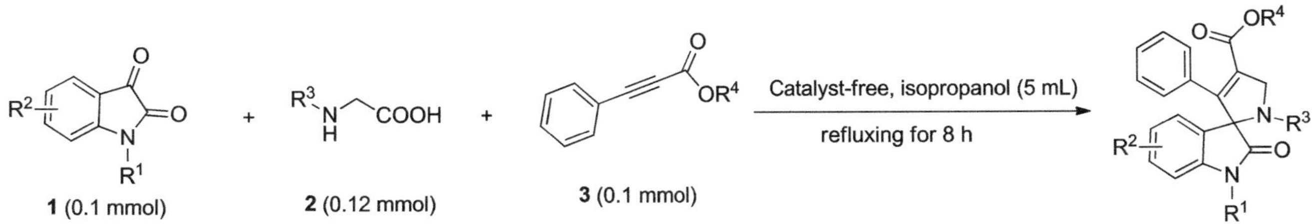

1 (0.1 mmol) **2** (0.12 mmol) **3** (0.1 mmol)

Spiro[indoline-3,2'-pyrroles] **4**
27 examples (yield: 50-90%)

R^1, R^2 = H, H; H, 5-CH_3; H, 5-C_2H_5; H, $(CH_3)_2CH$; H, 5-OCH_3; H, 5-F; H, 5-Cl; H, 5-Br; H, 5-I; H, 7-CH_3; H, 7-F; H, 6-Br; CH_3, H; CH_3, 5-C_2H_5; CH_3, 5-OCH_3

R^3 = H, CH_3, C_2H_5; R^4 = CH_3, C_2H_5

Scheme 3.11 Synthesis of spiro[indoline-3,2′-pyrroles] (**4**).

3.2.10 Entry-10: Synthesis of Spiro[indoline-3,2′-pyrroles]

Type of reaction: C–C and C–N bond formation
Reaction conditions: Catalyst-free, isopropanol, refluxing
Synthetic strategy: One-pot multicomponent reaction
Keywords: Isatins, α-amino acids, phenylpropiolic acid esters, catalyst-free, isopropanol, one-pot multicomponent reaction, regioselectivity, spiro[indoline-3,2′-pyrroles]

3.2.10.1 General Reaction Scheme

Xie and his group[56] reported a convenient catalyst-free method for preparing a series of spiro[indoline-3,2′-pyrroles] (**4**) from the one-pot three-component reaction of isatins (**1**), α-amino acids (**2**) and phenylpropiolic acid esters (**3**) in refluxing isopropanol with high regioselectivity and good yields (Scheme 3.11). The reaction is believed to be proceeded with the sequential generation of azomethine ylide and 1,3-dipolar cycloaddition reaction.

3.2.10.2 Representative Examples

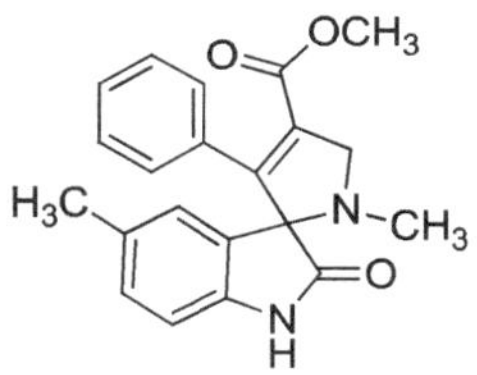

4a

Methyl 1',5-dimethyl-2-oxo-3'-phenyl-1',5'-dihydrospiro[indoline-3,2'-pyrrole]-4'-carboxylate (**4a**): white solid, mp 85–87 °C; yield: 83%

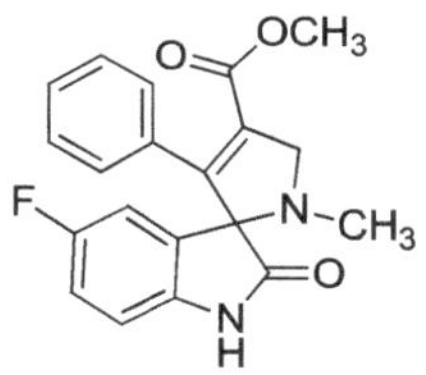

4b

Methyl 5-fluoro-1'-methyl-2-oxo-3'-phenyl-1',5'-dihydrospiro[indoline-3,2'-pyrrole]-4'-carboxylate (**4b**): white solid, mp 90–92 °C; yield: 90%

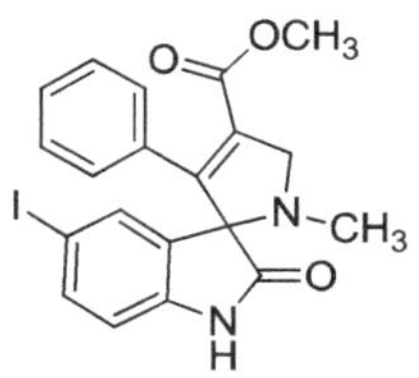

4c

Methyl 5-iodo-1'-methyl-2-oxo-3'-phenyl-1',5'-dihydrospiro[indoline-3,2'-pyrrole]-4'-carboxylate (**4c**): gray solid, mp 105–106 °C; yield: 86%

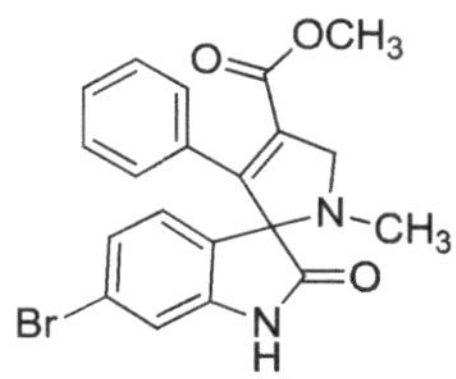

4d

Methyl 6-bromo-1'-methyl-2-oxo-3'-phenyl-1',5'-dihydrospiro[indoline-3,2'-pyrrole]-4'-carboxylate (**4d**): white solid, mp 89–90 °C; yield: 86%

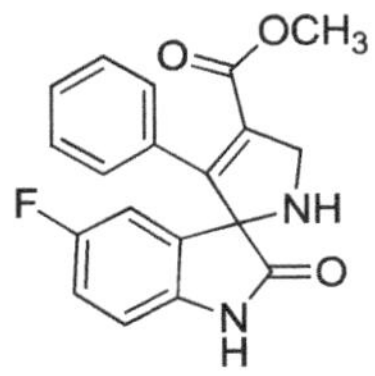

4e

Methyl 5-fluoro-2-oxo-3'-phenyl-1',5'-dihydrospiro[indoline-3,2'-pyrrole]-4'-carboxylate (**4e**): gray solid, mp 52–54 °C; yield: 88%

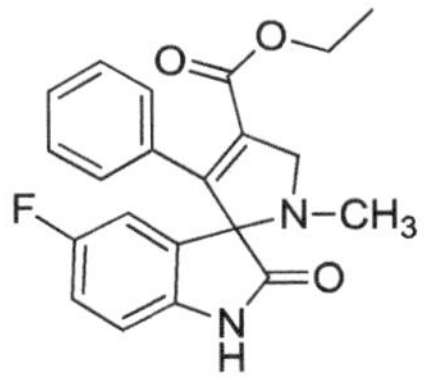

4f

Ethyl 5-fluoro-1'-methyl-2-oxo-3'-phenyl-1',5'-dihydrospiro[indoline-3,2'-pyrrole]-4'-carboxylate (**4f**): white solid, mp 66–67 °C; yield: 87%

3.2.10.3 Experimental Procedure

A mixture of isatins (**1**; 0.1 mmol), α-amino acid (**2**; 0.12 mmol) and phenylpropiolic acid esters (**3**; 0.1 mmol) in isopropanol (5 mL) was heated at reflux for 8 h in an oil bath. Upon completion of the reaction, the mixture was cooled to room temperature, and then evaporated in a vacuum. The residue was purified by flash column chromatography on silica gel with ethyl acetate and petroleum ether (3:1, v/v) as the eluent to furnish pure products of spiro[indoline-3,2′-pyrroles] (**4**), identified based on detailed spectral analyses.

3.2.10.4 Characterization Data of Two Representative Compounds

Methyl 6-bromo-1′-methyl-2-oxo-3′-phenyl-1′,5′-dihydrospiro[indoline-3,2′-pyrrole]-4′-carboxylate (**4d**): white solid, mp 89–90 °C; yield: 86%; IR (KBr): 3435, 1678, 1621, 1473, 1438, 1138, 1974, 989, 860, 545 cm^{-1}; ^{1}H NMR (DMSO-d_6, 400 MHz): δ 10.55 (1H, s), 7.25–7.13 (5H, m), 6.86–6.83 (3H, m), 4.13–4.06 (2H, m), 3.54 (3H, s), 2.20 (3H, s); ^{13}C NMR (DMSO-d_6, 100 MHz): δ 175.61, 172.25, 162.08, 159.06, 148.28, 142.86, 132.21, 131.65, 130.43, 129.49, 129.30, 128.40, 128.16, 127.97, 127.81, 126.98, 123.64, 123.10, 111.14, 82.96, 77.99, 52.45, 51.92, 51.90, 35.07; HRMS (ESI) calc. for $C_{20}H_{17}BrN_2O_3$ ($[M+H]^+$) 413.0501; found 413.0498

Ethyl 5-fluoro-1′-methyl-2-oxo-3′-phenyl-1′,5′-dihydrospiro[indoline-3,2′-pyrrole]-4′-carboxylate (**4f**): white solid, mp 66–67 °C; yield: 87%; IR (KBr): 3346, 1708, 1525, 1486, 1444, 1382, 1289, 1187, 1138, 801, 698 cm^{-1}; ^{1}H NMR (DMSO-d_6, 400 MHz): δ 10.51 (1H, s), 7.26–7.17 (4H, m), 7.04–6.99 (1H, m), 6.91–6.88 (2H, m), 6.71–6.68 (1H, m), 4.19–4.11 (2H, m), 4.02–3.96 (2H, m), 2.26 (3H, s), 0.99 (3H, t, $J=6.8$ Hz); ^{13}C NMR (DMSO-d_6, 100 MHz): δ 175.31, 162.90, 159.77, 158.64, 157.40, 151.14, 139.10, 139.07, 133.01, 131.87, 129.08, 128.65, 128.06, 127.89, 117.01, 116.76, 114.06, 113.82, 111.31, 111.25, 83.61,

83.59, 60.53, 60.21, 35.25, 13.99; HRMS (ESI) calc. for $C_{21}H_{19}FN_2O_3$ ($[M+H]^+$) 367.1458; found 367.1453.

3.2.10.5 *Critical Views*

The present protocol offers a facile and operationally simple route to access a series of diverse functionalized spiro[indoline-3,2′-pyrrole] derivatives as pharmaceutically-attractive building blocks in good yields without any catalysts or external additives. In comparison to the metal- and nonmetal catalyzed reactions,[57–65] this effective method appears to be an environmentally friendly process.

3.2.11 Entry-11: Synthesis of 4-(2-Substituted-3-iminoisoindolin-1-ylidene)-1-substituted-3-methyl-1*H*-pyrazol-5(4*H*)-ones

Type of reaction: C–C and C–N bond formation
Reaction conditions: Catalyst-free, ethanol, refluxing
Synthetic strategy: One-pot multicomponent reaction
Keywords: 2-Cyanobenzaldehyde, amine, 3-methyl-1*H*-pyrazol-5(4*H*)-one, 1,3-dimethyl-1*H*-pyrazol-5(4*H*)-one, catalyst-free, ethanol, one-pot multicomponent reaction, 4-(2-substituted-3-iminoisoindolin-1-ylidene)-1-substituted-3-methyl-1*H*-pyrazol-5(4*H*)-ones

3.2.11.1 *General Reaction Scheme*

Hu and coworkers[66] developed a one-pot three-component synthetic strategy for preparing a series of 4-(2-substituted-3-iminoisoindolin-1-ylidene)-1-substituted-3-methyl-1*H*-pyrazol-5(4*H*)-one derivatives (**4**) from the reaction between 2-cyanobenzaldehyde (**1**), amines (**2**), and 3-methyl-1*H*-pyrazol-5(4*H*)-one/1,3-dimethyl-1*H*-pyrazol-5(4*H*)-one (**3**) at reflux in ethanol with good to excellent yields (Scheme 3.12).

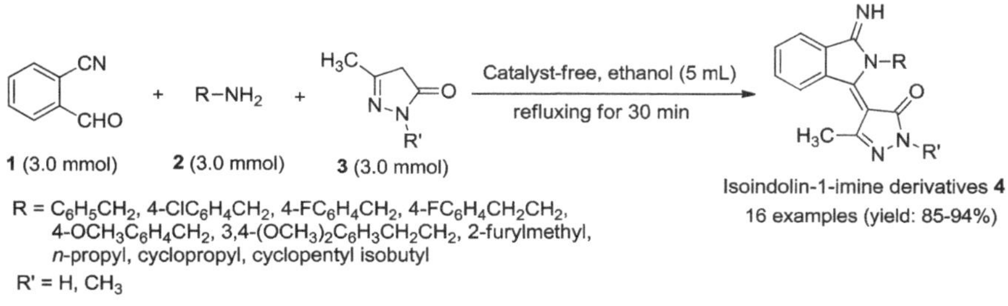

Scheme 3.12 Synthesis of isoindolin-1-imine derivatives (**4**).

3.2.11.2 Representative Examples

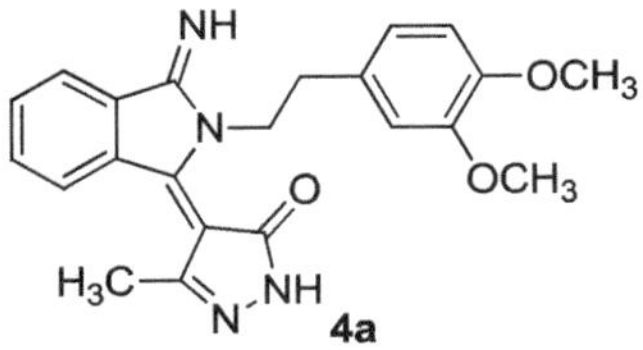

4a

(*Z*)-4-(2-(3,4-Dimethoxyphenethyl)-3-iminoisoindolin-1-ylidene)-3-methyl-1*H*-pyrazol-5(4*H*)-one (**4a**): red solid, mp 229.1–230.6 °C; yield: 87%

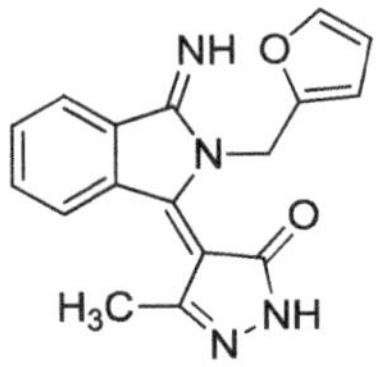

4b

(*Z*)-4-(2-(Furan-2-ylmethyl)-3-iminoisoindolin-1-ylidene)-3-methyl-1*H*-pyrazol-5(4*H*)-one (**4b**): red solid, mp 219.3–220.6 °C; yield: 85%

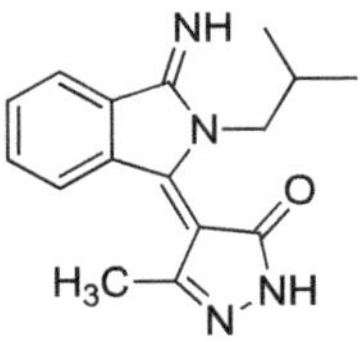

4c

(*Z*)-4-(3-Imino-2-isobutylisoindolin-1-ylidene)-3-methyl-1*H*-pyrazol-5(4*H*)-one (**4c**): red solid, mp 278.3–279.7 °C; yield: 90%

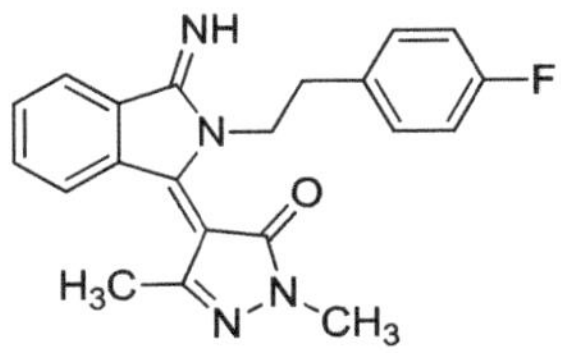

4d

(*Z*)-4-(2-(4-Fluorophenethyl)-3-iminoisoindolin-1-ylidene)-1,3-dimethyl-1*H*-pyrazol-5(4*H*)-one (**4d**): red solid, mp 146.7–148.3 °C; yield: 90%

3.2.11.3 Experimental Procedure

Amines (**2**; 3 mmol) were added to a stirred ethanol (5 mL) solution of 2-cyanobenzaldehyde (**1**; 3 mmol) and 3-methyl-1*H*-pyrazol-5(4*H*)-one/1,3-dimethyl-1*H*-pyrazol-5(4*H*)-one (**3**; 3 mmol) at room temperature. Then the mixture was heated to reflux with constant stirring for 30 min. After completion of the reaction, the precipitate was filtered off, washed with ethanol, and recrystallized to afford pure products of 4-(2-substituted-3-imino-isoindolin-1-ylidene)-1-substituted-3-methyl-1*H*-pyrazol-5(4*H*)-one derivatives (**4**). All the products were identified based on spectral studies.

3.2.11.4 Characterization Data of Two Representative Compounds

(*Z*)-4-(3-Imino-2-isobutylisoindolin-1-ylidene)-3-methyl-1*H*-pyrazol-5(4*H*)-one (**4c**): red solid, mp 278.3–279.7 °C; yield: 90%; ^{1}H NMR (DMSO-d_6, 400 MHz): δ 10.87 (1H, s), 9.74 (1H, t, $J = 6.1$ Hz), 9.58 (1H, dd, $J = 5.9$, 2.2 Hz), 8.02 (1H, dd, $J = 5.7$, 2.4 Hz), 7.64–7.55 (2H, m), 3.52 (2H, t, $J = 6.5$ Hz), 2.43 (3H, s), 2.09 (1H, dt, $J = 13.4$, 6.7 Hz), 0.97 (6H, d, $J = 6.7$ Hz); ^{13}C NMR (DMSO-d_6, 125 MHz): δ 169.8, 168.7, 164.9, 149.9, 140.2, 134.2, 131.8, 130.9, 128.4, 121.4, 112.2, 50.7, 28.7, 20.6 (2C), 19.2; MS (ESI): *m*/*z* 283.3 [M + H]$^+$; HRMS (ESI): calc. for $C_{16}H_{19}N_4O$ [M + H]$^+$: 283.1553; found: 283.1565.

(*Z*)-4-(2-(4-Fluorophenethyl)-3-iminoisoindolin-1-ylidene)-1,3-dimethyl-1*H*-pyrazol-5(4*H*)-one (**4d**): red solid, mp 146.7–148.3 °C; yield: 90%; ^{1}H NMR (DMSO-d_6, 400 MHz): δ 9.90 (1H, s), 9.60 (1H, dd, J = 5.9, 2.3 Hz), 7.96 (1H, dd, J = 5.2, 1.9 Hz), 7.63–7.59 (2H, m), 7.31 (2H, dd, J = 8.7, 5.6 Hz), 7.14 (2H, t, J = 8.9 Hz), 3.89 (2H, dd, J = 13.8, 6.8 Hz), 3.25 (3H, s), 3.06 (2H, t, J = 7.3 Hz), 2.47 (3H, s); ^{13}C NMR (DMSO-d_6, 125 MHz): δ 169.7, 165.9, 165.3, 162.29, and 160.36 ($^1J_{CF}$ = 241.3 Hz), 148.6, 140.0, 135.29, and 135.27 ($^4J_{CF}$ = 2.5 Hz), 134.2, 131.9, 131.1, 130.86, and 130.80 ($^3J_{CF}$ = 7.5 Hz) (2C), 128.5, 121.5, 115.60, and 115.43 ($^2J_{CF}$ = 21.2 Hz) (2C), 112.1, 44.8, 34.2, 31.1, 19.0; MS (ESI): *m/z* 363.3 $[M+H]^+$; HRMS (ESI): calc. for $C_{21}H_{20}FN_4O$ $[M+H]^+$: 363.1616; found: 363.1623.

3.2.11.5 Critical Views

The present protocol offers a straightforward synthesis of isoindolin-1-imine derivatives with excellent yields under catalyst-free conditions and within a short reaction time.

3.2.12 Entry-12: Synthesis of β-Lactam-triflones

Type of reaction: C–C and C–N bond formation
Reaction conditions: Catalyst-free, toluene, refluxing
Synthetic strategy: Wolff rearrangement followed by Staudinger [2 + 2]-cycloaddition
Keywords: 2-Diazo-1-aryl-2-(trifluoromethylsulfonyl)ethanones, imines, catalyst-free, toluene, Wolff rearrangement/Staudinger [2 + 2]-cycloaddition, stereoselectivity, β-lactam-triflones

3.2.12.1 General Reaction Scheme

Shibata and his group[67] developed a new strategy for the stereoselective synthesis of pharmaceutically interesting β-lactam-triflones (**3**) from the reaction between 2-diazo-1-aryl-2-(trifluoromethylsulfonyl)ethanones (**1**) and imines (**2**) at reflux in toluene under catalyst-free conditions (Scheme 3.13); the transformation involves a successive Wolff rearrangement and Staudinger [2 + 2]-cycloaddition reaction.

3.2.12.2 Representative Examples

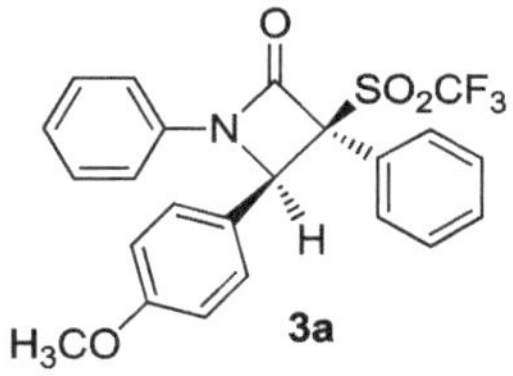

3a

(3*S*,4*R*)-4-(4-Methoxyphenyl)-1,3-diphenyl-3-((trifluoromethyl)sulfonyl)azetidin-2-one (**3a**): white solid, mp 165–166 °C; yield: 70% (d:r = 93:7)

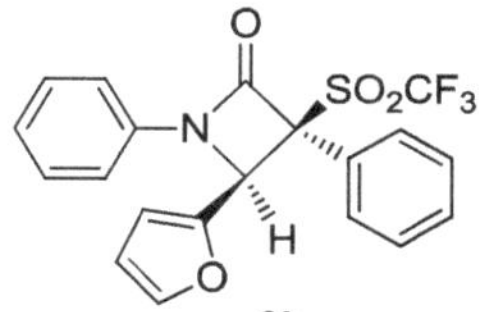

3b

(3*S*,4*R*)-4-(Furan-2-yl)-1,3-diphenyl-3-((trifluoromethyl)sulfonyl)azetidin-2-one (**3b**): white solid, mp 121–122 °C; yield: 69% (d:r = 94:6)

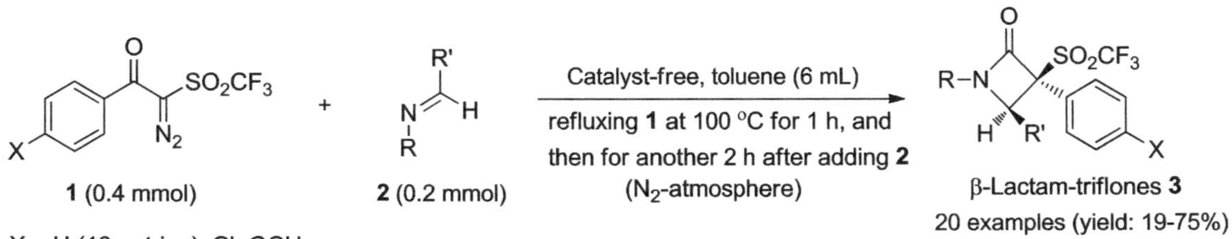

X = H (18 entries), Cl, OCH_3

R, R' = C_6H_5, C_6H_5; C_6H_5, 4-$OCH_3C_6H_4$; C_6H_5, 2-$OCH_3C_6H_4$; C_6H_5, 3-$OCH_3C_6H_4$; C_6H_5, 2,4-$(OCH_3)_2C_6H_4$; C_6H_5, 4-$NO_2C_6H_4$; C_6H_5, 4-ClC_6H_4; C_6H_5, 4-FC_6H_4; C_6H_5, 4-BrC_6H_4; C_6H_5, 4-$N(CH_3)_2C_6H_4$; C_6H_5, 2-furyl; C_6H_5, 2-thienyl; C_6H_5, $C_6H_5CH{=}CH_2$ (*trans*); $C_6H_5CH_2$, C_6H_5; 4-$OCH_3C_6H_4$, C_6H_5; 4-$OCH_3C_6H_4$, 2-naphthyl; isopropyl, 4-ClC_6H_5; 4-$OCH_3C_6H_4$, 4-$OCH_3C_6H_4$; 4-$OCH_3C_6H_4$, 4-ClC_6H_4

Scheme 3.13 Synthesis of β-lactam-triflones (3).

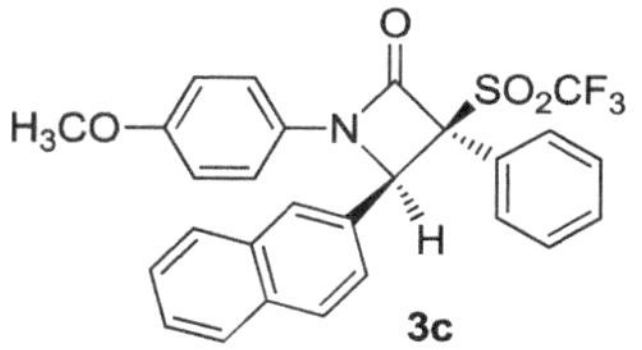

3c

(3*S*,4*R*)-1-(4-Methoxyphenyl)-4-(naphthalen-2-yl)-3-phenyl-3-((trifluoromethyl)sulfonyl)azetidin-2-one (**3c**): brown solid, mp 120–121 °C; yield: 52% (d:r = 90:10)

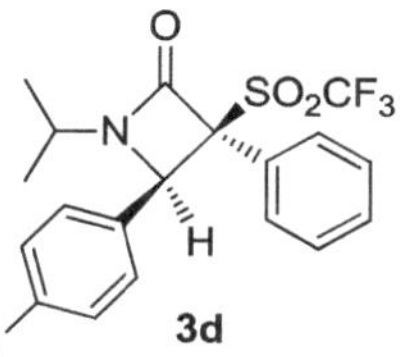

3d

(3*S*,4*R*)-4-(4-Chlorophenyl)-1-isopropyl-3-phenyl-3-((trifluoromethyl)sulfonyl)azetidin-2-one (**3d**): white solid, mp 154–155 °C; yield: 74% (d:r = 73:27)

3.2.12.3 Experimental Procedure

Diazotriflones (**1**; 0.4 mmol, 2.0 equiv.) were placed in a dry two-necked flask (50 mL) under a nitrogen atmosphere. Dry toluene (6.0 mL) was added to the flask, and the reaction mixture was stirred at 100 °C for 1 h. Then imines (**2**; 0.2 mmol, 1.0 equiv.) in toluene (1.0 mL) were added slowly to the mixture for 5 min, and the mixture was stirred at 100 °C until the imine was consumed (usually 2 h, monitored by TLC). Upon completion of the reaction, the solvent was removed under reduced pressure, and the crude product was purified by column chromatography with ethyl acetate–hexane as the eluent to afford β-lactam-triflones (**3**). All the compounds were characterized based on detailed spectral studies.

3.2.12.4 Characterization Data of Two Representative Compounds

(3*S*,4*R*)-4-(4-Methoxyphenyl)-1,3-diphenyl-3-((trifluoromethyl)sulfonyl)azetidin-2-one (**3a**): white solid, mp 165–166 °C; yield: 70% (d:r = 93:7); IR (KBr): 3071, 3047, 2962, 2933, 2838, 1769, 1615, 1514, 1366, 1194, 1112 cm^{-1}; ^{1}H NMR ($CDCl_3$, 500 MHz): δ 7.94–7.93 (2H, m), 7.57 (2H, d, $J = 7.9$ Hz), 7.53–7.52 (3H, m), 7.32–7.26 (4H, m), 7.13 (1H, t, $J = 7.0$ Hz), 6.98 (2H, d, $J = 8.2$ Hz), 5.54 (1H, s), 3.83 (3H, s); ^{13}C NMR ($CDCl_3$, 125 MHz): δ 160.9, 155.8, 136.1, 130.8, 130.7, 129.3, 129.2, 129.0, 128.2, 125.3, 120.4, 118.8 (q, $J = 331.2$ Hz), 117.9, 114.1, 84.2, 68.4, 55.3; ^{19}F NMR ($CDCl_3$, 282 MHz): δ −72.3; HRMS (ESI): calc. for $C_{23}H_{18}F_3NO_4SNa$ $[M + Na]^+$: 484.0806; found: 484.0813.

(3*S*,4*R*)-4-(4-Chlorophenyl)-1-isopropyl-3-phenyl-3-((trifluoromethyl)sulfonyl)-azetidin-2-one (**3d**): white solid, mp 154–155 °C; yield: 74% (d:r = 73:27); IR (KBr): 3065, 3032, 2984, 2933, 2877, 1771, 1361, 1199, 1114 cm^{-1}; ^{1}H NMR ($CDCl_3$, 500 MHz): δ 7.86–7.84 (2H, m), 7.63 (2H, d, $J = 8.6$ Hz), 7.52–7.47 (5H, m), 5.03 (1H, s), 3.74 (1H, heptet, $J = 6.8$ Hz), 1.42 (3H, d, $J = 6.9$ Hz), 1.20 (3H, d, $J = 6.7$ Hz); ^{13}C NMR ($CDCl_3$, 125 MHz): δ 158.6, 136.1, 130.6, 130.4, 129.5, 129.3, 129.0, 128.9, 128.2, 118.7 (q, $J = 331.2$ Hz), 84.3, 67.7, 47.1, 20.7, 20.0; ^{19}F NMR ($CDCl_3$, 282 MHz): δ −72.5; HRMS (ESI): calc. for $C_{19}H_{17}ClF_3NO_3SNa$ $[M + Na]^+$: 454.0467; found: 454.0474.

3.2.12.5 Critical Views

The investigators performed a highly stereoselective catalyst-free synthesis of a novel series of 3,3-aryltriflyl multisubstituted β-lactams from the reaction of diazotriflones with imines. The process is operationally simple and facile.

3.3 Concluding Remarks

Conventional heating can thus offer fruitful effects when performing many organic transformations even under catalyst-free conditions. Although conventional heating/refluxing requires much energy, in the cases of heating at higher temperatures, it should be given emphasis for designing those purposeful organic transformations which otherwise are not feasible without the aid of expensive catalysts and/or toxic solvents. In that sense, this technique would be of much benefit, and more developments are in progress.

References

1. B. Kaboudin and L. Malekzadeh, *Tetrahedron Lett.*, 2011, **52**, 6424.
2. R. Neidlein and L. Sheng, *Synth. Commun.*, 1995, **25**, 2379.
3. G. B. Liang and D. D. Feng, *Tetrahedron Lett.*, 1996, **37**, 6627.
4. R. Neidlein and L. Sheng, *J. Heterocyclic Chem.*, 1996, **33**, 1943.
5. J. R. Young and R. J. DeVita, *Tetrahedron Lett.*, 1998, **39**, 3931.
6. G. B. Liang and X. Qian, *Bioorg. Med. Chem. Lett.*, 1999, **9**, 2101.
7. A. G. Tyrkov, *Khim. Khimich. Tekhnol.*, 2000, **43**, 73.
8. H. C. Ryu, Y. T. Hong and S. K. Kang, *Heterocycles*, 2001, **54**, 985.
9. A. Pace and P. Pierro, *Org. Biomol. Chem.*, 2009, **7**, 4337.
10. C. Li and F. Zhang, *Tetrahedron Lett.*, 2017, **58**, 1572.
11. F. Wahid, M. Monneret and D. Dauzonne, *Chem. Pharm. Bull.*, 1999, **47**, 156.
12. L. Wu, R. Jiang, J.-M. Yang, S.-Y. Wang and S.-J. Ji, *Tetrahedron Lett.*, 2013, **54**, 2849.
13. Z. H. Skraup, *Ber. Dtsch. Chem. Ges.*, 1880, **13**, 2086.
14. O. Doebner and W. von Miller, *Ber. Dtsch. Chem. Ges.*, 1881, **14**, 2812.
15. W. H. Mills, J. E. G. Harris and H. Lambourne, *J. Chem. Soc.*, 1921, **119**, 1294.
16. M. Matsugi, F. Tabusa and J. Minamikawa, *Tetrahedron Lett.*, 2000, **41**, 8523.
17. B. C. Ranu, A. Hajra, S. S. Dey and U. Jana, *Tetrahedron*, 2003, **59**, 813.
18. Y. C. Wu, L. Liu, H. J. Li, D. Wang and Y. J. Chen, *J. Org. Chem.*, 2006, **71**, 6592.
19. S. Madapa, Z. Tusi and S. Batra, *Curr. Org. Chem.*, 2008, **12**, 1116.
20. J. Horn, S. P. Marsden, A. Nelson, D. House and G. G. Weingarten, *Org. Lett.*, 2008, **10**, 4117.
21. J. Marco-Contelles, E. Perez-Mayoral, A. Samadi, M. C. Carreiras and E. Soriano, *Chem. Rev.*, 2009, **109**, 2652.

22. P. P. Varma, A. Srinivasaa and K. M. Mahadevan, *Synth. Commun.*, 2011, **41**, 2186.
23. C. S. Kavitha, K. M. Hosamani and R. S. Harisha, *Can. J. Chem.*, 2010, **88**, 443.
24. C. S. Yi and S. Y. Yun, *Org. Lett.*, 2005, **7**, 2181.
25. B. Gabriele, R. Mancuso, G. Salerno, G. Ruffolo and P. Plastina, *J. Org. Chem.*, 2007, **72**, 6873.
26. Z. Zhang, J. Tan and Z. Wang, *Org. Lett.*, 2008, **10**, 173.
27. B. Gabriele, R. Mancuso, G. Salerno, E. Lupinacci, G. Ruffolo and M. Costa, *J. Org. Chem.*, 2008, **73**, 4971.
28. D. Choi, B. Yoo, L. Kimberly, G. Colson, E. Martin and K. Harold, *J. Org. Chem.*, 1995, **60**, 3391.
29. A. Srinivasaa, M. Kittappa and V. H. Mahadevan, *Synth. Commun.*, 2009, **39**, 93.
30. Y. Matsubara, S. Hirakawa, Y. Yamaguchi and Z. Yoshida, *Angew. Chem. Int. Ed.*, 2011, **50**, 7670.
31. Y. Chen and Y. R. Lee, *Bull. Korean Chem. Soc.*, 2011, **32**, 2485.
32. K. C. S. Achar, K. M. Hosamani and H. R. Seetharamareddy, *Synth. Commun.*, 2011, **41**, 33.
33. V. V. Kouznetsov, D. R. Merchan Arenas, C. J. Ortiz Areniz and C. M. Meléndez Gómez, *Synthesis*, 2011, 4011.
34. K. C. Kiran, K. M. Mahadevan, V. P. Prabhakara and A. Srinivasa, *Chin. J. Chem.*, 2012, **30**, 534.
35. M.-M. Zhang, J. Wang and X.-S. Wang, *J. Heterocyclic Chem.*, 2017, **54**, 1378.
36. M. Zhang, H. Jiang, H. Liu and Q. Zhu, *Org. Lett.*, 2007, **9**, 4111.
37. B. Carboni, L. Toupet and R. Carrié, *Tetrahedron*, 1987, **43**, 2293.
38. S. Ohta, Y. Hinata, M. Yamashita, I. Kawasaki, Y. Jinda and S. Horie, *Chem. Pharm. Bull.*; 1994, **42**, 1730.
39. A. Zamri, F. Sirockin and M. A. Abdallah, *Tetrahedron*, 1999, **55**, 5157.
40. H. Cho, K. Shima, M. Hayashimatsu, Y. Ohnaka, A. Mizuno and Y. Takeuchi, *J. Org. Chem.*, 1985, **50**, 4227.
41. K. Folkers and T. B. Johnson, *J. Am. Chem. Soc.*, 2002, **124**, 3784.
42. P. M. When and J. D. Bois, *J. Am. Chem. Soc.*, 2002, **124**, 12950.
43. D. Christian, C. S. W. Jacqueline, D. B. Mackay and M. A. L. Roch, *Tetrahedron Lett.*, 2004, **45**, 7197.
44. M. Adib, H. Yavari and M. Mollahosseini, *Tetrahedron Lett.*, 2004, **45**, 1803.
45. J. D. Clark, J. T. Collin, H. P. Kleine, G. A. Weisenburger and D. K. Anderson, *Org. Process Res. Dev.*, 2004, **8**, 571.
46. M. F. A. Adamo, J. E. Baldwin and R. M. Adlington, *J. Org. Chem.*, 2005, **70**, 3307.
47. S. H. Kang, S. Y. Kang, H. S. Lee and A. J. Buglass, *Chem. Rev.*, 2005, **105**, 4537.
48. M. Hagimori, Y. Murakami, N. Mizuyama and Y. Tominaga, *J. Heterocyclic Chem.*, 2016, **53**, 197.

49. J. Alen, L. Dobrzańska, W. M. De Borggraeve and F. Compernolle, *J. Org. Chem.*, 2007, **72**, 1055.
50. L. Crouillebois, L. Pantaine, J. Marrot, V. Coeffard, X. Moreau and C. Greck, *J. Org. Chem.*, 2015, **80**, 595.
51. D.-D. Zhu, X.-B. Chen, R. Huang, S.-J. Yan and J. Lin, *Tetrahedron*, 2015, **71**, 2363.
52. H.-Y. Huayou, Hu, Y. Liu, H. Zhong, Y.-L. Zhu, C. Wang and M. Ji, *Chem. - Asian J.*, 2012, **7**, 884.
53. C. He, J. Hao, H. Xu, Y.-P. Mo, H.-Y. Liu, J.-J. Han and A.-W. Lei, *Chem. Commun.*, 2012, 11073.
54. A. K. Bagdi, M. Rahman, S. Santra, A. Majee and A. Hajra, *Adv. Synth. Catal.*, 2013, **355**, 1741.
55. L. Dong, J.-R. Huang, C.-H. Qu, Q.-R. Zhang, W. Zhang, B. Han and C. Peng, *Org. Biomol. Chem.*, 2013, **11**, 6142.
56. G. Shi, X. He, Y. Shang and M. Xie, *RSC Adv.*, 2016, **6**, 10412.
57. S. L. Zhu, S. J. Jia and Y. Zhang, *Tetrahedron*, 2007, **63**, 9365.
58. A. A. Mohammadi, M. Dabiri and H. Qaraat, *Tetrahedron*, 2009, **65**, 3804.
59. Y. Han, Q. Wu, J. Sun and C.-G. Yan, *Tetrahedron*, 2012, **68**, 8539.
60. Y. Han, Q. Wu, J. Sun and C.-G. Yan, *Tetrahedron*, 2012, **68**, 539.
61. R. Sarkar and C. Mukhopadhyay, *Tetrahedron Lett.*, 2013, **54**, 3706.
62. T. Rajasekaran, G. Karthik, B. Sridhar and R. B. V. Subba, *Org. Lett.*, 2013, **15**, 1512.
63. A. Dandia, A. K. Jain and A. K. Laxkar, *Tetrahedron Lett.*, 2013, **54**, 3929.
64. N. Sharma, Z. Li, U. K. Sharma and E. V. Van der Eycken, *Org. Lett.*, 2014, **16**, 3884.
65. C.-S. Wang, R.-Y. Zhu, J. Zheng, F. Shi and S.-J. Tu, *J. Org. Chem.*, 2015, **80**, 512.
66. S. Shen, P. V. Khang, Y. Chen, M. Lei and L. Hu, *ARKIVOC*, 2013, **(iii)**, 413.
67. Z. Huang, C. Wang, E. Tokunaga, Y. Sumii and N. Shibata, *Org. Lett.*, 2015, **17**, 5610.

CHAPTER 4

Catalyst-free Reactions with Microwave Irradiation

4.1 Introduction

In current green chemistry practice, microwave irradiation is regarded as a '*green tool*' to reduce energy consumption,[1] and the revolutionary spread of this technique has resulted in an enormous development in synthetic organic chemistry.[2–11] The most common benefits from microwave irradiation are the considerable shortening of reaction times, thereby accelerating reaction rates, and the increase in the selectivities. However, the most valuable benefit is when a reaction can be carried out that is otherwise impossible under traditional thermal conditions. This may be the consequence of a so-called special microwave effect.[12,13] With this view, the present chapter offers detailed descriptions of more than 20 comprehensively screened catalyst-free organic synthetic protocols, which occur under the influence of microwave irradiation for synthesizing a number of synthetically and pharmaceutically useful organic scaffolds.

4.2 Catalyst-free Organic Transformations with Microwave Irradiation

A good deal of catalyst-free organic reactions leading to the synthesis of a variety of organic compounds under microwave irradiation have been reported so far. This section presents such useful organic transformations in an entry-based format and highlights the key aspects for each of them.

Green Chemistry Series No. 51
Catalyst-free Organic Synthesis
By Goutam Brahmachari

Published by the Royal Society of Chemistry, www.rsc.org

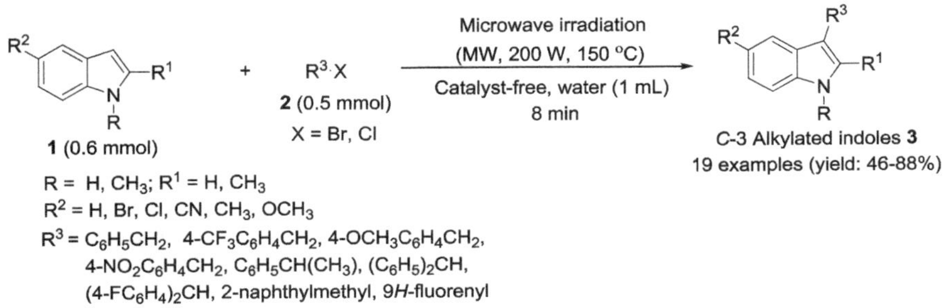

Scheme 4.1 Synthesis of *C*-alkylated indoles (**3**).

4.2.1 Entry-1: Synthesis of *C*-Alkylated Indoles

Type of reaction: C–C bond formation
Reaction conditions: Catalyst-free, water, microwave irradiation
Synthetic strategy: Alkylation reaction
Keywords: Indoles, alkyl halides, catalyst-free, water, Friedel–Crafts reaction, C-alkylation of indoles, substituted indoles

4.2.1.1 General Reaction Scheme

De Rosa and Soriente[14] developed an efficient and highly regioselective protocol for the synthesis of a series of functionalized indoles (**3**). These syntheses proceeded *via* an uncatalyzed Friedel–Crafts alkylation of the substrates (**1**) by alkyl halides (**2**) in the presence of water, as a solvent, under the influence of microwave irradiation (Scheme 4.1).

4.2.1.2 Representative Examples

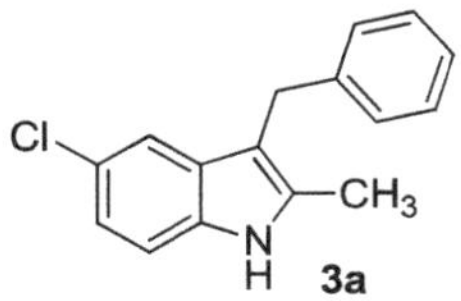

3a

3-Benzyl-5-chloro-2-methyl-1*H*-indole (**3a**): yellow solid, mp 95–97 °C; yield: 81%

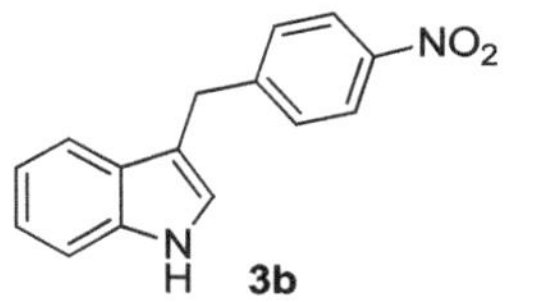

3b

3-(4-Nitrobenzyl)-1*H*-indole (**3b**): yellow solid, mp 121–123 °C; yield: 48%

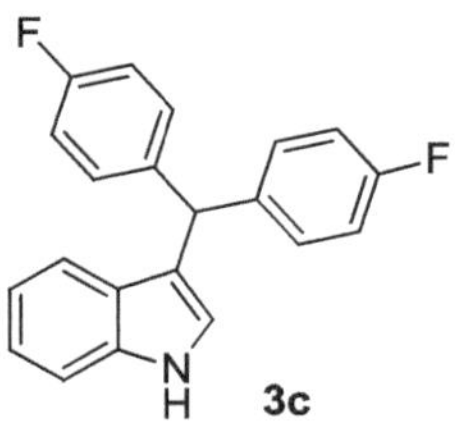

3c

3-(*Bis*(4-fluorophenyl)methyl)-1*H*-indole (**3c**): yellow oil; yield: 60%

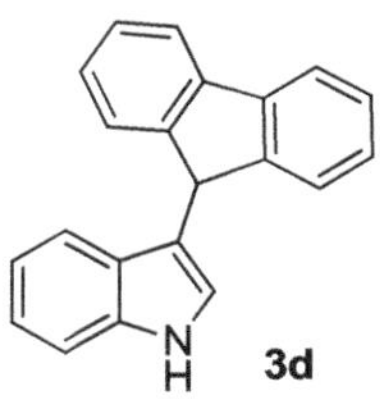

3d

3-(9*H*-Fluoren-9-yl)-1*H*-indole (**3d**): yellow solid, mp 198–200 °C; yield: 76%

4.2.1.3 Experimental Procedure

A sealed 10 mL glass tube containing a mixture of indoles (**1**; 0.6 mmol) and alkyl halides (**2**; 0.5 mmol) in distilled water (1 mL) was introduced in the cavity of a microwave reactor (CEM Co., Discover System, single-mode reactor; power 200 W) and irradiated at 150 °C with magnetic stirring for 8 min. After cooling the tube to room temperature, the reaction mixture was diluted with small amount of ethyl acetate and poured into a separating funnel. Water was added and the organic material was extracted (3×4 mL). The combined ethyl acetate extracts were then dried over magnesium sulfate and after removal of the solvent the mixture was purified by column chromatography (hexane–ethyl acetate as the eluent) to afford pure products of C-3 alkylated indole derivatives (**3**). All the products were characterized on the basis of their analytical and ^{1}H and ^{13}C NMR spectroscopic analysis.

4.2.1.4 Characterization Data of Two Representative Compounds

3-(*Bis*(4-fluorophenyl)methyl)-1*H*-indole (**3c**): yellow oil; yield: 60%; ^{1}H NMR ($CDCl_3$, 400 MHz): δ 7.96 (1H, br s), 7.37 (2H, d, *J* = 8.7 Hz), 7.26–7.15 (6H, m), 7.01–6.96 (5H, m), 6.54 (1H, s), 5.67 (1H, s); ^{13}C NMR ($CDCl_3$, 100 MHz): δ 163.3, 159.4, 139.3, 136.6, 130.3, 130.1, 126.6, 123.9, 122.2, 119.7, 119.5, 115.2, 114.9, 111.1, 47.2. Anal. calc. for $C_{21}H_{15}F_2N$: C, 78.98; H, 4.73; N, 4.39; found: C, 79.22; H, 4.48; N, 4.63.

3-(9*H*-Fluoren-9-yl)-1*H*-indole (**3d**): yellow solid, mp 198–200 °C; yield: 76%; ^{1}H NMR ($CDCl_3$, 400 MHz): δ 8.00 (1H, br s), 7.90–7.77 (5H, m), 7.49–7.08 (6H, m), 7.04 (1H, d, *J* = 2.3 Hz), 6.96 (1H, m), 5.39 (1H, s); ^{13}C NMR ($CDCl_3$, 100 MHz): δ 147.7, 140.7, 136.7, 127.1, 126.7, 125.2, 122.3, 122.1, 119.8, 119.4, 115.1, 111.2, 102.4, 45.9. Anal. calc. for $C_{21}H_{15}N$: C, 89.65; H, 5.37; N, 4.98; found: C, 89.90; H, 5.57 N, 5.23.

4.2.1.5 Critical Views

The investigators reported an efficient and microwave-assisted green alternative protocol for the Friedel–Crafts alkylation of indoles with different benzyl halides in water with moderate to good yields. In contrast to the classical conditions for Friedel–Crafts reactions, the key features of this procedure are the use of water as a green solvent, a shorter reaction time, catalyst-free conditions, operational simplicity and high selectivity.

4.2.2 Entry-2: Synthesis of Sulfonamides

Type of reaction: N–S bond formation
Reaction conditions: Catalyst-free, solvent-free, microwave irradiation
Synthetic strategy: Chemoselective sulfonylation of amines
Keywords: Amines, *p*-toluenesulfonyl chloride, catalyst- and solvent-free, chemoselective sulfonylation, sulfonamides

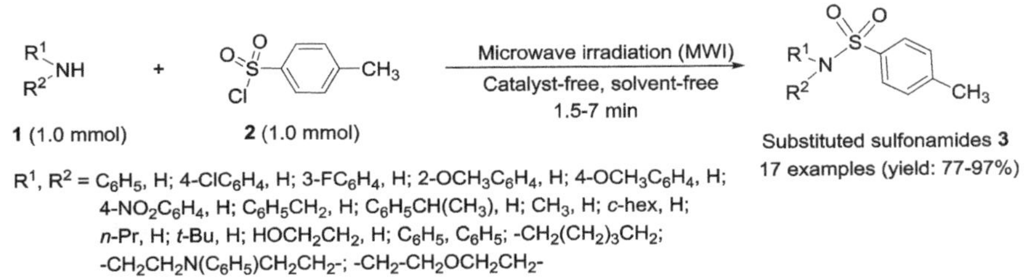

Scheme 4.2 Synthesis of sulfonamides (3).

4.2.2.1 General Reaction Scheme

A series of sulfonamides (**3**) was prepared by Aouf and his group[15] from the reaction between amines (**1**) and *p*-toluenesulfonyl chloride (**2**) under the influence of microwave irradiation in the absence of any catalyst and solvent with good to excellent yields within a very short time (Scheme 4.2).

4.2.2.2 Representative Examples

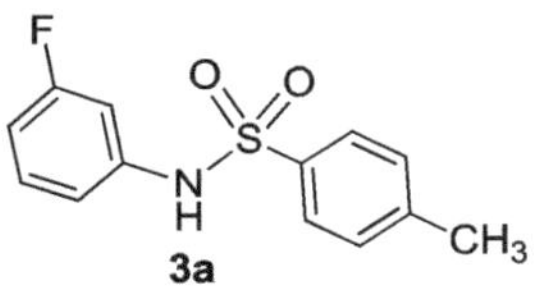

3a

N-(3-Fluorophenyl)-4-methylbenzenesulfonamide (**3a**): solid, mp 101.7–103.5 °C; time: 3 min; yield: 93%

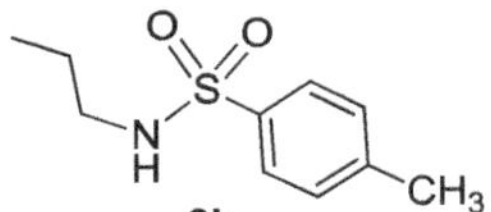

3b

4-Methyl *N*-propylbenzenesulfonamide (**3b**): time: 1.5 min; yield: 88%

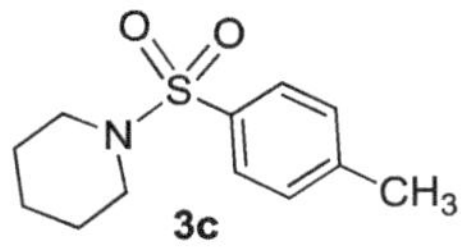

3c

1-Tosylpiperidine (**3c**): solid, mp 98.7–100.5 °C; time: 3.5 min; yield: 83%

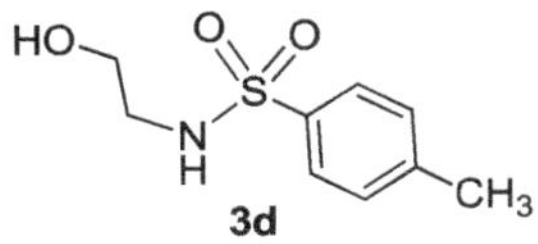

3d

N-(2-Hydroxyethyl)-4-methylbenzenesulfonamide (**3d**): time: 1.5 min; yield: 77%

4.2.2.3 Experimental Procedure

p-Toluenesulfonyl chloride (**2**; 1.0 mmol) was added to amines (**1**; 1.0 mmol) and the mixture was irradiated with microwaves for the appropriate time (1.5–7 min). Upon completion of the reaction (monitored by TLC), the reaction mixture was treated with *n*-hexane (15–20 mL), and was allowed to stand at room temperature for 7–10 h. The resulting crystals of pure sulfonamides **3** were collected by filtration, washed with *n*-hexane and dried. The products were identified based on their analytical and spectral studies.

4.2.2.4 Characterization Data of Two Representative Compounds

N-(3-Fluorophenyl)-4-methylbenzenesulfonamide (**3a**): solid, mp 101.7–103.5 °C; time: 3 min; yield: 93%; IR: 3457 (NH), 2941 (CH), 1594 (C=C), 1339 and 1107 (SO_2) cm^{-1}; ^1H NMR ($CDCl_3$, 250 MHz): δ 1.66 (1H, s, NH), 2.38 (3H, s, Ar–CH_3), 6.78–6.91 (3H, m, Ar–*H*), 7.15 (1H, m, Ar–*H*), 7.26 (2H, d, *J* = 7.8 Hz, Ar–*H*), 7.69 (2H, d, *J* = 7.8 Hz, Ar–*H*); ^{13}C NMR ($CDCl_3$, 60 MHz): δ 21.7, 108.5, 112.2, 116.4, 127.4, 130.0, 130.6, 130.8, 135.9, 138.3, 144.4. Anal. calc. for $C_{13}H_{12}FNO_2S$: C, 58.85; H, 4.56; N, 5.28; found: C, 58.63; H, 4.66; N, 5.34.

N-(2-Hydroxyethyl)-4-methylbenzenesulfonamide (**3d**): time: 1.5 min; yield: 77%; IR: 3597 (OH), 3219 (NH), 3065 (CH), 1605 (C=C), 1328, 1148 (SO_2) cm^{-1}; ^1H NMR ($CDCl_3$, 250 MHz): δ 2.33 (3H, s, Ar–CH_3), 2.94 (2H, q, *J* = 4.76 Hz, –CH_2–N), 3.58 (2H, t, *J* = 5.27 Hz, –CH_2–O), 6.25 (1H, t, *J* = 6.09 Hz, –NH), 7.22 (2H, d, *J* = 7.8 Hz, Ar–*H*), 7.68 (2H, d, *J* = 7.8 Hz, Ar–*H*); ^{13}C NMR ($CDCl_3$, 62.89 MHz): δ 21.4, 45.1, 60.9, 126.9, 129.7, 136.4, 143.4. Anal. calc. for $C_9H_{13}NO_3S$: C, 44.81; H, 5.39; N, 5.81; found: C, 44.69; H, 5.46; N, 5.67.

4.2.2.5 Critical Views

The present protocol offers an alternative microwave-assisted approach for *N*-sulfonylation of amines in a chemoselective manner without the use of any base or catalyst. This protocol is advantageous over the other earlier methods[16–24] in respect of shorter reaction times, good to excellent yields, avoidance of any catalyst or solvent and a high degree of chemoselctivity.

4.2.3 Entry-3: Synthesis of 4-Hydroxy-3-arylthiazolidine-2-thiones

Type of reaction: C–N and C–S bond formation
Reaction conditions: Catalyst-free, solvent-free, microwave irradiation
Synthetic strategy: Domino approach
Keywords: Aryl isothiocyanates, 1,4-dithiane-2,5-diol, catalyst- and solvent-free, domino reaction, 4-hydroxy-3-arylthiazolidine-2-thiones

4.2.3.1 General Reaction Scheme

Perumal and his group[25] developed a microwave-assisted protocol for the synthesis of 4-hydroxy-3-arylthiazolidine-2-thiones (**3**) through novel domino reactions of aryl isothiocyanates (**1**) and 1,4-dithiane-2,5-diol (**2**) under solvent- and catalyst-free conditions with excellent yields and atom-efficiency (Scheme 4.3). The authors assumed that the reaction proceeds *via* 2-mercaptoacetaldehyde generation/thia-Michael addition/regioselective hemi-aminalization domino sequence.

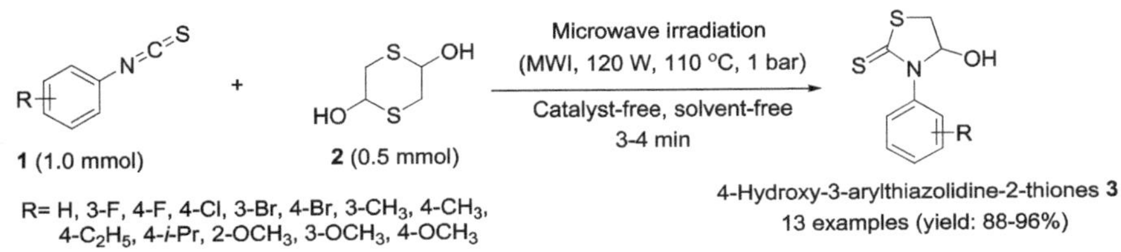

Scheme 4.3 Synthesis of 4-hydroxy-3-arylthiazolidine-2-thiones (**3**).

4.2.3.2 Representative Examples

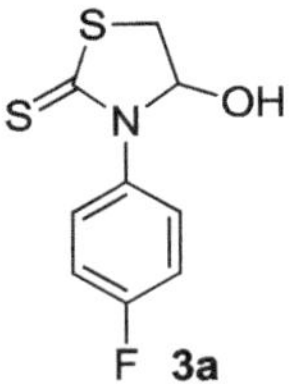

3a

3-(4-Fluorophenyl)-4-hydroxythiazolidine-2-thione (**3a**): off white solid, mp 166–167 °C; yield: 88%

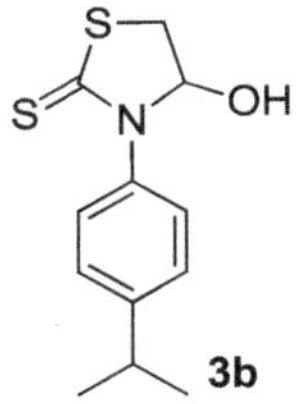

3b

4-Hydroxy-3-(4-isopropylphenyl)thiazolidine-2-thione (**3b**): white solid, mp 186–187 °C; yield: 95%

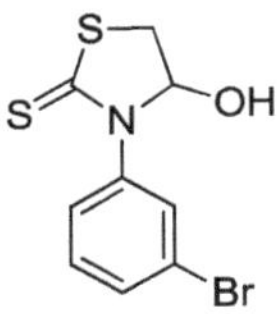

3c

3-(3-Bromophenyl)-4-hydroxythiazolidine-2-thione (**3c**): white solid, mp 124–125 °C; yield: 91%

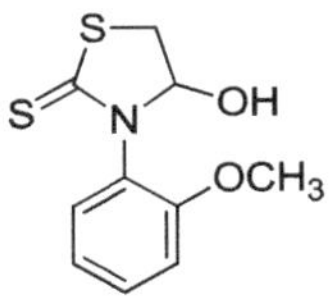

3d

4-Hydroxy-3-(2-methoxyphenyl)thiazolidine-2-thione (**3d**): white solid, mp 79–80 °C; yield: 96%

4.2.3.3 Experimental Procedure

A mixture of arylisothiocyanates (**1**; 1.0 mmol) and 1,4-dithiane-2,5-diol (**2**; 0.5 mmol) with a closed vial was irradiated with microwave for 3–4 min using a microwave synthesizer programmed at 110 °C, 120 W and 1 bar pressure. Upon completion of the reaction, the mixture was cooled to room temperature and then the crude reaction mixture was triturated with cold ethanol to give pure products of 4-hydroxy-3-arylthiazolidine-2-thiones (**3**). The products were identified based on their analytical and spectral studies.

4.2.3.4 Characterization Data of Two Representative Compounds

3-(4-Fluorophenyl)-4-hydroxythiazolidine-2-thione (**3a**): off white solid, mp 166–167 °C; yield: 88%; ^{1}H NMR (DMSO-d_6, 300 MHz): δ 3.20 (1H, d,

$J = 12.0$ Hz), 3.89 (1H , dd, $J = 6.3$, 12.0 Hz), 5.87 (1H, t, $J = 7.05$ Hz), 7.28–7.35 (2H, m), 7.38–7.43 (2H, m); ^{13}C NMR (DMSO-d_6, 75 MHz): δ 36.3, 91.9, 115.4 (d, $^2J_{CF} = 22.6$ Hz), 130.0 (d, $^3J_{CF} = 8.8$ Hz), 135.1, 160.9 (d, $^1J_{CF} = 243.3$ Hz), 196.9; ESI-MS: *m/z* 230.00 (M + 1). Anal. calc. for $C_9H_8FNOS_2$: C, 47.15; H, 3.52; N, 6.11; found: C, 47.10; H, 3.56; N, 6.19.

3-(3-Bromophenyl)-4-hydroxythiazolidine-2-thione (**3c**): white solid, mp 124–125 °C; yield: 91%; ^{1}H NMR (DMSO-d_6, 300 MHz): δ 3.29–3.33 (m, 1H), 3.82 (1H, d, $J = 6.3$, 12.0 Hz), 5.81 (1H, t, $J = 6.0$ Hz), 7.22 (1H, d, $J = 8.1$ Hz), 7.35–7.40 (1H, m), 7.50–7.53 (1H, m), 7.63 (1H, m); ^{13}C NMR (DMSO-d_6, 75 MHz): δ 36.9, 92.3, 121.2, 127.5, 130.8, 130.9, 140.7, 197.4; ESI-MS: *m/z* 289.92 (M + 1), 291.88 (M + 2). Anal. calc. for $C_9H_8BrNOS_2$: C, 37.25; H, 2.78; N, 4.83; found: C, 37.18; H, 2.89; N, 4.87.

4.2.3.5 Critical Views

The present protocol offers an expedient and solvent-free microwave-assisted alternative approach for the preparation of 4-hydroxy-3-arylthiazolidine-2-thiones as potentially useful synthons in excellent yields and atom-efficiency. This method seems to be superior to the earlier methods[26–30] that suffer from one or more disadvantages like limited diversity, polymeric nature of precursor (chloroacetaldehyde), need for excess amount of one of the reactants, longer reaction time, toxic solvents, moderate yields, mixture of products, need for chromatographic separation and laborious workup.[25]

4.2.4 Entry-4: Synthesis of 1,4-Dihydro-5-hydroxy-2-methyl-*N*,4-diphenylquinoline-3-carboxamides

Type of reaction: C–C and C–N bond formation
Reaction conditions: Catalyst-free, solvent-free, microwave irradiation
Synthetic strategy: One-pot multicomponent reaction via domino approach
Keywords: Resorcinol, aromatic aldehydes, acetoacetanilide, ammonium acetate, catalyst- and solvent-free, one-pot multicomponent reaction, domino reaction, substituted 1,4-dihydroquinolines, 1,4-dihydro-5-hydroxy-2-methyl-*N*,4-diphenylquinoline-3-carboxamide derivatives

4.2.4.1 General Reaction Scheme

Chidurala *et al.*[31] reported a catalyst- and solvent-free protocol for the synthesis of substituted 1,4-dihydroquinolines (**5**) through a one-pot multicomponent reaction between resorcinol (**1**), aromatic aldehydes (**2**), acetoacetanilide (**3**), and ammonium acetate (**4**) under microwave irradiation (Scheme 4.4). The reaction is supposed to proceed *via* a domino approach involving Knoevenagel condensation followed by subsequent Michael addition, intramolecular condensation and aromatization.

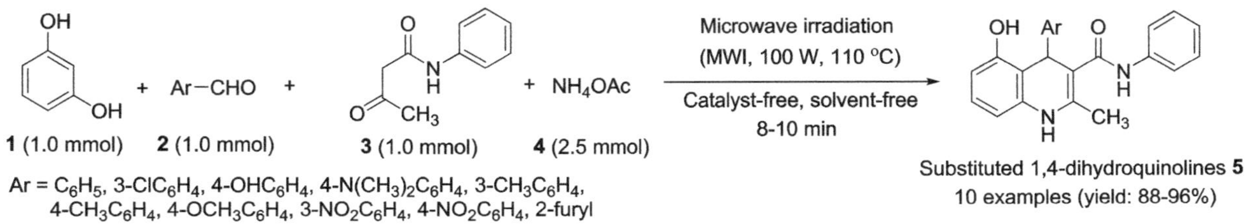

Scheme 4.4 Synthesis of 1,4-dihydro-5-hydroxy-2-methyl-*N*,4-diphenylquinoline-3-carboxamides (**5**).

4.2.4.2 Representative Examples

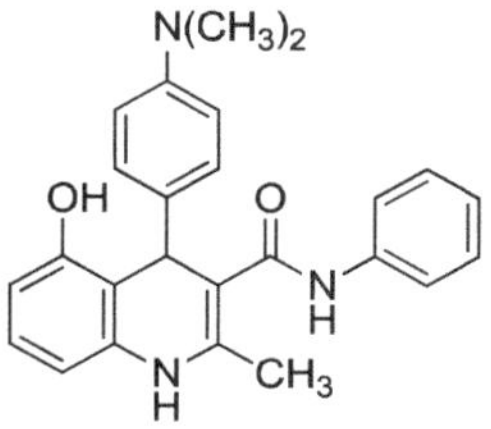

5a

4-(4-(Dimethylamino)phenyl)-5-hydroxy-2-methyl-*N*-phenyl-1,4-dihydroquinoline-3-carboxamide (**5a**): solid, mp 280 °C; yield: 93%

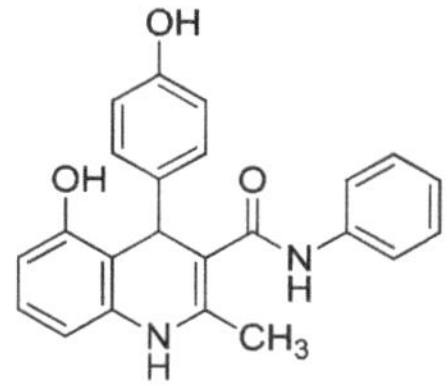

5b

5-Hydroxy-4-(4-hydroxyphenyl)-2-methyl-*N*-phenyl-1,4-dihydroquinoline-3-carboxamide (**5b**): solid, mp 263 °C; yield: 95%

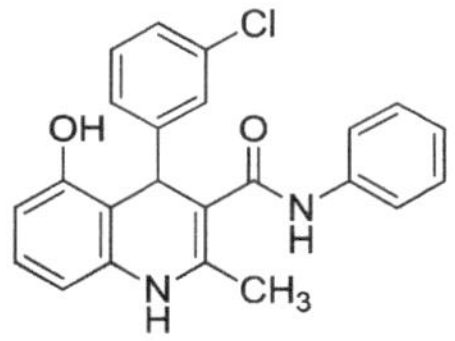

5c

4-(3-Chlorophenyl)-5-hydroxy-2-methyl-*N*-phenyl-1,4-dihydroquinoline-3-carboxamide (**5c**): solid, mp 260 °C; yield: 88%

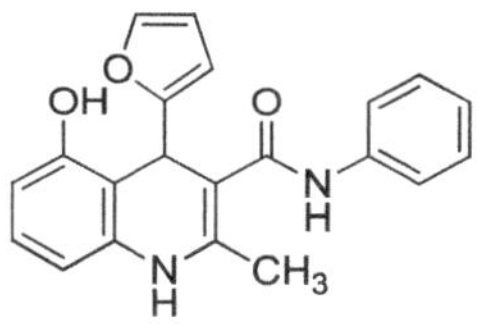

5d

4-(Furan-2-yl)-5-hydroxy-2-methyl-*N*-phenyl-1,4-dihydroquinoline-3-carboxamide (**5d**): solid, mp 235 °C; yield: 88%

4.2.4.3 Experimental Procedure

A mixture of resorcinol (**1**; 1.0 mmol), aromatic aldehydes (**2**; 1.0 mmol), acetoacetanilide (**3**; 1.0 mmol), and ammonium acetate (**4**; 2.5 mmol) in a loosely closed glass vessel was placed within a microwave oven (100 W) and irradiated at 110 °C for 8–10 min (TLC monitored). After completion of the reaction, the reaction mixture was cooled to room temperature and poured in ice water. The precipitated solid was filtered and purified using column chromatography (ethyl acetate and *n*-hexane mixture as the eluent) to give the desired products of substituted 1,4-dihydroquinolines (**5**) with excellent yields. The products were identified based on their analytical and spectral studies.

4.2.4.4 Characterization Data of Two Representative Compounds

5-Hydroxy-4-(4-hydroxyphenyl)-2-methyl-*N*-phenyl-1,4-dihydroquinoline-3-carboxamide (**5b**): solid, mp 263 °C; yield: 95%; IR (KBr): 3500 (O–H), 3300 (N–H), 1680 (C=O) cm^{-1}; ^1H NMR ($CDCl_3$, 400 MHz): δ 2.29 (3H, s, CH_3), 4.86 (1H, s), 5.53 (1H, br s, Ar–N*H*), 6.01–7.76 (12H, m, Ar–*H*), 8.42 (br s, 1H, –N*H*), 10.28 (1H, br s, Ar–O*H*), 10.92 (1H, br s, Ar–O*H*); ^{13}C NMR ($CDCl_3$,

100 MHz): δ 18.6 (–CH_3), 35.8 (–CH), 105.5, 109.5, 112.8, 117.8, 119.5, 123.7, 126.7, 129.9, 130.8, 132.6, 134.6, 137.8, 140.3, 145.8, 157.2, 159.3, 167.3 (C=O); Mass: *m/z* 372 (M^+). Anal. calc. for $C_{23}H_{20}N_2O_3$: C, 74.18; H, 5.41; N, 7.52; found: C, 74.16; H, 5.49; N, 7.52.

4-(Furan-2-yl)-5-hydroxy-2-methyl-*N*-phenyl-1,4-dihydroquinoline-3-carboxamide (**5d**): solid, mp 235 °C; yield: 88%; IR (KBr): 3510 (O–H), 3280 (N–H), 1680 (C=O) cm^{-1}; ^{1}H NMR ($CDCl_3$, 400 MHz): δ 2.25 (3H, s, CH_3), 4.67 (1H, s), 5.56 (1H, br s, Ar–NH), 6.00–7.84 (11H, m, Ar–*H*), 8.24 (1H, br s, –NH), 10.67 (1H, br s, Ar–OH); ^{13}C NMR ($CDCl_3$, 100 MHz): δ 19.1 (–CH_3), 35.2 (–CH), 105.3, 107.9, 109.6, 111.8, 113.7, 119.1, 122.8, 126.3, 130.1, 132.6, 138.2, 140.4, 143.7, 146.6, 153.8, 156.7, 166.2 (C=O); mass: *m/z* 346 (M^+). Anal. calc. for $C_{21}H_{18}N_2O_3$: C, 72.82; H, 5.24; N, 8.09; found: C, 72.76; H, 5.29; N, 8.02.

4.2.4.5 Critical Views

The present protocol offers an expedient and solvent-free microwave-assisted alternative approach for the preparation of biologically interesting 1,4-dihydro-5-hydroxy-2-methyl-*N*,4-diphenylquinoline-3-carboxamide derivatives in excellent yields.[31]

4.2.5 Entry-5: Synthesis of Azaarene-substituted 3-hydroxy-2-oxindoles

Type of reaction: C–C bond formation
Reaction conditions: Catalyst-free, water, microwave irradiation
Synthetic strategy: One-step process *via* functionalization of sp^3 C–H bond in methyl pyridine
Keywords: Isatins, 2- and 4-methyl pyridines, catalyst-free, water, azaarene-substituted 3-hydroxy-2-oxindoles

4.2.5.1 General Reaction Scheme

Meshram and his group[32] reported an alternative and efficient catalyst-free protocol for the synthesis of potentially useful azaarene-substituted 3-hydroxy-2-oxindoles (**3**) by the reaction of 2-methyl pyridines (**2**) with isatins (**1**) under microwave irradiation in aqueous medium *via* activation of sp^3 C–H bond in the methyl pyridines (Scheme 4.5). Use of no added catalyst, expedient synthesis, high yields and environmentally benign medium are the key features of the present protocol.

4.2.5.2 Plausible Mechanism

The authors proposed a tentative mechanistic path for the microwave-assisted conversion in water as depicted in Scheme 4.6.

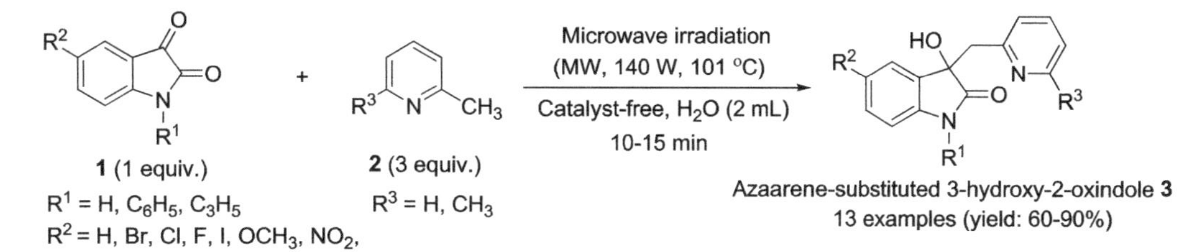

Scheme 4.5 Synthesis of azaarene-substituted 3-hydroxy-2-oxindoles (**3**).

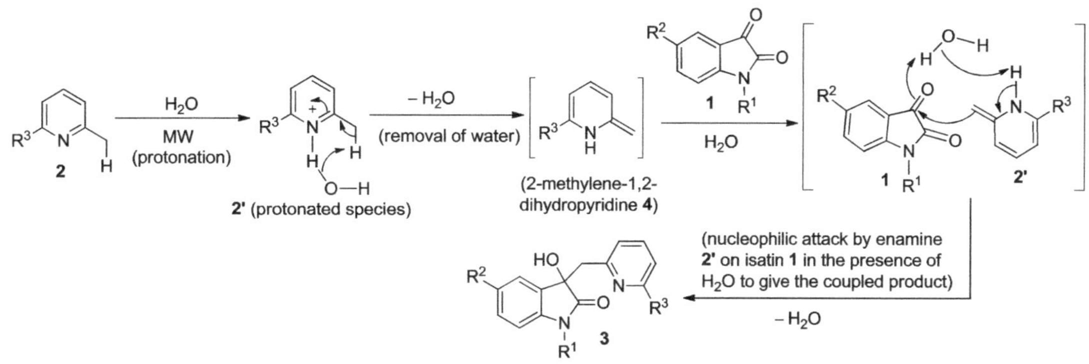

Scheme 4.6 Plausible mechanism for the synthesis of azaarene-substituted 3-hydroxy-2-oxindoles (**3**).

4.2.5.3 Representative Examples

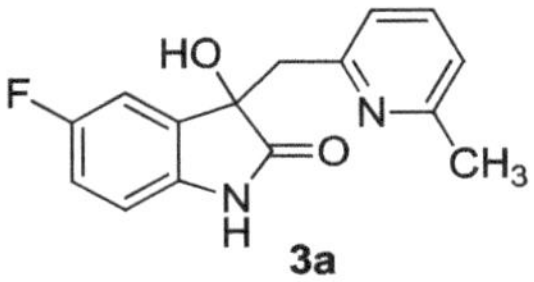

3a

5-Fluoro-3-hydroxy-3-((6-methylpyridin-2-yl)methyl)indolin-2-one (**3a**): solid, mp 169–171 °C, yield: 85%

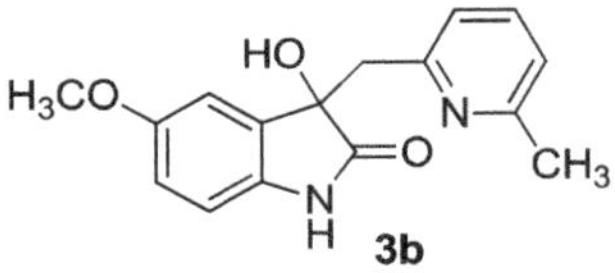

3b

3-Hydroxy-5-methoxy-3-((6-methylpyridin-2-yl)methyl)indolin-2-one (**3b**): solid, mp 175–177 °C, yield: 80%

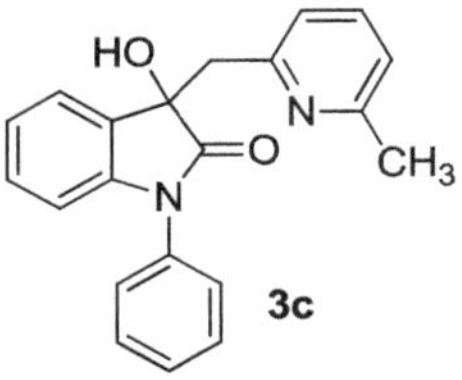

3c

3-Hydroxy-3-((6-methylpyridin-2-yl)methyl)-1-phenylindolin-2-one (**3c**): yield: 75%

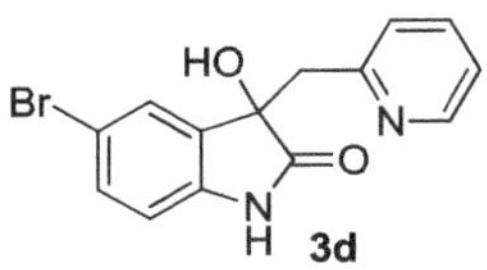

3d

5-Bromo-3-hydroxy-3-(pyridin-2-ylmethyl)indolin-2-one (**3d**): solid, mp 207–209 °C, yield: 90%

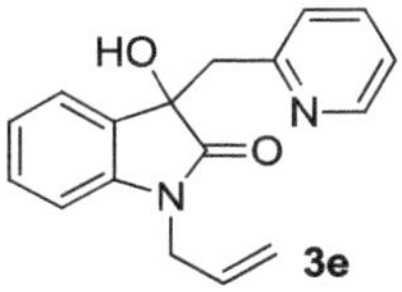

3e

1-Allyl-3-hydroxy-3-(pyridin-2-ylmethyl)indolin-2-one (**3e**): solid, mp 228–230 °C, yield: 80%

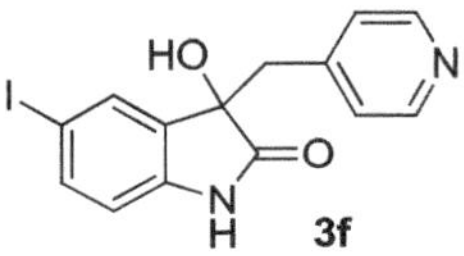

3f

3-Hydroxy-5-iodo-3-(pyridin-4-ylmethyl)indolin-2-one (**3f**): solid, mp 226–228 °C, yield: 80%

4.2.5.4 Experimental Procedure

A sealed 10 mL glass tube containing isatins (**1**; 1 equiv.), methyl pyridine (**2**; 3 equiv) and water (2 mL) placed in the cavity of a microwave reactor (power 140 W) was irradiated for 10–15 min at 101 °C. After cooling to room temperature, the tube was removed from the rotor, and the reaction mixture was diluted with water followed by extraction with ethyl acetate. The combined organic layer extracts were then dried over sodium sulfate, and the solvent removed under reduced pressure. The crude product obtained was then purified by short column chromatography (hexane–AcOEt as the eluent) to furnish pure 3-hydroxy-2-oxindole products **3** in good to excellent yields (60–90%). All the products were identified based on their detailed spectral studies including IR, ^{1}H and ^{13}C NMR, and MS.

4.2.5.5 Characterization Data of Three Representative Compounds

5-Fluoro-3-hydroxy-3-((6-methylpyridin-2-yl)methyl)indolin-2-one (**3a**): solid, mp 169–171 °C, yield: 85%; IR: ν_{max} 3280, 1718, 1597, 1486, 1145, 809 cm^{-1};

^{1}H NMR ($CDCl_3$ + DMSO-d_6, 300 MHz): δ 9.86 (1H, s), 7.56 (1H, t, J = 7.5 Hz), 7.13 (1H, d, J = 7.5 Hz), 6.93 (1H, d, J = 7.3 Hz), 6.75–6.88 (2H, m), 6.39 (1H, d, J = 7.1 Hz), 3.34 (1H, d, J = 14.5 Hz), 2.98 (1H, d, J = 14.5 Hz,), 2.56 (3H, s); ^{13}C NMR ($CDCl_3$ + DMSO-d_6, 75 MHz): δ 178.3, 156.4, 155.5, 136.5, 132.3, 121.2, 121.0, 114.7, 114.3, 111.5, 111.2, 110.2, 76.0, 42.3, 23.5; ESI: *m/z* 273 $[M+H]^+$; HRMS: *m/z* calc. for $C_{15}H_{14}N_2O_2F$: 273.10338; found: 273.10338.

1-Allyl-3-hydroxy-3-(pyridin-2-ylmethyl)indolin-2-one (**3e**): solid, mp 228–230 °C, yield: 80%; IR: ν_{max} 3349, 3088, 2986, 1706, 1603, 1255, 1208, 1124, 1005, 821, 802, 646 cm^{-1}; ^{1}H NMR ($CDCl_3$ + DMSO-d_6, 300 MHz): δ 8.47 (1H, d, J = 4.3 Hz); 7.55 (1H, td, J = 7.7, 1.5 Hz), 7.09–7.21 (2H, m), 6.96 (1H, d, J = 7.7 Hz), 6.83 (1H, t, J = 7.3 Hz), 6.72 (2H, dd, J = 10.7, 7.7 Hz), 5.65–5.79 (1H, m), 5.11 (2H, dd, J = 10.3, 6.4 Hz), 4.20 (2H, qd, J = 16.2, 5.2 Hz), 3.26 (1H, d, J = 14.7 Hz), 3.02 (1H, d, J = 14.7 Hz); ^{13}C NMR ($CDCl_3$ + DMSO-d_6, 75 MHz): δ 175.3, 154.6, 146.8, 140.7, 134.5, 129.9, 129.0, 127.4, 123.0, 122.7, 120.6, 120.2, 115.5, 107.2, 74.3, 43.7, 40.3; ESI: *m/z* 281 $[M+H]^+$; HRMS: *m/z* calc. for $C_{17}H_{17}N_2O_2$: 281.12868; found: 281.12845.

3-Hydroxy-5-iodo-3-(pyridin-4-ylmethyl)indolin-2-one (**3f**): solid, mp 226–228 °C, yield: 80%; IR: ν_{max} 3286, 1696, 1614, 1467, 1435, 1378, 1185, 1103, 999, 752 cm^{-1}; ^{1}H NMR ($CDCl_3$ + DMSO-d_6, 300 MHz): δ 10.17 (1H, s), 8.34 (2H, d, J = 4.7 Hz), 7.44 (1H, d, J = 8.1 Hz), 7.31(1H, s), 6.95 (2H, d, J = 4.7 Hz), 6.51 (1H, d, J = 8.1 Hz), 6.37 (1H, s), 3.19 (1H, d, J = 12.5 Hz), 3.00 (1H, d, J = 12.5 Hz); ^{13}C NMR ($CDCl_3$ + DMSO-d_6, 75 MHz): δ 175.5, 147.7, 142.8, 139.1, 131.2, 127.9, 125.4, 124.6, 123.6, 109.9, 75.3, 42.2; ESI: *m/z* 366 $[M+H]^+$; HRMS: *m/z* calc. for $C_{14}H_{12}N_2O_2I$: 366.99435; found: 366.99380.

4.2.5.6 Critical Views

The present method offers a rapid and eco-friendly way to access potentially useful methyl pyridine-substituted 3-hydroxy-2-oxindoles in good to excellent yields in water under the influence of microwave irradiation without using any catalyst in contrast to earlier methods with use Bronsted acid,[33] Lewis acids,[34–40] and transition metals.[41–45] Catalyst-free simple reaction conditions, rapid synthesis of the products with good yields and use of a green solvent are the advantages of the present method.[32] Moreover the present protocol provides catalyst-free easy access for the sp^3 C–H bond activation for the first time, which may find an application in organic synthesis.

4.2.6 Entry-6: Synthesis of Functionalized Quinoline Derivatives

Type of reaction: C–C bond formation
Reaction conditions: Catalyst-free, water, microwave irradiation
Synthetic strategy: Benzylic C–H functionalization
Keywords: 2-Methylquinolines, β-nitro-styrenes, catalyst-free, water, activation of methyl quinoline sp^3 C–H bond, Michael addition, functionalized quinoline derivatives

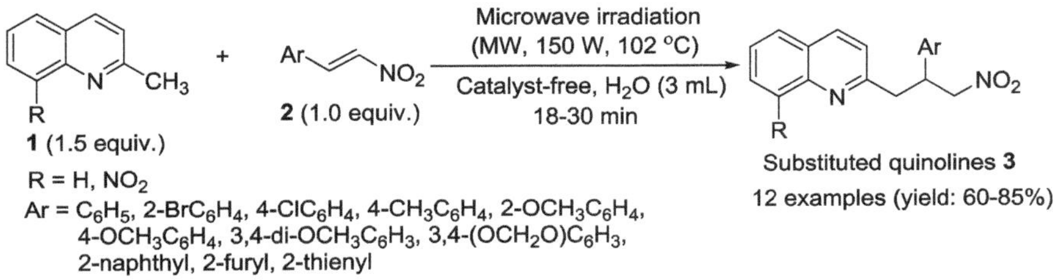

Scheme 4.7 Synthesis of functionalized quinolines (**3**).

4.2.6.1 *General Reaction Scheme*

A catalyst-free aqueous mediated C–H activation of methyl quinolines (**1**) followed by Michael addition to various β-nitro styrenes (**2**), under microwave irradiation, to generate a series of substituted quinolines (**3**) was accomplished by Rao and Meshram.[46] Catalyst- and additive-free clean reaction conditions and use of water as the environmentally benign medium are the notable features of the present protocol (Scheme 4.7).

4.2.6.2 *Representative Examples*

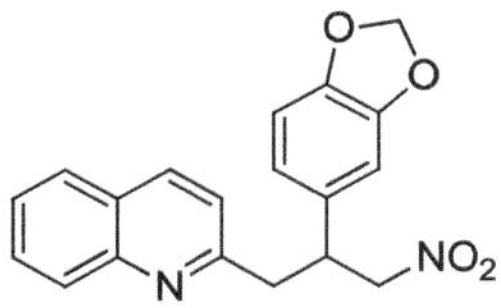

3a

2-(2-(Benzo[*d*][1,3]dioxol-5-yl)-3-nitropropyl)quinoline (**3a**): yield: 60%

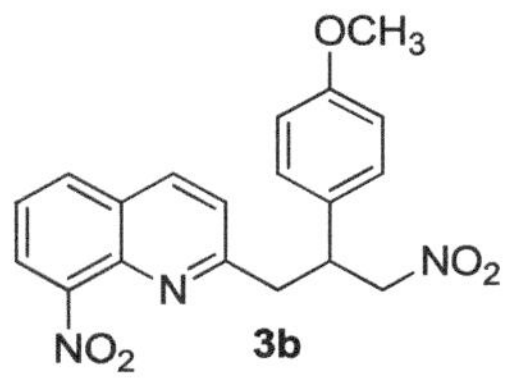

3b

2-(2-(4-Methoxyphenyl)-3-nitropropyl)-8-nitroquinoline (**3b**): yield: 75%

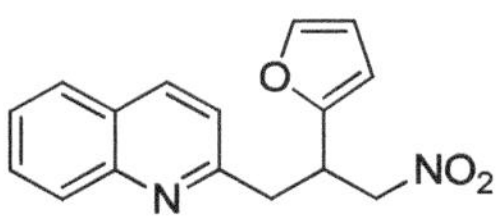

3c

2-(2-(Furan-2-yl)-3-nitropropyl)quinoline (**3c**): yield: 80%

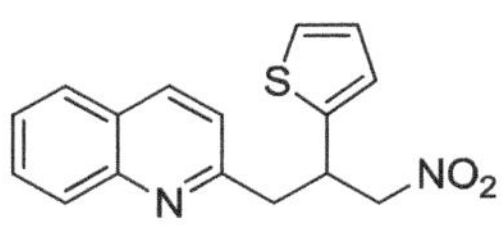

3d

2-(3-Nitro-2-(thiophen-2-yl)propyl)quinoline (**3d**): yield: 85%

4.2.6.3 *Experimental Procedure*

A mixture of 2-methylquinolines (**1**; 1.5 equiv.) and β-nitro styrenes (**2**; 1.0 equiv.) within a 10 mL closed microwave tube was placed inside the cavity of a microwave reactor, operated at 102 °C (temperature was monitored by a built-in infrared sensor; power 150 W) for 18–30 min. Upon completion of the reaction, the reaction tube was cooled at room temperature, and the reaction mixture was diluted with water and extracted with ethyl acetate. The combined ethyl acetate extracts were then dried over sodium sulfate, the solvent was removed, and then purified by column chromatography over silica gel using 20% ethyl acetate in hexane as the eluent to get the pure products of substituted quinolines (**3**). All the products were characterized based on their detailed spectral studies.

4.2.6.4 Characterization Data of Two Representative Compounds

2-(2-(Benzo[*d*][1,3]dioxol-5-yl)-3-nitropropyl)quinoline (**3a**): yield: 60%; IR: ν_{max} 2956, 2924, 2855, 1600, 1551, 1503, 1487, 1444, 1377, 1247, 1039, 934, 819, 751 cm^{-1}; ^1H NMR ($CDCl_3$, 300 MHz): δ 8.01–8.11 (2H, m), 7.65–7.81 (2H, m), 7.45–7.55 (1H, m), 7.28 (1H, d, $J=3.02$ Hz), 7.15 (1H, d, $J=8.3$ Hz), 6.75 (1H, s), 6.69 (1H, s), 5.90 (2H, s), 4.62–4.83 (2H, m), 4.06–4.23 (1H, m), 3.31 (2H, d, $J=7.5$ Hz); ^{13}C NMR ($CDCl_3$,75 MHz): δ 158.3, 147.9, 147.7, 147.5, 136.6, 129.6, 129.4, 128.8, 127.5, 126.2, 125.6, 121.6, 120.8, 108.4, 107.6, 101.0, 79.8, 43.6, 42.3; MS-ESI: *m/z* 307 $[M+H]^+$.

2-(3-Nitro-2-(thiophen-2-yl)propyl)quinoline (**3d**): yield: 85%; IR: ν_{max} 2924, 2854, 1558, 1550, 1502, 1428, 1376, 824, 701 cm^{-1}; ^1H NMR ($CDCl_3$, 300 MHz): δ 8.04 (2H, t, $J=8.1$ Hz), 7.66–7.79 (2H, m), 7.50 (1H, t, $J=7.1$ Hz), 7.13–7.20 (2H, m), 6.87 (2H, d, $J=3.3$ Hz), 4.67–4.90 (2H, m), 4.53–4.64 (1H, m), 3.34–3.48 (2H, m); ^{13}C NMR ($CDCl_3$, 75 MHz,): δ 157.8, 147.7, 142.2, 136.5, 129.6, 128.8, 127.4, 126.8, 126.2, 125.3, 124.4, 121.6, 80.0, 42.9, 38.9; MS-ESI: *m/z* 299 $[M+H]^+$.

4.2.6.5 Critical Views

The present protocol for microwave-assisted catalyst-free sp^3 C–H bond activation of methyl quinolines in water leading to the expedient synthesis of a series of substituted quinolines *via* Michael addition with β-nitro styrenes with moderate to good yields may find useful applications in organic transformations under green conditions.

4.2.7 Entry-7: Synthesis of Quinoxalines

Type of reaction: C–N bond formation
Reaction conditions: Catalyst-free, solvent-free, microwave irradiation
Synthetic strategy: Condensation
Keywords: Benzils, 1,2-diaminobenzenes, catalyst-free, solvent-free, quinoxalines

4.2.7.1 General Reaction Scheme

Zhou *et al.*[47] synthesized a series of functionalized quinoxaline derivatives (**3**) from microwave-assisted cyclocondensation reaction of benzils (**1**) and 1,2-diamines (**2**) under catalyst- and solvent conditions with good to excellent yields within short reaction time of minutes (Scheme 4.8).

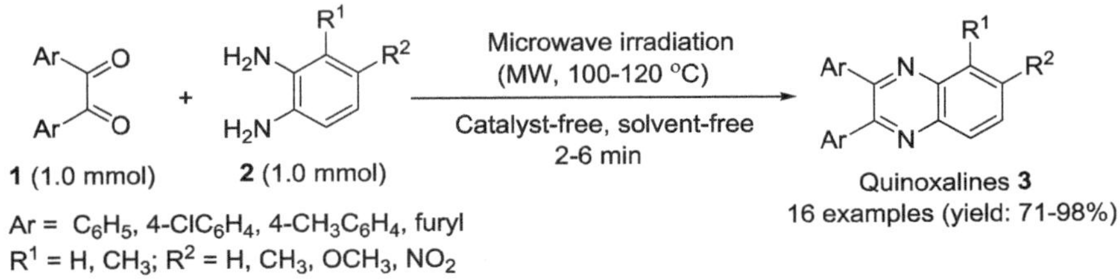

Scheme 4.8 Synthesis of quinoxalines (**3**).

4.2.7.2 *Representative Examples*

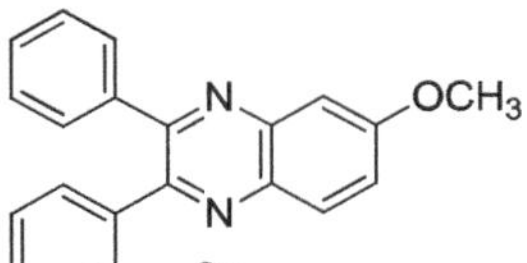

3a

6-Methoxy-2,3-diphenylquinoxaline (**3a**): solid, mp 156.4–156.5 °C; yield: 71%

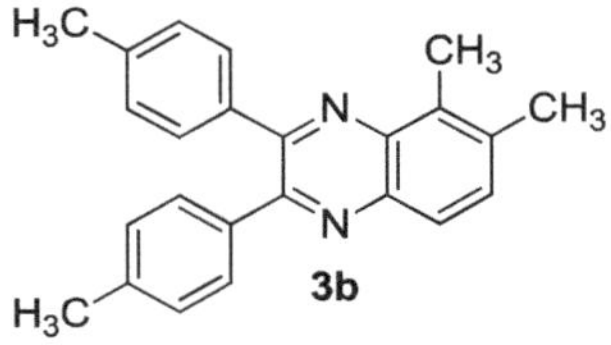

3b

5,6-Dimethyl-2,3-di-*p*-tolylquinoxaline (**3b**): solid, mp 173.4–173.6 °C; yield: 95%

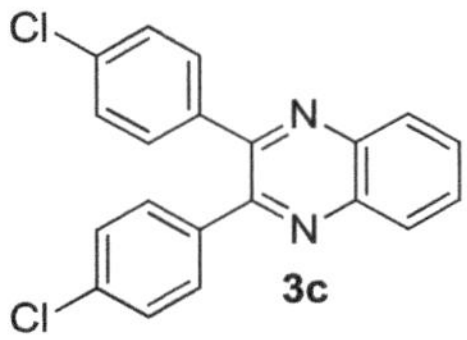

3c

2,3-*Bis*(4-chlorophenyl)quinoxaline (**3c**): solid, mp 174.9 °C; yield: 93%

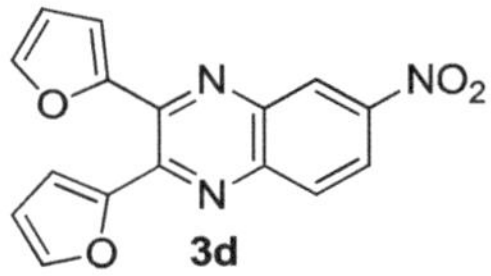

3d

2,3-Di(furan-2-yl)-6-nitroquinoxaline (**3d**): solid, mp 169.2–169.4 °C; yield: 98%

4.2.7.3 *Experimental Procedure*

A thoroughly ground mixture of benzils (**1**; 1 mmol) and 1,2-diaminobenzenes (**2**; 1 mmol) taken within a sealed vial was placed into the cavity of the microwave reactor and irradiated at 100–130 °C for 2–6 min (as indicated by TLC). After completion of the reaction, the mixture was then allowed to cool to room temperature when solid residues precipitated out. These crude products were filtered off, dried and purified by recrystallization from 95% ethanol to have the pure products of quinoxalines (**3**) with good yields ranging from 71–98%. All the products were characterized on the basis of their analytical and spectral studies.

4.2.7.4 *Characterization Data of Three Representative Compounds*

6-Methoxy-2,3-diphenylquinoxaline (**3a**): solid, mp 156.4–156.5 °C; yield: 71%; ^{1}H NMR ($CDCl_3$, 400 MHz): δ 8.09 (1H, d, $J = 9.2$ Hz), 7.51–7.54 (5H, m),

7.44–7.49 (1H, m), 7.34–7.40 (6H, m), 4.02 (3H, s); MS (%rel): *m/z* $[M+H]^+$ 313.7 (100).

2,3-Di(furan-2-yl)-6-nitroquinoxaline (**3d**): solid, mp 169.2–169.4 °C; yield: 98%; IR (KBr): 3094, 1566, 1522, 1241, 1059, 826, 741 cm^{-1}; ^1H NMR ($CDCl_3$, 400 MHz): δ 9.03 (1H, d, $J=2.4$ Hz), 8.52–8.50 (1H, m), 8.25 (2H, d, $J=9.2$ Hz), 7.69 (1H, d, $J=6.0$ Hz), 6.91 (1H, d, $J=3.2$ Hz), 8.86 (1H, d, $J=3.2$ Hz), 6.65–6.63 (2H, m); ^{13}C NMR ($CDCl_3$, 100 MHz): δ 150.2, 148.0, 145.4, 144. 8, 144.2, 143.0, 139.2, 130.4, 125.3, 123.6, 115.3, 114.4, 112.3; MS (%rel): *m/z* $[M+H]^+$ 308.3 (100). Anal. calc. for $C_{16}H_9N_3O_4$: C, 62.54; H, 2.95; N, 13.68; found: C, 65.31; H, 2.97; N, 13.58.

4.2.7.5 Critical Views

The present method offers an alternate way for the synthesis of a series of quinoxaline derivatives. Rapid and operational simplicity, high yields, use of no catalyst and solvent, easy workup, and eco-friendliness are the major advantages of this protocol over the other the huge number of previously reported methods involving so many catalytic systems and solvents.[48–73]

4.2.8 Entry-8: Synthesis of Coumarin-substituted Quinoxalines

Type of reaction: C–N bond formation
Reaction conditions: Catalyst-free, ethanol, microwave irradiation
Synthetic strategy: Condensation reaction
Keywords: 3-(ω-Bromoacetyl)coumarins, 1,2-diaminobenzenes, catalyst-free, ethanol, condensation, coumarin-substituted quinoxalines

4.2.8.1 General Reaction Scheme

A catalyst-free microwave-assisted synthetic protocol of a series of quinoxaline derivatives (**3**) from the reaction of 3-(ω-bromoacetyl)coumarins (**1**) with substituted 1,2-diaminobenzenes (**2**) in ethanol was developed by Zhou and his group[74] (Scheme 4.9).

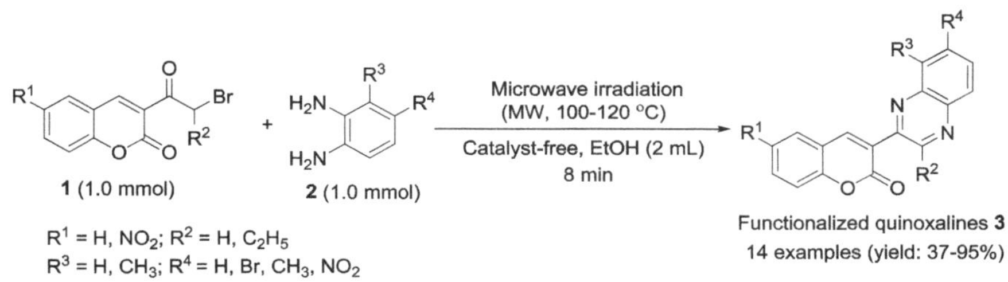

Scheme 4.9 Synthesis of coumarin-substituted quinoxalines (3).

4.2.8.2 Representative Examples

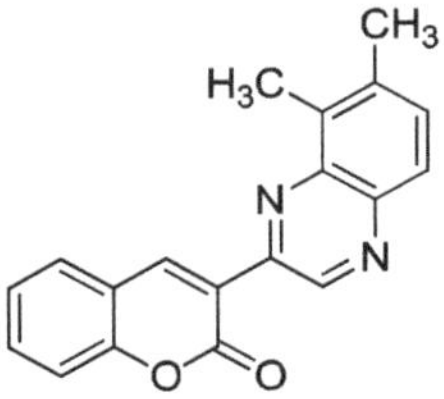

3a

3-(7,8-Dimethylquinoxalin-2-yl)-2*H*-chromen-2-one (**3a**): solid, mp 169.1–171 °C; yield: 95%

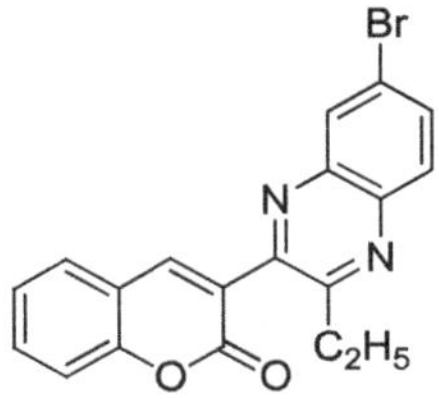

3b

3-(7-Bromo-3-ethylquinoxalin-2-yl)-2*H*-chromen-2-one (**3b**): solid, mp 224–226 °C; yield: 67%

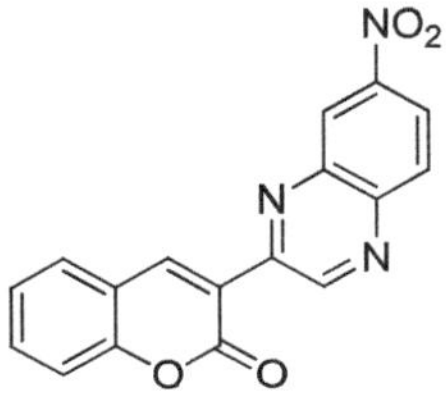

3c

3-(7-Nitroquinoxalin-2-yl)-2*H*-chromen-2-one (**3c**): solid, mp 240–242 °C; yield: 94%

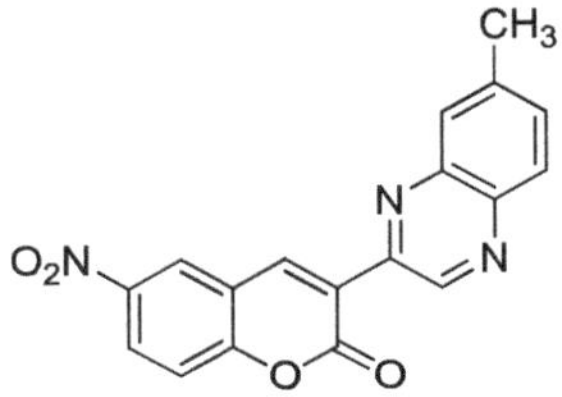

3d

3-(7-Methylquinoxalin-2-yl)-6-nitro-2*H*-chromen-2-one (**3d**): yellow solid, mp 202–203.5 °C; yield: 37%

4.2.8.3 Experimental Procedure

A mixture of 3-(ω-bromoacetyl)coumarin (**1**; 1 mmol) and 1,2-diaminobenzene (**2**; 1 mmol) in 2 mL ethanol placed within a sealed tube was irradiated with microwave using a microwave reactor at 100–120 °C for 8 minutes. After completion of the reaction (as indicated by TLC), the reaction mixture was then allowed to cool to room temperature, when solid product precipitated out, which was filtered off and dried. The isolated crude solids were purified by recrystallization from 95% ethanol to afford the pure products of quinoxalines (**3**). The products were characterized based on their spectral studies.

4.2.8.4 Characterization Data of Two Representative Compounds

3-(7-Bromo-3-ethylquinoxalin-2-yl)-2*H*-chromen-2-one (**3b**): solid, mp 224–226 °C; yield: 67%; IR (KBr): 3090, 2979, 1711, 1604, 1565, 1462, 1194, 998, 758 cm^{-1}; ^1H NMR ($CDCl_3$, 400 MHz): δ 1.41 (3H, t, $J = 14.8$ Hz), 3.06 (2H, q), 7.40 (1H, t, $J = 15.2$ Hz), 7.48 (1H, d, $J = 8.0$ Hz), 7.63–7.68 (2H, m), 7.86 (1H, s),

7.95–8.28 (2H, m), 8.34 (1H, s); ^{13}C NMR ($CDCl_3$, 100 MHz): δ 12.18, 28.76, 116.89, 118.92, 123.06, 124.90, 127.61, 128.59, 130.14, 131.36, 132.69, 133.80, 140.95, 141.18, 144.03, 150.04, 154.38, 158.01, 159.70; HRMS: calc. for $C_{19}H_{14}BrN_2O_2$ $[M+H]^+$: 382.2307; found: 382.9367.

3-(7-Nitroquinoxalin-2-yl)-2*H*-chromen-2-one (**3c**): solid, mp 240–242 °C; yield: 94%; IR (KBr): 3097, 1722, 1610, 1562, 1454, 1347, 1200, 1006, 747 cm^{-1}; ^{1}H NMR ($CDCl_3$, 400 MHz): δ 7.44 (1H, t, *J* = 14.8 Hz), 7.49 (1H, d, *J* = 8.4 Hz), 7.72 (1H, t, *J* = 15.2 Hz), 7.80 (1H, d, *J* = 7.2 Hz), 8.30–8.61 (2H, m), 8.99 (1H, s), 9.09 (1H, s), 10.05 (1H, s); ^{13}C NMR ($CDCl_3$, 100 MHz): δ 116.58, 119.34, 123.36, 125.62, 127.18, 128.69, 130.45, 131.37, 134.17, 141.12, 144.58, 147.55, 152.34, 154.01, 159.29, 160.46, 161.10 ; HRMS: calc. for $C_{17}H_{10}N_3O_4$ $[M+H]^+$: 320.0671; found: 319.9880.

4.2.8.5 Critical Views

The present method offers an alternative route for the synthesis of biologically useful coumarin containing quinoxaline derivatives under catalyst-free conditions. Although the method satisfies certain green chemistry aspects, it is not high yielding for all the substrates.

4.2.9 Entry-9: Synthesis of Functionalized 1,8-Naphthyridines and Quinolines

Type of reaction: C–C and C–N bond formation
Reaction conditions: Catalyst-free, solvent-free, microwave irradiation
Synthetic strategy: Cyclocondensation
Keywords: Aryl/heteroaryl amines, β-aryl vinyl ketones, catalyst-free, solvent-free, one-pot reaction, functionalized 1,8-naphthyridines and quinolines

4.2.9.1 General Reaction Scheme

Goswami *et al.*[75] synthesized a series of functionalized 1,8-naphthyridines/quinolines (**3**) from a microwave-assisted one-pot reaction of aryl/heteroaryl amines (**1**) and aryl vinyl ketones (**2**) under catalyst- and solvent conditions with moderate to good yields (Scheme 4.10).

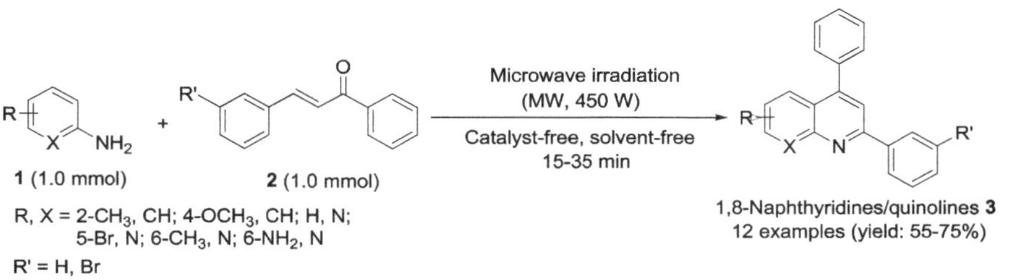

Scheme 4.10 Synthesis of functionalized 1,8-naphthyridines and quinolines (**3**).

4.2.9.2 Representative Examples

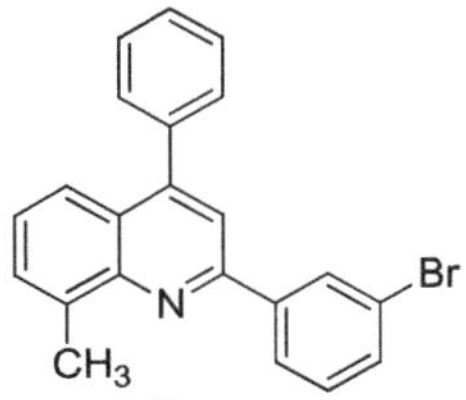
3a

2-(3-Bromophenyl)-8-methyl-4-phenylquinoline (**3a**): yield: 60%

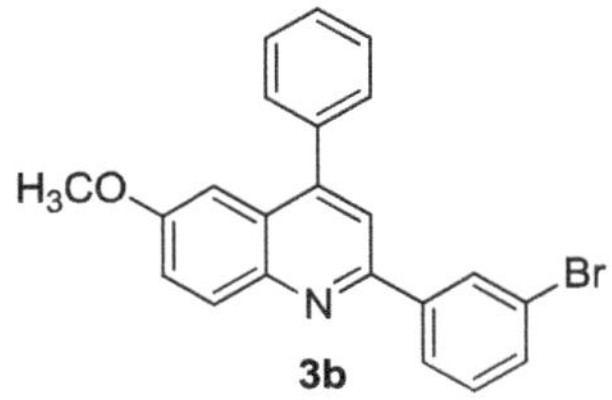
3b

2-(3-Bromophenyl)-6-methoxy-4-phenylquinoline (**3b**): solid, mp 110–112 °C; yield: 65%

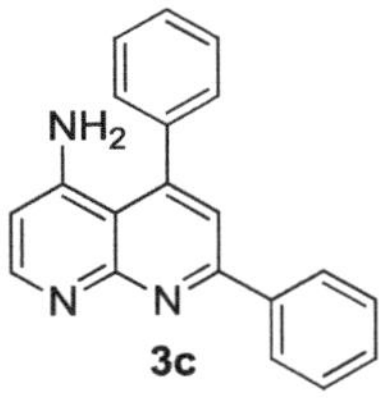
3c

5,7-Diphenyl-1,8-naphthyridin-4-amine (**3c**): yield: 70%

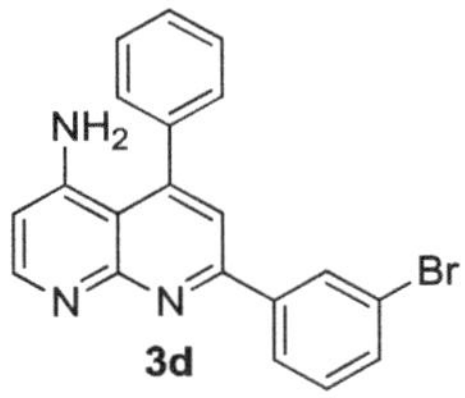
3d

7-(3-Bromophenyl)-5-phenyl-1,8-naphthyridin-4-amine (**3d**): off white solid, mp 243–246 °C; yield: 75%

4.2.9.3 Experimental Procedure

A thoroughly ground mixture of aryl/heteroaryl amines (**1**; 1 mmol) and aryl vinyl ketones (**2**; 1 mmol) was placed in an open-mouth conical flask and irradiated at 450 W with a domestic microwave oven for 15–35 min in the absence of any solvent and catalyst. Upon completion of the reaction, the solid residue was washed with water and extracted with ethyl acetate (4×20 mL), and the combined organic layers were then concentrated under reduced pressure to obtain the crude product. This crude product was purified by means of column chromatography (silica gel, 100–200 mesh; ethyl acetate–petroleum ether mixtures as the eluents) to furnish pure 1,8-naphthyridines/quinolines (**3**). All the products were characterized on the basis of their analytical and spectral studies.

4.2.9.4 Characterization Data of Three Representative Compounds

2-(3-Bromophenyl)-6-methoxy-4-phenylquinoline (**3b**): solid, mp 110–112 °C; yield: 65%; IR (KBr): 1623, 1589, 1253, 1224, 689 cm^{-1}; ^{1}H NMR ($CDCl_3$, 500 MHz): δ 8.15 (1H, d, $J=1.7$ Hz), 8.14(2H, d, $J=3.8$ Hz), 7.74(2H, s), 7.64 (1H, d, $J=7.9$ Hz), 7.51 (1H, t, $J=7.4$ Hz), 7.50 (2H, d, $J=7.6$ Hz), 7.46–7.39 (3H, m), 7.11 (1H, d, $J=2.4$ Hz), 3.82(3H, s); ^{13}C NMR ($CDCl_3$, 125 MHz): δ 158.41, 155.01, 146.49, 145.31, 143.35, 141.19, 139.92, 132.71, 132.15, 131.83,

130.61, 129.52, 129.25, 128.42, 127.70, 123.25, 122.41, 119.96, 103.72, 55.92; MS (ESI; rel%): *m/z* 390.1 (M^+, 94), 391($[M+1]^+$, 26), 392($[M+2]^+$, 100), 393($[M+3]^+$, 24). Anal. calc. for $C_{22}H_{16}BrNO$: C, 67.71; H, 4.13; N, 3.59; found: C, 67.75; H, 4.10; N, 3.54.

7-(3-Bromophenyl)-5-phenyl-1,8-naphthyridin-4-amine (**3d**): off white solid, mp 243–246 °C; yield: 75%; IR (KBr): 3470, 3051, 1644, 1564, 1357, 771, 705 cm^{-1}; ^1H NMR (DMSO-d_6, 500 MHz): δ 8.46 (1H, t, $J=1.7$ Hz), 8.25 (1H, d, $J=8.0$ Hz), 7.81(1H, d, $J=9.0$ Hz), 7.23 (1H, s), 7.64 (1H, d, $J=7.5$ Hz), 7.55 (4H, d, $J=4.3$ Hz), 7.53–7.50 (1H, m), 7.45 (1H, t, $J=7.9$ Hz), 6.91 (2H, bs), 6.81 (1H, d, $J=9.0$ Hz); ^{13}C NMR (DMSO-d_6, 125 MHz): δ 161.78, 157.95, 156.08, 150.30, 142.17, 138.23, 135.66, 132.90, 131.71, 130.63, 130.38, 129.54, 129.41, 126.96, 123.18, 115.38, 114.87, 114.58; MS (ESI; rel%): *m/z* 378.1 ($[M+2]^+$, 100), 376.1(M^+, 94.7). Anal. calc. for $C_{20}H_{14}BrN_3$: C, 63.85; H, 3.75; N, 11.17; found: C 63.91; H, 3.71; N, 11.12.

4.2.9.5 Critical Views

The present method offers an alternate way for the synthesis of 1,8-naphthyridines and quinolines. Use of no catalyst or solvent, operational simplicity and clean reaction profiles are the major advantages of this protocol.

4.2.10 Entry-10: Synthesis of Substituted Pyrazoles

Type of reaction: C–N bond formation
Reaction conditions: Catalyst-free, neat, microwave irradiation
Synthetic strategy: Condensation reaction
Keywords: Aryl/heteroaryl hydrazines, 1,3-diketones, catalyst-free, neat, condensation, pyrazoles

4.2.10.1 General Reaction Scheme

A microwave-assisted catalyst-free alternative method for the facile condensation of hydrazines (**1**) with 1,3-diketones (**2**) leading to the preparation of pyrazoles (**3**) under neat conditions was designed by Reddy Vaddula *et al.* (Scheme 4.11).[76] Rapid synthesis, high yields and eco-friendly conditions are the notable advantages of this protocol.

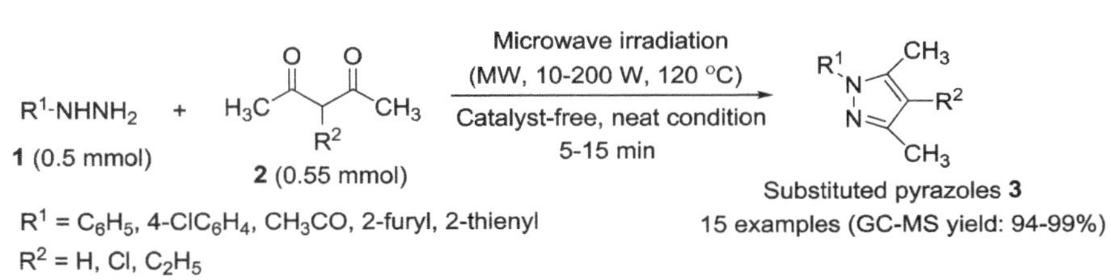

Scheme 4.11 Synthesis of substituted pyrazoles (**3**).

4.2.10.2 Representative Examples

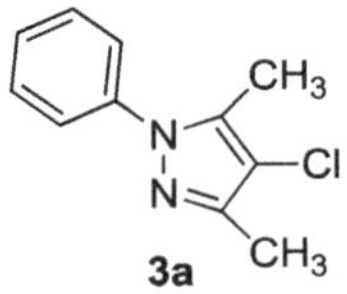

3a

4-Chloro-3,5-dimethyl-1-phenyl-1*H*-pyrazole (**3a**): yield: 98% (GCMS)

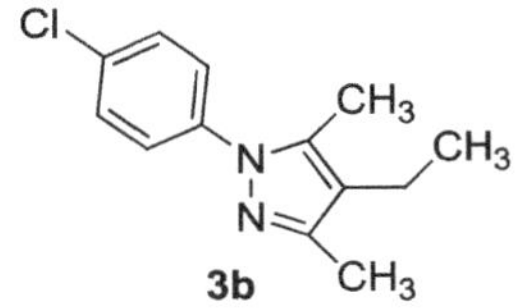

3b

1-(4-Chlorophenyl)-4-ethyl-3,5-dimethyl-1*H*-pyrazole (**3b**): yield: 98% (GCMS)

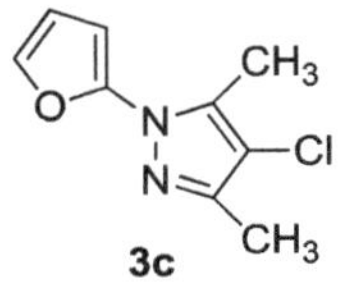

3c

4-Chloro-1-(furan-2-yl)-3,5-dimethyl-1H-pyrazole (**3c**): yield: 98% (GCMS)

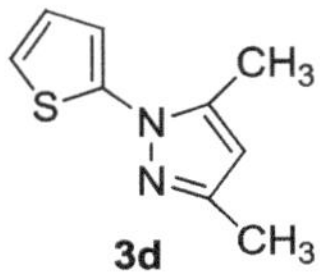

3d

3,5-Dimethyl-1-(thiophen-2-yl)-1H-pyrazole (**3d**): yield: 98% (GCMS)

4.2.10.3 Experimental Procedure

A mixture of hydrazines (**1**; 0.5 mmol) and 1,3-diketones (**2**; 0.55 mmol) within a closed microwave tube was placed inside the cavity of a CEM Discover focused microwave synthesis system, operated at 120 °C (temperature was monitored by a built-in infrared sensor), power 10–200 W for 5–15 min. Upon completion of the reaction, crude products were purified by column chromatography over silica gel using 20% ethyl acetate in hexane as the eluent to afford the pure products of pyrazoles (**3**). All the products are known in literature.

4.2.10.4 Critical Views

The present protocol is advantageous over earlier methods[77–81] reported in literature as accessing pyrazole derivatives, particularly in regard to its fastness, high yields and use of no catalyst.

4.2.11 Entry-11: Synthesis of Substituted Tetrahydropyrimidines

Type of reaction: C–C and C–N bond formation
Reaction conditions: Catalyst-free, solvent-free, microwave irradiation
Synthetic strategy: One-pot multicomponent reaction

Keywords: Ethyl 3-oxo-4-(arylsulfonyl)butanoate, aromatic aldehydes, urea/thiourea, catalyst-free, solvent-free, Biginelli reaction, regioselectivity, ethyl 2-oxo/thioxo-4-aryl-6-(arylsulfonylmethyl)-1,2,3,4-tetrahydropyrimidine-5-carboxylates, substituted tetrahydropyrimidines (THPMs)

4.2.11.1 General Reaction Scheme

Harikrishnan *et al.*[82] designed a microwave-mediated catalyst- and solvent-free regioselective Biginelli reaction for the synthesis of a new series of highly functionalized tetrahydropyrimidines, ethyl 2-oxo/thioxo-4-aryl-6-(arylsulfonylmethyl)-1,2,3,4-tetrahydropyrimidine-5-carboxylates (**4**) *via* a one-pot three-component reaction of ethyl 3-oxo-4-(arylsulfonyl)butanoate (**1**), aromatic aldehyde (**2**) and urea/thiourea (**3**) (Scheme 4.12).

4.2.11.2 Plausible Mechanism

The investigators proposed a unique mechanistic pathway for this catalyst-free Biginelli reaction based on their experimental findings, the ouline of which is in Scheme 4.12. Interestingly, the reaction proceeds in a regioselective fashion as the product is formed by the reaction of a methylene group flanked by the keto and ester function of (**1**) leading to the generation of regioisomer (**4**) exclusively. The other methylene group adjacent to the arylsulfonyl moiety of **1** does not take part in the process, possibly due to the steric interaction between the arylsulfonyl group in the enol (**1′**) and the aryl ring of the imine (**6**), thereby restricting the formation of another regioisomer (**4′**) (Scheme 4.13).

4.2.11.3 Representative Examples

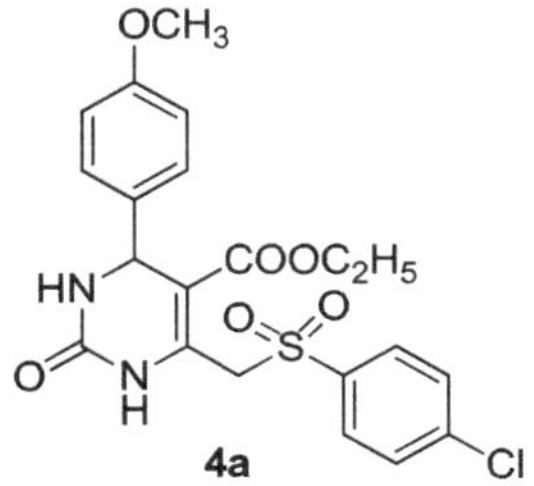

Ethyl 6-(((4-chlorophenyl)sulfonyl)methyl)-4-(4-methoxyphenyl)-2-oxo-1,2,3,4-tetrahydropyrimidine-5-carboxylate (**4a**): pale yellow solid, mp 222–223 °C; yield: 88%

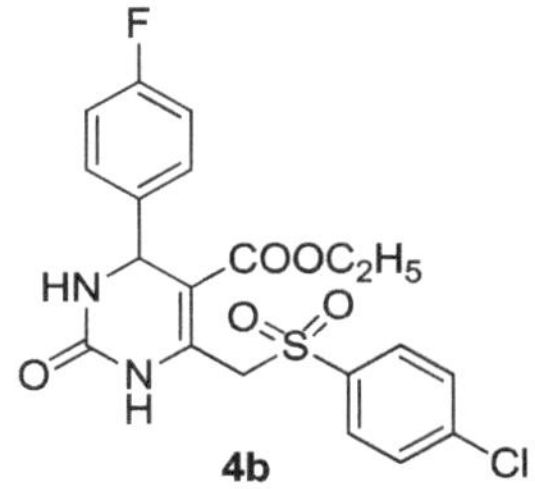

Ethyl 6-(((4-chlorophenyl)sulfonyl)methyl)-4-(4-fluorophenyl)-2-oxo-1,2,3,4-tetrahydropyrimidine-5-carboxylate (**4b**): pale yellow solid, mp 209–210 °C; yield: 80%

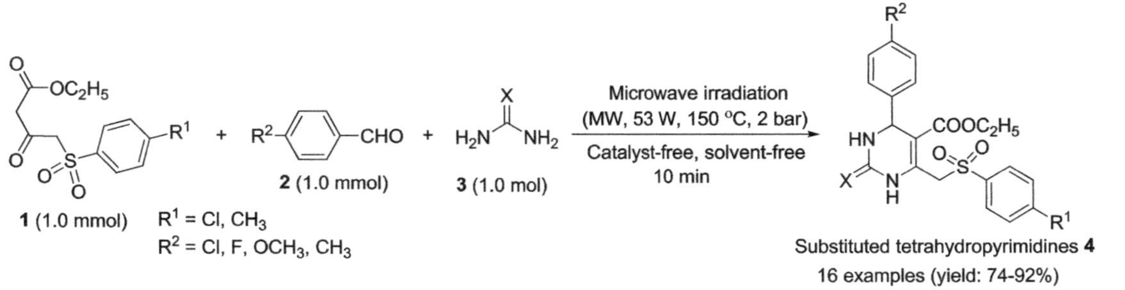

Scheme 4.12 Synthesis of substituted tetrahydropyrimidines (**4**).

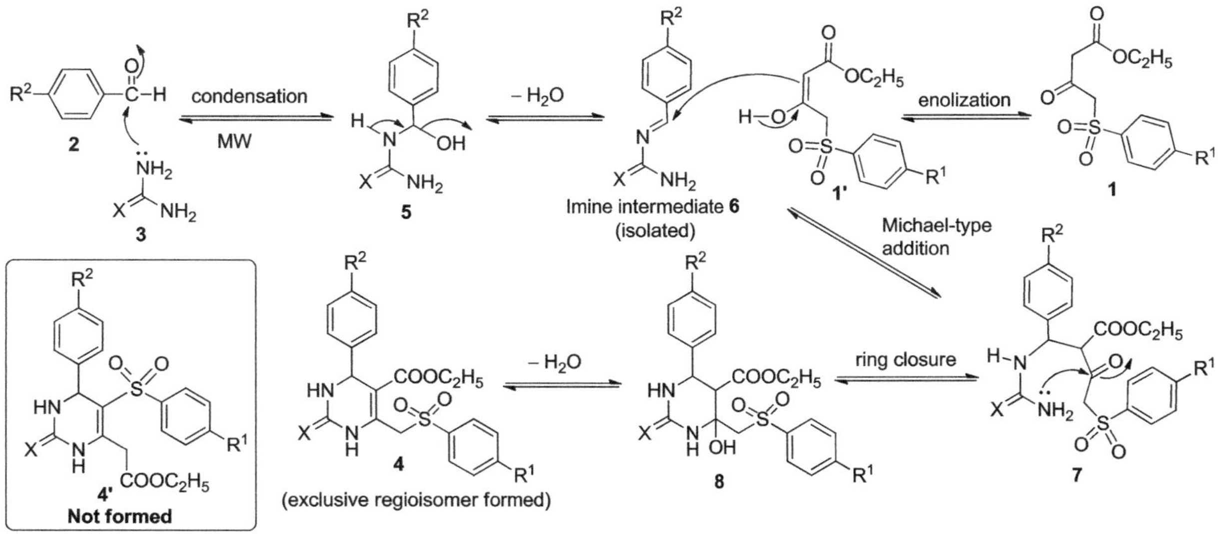

Scheme 4.13 Plausible mechanism for the formation of substituted tetrahydropyrimidines (**4**).

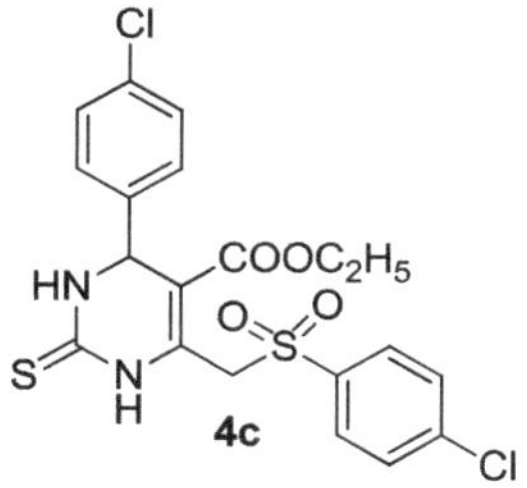

Ethyl 4-(4-chlorophenyl)-6-(((4-chlorophenyl)sulfonyl)methyl)-2-thioxo-1,2,3,4-tetrahydropyrimidine-5-carboxylate (**4c**): pale yellow solid, mp 201–202 °C; yield: 75%

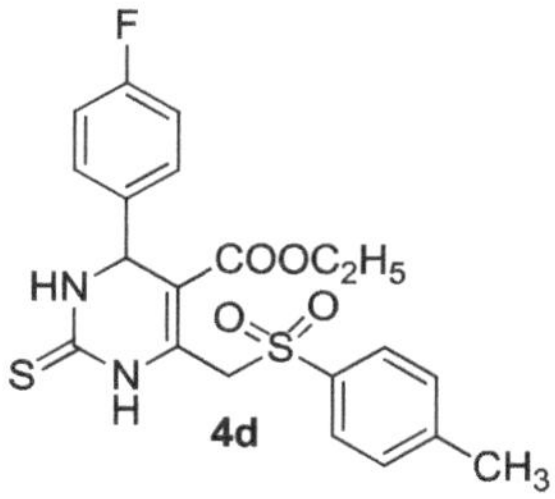

Ethyl 4-(4-fluorophenyl)-2-thioxo-6-(tosylmethyl)-1,2,3,4-tetrahydropyrimidine-5-carboxylate (**4d**): pale yellow solid, mp 204–205 °C; yield: 92%

4.2.11.4 Experimental Procedure

A mixture of ethyl 3-oxo-4-(arylsulfonyl)butanoate (**1**; 1 mmol), aromatic aldehyde (**2**; 1 mmol) and urea/thiourea (**3**; 1 mol) was taken in a 10 mL quartz vial, sealed and placed in a Biotage microwave oven (power 53 W), followed by microwave irradiation at 150 °C and 2 bar pressure for 10 min. Upon completion of the reaction, the reaction tube was cooled at room temperature, and the reaction mixture was diluted with water and extracted with dichloromethane (2×5 mL). The combined dichloromethane extracts were then dried over sodium sulfate, the solvent was removed under reduced pressure, and then purified by column chromatography over silica gel using ethyl acetate–petroleum ether (1:4 v/v) as the eluent to obtain the pure product (**4**). All the products were characterized based on their analytical and spectral studies.

4.2.11.5 Characterization Data of Two Representative Compounds

Ethyl 6-(((4-chlorophenyl)sulfonyl)methyl)-4-(4-methoxyphenyl)-2-oxo-1,2,3,4-tetrahydropyrimidine-5-carboxylate (**4a**): pale yellow solid, mp 222–223 °C; yield: 88%; ^{1}H NMR ($CDCl_3$, 300 MHz): δ 1.08 (3H, t, J = 7.2 Hz, –CH_3), 3.79 (3H, s, –OCH_3), 3.90 (2H, q, J = 7.2 Hz, –CH_2), 4.54 (1H, d, J = 13.8 Hz), 5.13 (1H, d, J = 13.8 Hz), 5.28 (1H, s), 5.99 (1H, s, –NH), 6.81 (2H, d, J = 8.7 Hz, Ar–*H*), 7.07 (2H, d, J = 8.7 Hz, Ar–*H*), 7.33 (2H, d, J = 8.7 Hz, Ar–*H*), 7.75 (2H, d, J = 8.7 Hz, Ar–*H*), 8.24 (1H, s, –NH); ^{13}C NMR ($CDCl_3$, 75 MHz): δ 13.9, 54.8, 55.3, 55.7, 60.7, 106.7, 114.0, 127.7, 129.1, 130.3, 134.7, 135.8, 136.2, 140.8, 152.5, 159.4, 164.1. Anal. calc. for $C_{21}H_{21}ClN_2O_6S$: C, 54.25; H, 4.55; N, 6.03%; found C, 54.32; H, 4.47; N, 6.13%.

Ethyl 4-(4-fluorophenyl)-2-thioxo-6-(tosylmethyl)-1,2,3,4-tetrahydropyrimidine-5-carboxylate (**4d**): pale yellow solid, mp 204–205 °C; yield: 92%; ^{1}H NMR

($CDCl_3$, 300 MHz): δ 1.08 (3H, t, $J = 7.2$ Hz, $-CH_3$), 2.26 (3H, s, $-CH_3$), 3.96 (2H, q, $J = 7.2$ Hz, $-CH_2$), 4.50 (1H, d, $J = 13.8$ Hz), 5.18 (1H, d, $J = 13.8$ Hz), 5.26 (1H, s), 6.04 (1H, s, –NH), 6.89 (2H, d, $J = 8.7$ Hz, Ar–*H*), 7.15 (2H, d, $J = 8.7$ Hz, Ar–*H*), 7.38 (2H, d, $J = 8.7$ Hz, Ar–*H*), 7.70 (2H, d, $J = 8.7$ Hz, Ar–*H*), 8.20 (1H, s, –NH); ^{13}C NMR ($CDCl_3$,75 MHz): δ 13.4, 21.6 54.5, 55.6, 60.2, 105.1, 114.6, 127.8, 129.6, 130.5, 134.6, 135.9, 136.9, 140.2, 152.7, 159.1, 164.5. Anal. calc. for $C_{21}H_{21}FN_2O_4S_2$: C, 56.23; H, 4.72; N, 6.25%; found C, 56.17; H, 4.80; N, 6.33%.

4.2.11.6 Critical Views

A novel series of biologically relevant tetrahydropyrimidine derivatives were successfully synthesized *via* a modified Biginelli reaction under catalyst- and solvent-free conditions under microwave irradiation. The process is expedient, high yielding, and clean. However, the scope of the products is limited.

4.2.12 Entry-12: Synthesis of 1-Carboxymethyl-5-trifluoromethyl-5-hydroxy-4,5-dihydro-1*H*-pyrazoles

Type of reaction: C–N bond formation
Reaction conditions: Catalyst-free, solvent-free, microwave irradiation
Synthetic strategy: Cyclocondensation reaction
Keywords: Enones, methyl hydrazinocarboxylate, catalyst-free, ethanol, condensation, 1-carboxymethyl-5-trifluoromethyl-5-hydroxy-4,5-dihydro-1*H*-pyrazoles

4.2.12.1 General Reaction Scheme

Martins *et al.*[83] accomplished the synthesis of a series of substituted 1-carboxymethyl-5-trifluoromethyl-5-hydroxy-4,5-dihydro-1*H*-pyrazoles (**3**) from the cyclocondensation reaction between enones (**1**) and methyl hydrazinocarboxylate (**2**) under the influence of microwave irradiation in the absence of any catalyst and solvent (Scheme 4.14).

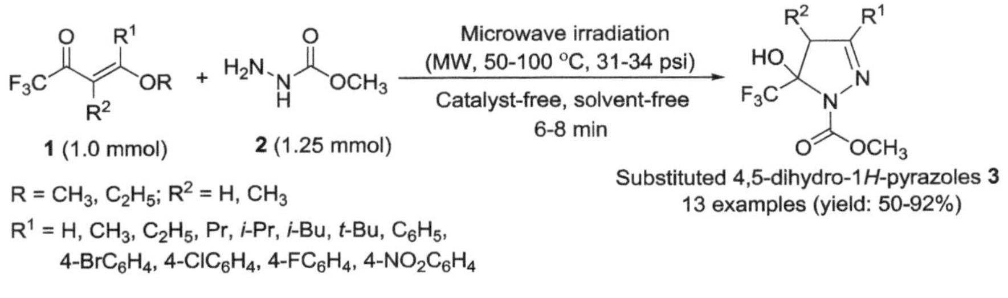

Scheme 4.14 Synthesis of 1-carboxymethyl-5-trifluoromethyl-5-hydroxy-4,5-dihydro-1*H*-pyrazoles (**3**).

4.2.12.2 Representative Examples

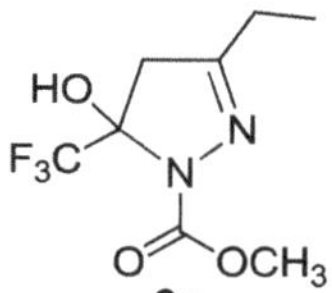

3a

Methyl 3-ethyl-5-hydroxy-5-(trifluoromethyl)-4,5-dihydro-1*H*-pyrazole-1-carboxylate (**3a**): solid, mp 82–84 °C; yield: 92%

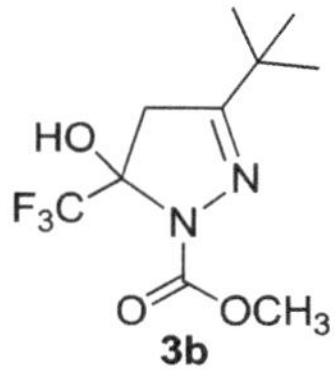

3b

Methyl 3-(*tert*-butyl)-5-hydroxy-5-(trifluoromethyl)-4,5-dihydro-1*H*-pyrazole-1-carboxylate (**3b**): solid, mp 97–99 °C; yield: 80%

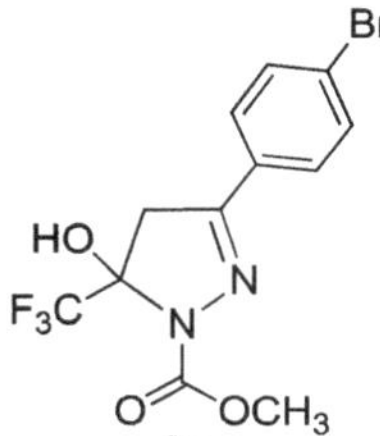

3c

Methyl 3-(4-bromophenyl)-5-hydroxy-5-(trifluoromethyl)-4,5-dihydro-1*H*-pyrazole-1-carboxylate (**3c**): solid, mp 148–151 °C; yield: 85%

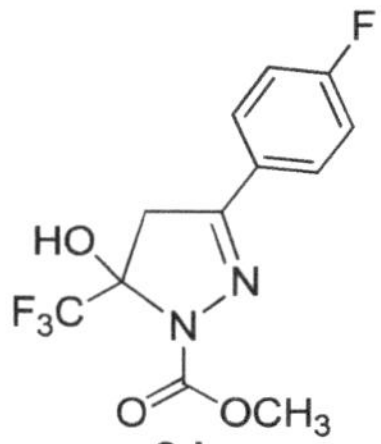

3d

Methyl 3-(4-fluorophenyl)-5-hydroxy-5-(trifluoromethyl)-4,5-dihydro-1*H*-pyrazole-1-carboxylate (**3d**): yellow solid, mp 145–147 °C; yield: 80%

4.2.12.3 Experimental Procedure

A mixture of enone derivatives (**1**; 1 mmol) and methyl hydrazinocarboxylate (**2**; 1.25 mmol) placed into a 10 mL reaction vessel equipped with a stirrer bar was irradiated at 50–100 °C using a microwave reactor (power 45–80 W; internal pressure 31–34 psi) for 6–8 min. After completion of the reaction, dichloromethane (10 mL) was added and the solution was washed with water (3×10 mL). The organic layer was dried with sodium sulfate and the solvent was removed under reduced pressure to furnish the product (**3**) in pure form without further purification. The products were characterized based on their spectral studies.

4.2.12.4 Characterization Data of Two Representative Compounds

Methyl 3-(tert-butyl)-5-hydroxy-5-(trifluoromethyl)-4,5-dihydro-1H-pyrazole-1-carboxylate (**3b**): solid, mp 97–99 °C; yield: 80%; ^{1}H NMR ($CDCl_3$, 400 MHz): δ 1.21 (9H, s, 3×-CH_3, H-10), 3.13 (1H, d, $^2J = 18$ Hz, H-4b), 3.29

(1H, d, $^2J = 18$ Hz, H-4a), 3.88 (3H, s, $-OCH_3$); ^{13}C NMR ($CDCl_3$, 100 MHz): δ 27.6 ($3\times-CH_3$), 42.5 (C4), 58.0 ($-OCH_3$), 90.9 (q, $^2J = 34$ Hz, C-5), 123.1 (q, $^1J = 286$ Hz, $-CF_3$), 153.9 (C-3), 164.2 (C=O); GCMS (*m*/*z*, rel%): 268 (M^+, 19), 199 (100), 167 (19), 140 (10).

Methyl 3-(4-fluorophenyl)-5-hydroxy-5-(trifluoromethyl)-4,5-dihydro-1*H*-pyrazole-1-carboxylate (**3d**): yellow solid, mp 145–147 °C; yield: 80%; ^1H NMR ($CDCl_3$, 400 MHz): δ 3.52 (1H, d, $^2J = 18$ Hz, H-4a), 3.66 (1H,d, $^2J = 18$ Hz, H-4b), 3.94 (3H, s, $-OCH_3$), 7.11 (2H, d, Ar–*H*), 7.72 (2H, d, Ar–*H*); ^{13}C NMR ($CDCl_3$, 100 MHz): δ 43.3 (C-4), 53.8 ($-OCH_3$), 91.3 (q, $^2J = 34$ Hz, C-5), 122.5 (q, $^1J = 286$ Hz, $-CF_3$), 165.6, 163.1, 128.8, 115.9, 151.8 (C-3), 153.9 (C=O); GCMS (*m*/*z*, rel%): 306 (M^+, 75), 237 (100), 218 (6).

4.2.12.5 Critical Views

The present method is an expedient alternative method for easy access to 1-carboxymethyl-5-trifluoromethyl-5-hydroxy-4,5-dihydro-1*H*-pyrazoles as useful synthons under microwave irradiation. Rapid synthesis, use of no organic solvent and catalyst, easy workup, and good to excellent yields are the key advantages of this method.

4.2.13 Entry-13: Synthesis of 5,6-Dihydropyrido[4,3-*d*]pyrimidines and Pyrido[4,3-*d*]pyrimidines

Type of reaction: C–C and C–N bond formation
Reaction conditions: Catalyst-free, solvent-free, microwave irradiation
Synthetic strategy: One-pot multicomponent reaction involving aza-Diels–Alder coupling
Keywords: 6-[2-(Dimethylamino)vinyl]-1,3-dimethyluracil, aryl/heteroaryl aldehydes, urea, *in situ* generated aldimines, catalyst-free, solvent-free, aza-Diels–Alder reaction, 5,6-dihydropyrido[4,3-*d*]pyrimidines, pyrido[4,3-*d*]pyrimidines

4.2.13.1 General Reaction Scheme

A microwave-assisted alternative route to access a series of 5,6-dihydropyrido[4,3-*d*]pyrimidine derivatives (**4**) as well as pyrido[4,3-*d*]pyrimidines (**5**) from a one-pot three-component reaction between 6-[2-(dimethylamino)vinyl]-1,3-dimethyluracil (**1**), aldehydes (**2**) and urea (**3**) in the absence of any catalyst or solvent was designed by Prajapati and his group[84] (Scheme 4.15). The key step in the reaction is *in situ* generation and trapping of the reactive aldimine formed from urea and aldehyde by the diene system of the uracil *via* an aza-Diels – Alder reaction.

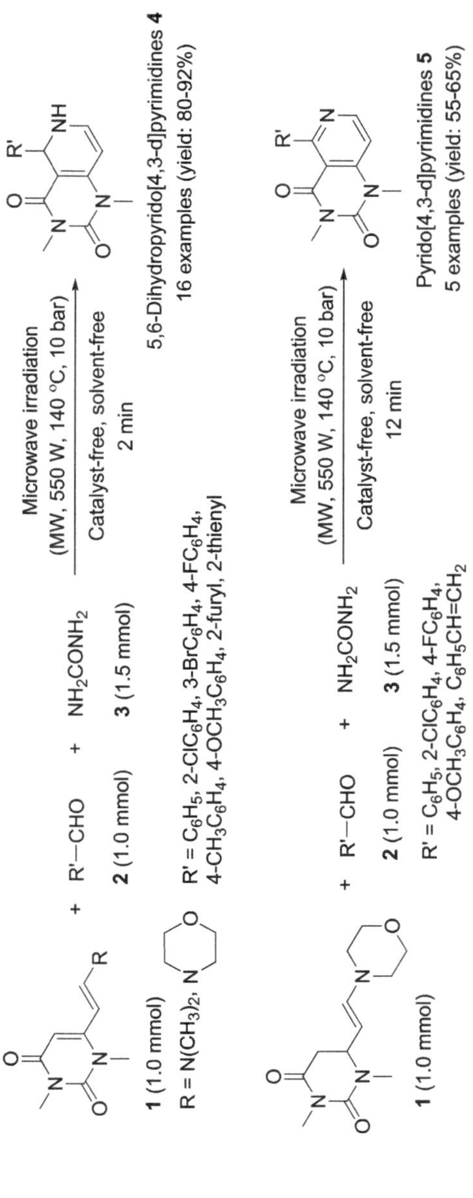

Scheme 4.15 Synthesis of 5,6-dihydropyrido[4,3-*d*]pyrimidines (**4**) and pyrido[4,3-*d*]pyrimidines (**5**).

4.2.13.2 Representative Examples

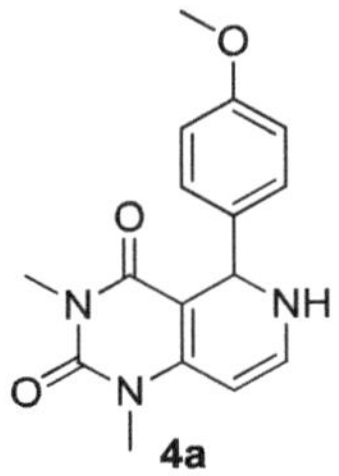

4a

5-(4-Methoxyphenyl)-1,3-dimethyl-5,6-dihydropyrido[4,3-*d*]pyrimidine-2,4(1*H*,3*H*)-dione (**4a**): yellow solid, mp 178–180 °C; yield: 90%

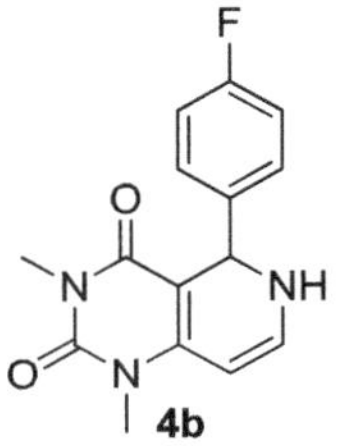

4b

5-(4-Fluorophenyl)-1,3-dimethyl-5,6-dihydropyrido[4,3-*d*]pyrimidine-2,4(1*H*,3*H*)-dione (**4b**): yellow solid, mp 280–282 °C; yield: 90%

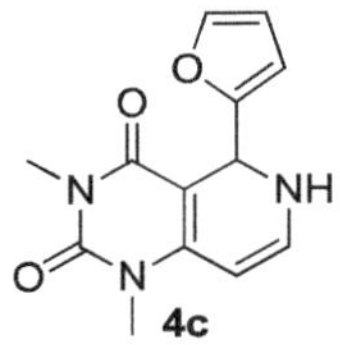

4c

5-(Furan-2-yl)-1,3-dimethyl-5,6-dihydropyrido[4,3-*d*]pyrimidine-2,4(1*H*,3*H*)-dione (**4c**): brown solid, mp 144–147 °C; yield: 80%

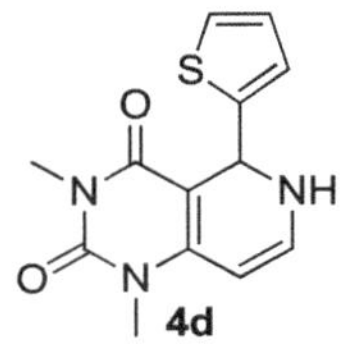

4d

1,3-Dimethyl-5-(thiophen-2-yl)-5,6-dihydropyrido[4,3-*d*]pyrimidine-2,4(1*H*,3*H*)-dione (**4d**): yellow solid, mp 125–126 °C; yield: 82%

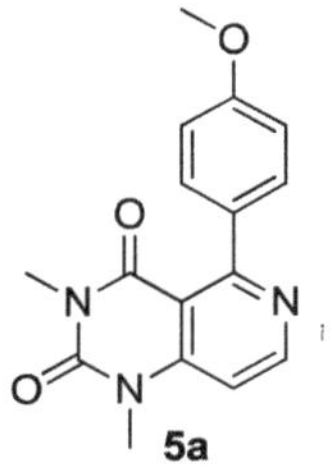

5a

5-(4-Methoxyphenyl)-1,3-dimethylpyrido[4,3-*d*]pyrimidine-2,4(1*H*,3*H*)-dione (**5a**): solid; yield: 55%

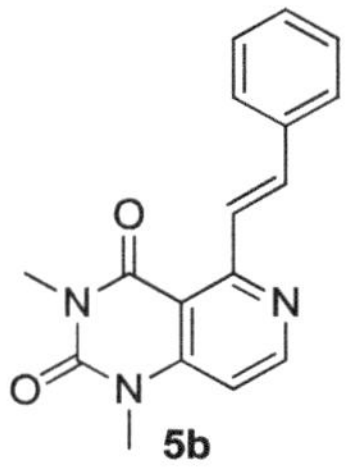

5b

1,3-Dimethyl-5-styrylpyrido[4,3-*d*]pyrimidine-2,4(1*H*,3*H*)-dione (**5b**): solid; yield: 65%

4.2.13.3 Experimental Procedures

A mixture of 6-[2-(dimethylamino)vinyl]-1,3-dimethyluracil (**1**; 1 mmol), aldehydes (**2**; 1 mmol) and urea (**3**; 1.5 mmol) was irradiated in a closed vessel in the absence of any solvent in a Synthos 3000 microwave reactor at 550 W, 140 °C, and 10 bar for 2 min. The crude product mixture was dissolved in ethyl acetate and directly column chromatographed using 9:1 ethyl acetate–hexane to afford pure products of 5,6-dihydropyrido[4,3-*d*]pyrimidines (**4**) in good to excellent yields. When the irradiation was continued up to 12 min (instead of 2 min), the dehydro-derivatives, pyrido[4,3-*d*]pyrimidines (**5**) were obtained *via* oxidative aromatization keeping the other reaction conditions unchanged. All the products were characterized on the basis of their analytical and spectral studies.

4.2.13.4 Characterization Data of Three Representative Compounds

5-(Furan-2-yl)-1,3-dimethyl-5,6-dihydropyrido[4,3-*d*]pyrimidine-2,4(1*H*,3*H*)-dione (**4c**): brown solid, mp 144–147 °C; yield: 80%; IR ($CHCl_3$): 3298.8, 2925.2, 1681.2, 1642.9, 1485.2 cm^{-1}; ^1H NMR ($CDCl_3$, 300 MHz): δ 7.34 (1H, d, *J* = 0.9 Hz), 6.75 (1H, t, *J* = 6.8 Hz), 6.29 (1H, m), 6.16 (1H, d, *J* = 3.1 Hz), 5.90 (1H, d, *J* = 2.5 Hz), 5.18 (1H, br s), 5.06 (1H, dd, *J* = 7.3 Hz), 3.43 (3H, s), 3.34 (3H, s); ^{13}C NMR ($CDCl_3$, 75 MHz): δ 160.7, 155.2, 152.1, 146.9, 143.0, 142.1, 110.3, 106.3, 91.5, 85.0, 46.7, 30.9, 28.0; MS (GCMS, *m/z*): 259 $[M]^+$. Anal. calc. for $C_{13}H_{13}N_3O_3$: C, 60.22; H, 5.05; N, 16.21; found: C, 60.30; H, 5.03; N, 16.18.

1,3-Dimethyl-5-(thiophen-2-yl)-5,6-dihydropyrido[4,3-*d*]pyrimidine-2,4(1*H*,3*H*)-dione (**4d**): yellow solid, mp 125–126 °C; yield: 82%; IR ($CHCl_3$): 3330.4, 2926.7, 1710.7, 1664.1, 1642.7, 1478.6 cm^{-1}; ^1H NMR ($CDCl_3$, 300 MHz): δ 7.16 (1H, m), 7.05 (1H, m), 6.93 (1H, m), 6.82 (1H, t, *J* = 6.8 Hz), 6.13 (1H, d, *J* = 2.8 Hz), 5.26 (1H, br s), 5.08 (1H, dd, *J* = 7.2 Hz), 3.42 (3H, s), 3.33 (3H, s); ^{13}C NMR ($CDCl_3$, 75 MHz): δ 160.2, 151.9, 150.6, 148.0, 141.6, 130.4, 128.6, 127.1, 108.6, 94.9, 56.5, 31.0, 28.9; MS (GCMS): *m/z* 275 $[M]^+$. Anal. calc. for $C_{13}H_{13}N_3O_2S$: C, 56.71; H, 4.76; N, 15.26; found: C, 56.76; H, 4.72; N, 15.22.

1,3-Dimethyl-5-styrylpyrido[4,3-*d*]pyrimidine-2,4(1*H*,3*H*)-dione (**5b**): solid; yield: 65%; IR ($CHCl_3$): 3024.9, 2924.9, 1707.8, 1660.1, 1577.5, 1473.4 cm^{-1}; ^1H NMR ($CDCl_3$, 300 MHz): δ 8.85 (1H, d, *J* = 15.6 Hz), 8.67 (1H, d, *J* = 5.7 Hz), 8.01 (1H, d, *J* = 15.6 Hz), 7.71 (1H, d, *J* = 6.9 Hz,), 7.41–7.32 (4H, m), 6.96 (1H, d, *J* = 5.7 Hz), 3.60 (3H, s), 3.50 (3H, s); ^{13}C NMR ($CDCl_3$, 75 MHz): δ 161.5, 158.1, 152.4, 150.6, 147.5, 138.0, 136.6, 129.0, 128.7, 127.9, 125.7, 107.6, 106.2, 31.1, 28.7; MS (GCMS): *m/z* 293 $[M]^+$. Anal. calc. for $C_{17}H_{15}N_3O_2$: C, 69.61; H, 5.15; N, 14.33; found: C, 69.56; H, 5.28; N, 14.25.

4.2.13.5 Critical Views

The investigators constructed a good deal of dihydropyrido[4,3-*d*]pyrimidines and their aromatized analogs (pyrido[4,3-*d*]pyrimidines) by exploiting an efficient aza-Diels–Alder strategy in an improved and more eco-friendly protocol than previously reported procedures.[85–93] This was without the aid of any catalyst or solvent and under the influence of microwave irradiation. The present protocols offer an expedient synthesis, good yields and a clean reaction profile.

4.2.14 Entry-14: Synthesis of 2,4,5-Triarylimidazoles

Type of reaction: C–N bond formation
Reaction conditions: Catalyst-free, solvent-free, microwave irradiation
Synthetic strategy: One-pot multicomponent reaction
Keywords: Benzils, aromatic aldehydes, ammonium acetate, catalyst-free, solvent-free, one-pot multicomponent reaction, 2,4,5-triarylimidazoles

4.2.14.1 General Reaction Scheme

Zhou *et al.*[94] synthesized a series of 2,4,5-triarylimidazole derivatives (**4**) from a microwave-assisted one-pot condensation reaction between benzils (**1**), aromatic aldehydes (**2**) and ammonium acetate (**2**) under catalyst- and solvent-free conditions with excellent yields and within a short reaction time of minutes (Scheme 4.16).

4.2.14.2 Representative Examples

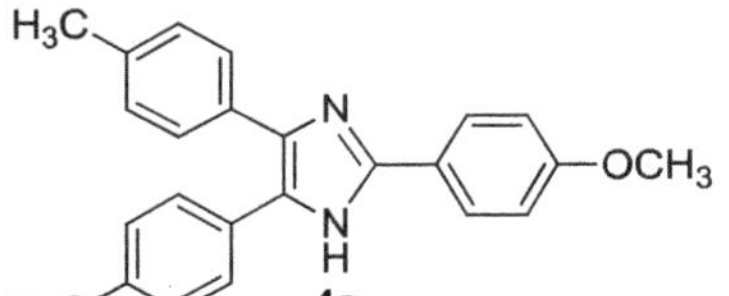

4a

2-(4-Methoxyphenyl)-4,5-di-*p*-tolyl-1*H*-imidazole (**4a**): solid, mp 250–251 °C; yield: 96%

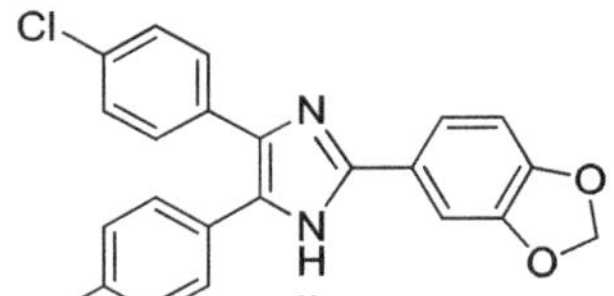

4b

2-(Benzo[*d*][1,3]dioxol-5-yl)-4,5-*bis*(4-chlorophenyl)-1*H*-imidazole (**4b**): solid, mp 274–276 °C; yield: 95%

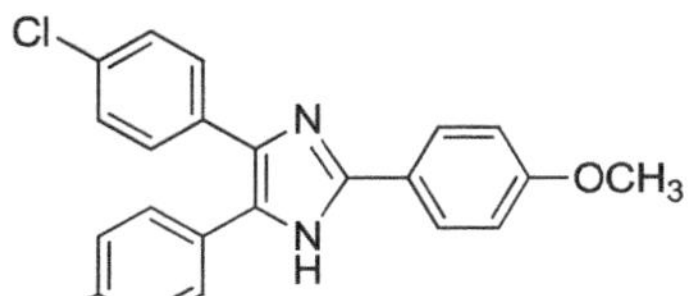

4c

4,5-*Bis*(4-chlorophenyl)-2-(4-methoxyphenyl)-1*H*-imidazole (**4c**): solid, mp 258–259 °C; yield: 91%

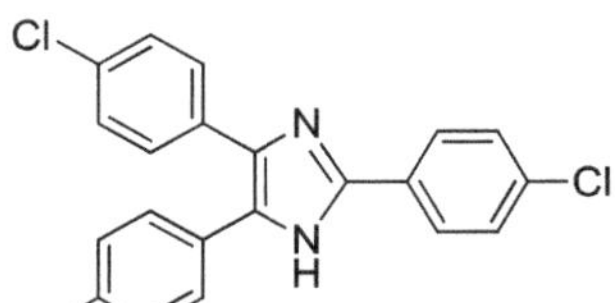

4d

2,4,5-*Tris*(4-chlorophenyl)-1*H*-imidazole (**4d**): solid, mp 271–272 °C; yield: 80%

4.2.14.3 Experimental Procedure

A mixture of benzils (**1**; 1 mmol), aromatic aldehyde (**2**; 1 mmol) and ammonium acetate (**3**; 3 mmol) taken within a sealed vial was placed into the cavity of the microwave reactor (power 150 W) and irradiated at 120 °C for 3–5 min. After completion of the reaction (as indicated by TLC), the reaction mixture was then allowed to cool to room temperature when solid residues precipitated out. These crude products were filtered off, dried and purified by recrystallization from 95% ethanol to give the pure products of substituted triarylimidazoles (**3**). All the products were characterized on the basis of their analytical and spectral studies.

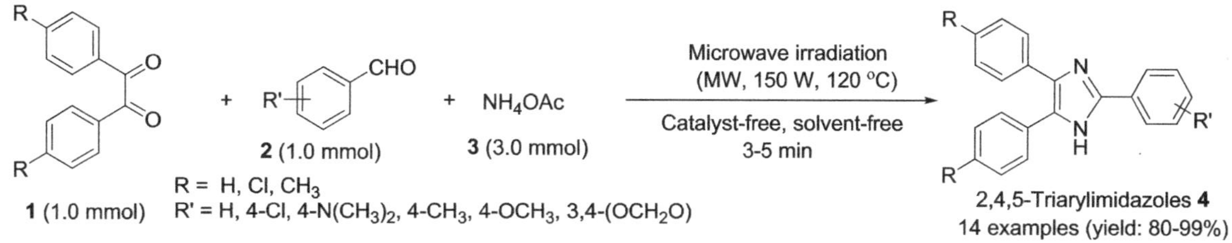

Scheme 4.16 Synthesis of 2,4,5-triarylimidazoles (**4**).

4.2.14.4 Characterization Data of Two Representative Compounds

2-(Benzo[*d*][1,3]dioxol-5-yl)-4,5-*bis*(4-chlorophenyl)-1*H*-imidazole (**4b**): solid, mp 274–276 °C; yield: 95%; IR (KBr): 3425, 3078, 2891, 1479, 1237, 1091, 1039, 833, 740 cm^{-1}; ^1H NMR ($CDCl_3$, 400 MHz): δ 7.45 (4H, d, $J=7.20$ Hz), 7.42 (1H, s), 7.35–7.32 (5H, m), 6.88 (1H, d, $J=8.00$ Hz), 6.04 (2H, s, $-OCH_2O-$); ^{13}C NMR (DMSO-d_6, 100 MHz): δ 148.1, 146.4, 136.6, 134.3, 132.9, 131.6, 130.5, 130.1, 129.3, 129.2, 128.8, 127.4, 124.9, 120.0, 109.0, 106.1, 101.8; MS (%rel): *m/z* $(M)^+$ 409.5 (100). Anal. calc. for $C_{22}H_{14}Cl_2N_2O_2$: C, 64.56; H, 3.45; N, 6.84; found: C, 64.30; H, 3.46; N, 6.77.

4,5-*Bis*(4-chlorophenyl)-2-(4-methoxyphenyl)-1*H*-imidazole (**4c**): solid, mp 258–259 °C; yield: 91%; IR (KBr): 3427, 3068, 2961, 1614, 1496, 1441, 1248, 830, 742 cm^{-1}; ^1H NMR ($CDCl_3$, 400 MHz): δ 7.86 (2H, d, $J=8.4$ Hz), 7.46 (4H, d, $J=8.0$ Hz), 7.32 (4H, d, $J=8.0$ Hz), 6.99 (2H, d, $J=8.8$ Hz), 3.88 (3H, s, $-OCH_3$); ^{13}C NMR (DMSO-d_6, 100 MHz): δ 160.1, 146.6, 136.5, 136.5, 134.4, 132.8, 131.6, 130.5, 130.2, 129.2, 129.2, 128.8, 127.7, 127.3, 123.3, 114.6; MS (%rel): *m/z* $(M)^+$ 395.6 (100). Anal. calc. for $C_{22}H_{16}Cl_2N_2O$: C, 66.85; H, 4.08; N, 7.09; found: C, 66.59; H, 4.10; N, 7.02.

4.2.14.5 Critical Views

The present method offers an alternative way for the synthesis of 2,4,5-triarylimidazole derivatives. Rapid and operational simplicity, high yields, use of no catalyst or solvent, easy workup, and eco-friendliness are the major advantages of this protocol over the other previously reported methods involving a variety of catalysts, solvents and harsh reaction conditions.[95–102]

4.2.15 Entry-15: Synthesis of 7-Amino-substituted Pyrazolo[1,5-*a*][1,3,5]triazine-8-carbonitriles

Type of reaction: C–N bond formation
Reaction conditions: Catalyst-free, methanol, microwave irradiation
Synthetic strategy: One-pot multicomponent reaction
Keywords: 3-Amino-substituted 5-aminopyrazole-4-carbonitriles, cyanamide, triethyl orthoformate, catalyst-free, methanol, one-pot multicomponent reaction, 7-amino-substituted pyrazolo[1,5-*a*][1,3,5]triazine-8-carbonitriles

4.2.15.1 General Reaction Scheme

Lim *et al.*[103] synthesized a series of 7-amino-substituted pyrazolo[1,5-*a*][1,3,5]triazine-8-carbonitriles (**4**) *via* a three-component reaction of 3-amino-substituted 5-aminopyrazole-4-carbonitriles (**1**), triethyl orthoformate (**2**) and cyanamide (**3**) under microwave irradiation in methanol without the aid of any added catalyst (Scheme 4.17).

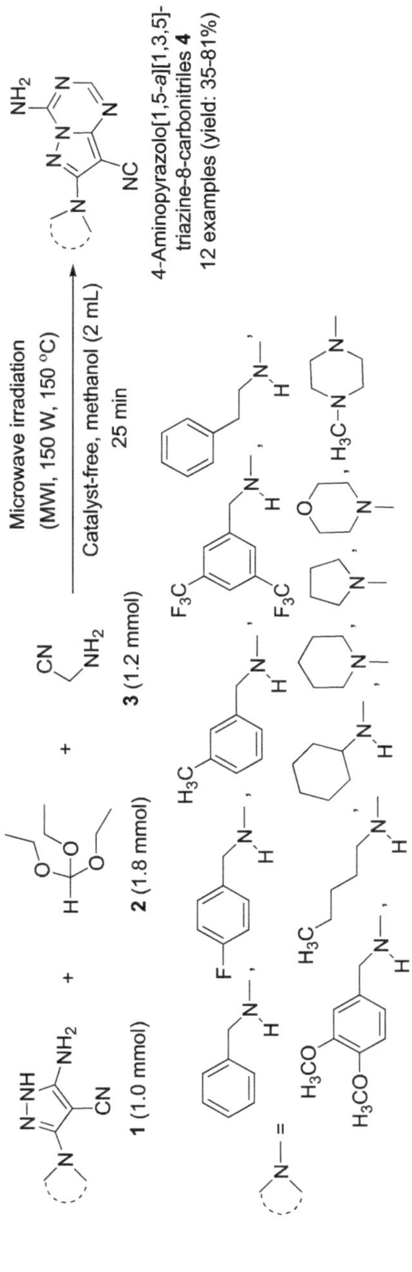

Scheme 4.17 Synthesis of 7-amino-substituted pyrazolo[1,5-*a*][1,3,5]triazine-8-carbonitriles (**4**).

4.2.15.2 Representative Examples

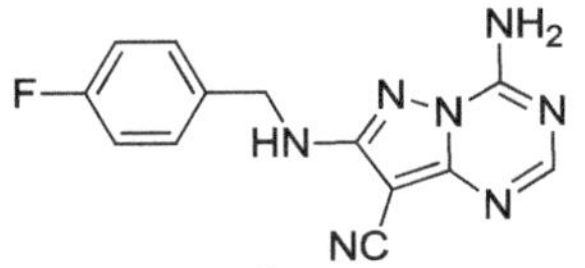

4a

4-Amino-7-((4-fluorobenzyl)amino)pyrazolo[1,5-*a*][1,3,5]triazine-8-carbonitrile (**4a**): solid, mp 258–260 °C; yield: 63%

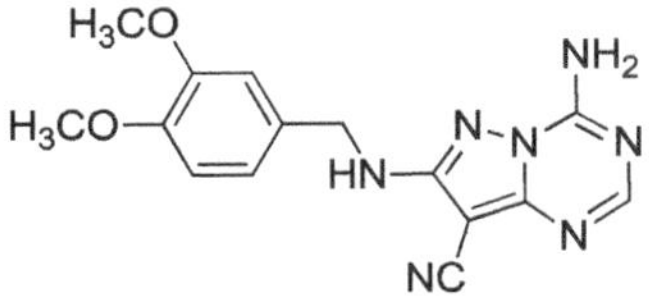

4b

4-Amino-7-((3,4-dimethoxybenzyl)amino)pyrazolo[1,5-*a*][1,3,5]triazine-8-carbonitrile (**4b**): solid, mp 266–268 °C; yield: 71%

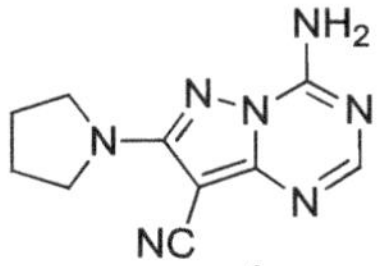

4c

4-Amino-7-(pyrrolidin-1-yl)pyrazolo[1,5-*a*][1,3,5]triazine-8-carbonitrile (**4c**): solid, mp 273–275 °C; yield: 81%

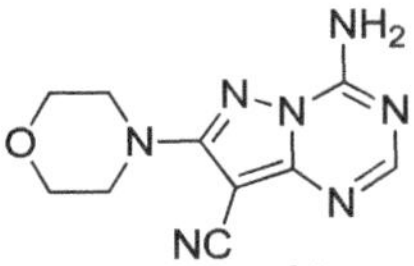

4d

4-Amino-7-morpholinopyrazolo[1,5-*a*][1,3,5]triazine-8-carbonitrile (**4d**): solid, mp 305–307 °C; yield: 40%

4.2.15.3 Experimental Procedure

A mixture of 3-amino-substituted 5-aminopyrazole-4-carbonitriles (**1**; 1.0 mmol), triethyl orthoformate (**2**; 0.3 mL, 1.8 mmol) and cyanamide (**3**; 51 mg, 1.2 mmol) in methanol (2 mL) were irradiated in a 10 mL seamless pressure vial using a microwave system operating at maximal microwave power up to 150 W at 150 °C for 25 min. Upon completion of the reaction, the reaction mixture was cooled to room temperature. The solvent was then evaporated and the product was isolated by column chromatography on silica gel using an ethyl acetate–hexane mixture as eluents to afford pure products of 7-amino-substituted pyrazolo[1,5-*a*][1,3,5]triazine-8-carbonitriles (**4**). All the products were identified based on analytical and spectral studies.

4.2.15.4 Characterization Data of Two Representative Compounds

4-Amino-7-((4-fluorobenzyl)amino)pyrazolo[1,5-*a*][1,3,5]triazine-8-carbonitrile (**4a**): solid, mp 258–260 °C; yield: 63%; IR (ATR): 3460 (N–H), 3359 (N–H) 3027 (C–H), 2219 (C≡N), 1669, 1583, 1558, 1441, 1357 cm^{-1}; ^{1}H NMR (DMSO-d_6, 400 MHz): δ 4.46 (2H, d, $J = 6.2$ Hz, –CH_2), 7.13 (2H, dd, $J = 8.9$ Hz, $^{3}J_{HF} = 8.9$ Hz, H-3′ and H-5′), 7.48 (2H, dd, $J = 5.7$ Hz, $^{3}J_{HF} = 8.6$ Hz, H-2′ and H-6′), 7.63 (1H, t, $J = 6.2$ Hz, –NH), 8.04 (1H, br s, –NH), 8.07 (1H, s, H-2), 8.77 (1H, br s, –NH); ^{13}C NMR (DMSO-d_6, 100 MHz): δ 44.7 (CH_2), 66.4 (C-8), 113.5 (C≡N), 114.8 (d, $^{2}J_{CF} = 21.2$ Hz, C-3′ and C-5′), 129.9 (d,

$^3J_{CF}=8.1$ Hz, C-2′ and C-6′), 135.7 (d, $^4J_{CF}=3.0$ Hz, C-1′), 149.5 (C-8a), 153.4 (C-4), 156.6 (C-2), 158.8 (C-7), 161.2 (d, $^1J_{CF}=242.1$ Hz, C-4′). Anal. calc. for $C_{13}H_{10}FN_7$: C, 55.12; H, 3.56; N, 34.61; found: C, 55.02; H, 3.64; N, 34.50.

4-Amino-7-morpholinopyrazolo[1,5-*a*][1,3,5]triazine-8-carbonitrile (**4d**): solid, mp 305–307 °C; yield: 40%; IR (ATR): 3328 (N–H), 3257 (N–H), 3169 (C–H), 2219 (C≡N), 1657, 1570, 1508, 1446, 1372 cm^{-1}; ^{1}H NMR (DMSO-d_6, 400 MHz): δ 3.54–3.56 (4H, m, $(CH_2)_2$), 3.74–3.77 (4H, m, $(CH_2)_2$), 8.14 (1H, s, H-2), 8.27 (1H, br s, –NH), 8.86 (1H, br s, –N*H*); ^{13}C NMR (DMSO-d_6, 100 MHz): δ 46.6 (CH_2NCH_2), 65.4 (CH_2OCH_2), 67.1 (C-8), 114.5 (C≡N), 149.6 (C-8a), 154.4 (C-4), 156.9 (C-2), 159.5 (C-7). Anal. calc. for $C_{10}H_{11}N_7O$: C, 48.98; H, 4.52; N, 39.98; found: C, 48.85; H, 4.63; N, 39.86.

4.2.15.5 Critical Views

The investigators offered a route for catalyst-free synthesis of a novel series of 7-amino-substituted 4-aminopyrazolo[1,5-*a*][1,3,5]triazines as useful building blocks *via* a one-pot, three-component reaction under microwave irradiation with moderate yields within a short reaction time. The structural similarity of the prepared compounds to substituted 8-aminoadenines may invoke interest in exploring their biological profiles.[103]

4.2.16 Entry-16: Synthesis of Spiroindenotetrahydropyridines

Type of reaction: C–C and C–N bond formation
Reaction conditions: Catalyst-free, solvent-free, microwave irradiation
Synthetic strategy: One-pot multicomponent reaction *via* tandem synthesis
Keywords: 1,3-Indanedione, aromatic aldehydes, ammonium acetate, catalyst-free, solvent-free, tandem Knoevenagel/aza-Diels–Alder reaction, spiroindenotetrahydropyridine derivatives

4.2.16.1 General Reaction Scheme

Bhuyan *et al.*[104] reported on a microwave-assisted, one-pot pseudo-five-component reaction for the synthesis of some novel spiroindenotetrahydropyridines (**4**) from 1,3-indanedione (**1**), aromatic aldehydes (**2**), and ammonium acetate (**3**) under catalyst- and solvent-free conditions *via* a tandem Knoevenagel/aza-Diels–Alder reaction strategy (Scheme 4.18).

4.2.16.2 Plausible Mechanism

The investigators proposed a mechanism for this microwave-assisted catalyst-free transformation (Scheme 4.19) that is supposed to be involved in a tandem Knoevenagel/aza-Diels–Alder reaction strategy where both diene and dienophile are generated *in situ*. This type of catalyst- and solvent-free aza-Diels–Alder pathway is rare in the literature.[104]

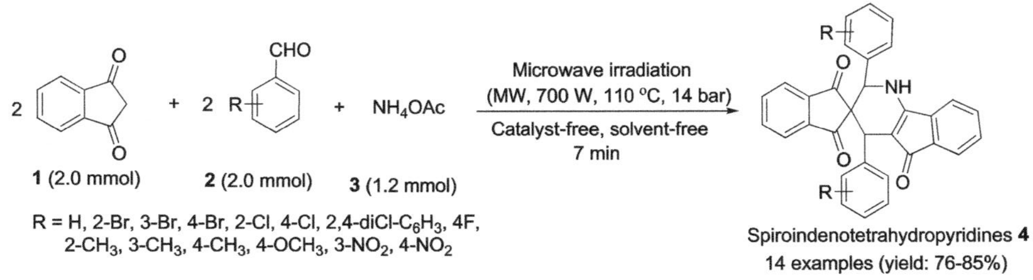

Scheme 4.18 Synthesis of spiroindenotetrahydropyridines (**4**).

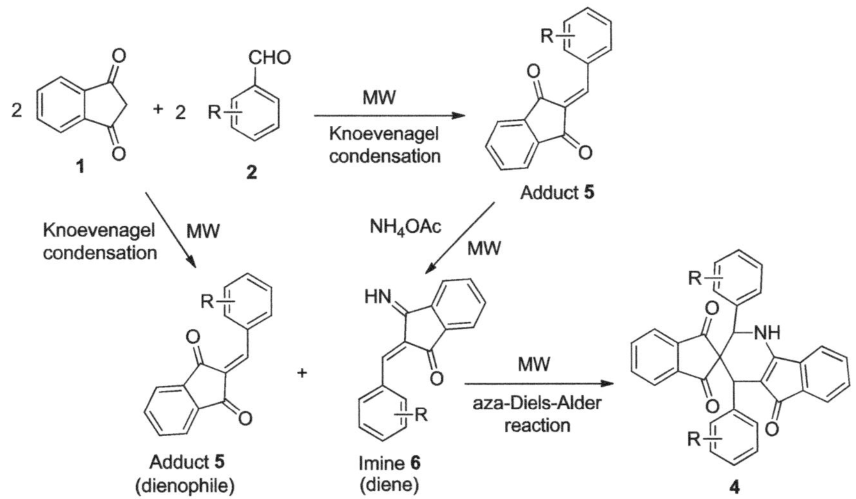

Scheme 4.19 Plausible mechanism for the synthesis of spiroindenotetrahydropyridines (**4**).

4.2.16.3 Representative Examples

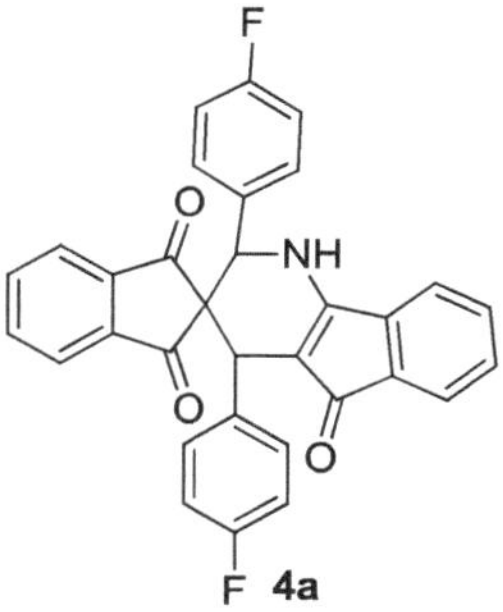

4a

2',4'-*Bis*(4-fluorophenyl)-1',2'-dihydrospiro[indene-2,3'-indeno[1,2-*b*]pyridine]-1,3,5'(4'*H*)-trione (**4a**): orange solid, mp 258–259 °C; yield: 85%

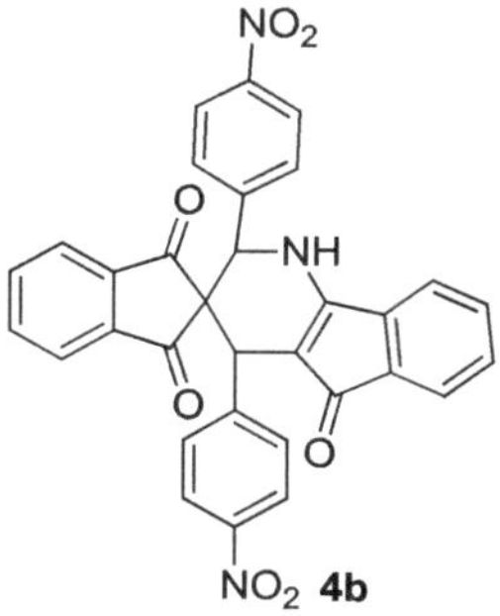

4b

2',4'-*Bis*(4-nitrophenyl)-1',2'-dihydrospiro[indene-2,3'-indeno[1,2-*b*]pyridine]-1,3,5'(4'*H*)-trione (**4b**): yield: 78%

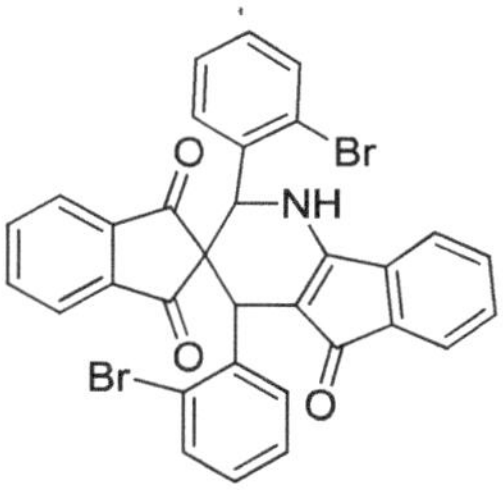

4c

Bis(2-bromophenyl)-1',2'-dihydrospiro[indene-2,3'-indeno[1,2-*b*]pyridine]-1,3,5'(4'*H*)-trione (**4c**): yield: 83%

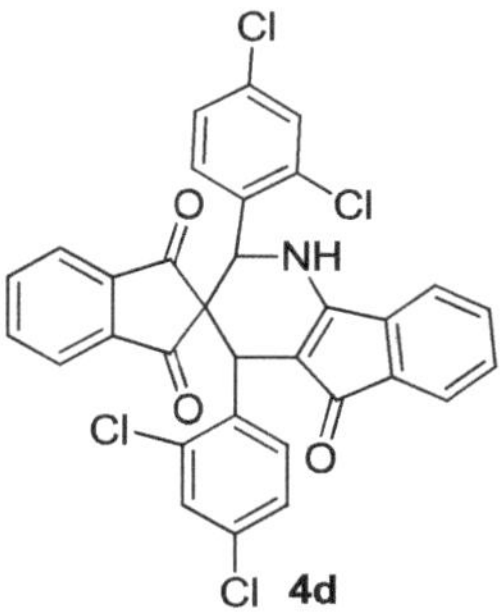

4d

Bis(2,4-dichlorophenyl)-1',2'-dihydrospiro[indene-2,3'-indeno[1,2-*b*]pyridine]-1,3,5'(4'*H*)-trione (**4d**): yield: 76%

4.2.16.4 Experimental Procedure

A mixture of 1,3-indanedione (**1**; 2.0 mmol), aromatic aldehydes (**2**; 2 mmol) and ammonium acetate (**3**; 1.2 mmol) taken in a closed vessel was irradiated within a Synthos 3000 microwave reactor (700 W, 14 bar) at 110 °C for 7 min. Upon completion of the reaction, the reaction tube was cooled at room temperature, and the crude product mixture was subjected to column chromatographic purification using 3:7 ethyl acetate–hexane as the eluent to afford pure products of spiroindenotetrahydropyridines (**4**). All the products were characterized based on their analytical and spectral studies.

4.2.16.5 Characterization Data of One Representative Compound

2′,4′-*Bis*(4-fluorophenyl)-1′,2′-dihydrospiro[indene-2,3′-indeno[1,2-*b*]pyridine]-1,3,5′(4′*H*)-trione (**4a**): orange solid, mp 258–259 °C; yield 85%; IR

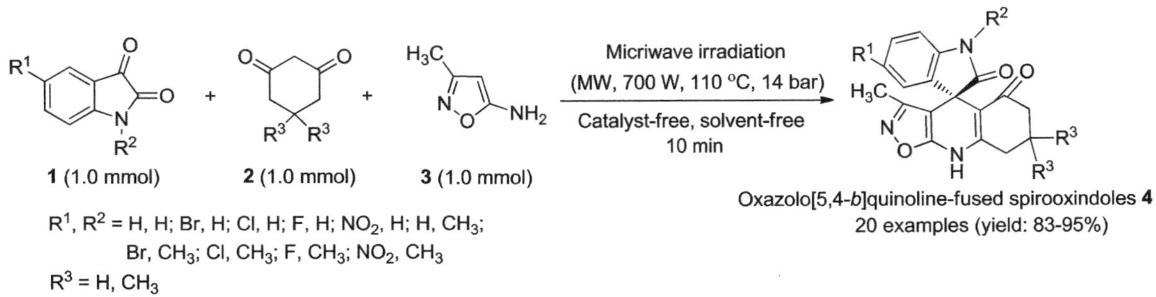

Scheme 4.20 Synthesis of oxazolo[5,4-*b*]quinoline-fused spirooxindoles (**4**).

($CHCl_3$): 3415, 1735, 1703, 1659, 1547, 1475, 1429 cm^{-1}; ^{1}H NMR (DMSO-d_6, 300 MHz): δ 9.37 (1H, br s), 7.73–6.77 (16H m, Ar–*H*), 5.33 (1H, s), 4.54 (1H, s); ^{13}C NMR (DMSO-d_6, 75 MHz): δ 201.3, 198.4, 187.3, 163.4, 143.1, 141,7, 136.7, 136.4, 136.2, 132.6, 132.5, 131.4, 131.3, 131.2, 131.1, 130.9, 130.8, 130.7, 122.5, 122.3, 120.0, 119.5, 115.7, 115.5, 114.6, 114.4, 100.8, 61.6, 60.5, 44.2; GCMS: *m/z* 503 [M]$^+$. Anal. calc. for $C_{32}H_{19}F_2NO_3$: C, 76.33; H, 3.80; N, 2.78; found: C, 76.28; H, 3.82; N, 2.80.

4.2.16.6 Critical Views

The present protocol offers an easy route for the rapid synthesis of a novel series of biologically relevant spiroindenotetrahydropyridines *via* tandem Knoevenagel/aza-Diels–Alder reaction strategy. Expedient conversion, use of no catalyst and solvent, moderate to good yields and clean reaction profiles are the key advantages of this method.

4.2.17 Entry-17: Synthesis of Oxazolo[5,4-*b*]quinoline-fused Spirooxindoles

Type of reaction: C–C and C–N bond formation
Reaction conditions: Catalyst-free, solvent-free, microwave irradiation
Synthetic strategy: One-pot multicomponent reaction *via* tandem synthesis
Keywords: Isatins, 1,3-diketones, 5-amino-3-methylisoxazole, catalyst-free, solvent-free, tandem Knoevenagel/Michael addition, chemoselectivity, oxazolo[5,4-*b*]quinoline-fused spirooxindoles

4.2.17.1 General Reaction Scheme

Reddy and his group[105] developed a microwave-assisted synthesis of a novel series of diversely substituted oxazolo[5,4-*b*]quinoline-fused spirooxindoles (**4**) *via* three-component tandem Knoevenagel/Michael addition reaction of isatins (**1**), β-diketones (**2**) and 5-amino-3-methylisoxazole (**3**) in a chemoselective fashion with good to excellent yields under catalyst- and solvent-free conditions (Scheme 4.20).

4.2.17.2 Representative Examples

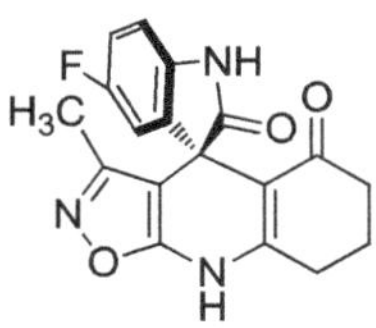

4a

(4*S*)-5'-Fluoro-3-methyl-1',2',6,7,8,9-hexahydro-5*H*-spiro[[1,2]oxazolo[5,4-*b*]quinoline-4,3'-indole]-2',5-dione (**4a**): white solid, mp 212–215 °C; yield: 88%

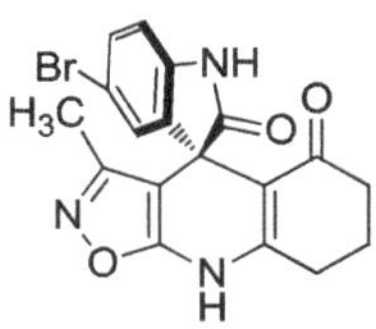

4b

(4*S*)-5'-Bromo-3-methyl-1',2',6,7,8,9-hexahydro-5*H*-spiro[[1,2]oxazolo[5,4-*b*]quinoline-4,3'-indole]-2',5-dione (**4b**): white solid, mp 209–212 °C; yield: 89%

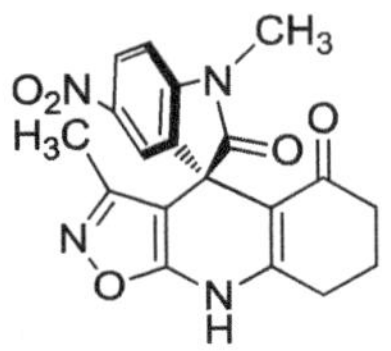

4c

[(4*S*)-1',3-Dimethyl-2',5-dioxo-1',2',6,7,8,9-hexahydro-5*H*-spiro[[1,2]oxazolo[5,4-*b*]quinoline-4,3'-indole]-5'-yl]azinic acid (**4c**): white solid, mp 220–223 °C; yield: 92%

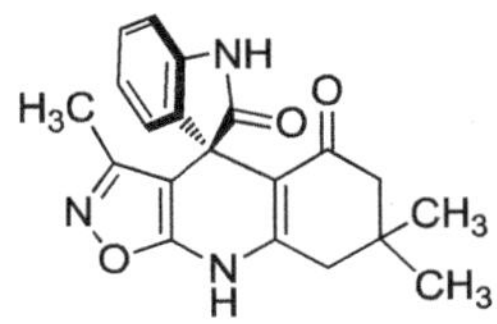

4d

(4*S*)-3,7,7-Trimethyl-1',2',6,7,8,9-hexahydro-5*H*-spiro[[1,2]oxazolo[5,4-*b*]quinoline-4,3'-indole]-2',5-dione (**4d**): white solid, mp 210–213 °C; yield: 87%

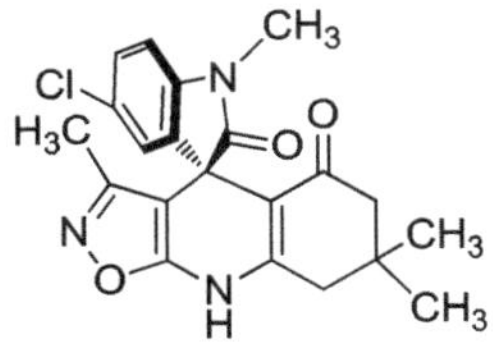

4e

(4*S*)-5′-Chloro-1′,3,7,7-tetramethyl-1′,2′,6,7,8,9-hexahydro-5*H*-spiro[[1,2]oxazolo[5,4-*b*]quinoline-4,3′-indole]-2′,5-dione (**4e**): white solid, mp 230–232 °C; yield: 86%

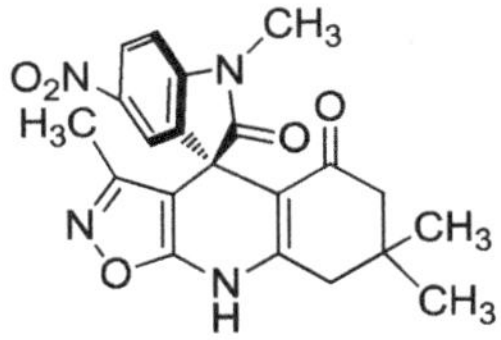

4f

[(4*S*)-1',3,7,7-Tetramethyl-2',5-dioxo-1',2',6,7,8,9-hexahydro-5*H*-spiro[[1,2]oxazolo [5,4-*b*]quinoline-4,3'-indole]-5'-yl]azinic acid (**4f**): white solid, mp 215–217 °C; yield: 89%

4.2.17.3 Experimental Procedure

A mixture of isatin (**1**; 1 mmol), β-diketones (**2**; 1 mmol) and 5-amino-3-methylisoxazole (**3**; 1 mmol) was irradiated in a closed vessel in the absence of any solvent in a Synthos 3000 microwave reactor at 700 W, 14 bar, and 110 °C for 10 min. Upon completion of the reaction (TLC monitored), the reaction mixture was filtered hot and the resulting solid products were washed with ethanol, dried in air and recrystallized from ethanol. All the products were characterized based on their spectral studies.

4.2.17.4 Characterization Data of Two Representative Compounds

(4*S*)-5′-Fluoro-3-methyl-1′,2′,6,7,8,9-hexahydro-5*H*-spiro[[1,2]oxazolo[5,4-*b*]quinoline-4,3′-indole]-2′,5-dione (**4a**): white solid, mp 212–215 °C; yield: 88%; ^{1}H NMR (DMSO-d_6, 400 MHz): δ 1.53 (3H, s), 1.89 (2H, d, $J = 7$ Hz), 2.15 (2H, t, $J = 14.5$ Hz), 2.63 (2H, d, $J = 7.0$ Hz), 6.80–6.90 (3H, m), 7.11–7.15 (1H, m), 10.42 (1H, s), 11.12 (1H, s,); ^{13}C NMR (DMSO-d_6, 100 MHz): δ 9.3, 21.4, 27.9, 37.3, 49.2, 94.3, 109.7, 110.3, 122.2, 123.8, 128.3, 136.5, 141.9, 154.7, 157.6, 160.0, 179.4, 194.8; ESI-MS: calc. for $C_{18}H_{15}N_3O_3$: *m/z* at 321.41; found: *m/z* at 322.07 [M + 1].

[(4*S*)-1′,3,7,7-Tetramethyl-2′,5-dioxo-1′,2′,6,7,8,9-hexahydro-5*H*-spiro[[1,2]-oxazolo [5,4-*b*]quinoline-4,3′-indole]-5′-yl]azinic acid (**4f**): white solid, mp 215–217 °C; yield: 89%; ^{1}H NMR (DMSO-d_6, 400 MHz): δ 1.03 (3H, s), 1.04 (3H, s), 1.59 (3H, s), 2.04–2.11 (2H, m), 2.31–2.51 (2H, m), 2.76 (3H, s), 6.79–7.08 (2H, m), 7.22–7.34 (1H, m), 11.18 (1H, s); ^{13}C NMR (DMSO-d_6, 100 MHz,): δ 9.3, 27.7, 28.0, 30.6, 32.8, 49.3, 50.6, 93.6, 108.6. 111.3, 113.8, 123.8, 126.5, 131.0, 138.6, 141.3, 153.1, 160.3, 178.9, 194.4; ESI-MS: calc. for $C_{21}H_{20}N_4O_5$: *m/z* at 408.05; found: *m/z* at 409.24 [M + 1].

4.2.17.5 Critical Views

The present protocol offers a facile and efficient green strategy for the rapid synthesis of a novel series of biologically relevant oxazolo[5,4-*b*]quinoline-fused spirooxindoles *via* tandem Knoevenagel/Michael addition. Expedient conversion, use of no catalyst or solvent, moderate to good yields, clean reaction profiles are the key advantages of this method.

4.2.18 Entry-18: Synthesis of Substituted Fused Pyrans

Type of reaction: C–C and C–O bond formation
Reaction conditions: Catalyst-free, ethanol, microwave irradiation
Synthetic strategy: One-pot multicomponent reaction involving tandem Knoevenagel condensation and intramolecular cyclization
Keywords: Arylglyoxal monohydrate, malononitrile, quinolones, coumarins, naphthoquinone, catalyst-free, ethanol, one-pot multicomponent reaction, substituted fused pyrans

4.2.18.1 General Reaction Scheme

Misra and Choudhury[106] reported on a catalyst-free microwave-induced protocol for the synthesis of substituted fused pyrans (**4**/**6**) from the one-pot three-component reaction between arylglyoxal monohydrate (**1**), malononitrile (**2**), and cyclic 1,3-dicarbonyls (coumarin **3a**, quinolone **3b** or 2-hydroxynaphthoquinone **5**) in ethanol (Scheme 4.21).

4.2.18.2 Representative Examples

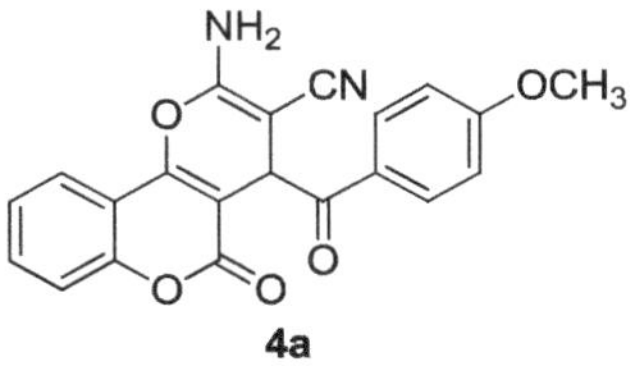

4a

2-Amino-4-(4-methoxybenzoyl)-5-oxo-4,5-dihydropyrano[3,2-*c*]chromene-3-carbonitrile (**4a**): white solid, mp 262–264 °C; yield: 75%

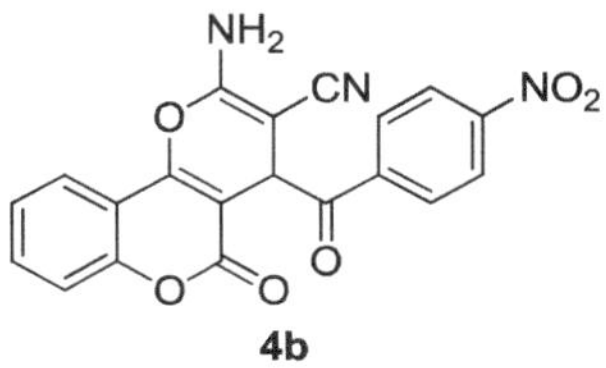

4b

2-Amino-4-(4-nitrobenzoyl)-5-oxo-4,5-dihydropyrano[3,2-*c*]chromene-3-carbonitrile (**4b**): yellow solid, mp 248–250 °C; yield: 70%

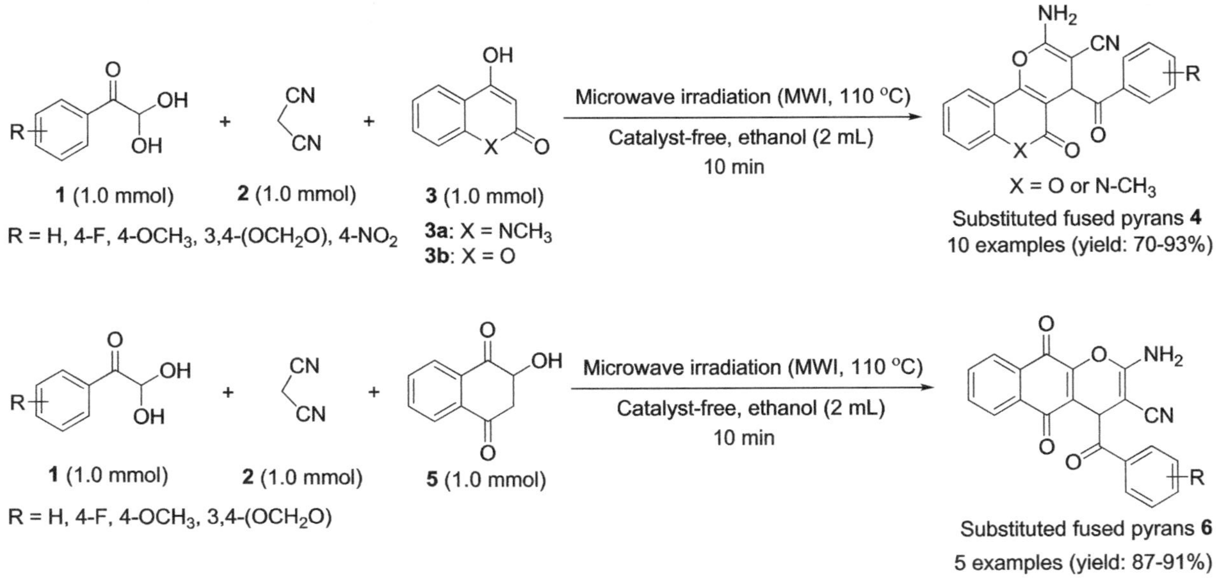

Scheme 4.21 Synthesis of substituted fused pyrans (**4**/**6**).

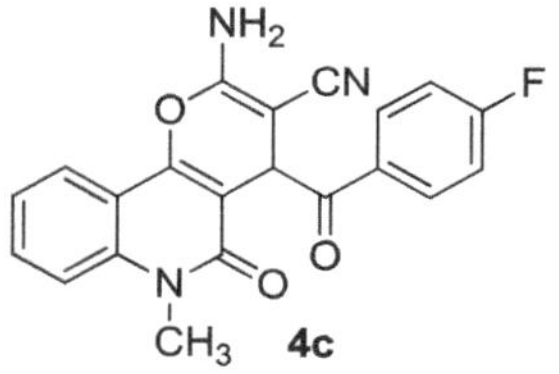

4c

2-Amino-4-(4-fluorobenzoyl)-6-methyl-5-oxo-5,6-dihydro-4*H*-pyrano[3,2-*c*]quinoline-3-carbonitrile (**4c**): white solid, mp 264–266 °C; yield: 85%

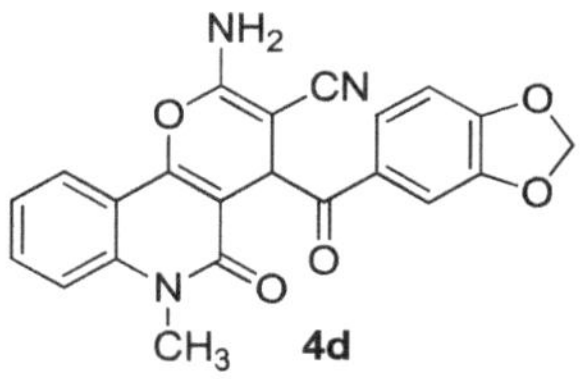

4d

2-Amino-4-(benzo[*d*][1,3]dioxole-5-carbonyl)-6-methyl-5-oxo-5,6-dihydro-4*H*-pyrano[3,2-*c*]quinoline-3-carbonitrile (**4d**): white solid, mp 264–266 °C; yield: 93%

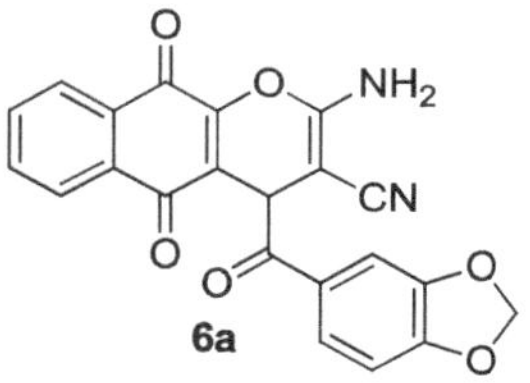

6a

2-Amino-4-(benzo[*d*][1,3]dioxole-5-carbonyl)-5,10-dioxo-5,10-dihydro-4*H*-benzo[g]chromene-3-carbonitrile (**6a**): orange solid, mp 246–248 °C; yield: 87%

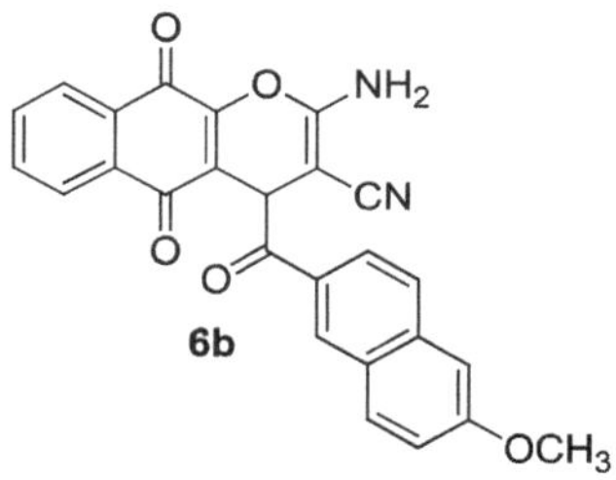

6b

2-Amino-4-(6-methoxy-2-naphthoyl)-5,10-dioxo-5,10-dihydro-4*H*-benzo[*g*]chromene-3-carbonitrile (**6b**): pale yellow solid, mp 238–240 °C; yield: 91%

4.2.18.3 Experimental Procedure

A mixture of arylglyoxal monohydrate (**1**; 1.0 mmol) and malononitrile (**2**; 1.0 mmol) in ethanol (2 mL) was stirred at room temperature for 5 minutes. To this mixture, cyclic 1,3-dicarbonyls (coumarin **3a**, quinolone **3b** or 2-hydroxynaphthoquinone **5**; 1.0 mmol) was added and the resultant mixture was kept under microwave under sealed and stirring conditions for 10 minutes, at 110 °C. Upon completion of the reaction, the reaction mixture was cooled to room temperature and the crude product was obtained by simple filtration, followed by purification by recrystallization from a mixture of ethanol and THF to give the desired products of fused pyrans (**4**/**6**). All the products were characterized with the help of elemental analyses and spectral studies.

4.2.18.4 Characterization Data of Three Representative Compounds

2-Amino-4-(4-methoxybenzoyl)-5-oxo-4,5-dihydropyrano[3,2-*c*]chromene-3-carbonitrile (**4a**): white solid, mp 262–264 °C; yield: 75%; IR: 3456, 3311, 2192, 1668, 1594, 1379, 1258, 1158, 1025, 945, 802, 782 cm^{-1}; ^{1}H NMR (DMSO-d_6 + $CDCl_3$, 400 MHz): δ 8.12 (2H, d, $J = 8.92$, Ar–H), 7.88 (1H, d,

J = 7.72 Hz, Ar–H), 7.73–7.69 (1H, m, Ar–H), 7.57 (2H, bs, –NH_2), 7.47 (2H, t, J = 8.8 Hz, Ar–H), 7.09 (2H, d, J = 8.8 Hz, Ar–H), 5.29 (1H, s, –CH), 3.88 (3H, s, Ar–OCH_3); ^{13}C NMR (DMSO-d_6 + $CDCl_3$, 100 MHz): δ 196.0, 163.8, 159.9, 159.5, 154.7, 152.1, 133.0, 131.5, 128.1, 124.7, 122.1, 118.5, 116.6, 113.9, 112.6, 102.0, 55.5, 52.2, 36.7. Anal. calc. for $C_{21}H_{14}N_2O_5$: C, 67.38; H, 3.77; N, 7.48; found: C 67.36; H, 3.78; N, 7.51.

2-Amino-4-(4-fluorobenzoyl)-6-methyl-5-oxo-5,6-dihydro-4*H*-pyrano[3,2-*c*]quinoline-3-carbonitrile (**4c**): white solid, mp 264–266 °C; yield: 85%; IR (ATR) 3779, 3435, 3259, 2188, 1825, 1654, 1590, 1418, 1367, 1268, 1217, 1201, 1173, 1043 cm^{-1}; ^1H NMR (DMSO-d_6 + $CDCl_3$, 400 MHz): δ 8.25–8.22 (2H, m, Ar–H), 7.98 (1H, dd, J = 7.96, 1.36 Hz, Ar–H), 7.74 (1H, t, J = 7.16 Hz, Ar–H), 7.62 (1H, d, J = 8.52 Hz, Ar–H), 7.51 (2H, br s, –NH_2), 7.4 (3H, t, J = 8.76 Hz, Ar–H), 5.40 (1H, s, –CH), 3.56 (3H, s, –NCH_3); ^{13}C NMR (DMSO-d_6 + $CDCl_3$, 100 MHz): δ 197.7, 166.6, 164.1, 160.1, 151.2, 138.1, 132.8, 132.7, 132.1, 132.0, 131.9, 122.4, 121.8, 119.1, 115.9, 115.2, 112.5, 106.0, 51.7, 40.1, 37.6, 29.3; Anal. calc. for $C_{21}H_{14}FN_3O_3$: C, 67.20; H, 3.76; N, 11.19; found: C, 67.19; H, 3.74; N, 11.22.

2-Amino-4-(benzo[*d*][1,3]dioxole-5-carbonyl)-5,10-dioxo-5,10-dihydro-4*H*-benzo[g]chromene-3-carbonitrile (**6a**): orange solid, mp 246–248 °C; yield: 87%; IR 3470, 3335, 3088, 2195, 1677, 1647, 1062, 1589, 1503, 1438, 1353, 1268, 1249, 1209, 1100, 1038, 949, 869, 797, 726 cm^{-1}; ^1H NMR (DMSO-d_6 + $CDCl_3$, 400 MHz): δ 8.09–8.07 (1H, m, Ar–H), 7.94–7.91 (1H, m, Ar–H), 7.89–7.83 (3H, m, Ar–H), 7.58–7.57 (1H, m, Ar–H), 7.54 (2H, br s, –NH_2), 7.11 (1H, d, J = 8.56 Hz, Ar–H), 6.19 (2H, s, –CH_2), 5.4 (1H, s, –CH); ^{13}C NMR (DMSO-d_6 + $CDCl_3$, 100 MHz): δ 195.5, 182.4, 176.5, 159.8, 152.3, 149.8, 148.1, 134.7, 134.4, 130.5, 130.3, 129.6, 126.3, 126.2, 125.9, 121.2, 118.6, 108.2, 108.1, 102.2, 51.8, 36.9; Anal. calc. for $C_{22}H_{12}N_2O_6$: C, 66.00; H, 3.02; N, 7.00; found: C, 66.02; H, 3.05; N, 7.08.

4.2.18.5 Critical Views

The present protocol offers a facile way for the preparation of a wide range of fused pyrans tethered with amino, nitrile and benzoyl groups under microwave irradiation that requires a short reaction time. It also avoids column chromatographic purification and catalysts unlike other methods available in the literature for such scaffolds.[107–114]

4.2.19 Entry-19: Synthesis of Ring-fused Aminals

Type of reaction: C–N bond formation
Reaction conditions: Catalyst-free, solvent-free, microwave irradiation
Synthetic strategy: α-Amination of nitrogen heterocycles
Keywords: α-Aminoaldehydes, cyclic amines, solvent- and catalyst-free, α-amination, ring-fused aminals

4.2.19.1 General Reaction Scheme

Polshettiwar and Varma[115] accomplished an expedient synthesis of ring-fused aminals (**3**) from the reaction between α-aminoaldehydes (**1**) and cyclic

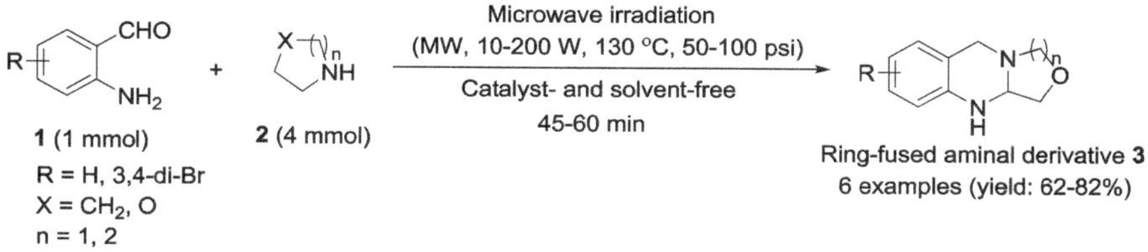

Scheme 4.22 Synthesis of ring-fused aminals (**3**).

amines (**2**) under microwave irradiation in the absence of any added catalyst and solvent (Scheme 4.22). This protocol is faster with moderate yields, thus offering an alternative way for the previously reported sluggish method by Seidel *et al.*[116] for the synthesis of such compounds.

4.2.19.2 Representative Examples

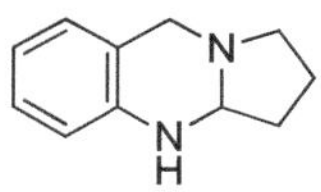

3a

1,2,3,3a,4,9-Hexahydropyrrolo[2,1-*b*]quinazoline (**3a**): yield: 75%

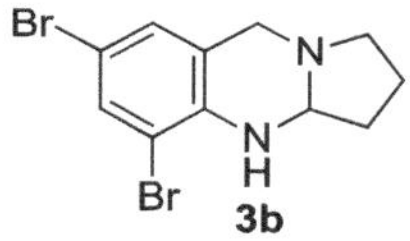

3b

5,7-Dibromo-1,2,3,3a,4,9-hexahydropyrrolo[2,1-*b*]quinazoline (**3b**): yield: 82%

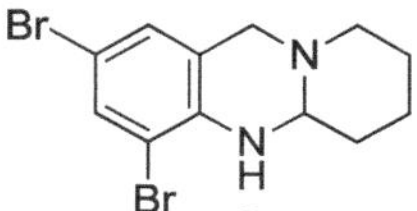

3c

2,4-Dibromo-5a,6,7,8,9,11-hexahydro-5*H*-pyrido[2,1-*b*]quinazoline (**3c**): yield: 62%

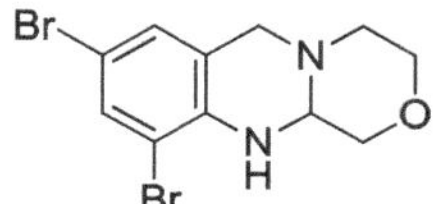

3d

8,10-dibromo-1,3,4,6,11,11a-hexahydro-[1,4]oxazino[3,4-*b*]quinazoline (**3d**): yield: 65%

4.2.19.3 Experimental Procedure

A mixture of aminoaldehydes (**1**; 1 mmol) and cyclic amines (**2**; 4 mmol) within a 10 mL crimp-sealed thick-walled glass tube (equipped with a pressure sensor and a magnetic stirrer) was placed inside the cavity of a CEM Discover focused microwave synthesis system, operated at 130 ± 5 °C (temperature was monitored by a built-in infrared sensor), power 10–200 W, and pressure 50–100 psi for 45–60 min. Upon completion of the reaction, crude products were purified by column chromatography to give pure ring-fused aminals (**3**), followed by their characterization by spectral methods.

4.2.19.4 Characterization Data of One Representative Compound

5,7-Dibromo-1,2,3,3a,4,9-hexahydropyrrolo[2,1-*b*]quinazoline (**3a**): yield: 82%; ^{1}H NMR ($CDCl_3$): δ 7.39 (1H, s), 7.00 (s, 1H), 4.41 (1H, m), 4.25 (1H, s),

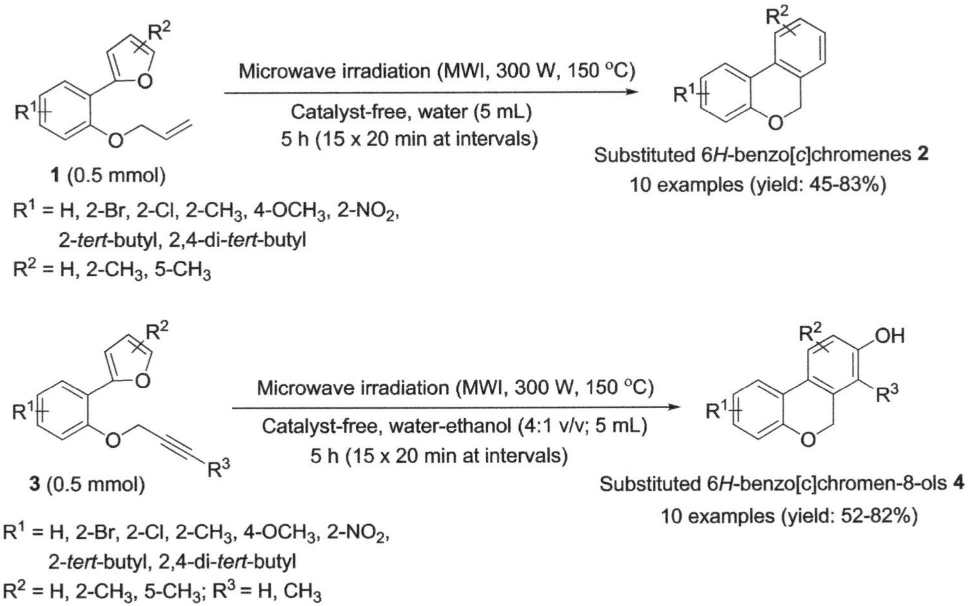

Scheme 4.23 Synthesis of substituted 6*H*-benzo[*c*]chromenes (**2**) and 6*H*-benzo[*c*]-chromen-8-ols (**4**).

4.12 (1H, d), 3.80 (1H, d), 2.80 (2H, m), 2.20 (1H, m), 1.98 (2H, m), 1.75 (1H, m); ^{13}C NMR ($CDCl_3$): δ 21.5, 32.5, 50.0, 70.9, 108.2, 109.1, 121.1, 128.6, 131.9, 139.5; MS: (M^+) 332.

2,4-Dibromo-5a,6,7,8,9,11-hexahydro-5*H*-pyrido[2,1-*b*]quinazoline (**3c**): yield: 62%; ^{1}H NMR ($CDCl_3$): δ 7.50 (1H, s), 6.98 (1H, s), 4.23 (1H, s), 3.82 (1H, s), 3.78 (2H, m), 2.98 (1H, m), 2.24 (1H, m), 1.92 (1H, m), 1.78 (1H, m), 1.65 (2H, m), 1.60 (1H, m), 1.48 (1H, m); ^{13}C NMR ($CDCl_3$): δ 21.1, 25.8, 32.0, 51.6, 56.1, 70.0, 108.2, 108.4, 121.9, 128.3, 131.9, 139.2; MS: (M^+) 346.

4.2.19.5 *Critical Views*

The investigators[115] developed a rapid and one-step protocol for pharmaceutically relevant ring-fused aminals with the use of microwaves under catalyst- and solvent-free conditions. The yields were obtained in moderate to good amounts but the main advantages of this protocol are the clean reaction profile and the rapidness compared to the previously reported sluggish (long reaction time of 80 h) protocol.[116]

4.2.20 Entry-20: Synthesis of Substituted 6*H*-Benzo[*c*]chromenes and 6*H*-Benzo[*c*]-chromen-8-ols

Type of reaction: C–C bond formation
Reaction conditions: Catalyst-free, water/ethanol, microwave irradiation
Synthetic strategy: Intramolecular Diels–Alder cycloaddition *via* cascade reaction
Keywords: 2-(2-(Allyloxy)phenyl)furans, 2-(2-(prop-2-ynyloxy)phenyl)furans, catalyst-free, water/ethanol, substituted 6*H*-benzo[*c*]chromenes, 6*H*-benzo[*c*]-chromen-8-ols

4.2.20.1 *General Reaction Scheme*

He *et al.*[117] synthesized diversely substituted 6*H*-benzo[*c*]chromenes (**2**) and 6*H*-benzo[*c*]-chromen-8-ols (**4**) *via* cascade reactions of 2-(2-(allyloxy)phenyl)furans (**1**) and 2-(2-(prop-2-ynyloxy)phenyl)furans (**2**), respectively, involving intramolecular Diels–Alder reactions in aqueous/aqua-organic media under microwave irradiation without the aid of a catalyst (Scheme 4.23).

4.2.20.2 *Representative Examples*

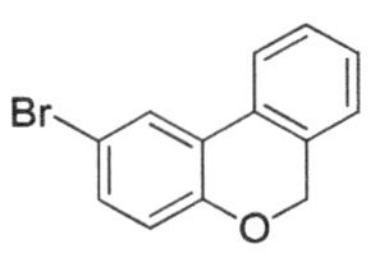

2a

2-Bromo-6*H*-benzo[*c*]chromene (**2a**): yield: 71%

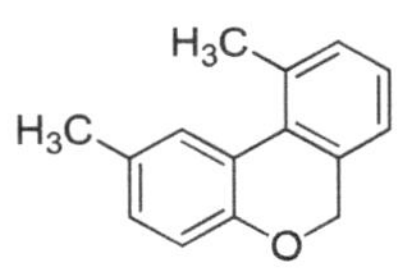

2b

2,10-Dimethyl-6*H*-benzo[*c*]chromene (**4b**): yield: 74%

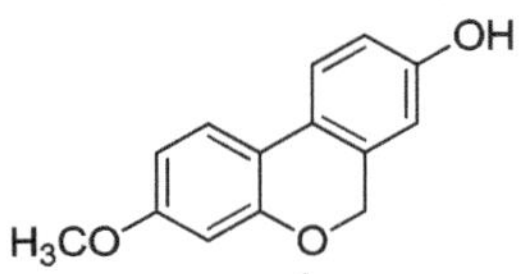

4a

3-Methoxy-6*H*-benzo[c]chromen-8-ol (**4a**): yield: 82%

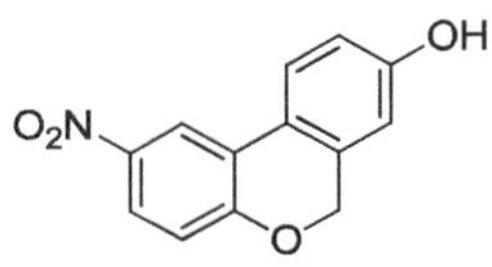

4b

2-Nitro-6*H*-benzo[*c*]chromen-8-ol (**4b**): yield: 52%

4.2.20.3 Experimental Procedures

A sealed 10 mL glass tube (equipped with a magnetic stirring bar) containing 2-(2-(allyloxy)phenyl)furans (**1**; 0.5 mmol) and water (5 mL) was irradiated with microwaves with the help of microwave synthesis apparatus (300 W) at 150 °C for 15×20 min at intervals. Between two irradiations, the reaction temperature was allowed to cool to 60 °C. Upon completion of the reaction, the vessel was cooled to room temperature and the crude product was extracted with ethyl acetate (3×5 mL). The combined organic phases were dried, filtered and concentrated under vacuum. The residue was purified by column chromatography with silica gel using ethyl acetate–hexane mixtures as the eluents to afford pure 6*H*-benzo[*c*]chromenes (**2**). A similar procedure was followed for synthesizing 6*H*-benzo[*c*]-chromen-8-ols (**4**) but replacing water with water–ethanol (4:1 v/v) as the solvent. All the products were characterized on the basis of their NMR and mass spectral studies.

4.2.20.4 Characterization Data of Two Representative Compounds

2-Bromo-6*H*-benzo[*c*]chromene (**2a**): yield: 71%; ^{1}H NMR ($CDCl_3$, 400 MHz): δ 5.11 (s, 3H), 6.88 (1H, d, J = 8.4 Hz), 7.15 (1H, d, J = 7.2 Hz), 7.31–7.33 (2H, m), 7.39 (1H, t, J = 8.0 Hz), 7.64 (1H, d, J = 7.6 Hz), 7.83 (1H, d, J = 2.0 Hz); ^{13}C NMR ($CDCl_3$, 100 MHz): δ 68.5, 114.5, 119.2, 122.1, 124.7, 124.9, 126.1, 128.3, 128.6, 128.9, 131.2, 132.0, 153.8; MS: *m/z* 261 $[M+H]^+$; HRMS calc. for $C_{13}H_{10}BrO$: 260.9916 $[M+H]^+$; found: 260.9901.

3-Methoxy-6*H*-benzo[c]chromen-8-ol (**4a**): yield: 82%; ^{1}H NMR ($CDCl_3$, 400 MHz): δ 3.81 (3H, s), 5.04 (2H, s), 5.11 (1H, s), 6.55 (1H, d, J = 2.4 Hz), 6.61–6.64 (2H, m), 6.81 (1H, dd, J = 2.0, 8.0 Hz), 7.48 (1H, d, J = 8.4 Hz), 7.54 (1H, d, J = 8.4 Hz); ^{13}C NMR ($CDCl_3$, 100 MHz): δ 55.4, 68.5, 102.4, 108.7, 111.5, 115.4, 115.9, 122.8, 123.4, 132.0, 154.6, 155.0, 160.1; MS: *m/z* 229 $[M+H]^+$; HRMS calc. for $C_{14}H_{13}O_3$: 229.0865 $[M+H]^+$, found: 229.0861.

4.2.20.5 Critical Views

The investigators developed an alternative method for the synthesis of biologically interesting 6*H*-benzo[*c*]chromenes and 6*H*-benzo[*c*]-chromen-8-ols under catalyst-free conditions in water or water–ethanol media. They also extended their reaction protocol in preparing some useful derivatives of these compounds as well in this communication.[117]

4.2.21 Entry-21: Synthesis of Functionalized 1,4-Pyranonaphthoquinones

Type of reaction: C–C and C–O bond formation
Reaction conditions: Catalyst-free, solvent-free, microwave irradiation
Synthetic strategy: One-pot multicomponent reaction *via* tandem synthesis
Keywords: Anilines, diethyl acetylenedicarboxylate, 2-hydroxynaphthalene-1,4-dione, aromatic aldehydes, catalyst-free, solvent-free, Michael addition, stereoselectivity, functionalized 1,4-pyranonaphthoquinones, 4-aryl-5,10-dioxo-2-aryl-3,4,5,10-tetrahydro-2*H*-benzo[*g*]chromene-2,3-dicarboxylates

4.2.21.1 General Reaction Scheme

Perumal and his group[118] accomplished stereoselective synthesis of a series of novel functionalized 4-aryl-5,10-dioxo-2-aryl-3,4,5,10-tetrahydro-2*H*-benzo[*g*]chromene-2,3-dicarboxylates from one-pot sequential reactions of anilines (**3**), diethyl acetylenedicarboxylate (**4**), 2-hydroxynaphthalene-1,4-dione (**1**) and benzaldehydes (**2**) under microwave irradiation in the absence of any catalyst or solvent (Scheme 4.24), presumably involving two Michael additions, aldol condensation and annulation reactions.

4.2.21.2 Representative Examples

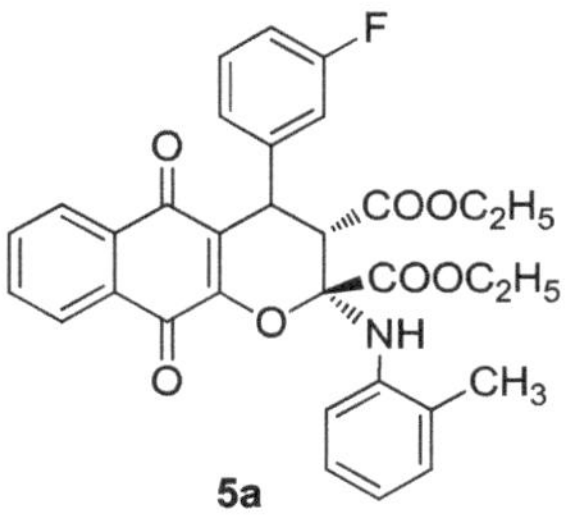

5a

(2*S*,3*S*)-Diethyl 4-(3-fluorophenyl)-5,10-dioxo-2-(o-tolylamino)-3,4,5,10-tetrahydro-2*H*-benzo[*g*]chromene-2,3-dicarboxylate (**5a**): yellow solid, mp 142 °C; yield: 79%

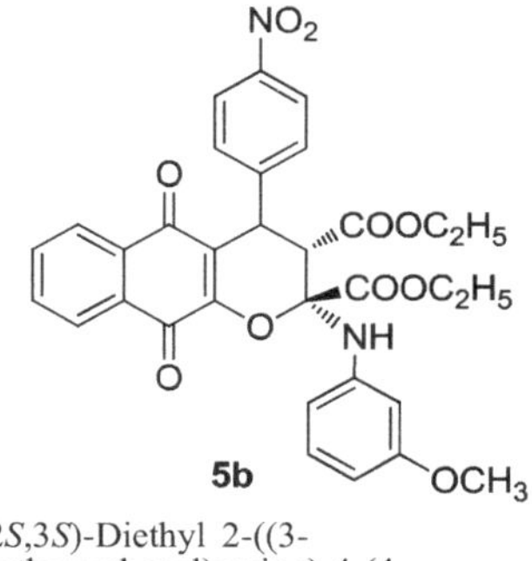

5b

(2*S*,3*S*)-Diethyl 2-((3-methoxyphenyl)amino)-4-(4-nitrophenyl)-5,10-dioxo-3,4,5,10-tetrahydro-2*H*-benzo[*g*]chromene-2,3-dicarboxylate (**5b**): yellow solid, mp 173 °C; yield: 72%

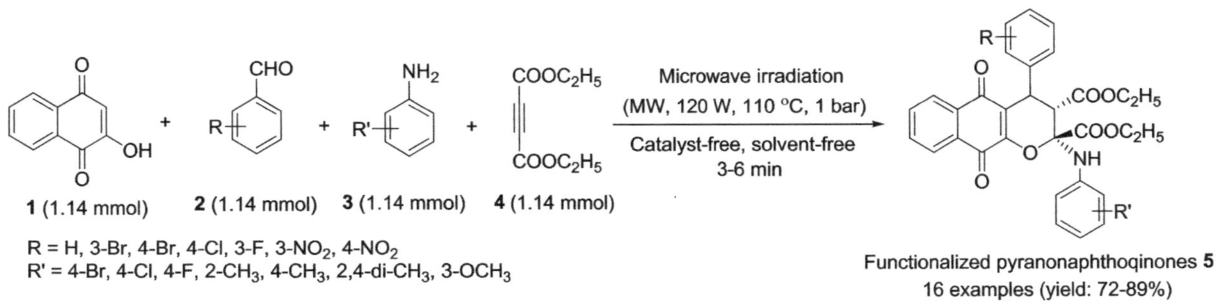

Scheme 4.24 Synthesis of functionalized 1,4-pyranonaphthoquinones (5).

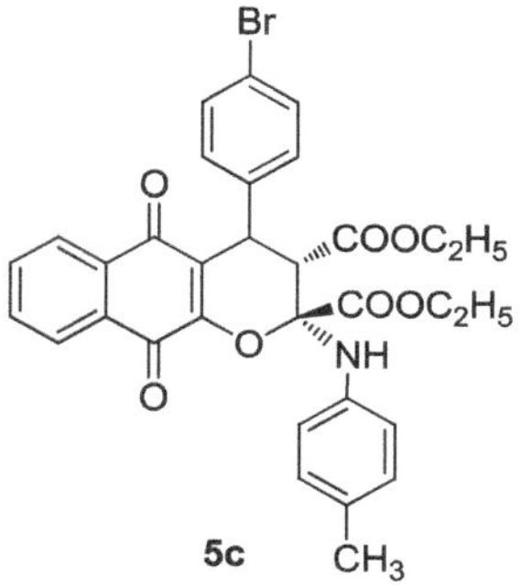

5c

(2*S*,3*S*)-Diethyl 4-(4-bromophenyl)-5,10-dioxo-2-(*p*-tolylamino)-3,4,5,10-tetrahydro-2*H*-benzo[*g*]chromene-2,3-dicarboxylate (**5c**): yellow solid, mp 153 °C; yield: 89%

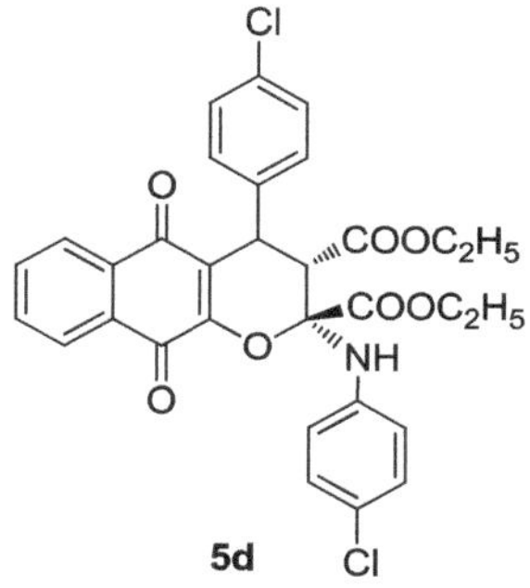

5d

(2*S*,3*S*)-Diethyl 4-(4-chlorophenyl)-2-((4-chlorophenyl)amino)-5,10-dioxo-3,4,5,10-tetrahydro-2*H*-benzo[*g*]chromene-2,3-dicarboxylate (**5d**):yellow solid, mp 157 °C; yield: 78%

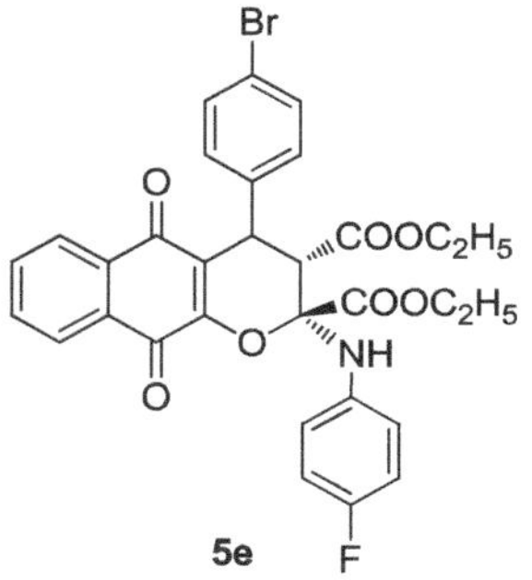

5e

(2*S*,3*S*)-Diethyl 4-(4-bromophenyl)-2-((4-fluorophenyl)amino)-5,10-dioxo-3,4,5,10-tetrahydro-2*H*-benzo[*g*]chromene-2,3-dicarboxylate (**5e**): yellow solid, mp 171 °C; yield 84%

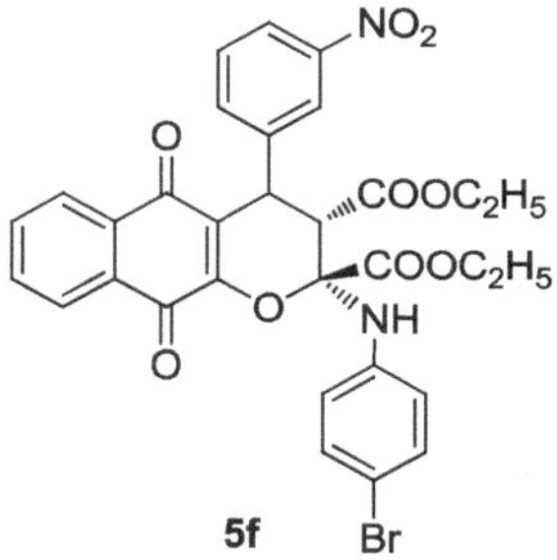

5f

(2*S*,3*S*)-Diethyl 2-((4-bromophenyl)amino)-4-(3-nitrophenyl)-5,10-dioxo-3,4,5,10-tetrahydro-2*H*-benzo[*g*]chromene-2,3-dicarboxylate (**5f**): yellow solid, mp 169 °C; yield 87%

4.2.21.3 Experimental Procedure

A mixture of anilines (**3**; 1.14 mmol) and diethyl acetylenedicarboxylate (**4**; 1.14 mmol) was first taken in a 10 mL quartz vial, and ground thoroughly for about 30 s to form a yellowish Michael adduct, which, without isolation, was treated with 2-hydroxynaphthalene-1,4-dione (**1**; 1.14 mmol) and aromatic aldehyde (**2**; 1.14 mmol) and sealed and placed in a Biotage microwave synthesizer (power 120 W). Then the vial was subjected to microwave irradiation at 100 °C and 1 bar pressure for 3–6 min. Upon completion of the reaction (as monitored by TLC), the resulting brownish solid obtained was recrystallized from ethanol to obtain the pure yellow products of 1,4-pyranonaphthoquinones (**5**). All the products were characterized based on their elemental, as well as their spectral studies.

4.2.21.4 Characterization Data of Two Representative Compounds

(2*S*,3*S*)-Diethyl 2-((3-methoxyphenyl)amino)-4-(4-nitrophenyl)-5,10-dioxo-3,4,5,10-tetrahydro-2*H*-benzo[*g*]chromene-2,3-dicarboxylate (**5b**): yellow solid, mp 173 °C; yield: 72%; ^{1}H NMR ($CDCl_3$, 300 MHz): δ 1.03–1.11 (6H, m), 3.20 (1H, d, *J* = 11.4 Hz), 3.80 (3H, s), 4.06–4.22 (4H, m), 4.68 (1H, d, *J* = 11.4 Hz), 5.64 (1H, s), 6.43 (1H, dd, *J* = 8.1, 1.6 Hz), 6.56 (1H, dd, *J* = 8.1, 1.8 Hz), 6.64–6.65 (1H, m), 7.08 (1H, t, *J* = 8.1 Hz), 7.41 (2H, d, *J* = 8.7 Hz), 7.66–7.70 (2H, m), 7.84–7.87 (1H, m), 8.10–8.12 (1H, m), 8.18 (2H, d, *J* = 8.7 Hz); ^{13}C NMR ($CDCl_3$,75 MHz): δ 13.8, 13.9, 40.2, 53.1, 55.2, 62.5, 62.9, 88.5, 102.0, 108.9, 123.0, 123.9, 124.2, 126.4, 126.5, 128.4, 129.2, 131.0, 131.8, 133.6, 134.3, 144.3, 147.3, 147.7, 152.3, 166.9, 170.3, 178.2, 182.6. ESI-MS: *m/z* 601.16 (M^+). Anal. calc. for $C_{32}H_{28}N_2O_{10}$: C, 64.00; H, 4.70; N, 4.66; found C, 63.93; H, 4.81; N, 4.72.

(2*S*,3*S*)-Diethyl 4-(4-bromophenyl)-2-((4-fluorophenyl)amino)-5,10-dioxo-3,4,5,10-tetrahydro-2*H*-benzo[*g*]chromene-2,3-dicarboxylate (**5e**): yellow solid, mp 171 °C; yield: 84%; ^{1}H NMR ($CDCl_3$, 300 MHz): δ 1.02–1.28 (6H, m), 3.19 (1H, d, *J* = 11.1 Hz), 4.04–4.18 (4H, m), 4.52 (1H, d, *J* = 11.1 Hz), 5.60 (1H, s), 6.86–6.98 (4H, m), 7.09 (2H, d, *J* = 8.2 Hz), 7.43 (2H, d, *J* = 8.2 Hz), 7.67–7.70 (2H, m), 7.88–7.91 (1H, m), 8.10–8.13 (1H, m); ^{13}C NMR ($CDCl_3$,75 MHz): δ 13.7, 13.8, 39.8, 53.4, 62.3, 62.7, 88.9, 115.8, 118.1, 121.4, 123.8, 126.4, 129.2, 131.0, 132.0, 132.1, 133.4, 134.2, 139.0, 139.2, 139.3, 152.0, 158.0, 167.0, 170.8, 178.5, 182.7. Anal. calc. for $C_{31}H_{25}BrFNO_7$: C, 59.82; H, 4.05; N, 2.25; found C, 59.73; H, 3.98; N, 2.36.

4.2.21.5 Critical Views

The investigators designed a microwave-assisted, catalyst- and solvent-free four-component reaction protocol to assess a new series of 1,4-pyranonaphthoquinones from the sequential reaction of anilines, diethyl acetylenedicarboxylate, 2-hydroxynaphthalene-1,4-dione and benzaldehydes.[118] Simple starting materials, catalyst- and solvent-free reaction conditions, generation of water as the green waste and a simple workup are the key advantages of this method.

4.2.22 Entry-22: Synthesis of α-Aminophosphonates

Type of reaction: C–N and C–P bond formation
Reaction conditions: Catalyst-free, solvent-free, microwave irradiation
Synthetic strategy: One-pot three-component reaction
Keywords: Aldehydes, amines, dimethyl phosphite, solvent- and catalyst-free, one-pot multicomponent reaction, Kabachnik–Fields reaction, α-aminophosphonates

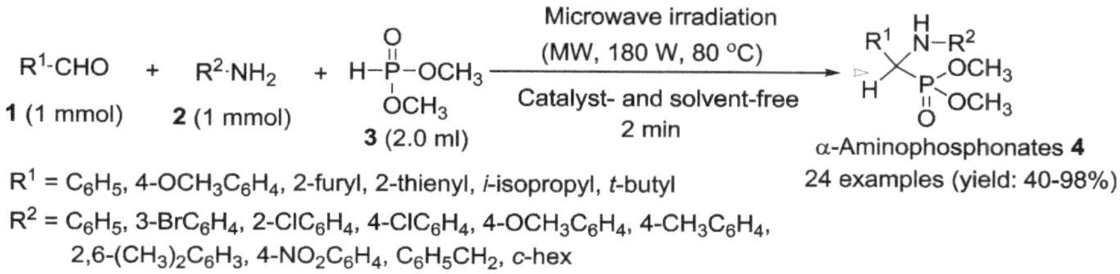

Scheme 4.25 Synthesis of α-aminophosphonates (**4**).

4.2.22.1 General Reaction Scheme

A microwave-assisted highly efficient solvent-free and catalyst-free method for the synthesis of α-amino phosphonates (**4**) *via* a one-pot three-component Kabachnik–Fields reaction involving aldehyde (**1**), amine (**2**), and dimethyl phosphite (**3**) was developed by Mu *et al.*[119] (Scheme 4.25).

4.2.22.2 Representative Examples

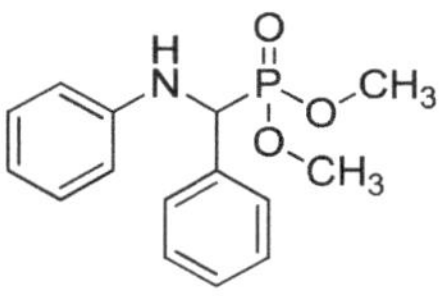

4a

Dimethyl (phenyl(phenylamino)methyl)phosphonate (**4a**): yield: 98%

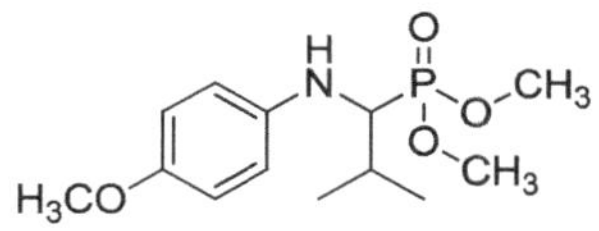

4b

Dimethyl (1-((4-methoxyphenyl)amino)-2-methylpropyl)phosphonate (**4b**): yield: 85%

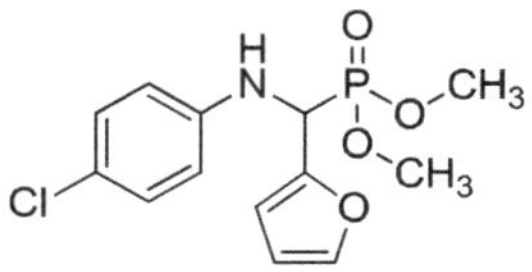

4c

Dimethyl (((4-chlorophenyl)amino)(furan-2-yl)methyl)phosphonate (**4c**): yield: 76%

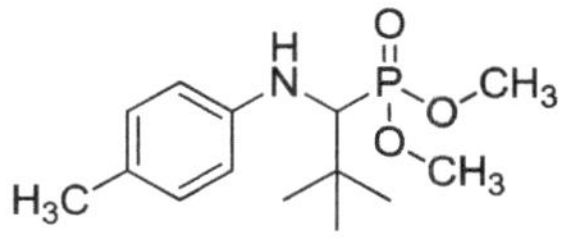

4d

Dimethyl (2,2-dimethyl-1-(p-tolylamino)propyl)phosphonate (**4d**): yield: 53%

4.2.22.3 Experimental Procedure

A 25 mL three-necked flask was charged with aldehyde (**1**; 1 mmol), amine (**2**; 1 mmol), and dimethyl phosphite (2 mL), and the mixture was then heated at 80 °C for 2 min at 180 W in a multimode microwave oven. Upon completion of the reaction, the resulting mixture was diluted with water and extracted with dichloromethane (20 mL). The organic layer was washed with water (3×10 mL), dried over anhydrous sodium sulfate and filtered. The filtrate was concentrated under reduced pressure to get the crude product, which was then purified by silica gel flash column chromatography using with 2:1 petroleum ether–acetone as eluent to afford pure α-aminophosphonate (**4**). All the products were characterized based on NMR and HRMS measurements.

4.2.22.4 Characterization Data of One Representative Compound

Dimethyl (phenyl(phenylamino)methyl)phosphonate (**4a**): solid, mp 90–91 °C; yield: 98%; ^{1}H NMR ($CDCl_3$, 400 MHz): δ 7.48–6.59 (10H, m, 2×C_6H_5), 4.81 (1H, d, $^{1}J_{PH}$ = 24.40 Hz, -C*H*), 3.76 (3H, d, $^{2}J_{PH}$ = 10.60 Hz, -OCH_3), 3.46 (3H, d, $^{2}J_{PH}$ = 10.60 Hz, -OCH_3); ^{13}C NMR ($CDCl_3$, 100 MHz): δ 146.4, 135.9, 129.6, 129.2, 128.5, 128.2, 118.9, 114.2, 55.5 (d, $^{1}J_{CP}$ = 150.2 Hz, -CH), 54.3, 54.2; HRMS calc. for $C_{15}H_{18}NO_3P$ 291.1024; found 291.1022.

4.2.22.5 Critical Views

The present method offers a microwave-assisted rapid access of a series of α-aminophosphonates in the absence of any catalyst or solvent. The reaction proceeds well with aryl and heteroaryl aldehydes with good to excellent yields ranging from 78–98%; however, reaction with aliphatic aldehydes (*viz.* pivalaldehyde) resulted with poor yields (40–53%).

4.3 Concluding Remarks

Microwave (MW) heating is now regarded as a tool for sustainable chemistry. MW heating is a multiphysics phenomenon that involves electromagnetic waves and heat transfer to a target material. It has been well-demonstrated that microwave heating, with a number of quantitative and qualitative advantages over conventional heating techniques, can be used to facilitate the clean and sustainable synthesis of organic compounds. Shorter reaction times, higher product yields, and enhanced selectivity are some of the notable advantages of microwave heating over conventional methods. This motivates the researchers considerably with its continued applications as a green tool in diverse fields of chemical sciences, including the area of catalyst-free synthesis.

References

1. J. D. Moseley, Microwave Heating as a Tool for Process Chemistry, in *Microwave Heating as a Tool for Sustainable Chemistry*, ed. N. Leadbeater, CRC Press, New York, 2010, pp. 105–147.
2. G. Keglevich, Application of Microwave Irradiation in the Synthesis of P-heterocycles, in *Green Synthetic Approaches for Biologically Relevant Heterocycles*, ed. G. Brahmachari, Elsevier, Amsterdam, The Netherland, 2015, pp. 559–570.

3. *Microwaves in Organic Synthesis*, ed. A. De La Hoz and A. Loupy, 2012, vol. 1. 3rd edn, Wiley-VCH, Weinheim, Germany.
4. C. O. Kappe, A. Stadler and D. Dallinger, in *Microwaves in Organic and Medicinal Chemistry*, ed. R. Mannhold, H. Kubinyi and G. Folkers, 2nd edn, 2012, Wiley-VCH, Weinheim, Germany.
5. *Microwave-assisted Organic Synthesis: A Green Chemical Approach*, ed. S. C. Ameta, P. B. Punjabi, R. Ameta and C. Ameta, 2014, 1st edn, CRC Press, New York.
6. *Milestones in Microwave Chemistry*, ed. G. Keglevich, 2016, 1st edn, Springer, Germany.
7. R. Gedye, F. Smith, K. Westaway, H. Ali, L. Baldisera, L. Laberge and J. Rousell, *Tetrahedron Lett.*, 1986, **27**, 279.
8. R. A. Abramovitch, *Org. Prep. Proced. Int.*, 1991, **23**, 683.
9. B. L. Hayes, *Aldrichimica Acta*, 2004, **37**, 66.
10. K. Kranjc and M. Kočevar, *Curr. Org. Chem.*, 2010, **14**, 1050.
11. K. Kempe, C. R. Becer and U. S. Schubert, *Macromol.*, 2011, **44**, 5825.
12. D. Bandyopadhyay and B. K. Banik, Microwave-induced Synthesis of Heterocycles of Medicinal Interests, in *Green Synthetic Approaches for Biologically Relevant Heterocycles*, ed. G. Brahmachari, Elsevier, Amsterdam, The Netherland, 2015, pp. 517–557.
13. G. Brahmachari, *ChemText*, 2015, **1**, 15, DOI: 10.1007/s40828-015-0017-x.
14. M. De Rosa and A. Soriente, *Eur. J. Org. Chem.*, 2010, 1029.
15. S. Lakrout, H. K'tir, A. Amira, M. Berredjem and N.-E. Aouf, *RSC Adv.*, 2014, **4**, 16027.
16. D. A. Alonso and P. G. Andersson, *J. Org. Chem.*, 1998, **63**, 9455.
17. C. S. Pak and D. S. Lim, *Synth. Commun.*, 2001, **31**, 2209.
18. H. H. Kang, H. S. Rho, D. H. Kim and S.-G. Oh, *Tetrahedron Lett.*, 2003, **44**, 7225.
19. A. Kamal, J. S. Reddy, E. V. Bharathi and D. Dastagiri, *Tetrahedron Lett.*, 2008, **49**, 348.
20. G. A. Meshram and V. D. Patil, *Tetrahedron Lett.*, 2009, **50**, 1117.
21. F. Tamaddon, A. Nasiri and S. Farokhi, *Catal. Commun.*, 2011, **12**, 1477.
22. F. Tamaddon, M. Reza Sabeti, A. Ali Jafari, F. Tigir and E. Keshavarz, *J. Mol. Catal. A: Chem.*, 2011, **41**, 41.
23. M. Mokhtar, T. S. Saleh, N. S. Ahmed, S. A. Al-Thabaiti and R. A. Al-Shareef, *Ultrason. Sonochem.*, 2011, **18**, 172.
24. C. Barbey, R. Bouasla, M. Berredjem, N. Dupont, B. Retailleau, M. Lecouvey and N.-E. Aouf, *Tetrahedron*, 2012, **68**, 9125.
25. S. V. Kumar, S. Muthusubramanian and S. Perumal, *RSC Adv.*, 2015, **5**, 90451.
26. S. Hünig and G. Sauer, *Liebigs Ann. Chem.*, 1971, **748**, 173.
27. Y. Yavari, S. Seyfi, Z. Hossaini, M. Sabbaghan and F. Shirgahi-Talari, *Monatsh. Chem.*, 2008, **139**, 1479.
28. S.-F. Gan, J.-P. Wan, Y.-J. Pan and C.-R. Sun, *Synlett*, 2010, **6**, 973.

29. A. Alizadeh, N. Zohreh, H. Sabahnoo and Z. Noaparast, *Tetrahedron*, 2011, **67**, 1709.
30. A. Hassanabadi and K. Khandan-Barani, *J. Chem. Res.*, 2013, **37**, 71.
31. P. Chidurala, V. Jetti, R. Pagadala, J. S. Meshram and S. B. Jonnalagadda, *J. Heterocyclic Chem.*, 2015, **52**, 1302.
32. H. M. Meshram, N. N. Rao, L. C. Rao and N. S. Kumar, *Tetrahedron Lett.*, 2012, **53**, 3963.
33. R. Niu, J. Xiao, T. Liang and X. Li, *Org. Lett.*, 2012, **14**, 676.
34. M. Tobisu and N. Chatni, *Angew. Chem. Int. Ed.*, 2006, **45**, 1683.
35. K. R. Campos, *Chem. Soc. Rev.*, 2007, **36**, 1069.
36. Y. Nakao, Y. Yamada, N. Kashihara and T. Hiyama, *J. Am. Chem. Soc.*, 2010, **132**, 13666.
37. C. C. Tsai, W. C. Shih, C. H. Fang, T. G. Ong and G. P. A. Yap, *J. Am. Chem. Soc.*, 2010, **132**, 11887.
38. M. Rueping and N. Tolstoluzhsky, *Org. Lett.*, 2011, **13**, 1095.
39. H. Komai, T. Yoshino, S. Matsuguna and M. Kanai, *Org. Lett.*, 2011, **13**, 1706.
40. T. A. Ramirez, B. Zhao and Y. Shi, *Chem. Soc. Rev.*, 2012, **41**, 931.
41. X. Chen, K. M. Engle, D. H. Wang and J. Q. Yu, *Angew. Chem. Int. Ed.*, 2009, **121**, 5196.
42. T. W. Lyons and M. S. Sanford, *Chem. Rev.*, 2010, **110**, 1147.
43. O. Baudoin, *Chem. Soc. Rev.*, 2011, **40**, 4902.
44. A. Iglesias, R. Alvarez, A. de Lera and K. Muniz, *Angew. Chem. Int. Ed.*, 2012, **51**, 1.
45. J. Xie, H. Li, J. Zhou, Y. Cheng and C. Zhu, *Angew. Chem. Int. Ed.*, 2012, **51**, 1252.
46. N. N. Rao and H. M. Meshram, *Tetrahedron Lett.*, 2013, **54**, 1315.
47. J.-F. Zhou, G.-X. Gong, S.-J. Zhi and X.-L. Duan, *Synth. Commun.*, 2009, **39**, 3743.
48. J. Feng, L. Yang, Q. Meng and B. Liu, *Synth. Commun.*, 1998, **28**, 193.
49. Z. Wu and N. J. Ede, *Tetrahedron Lett.*, 2001, **42**, 8115.
50. S. Rostamizadeh and S. Jafari, *Indian J. Heterocyclic Chem.*, 2001, **10**, 303.
51. S. K. Singh, P. Gupta, S. Duggineni and B. Kundu, *Synlett*, 2003, 2147.
52. L. Mao, H. Sakurai and T. Hirao, *Synthesis*, 2004, 2535.
53. Z. Zhao, D. D. Wisnoski, S. E. Wolkenberg, W. H. Leister, Y. Wang and C. W. Lindsley, *Tetrahedron Lett.*, 2004, **45**, 4873.
54. S. A. Raw, C. D. Wilfred and R. J. K. Taylor, *Org. Biomol. Chem.*, 2004, **2**, 788.
55. S. Y. Kim, K. H. Park and Y. K. Chung, *Chem. Commun.*, 2005, **35**, 1321.
56. R. S. Bhosale, S. R. Sarda, S. S. Andhapure, W. N. Jadhav, S. R. Bhusare and R. P. Pawar, *Tetrahedron Lett.*, 2005, **46**, 7183.
57. R. S. Robinson and R. J. K. Taylor, *Synlett*, 2005, 1003.
58. S. V. More, M. N. V. Sastry, C. C. Wang and C. F. Yao, *Tetrahedron Lett.*, 2005, **46**, 6345.

59. C. Venkatesh, B. Singh, P. K. Mahata, H. Junjappa and H. Iia, *Org. Lett.*, 2005, **7**, 2169.
60. D. Aparicio, O. A. Attanasi, P. Filippone, R. Ignacio, S. Lillini, F. Mantellini, F. Palacios and J. M. de los Santos, *J. Org. Chem.*, 2006, **71**, 5897.
61. S. V. More, M. N. V. Sastry and C. F. Yao, *Green Chem.*, 2006, **8**, 91.
62. F. Mohsenzadeh, K. Aghapoor and H. R. Darabi, *J. Braz. Chem. Soc.*, 2007, **18**, 297.
63. H. R. Darabi, S. Mohandessi, K. Aghapoor and F. Mohsenzadeh, *Catal.Commun.*, 2007, **8**, 389.
64. B. Das, K. Venkateswarlu, K. Suneel and A. Majhi, *Tetrahedron Lett.*, 2007, **48**, 5371.
65. M. M. Heravi, K. Bakhtiari, F. F. Bamoharram and M. H. Tehrani, *Monatsh Chem.*, 2007, **138**, 465.
66. M. M. Heravi, *Catal. Commun.*, 2007, **8**, 1341.
67. P. K. Dubey, P. V. V. Prasada Reddy and K. Srinivas, *Synth. Commun.*, 2008, **38**, 613.
68. M. M. Heravi, S. Taheri, K. Bakhtiari and H. A. Oskooie, *Catal. Commun.*, 2007, **8**, 211.
69. A. Hossein, M. M. Oskooie, K. B. Heravi and T. Shima, *Monatsh. Chem.*, 2007, **138**, 875.
70. T. Huang, R. Wang, L. Shi and X. Lu, *Catal. Commun.*, 2008, **9**, 1143.
71. K. Ajeet, K. Santosh, S. Amit, D. Arnab and M. Subho, *Catal. Commun.*, 2008, **9**, 778.
72. D. Q. Shi and G. L. Dou, *Synth. Commun.*, 2008, **38**, 3329.
73. G. Brahmachari, S. Lascar and P. Barik, *RSC Adv.*, 2013, **3**, 14245.
74. J.-F. Zhou, G.-X. Gong, L.-T. An, Y. Liu, F.-X. Zhu, Y.-L. Zhu and S.-J. Ji, *Synlett.*, 2008, 3163.
75. S. Goswami, S. Jana, A. Hazra and A. K. Adak, *J. Heterocyclic Chem.*, 2007, **44**, 1191.
76. B. Reddy Vaddula, R. S. Varma and J. Leazer, *Tetrahedron Lett.*, 2013, **54**, 1538.
77. F. A. Rosa, P. Machado, P. S. Vargas, H. G. Bonacorso, N. Zanatta and M. A. P. Martins, *Synlett*, 2008, 1673.
78. X. Chen, J. She, Z. Shang, J. Wu, H. Wu and P. Zhang, *Synthesis*, 2008, 3478.
79. V. Polshettiwar and R. S. Varma, *Tetrahedron Lett.*, 2008, **49**, 397.
80. X. Chen, J. She, Z.-C. Shang, J. Wu and P. Zhang, *Synth. Commun.*, 2009, **39**, 947.
81. B. Gutmann, D. Obermayer, B. Reichart, B. Prekodravac, M. Irfan, J. M. Kremsner and C. O. Kappe, *Chem. Eur. J.*, 2010, **16**, 12182.
82. P. S. Harikrishnan, S. M. Rajesh, S. Perumal and A. I. Almansour, *Tetrahedron Lett.*, 2013, **54**, 1076.
83. M. A. P. Martins, P. H. Beck, D. N. Moreira, L. Buriol, C. P. Frizzo, N. Zanatta and H. G. Bonacorso, *J. Heterocyclic Chem.*, 2010, **47**, 301.

84. R. Sarma, M. M. Sarmah and D. Prajapati, *J. Org. Chem.*, 2012, **77**, 2018.
85. A. M. Thompson, A. J. Bridges, D. W. Fry, A. J. Kraker and W. A. Denny, *J. Med. Chem.*, 1995, **38**, 3780.
86. A. M. Thompson, D. K. Murray, W. L. Elliott, D. W. Fry, J. A. Nelson, H. D. Hollis Showalter, B. J. Roberts, P. W. Vincent and W. A. Denny, *J. Med. Chem.*, 1997, **40**, 3915.
87. H. I. El-Subbagh, S. M. Abu-Zaid, M. A. Mahran, F. A. Badria and A. M. Al-Obaid, *J. Med. Chem.*, 2000, **43**, 2915.
88. M.-Y. Jang, S. D. Jonghe, L.-J. Gao, J. Rozenski and P. Herdewijn, *Eur. J. Org. Chem.*, 2006, 4257.
89. A. Herrera, R. Martınez-Alvarez, R. Chiouab and J. Almy, *Tetrahedron Lett.*, 2006, **47**, 5463.
90. E. B. Tsupak, M. A. Shevchenko, V. V. Kostrub and Y. N. Tkachenko, *Russ. Chem. Bull., Int. Ed.*, 2007, **56**, 2330.
91. V. V. Kostrub, E. B. Tsupak, Y. N. Tkachenko and M. A. Shevchenko, *Russ. Chem. Bull., Int. Ed.*, 2008, **57**, 1754.
92. N. A. Kheder, Y. N. Mabkhot and A. M. Farag, *ARKIVOC*, 2008, **xvii**, 107.
93. I. Cikotiene, V. Kairys, R. Buksnaitiene, M. Morkunas, S. Rudys, A. Brukstus and M. X. Fernandes, *Tetrahedron*, 2009, **65**, 5752.
94. J.-F. Zhou, G.-X. Gong, X.-J. Sun and Y.-L. Zhu, *Synth. Commun.*, 2010, **40**, 1134.
95. S. Balalaie, A. Arabanian and M. S. Hashtroudi, *Monatsh. Chem.*, 2000, **131**, 945.
96. J. F. Zhou, Y. Z. Song, Y. L. Yang, Y. L. Zhu and S. J. Tu, *Synth. Commun.*, 2005, **35**, 1369.
97. S. A. Siddiqui, U. C. Narkhede, S. S. Palimkar, T. Daniel, R. J. Lahoti and K. V. Srinivasan, *Tetrahedron*, 2005, **61**, 3539.
98. G. Sharma, Y. Jyothi and P. Lakshmi, *Synth. Commun.*, 2006, **36**, 2991.
99. M. Kidwai, P. Mothsra, V. Bansal and R. Goyal, *Monatsh. Chem.*, 2006, **137**, 1189.
100. A. Mohammed, N. Lokare, J. Sangshetti and D. Shinde, *J. Kor. Chem. Soc.*, 2007, **51**, 418.
101. J. Sangshetti, N. Kokare, A. Kotharkar and D. Shinde, *Monatsh. Chem.*, 2008, **139**, 125.
102. S. Ahmad, A. Maleki and M. Benham, *Synth. Commun.*, 2009, **39**, 102.
103. F. P. L. Lim, G. Luna and A. V. Dolzhenko, *Tetrahedron Lett.*, 2015, **56**, 7016.
104. D. Bhuyan, M. M. Sarmah, Y. Dommaraju and D. Prajapati, *Tetrahedron Lett.*, 2014, **55**, 5133.
105. P. Yuvaraj, K. Manivannan and B. S. R. Reddy, *Tetrahedron Lett.*, 2015, **56**, 78.
106. R. Misra and L. H. Choudhury, *RSC Adv.*, 2016, **6**, 24464.
107. V. F. Ferreira, S. B. Ferreira and F. C. da Silva, *Org. Biomol. Chem.*, 2010, **8**, 4793.
108. S. Ahadi, M. Zolghadr, H. R. Khavasi and A. Bazgir, *Org. Biomol. Chem.*, 2013, **11**, 279.

109. S. Khodabakhshi, B. Karami, K. Eskandari and M. Farahi, *Tetrahedron Lett.*, 2014, **55**, 3753.
110. S. Khodabakhsi and B. Karami, *New J. Chem.*, 2014, **38**, 3586.
111. A. Saha, S. Payra and S. Banerjee, *RSC Adv.*, 2015, **5**, 101664.
112. M. V. Reddy, B. S. Kumar, K. T. Lim, B. G. Cho and Y. T. Jeong, *Tetrahedron Lett.*, 2016, **57**, 476.
113. P. T. Moshapo, M. Sokamisa, E. M. Mmutlane, R. M. Mampa and H. H. Kinfe, *Org. Biomol. Chem.*, 2016, **14**, 5627.
114. A. A. Esmaeili, F. Salehan, A. Habibi and A. R. Fakhari, *Tetrahedron Lett.*, 2016, **57**, 100.
115. V. Polshettiwar and R. S. Varma, *Tetrahedron Lett.*, 2008, **49**, 7165.
116. C. Zhang, C. K. De, R. Mal and D. Seidel, *J. Am. Chem. Soc.*, 2008, **130**, 416.
117. Y. He, X. Zhang, L. Cui, J. Wang and X. Fan, *Green Chem.*, 2012, **14**, 3429.
118. B. D. Bala, S. M. Rajesh and S. Perumal, *Green Chem.*, 2012, **14**, 2484.
119. X.-J. Mu, M.-Y. Lei, J.-P. Zou and W. Zhang, *Tetrahedron Lett.*, 2006, **47**, 1125.

CHAPTER 5

Catalyst-free Organic Reactions with Ultrasound Irradiation

5.1 Introduction

Ultrasound irradiation is now regarded as a very useful tool in enhancing the reaction rates in a variety of reacting systems and *sonochemistry* (the term used when ultrasound induces 'true' chemical effects on a reaction system) has already established itself as a distinct branch of organic chemistry.[1–12] Ultrasound irradiation is associated with a unique physical process, known as cavitations, that creates, enlarges, and implodes gaseous and vaporous cavities in an irradiated liquid, thereby inducing very high local temperatures and pressures inside the bubbles (cavities) and leads to turbulent flow of the liquid and enhanced mass transfer. Ultrasound is widely used in chemistry as it is associated with a series of key characteristics such as safety, energy savings, waste prevention, the use of ambient conditions, and the improvement in the mass transfer and product selectivity. Sonochemistry normally uses frequencies between 20 and 40 kHz because this is the range employed in common laboratory equipment. In order to meet the green chemistry goals, ultrasound-assisted organic synthesis has attracted considerable interest during recent decades in terms of accelerating reaction rate, product selectivity, enhancing catalyst activity and recyclability and operational simplicity. Such unique features make ultrasound-assisted organic synthesis (UAOS) an interesting alternative technique to synthesize structurally diverse organic compounds. There are a handful of useful reviews, books and book chapters on this promising subject delineating the impact, chemistry and applications of ultrasound in organic synthesis.[13–22] It has been observed that ultrasonic irradiation can also be effective in accomplishing certain organic transformations in the absence of any catalyst as well. With this view, the present chapter offers detailed

Green Chemistry Series No. 51
Catalyst-free Organic Synthesis
By Goutam Brahmachari

Published by the Royal Society of Chemistry, www.rsc.org

descriptions of 16 comprehensively screened catalyst-free ultrasound-assisted organic synthetic protocols.

5.2 Catalyst-free Organic Transformations with Ultrasound Irradiation

A number of catalyst-free organic reactions leading to the synthesis of useful organic compounds under ultrasound irradiation has been reported so far. This section presents these organic transformations as entry-based format highlighting the key aspects for each of them.

5.2.1 Entry-1: Synthesis of *N*-formylated Derivatives

Type of reaction: C–N bond formation
Reaction conditions: Catalyst-free, solvent-free, room temperature, ultrasound irradiation
Synthetic strategy: *N*-Formylation of amines
Keywords: Amines, formic acid, solvent- and catalyst-free, room temperature, ultrasonication, *N*-formylation

5.2.1.1 General Reaction Scheme

Habibi and Nasrollahzadeh[23] developed an ultrasound-assisted fast and efficient process for *N*-formylation of primary and secondary amines (**1**) in the presence of formic acid (**2**) The reaction was solvent- and catalyst-free and under ambient conditions to afford a series of *N*-formylated derivatives (**3**) with excellent yields and atom economy (Scheme 5.1).

5.2.1.2 Representative Examples

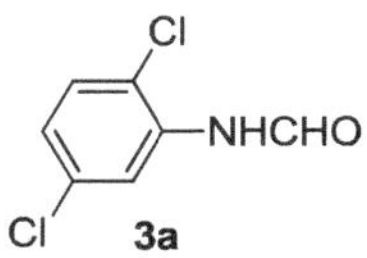

3a

N-(2,4-Dimethylphenyl)formamide (**3a**): solid, mp 148–151 °C; time: 240 min; yield: 95%

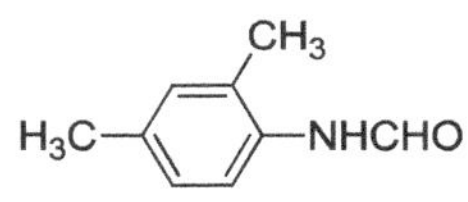

3b

N-(2,5-Dichlorophenyl)formamide (**3b**): solid, mp 114–118 °C; time: 14 min; yield: 94%

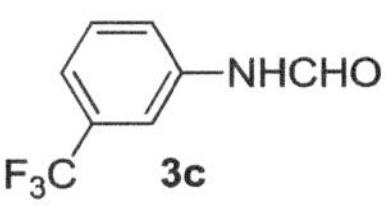

3c

N-(3-(Trifluoromethyl)phenyl)formamide (**3c**): solid, mp 59–61 °C; time: 18 min; yield: 92%

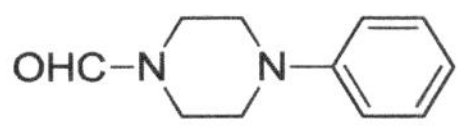

3d

4-Phenylpiperazine-1-carbaldehyde (**3d**): solid, mp 82–84 °C; time: 80 min; yield: 92%

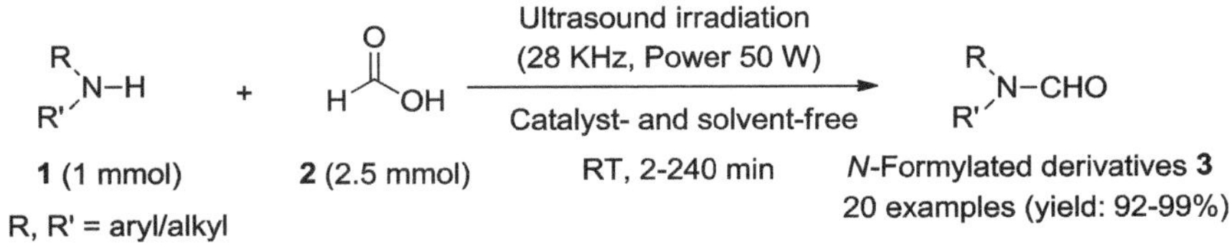

Scheme 5.1 Synthesis of *N*-formylated derivatives (**3**).

5.2.1.3 *Experimental Procedure*

A mixture of amine (**1**; 2 mmol) and formic acid (**2**; 2.5 mM) was irradiated by ultrasound for an appropriate time at room temperature. After completion (as monitored by TLC), water was added and the organic layer was extracted with ethyl acetate to afford the pure product of the *N*-formylated derivative (**3**). The synthesized compounds were identified by comparison of their physical and spectral data with those reported in the literature.

5.2.1.4 *Characterization Data of Two Representative Compounds*

N-(3-(Trifluoromethyl)phenyl)formamide (**3c**): solid, mp 59–61 °C; yield: 92%; IR (KBr): 3279, 3221, 3163, 3092, 3055, 3010, 2881, 1693, 1675, 1608, 1560, 1486, 1452, 1400, 1339, 1319, 1283, 1260, 1180, 1151, 1125, 1069, 1000, 975, 911, 892, 818, 797, 736, 694, 655 cm^{-1}; ^{1}H NMR ($CDCl_3$, 90 MHz): δ 7.42–8.33 (4H, m), 8.89 (1H, d, $J = 10.6$,), 10.49 (1H, s).

4-Phenylpiperazine-1-carbaldehyde (**3d**): solid, mp 82–84 °C; yield: 92%; IR (KBr): 2916, 2814, 1654, 1632, 1599, 1499, 1447, 1405, 1385, 1366, 1350, 1334, 1283, 1254, 1239, 1197, 1149, 1092, 1057, 1033, 1007, 916, 766, 693 cm^{-1}; ^{1}H NMR ($CDCl_3$, 300 MHz): δ 10.35 (1H, s), 8.55 (1H, s), 6.92–7.33 (5H, m, Ar–H), 3.72–3.16 (8H, m); ^{13}C NMR ($CDCl_3$,125 MHz): δ 161.2, 151.4, 129.7, 121.4, 117.6, 50.9, 49.8, 46.0, 40.4.

5.2.1.5 *Critical Views*

The investigators claimed their method as the first example of an efficient ultrasound-promoted approach for the synthesis of *N*-formamides. This process offers several advantages, such as high atom efficiency and yields, rapid and cleaner reaction profile, easy workup and excellent chemo-selectivity over earlier methods.[24–33]

5.2.2 Entry-2: Synthesis of Silyl Ethers

Type of reaction: O–Si bond formation
Reaction conditions: Catalyst-free, solvent-free, room temperature, ultra-sound irradiation
Synthetic strategy: Protection of alcohols/phenols *via O*-silyation

Keywords: Alcohols/phenols, 1,1,1,3,3,3-hexamethyldisilazane (HMDS), solvent- and catalyst-free, room temperature, ultrasonication, silyl ethers

5.2.2.1 *General Reaction Scheme*

Mojtahedi and his group[34] reported a facile and efficient protection of alcohols/phenols (**1**) with 1,1,1,3,3,3-hexamethyldisilazane (HMDS; **2**) so as to prepare varying silyl ethers (**3**) in good to excellent yields at ambient conditions upon high-intensity ultrasound irradiation without the use of any solvent and catalyst (Scheme 5.2).

5.2.2.2 *Representative Examples*

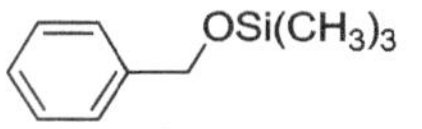

3a

(Benzyloxy)trimethylsilane (**3a**): colorless liquid; time: 10 min; yield: 99%

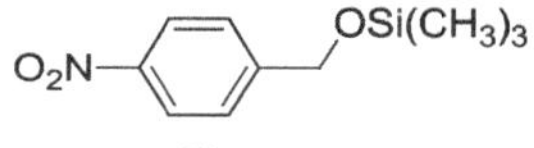

3b

Trimethyl((4-nitrobenzyl)oxy)silane (**3b**): colorless liquid; time: 10 min; yield: 84%

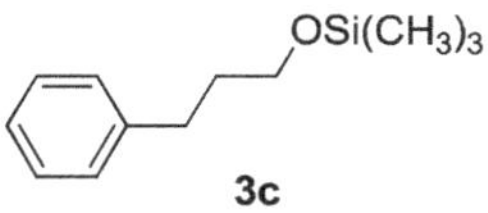

3c

Trimethyl(3-phenylpropoxy)silane (**3c**): colorless liquid; time: 5 min; yield: 99%

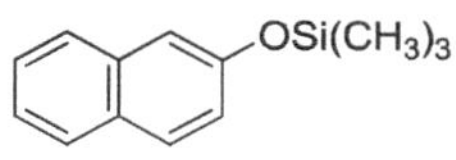

3d

Trimethyl(naphthalen-2-yloxy)silane (**3d**): colorless liquid; time: 10 min; yield: 97%

5.2.2.3 *Experimental Procedure*

A mixture of alcohol/phenol (**1**; 5 mmol) and hexamethyldisilazane (**2**; 3.0 mmol) was sonicated for an appropriate length of time under ambient conditions. Upon completion of the reaction (as monitored by TLC), the crude product was obtained after evaporation of the volatile portion and was then purified using column chromatography to obtain pure products of silyl ether (**3**) with good to excellent yields. The synthesized compounds were identified by comparison of their physical and spectral data with those reported in the literature.

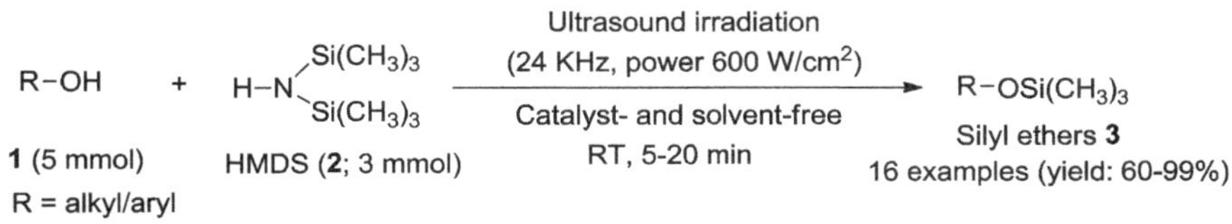

Scheme 5.2 Synthesis of silyl ethers (**3**).

5.2.2.4 Characterization Data of Two Representative Compounds

Trimethyl((4-nitrobenzyl)oxy)silane (**3b**): colorless liquid; yield: 84%; IR (KBr): 1253, 1095, 844 cm^{-1}; ^1H NMR ($CDCl_3$, 80 MHz): δ 7.31–7.19 (4H, m), 4.72 (2H, s), 0.10 (9H, s).

Trimethyl(3-phenylpropoxy)silane (**3c**): colorless liquid; yield: 99%; IR (KBr): 1251, 1100, 841 cm^{-1}; ^1H NMR ($CDCl_3$, 80 MHz): δ 7.22–7.18 (5H, m), 3.60 (2H, t, J=6.5 Hz), 2.68 (2H, t, J=8 Hz,), 1.85 (2H, m), 0.11 (9H, s).

5.2.2.5 Critical Views

Protection of alcohols and phenols is an important task for organic chemists, and the present protocol is superior to the earlier methods[35–48] in terms of the use of no catalysts or solvents. In addition, this protocol offers varying types of silyl ethers with good yields and shows good chemoselectivity for competitive protection of hydroxyl groups in favor of sterically less hindered alcohols. Moreover, this ultrasound-promoted method has been reported to exert exclusive protection of phenols in the presence of aromatic amines as well.[34]

5.2.3 Entry-3: Synthesis of Substituted Thiourea Derivatives

Type of reaction: C–N bond formation
Reaction conditions: Catalyst-free, water–polyethylene glycol, ultrasound irradiation
Synthetic strategy: Condensation reaction
Keywords: Amines, carbon disulfide, water–polyethylene glycol, catalyst-free, ultrasonication, substituted thioureas

5.2.3.1 General Reaction Scheme

Azizi *et al.*[49] reported a rapid and convenient ultrasound-assisted catalyst-free synthesis of both unsymmetrical and symmetrical thiourea derivatives (**3** and **5**) from condensation between aliphatic amines (**1**/**4**) and carbon disulfide (**2**) in water and polyethylene glycol in good yield without any tedious workup (Scheme 5.3).

5.2.3.2 Representative Examples

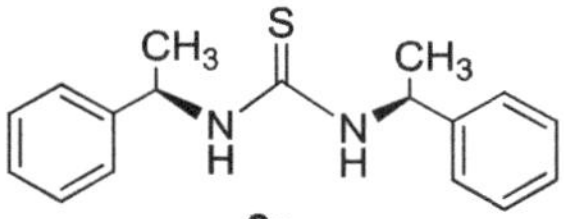

3a

1-((*R*)-1-Phenylethyl)-3-((*S*)-1-phenylethyl)thiourea (**3a**): solid, mp 178–179 °C; yield: 97% (both in water and PEG)

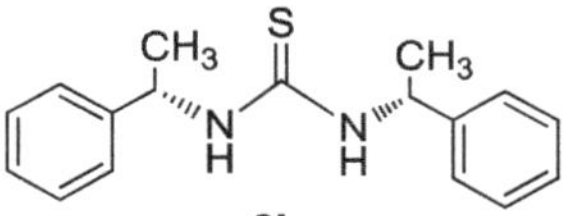

3b

1-((*R*)-1-Phenylethyl)-3-((*S*)-1-phenylethyl)thiourea (**3b**): solid, mp 101–103 °C; yield: 97% (both in water and PEG)

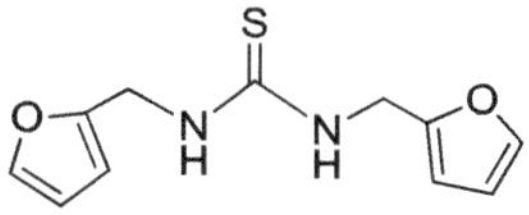

3c

1,3-*Bis*(furan-2-ylmethyl)thiourea (**3c**): solid, mp 69–72 °C; yield: 80% (PEG); 45% (water)

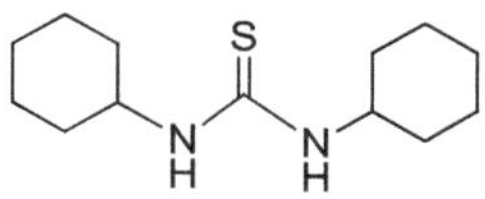

3d

1,3-Dicyclohexylthiourea (**3d**): solid, mp 178–180 °C; yield: 97% (PEG); 90% (water)

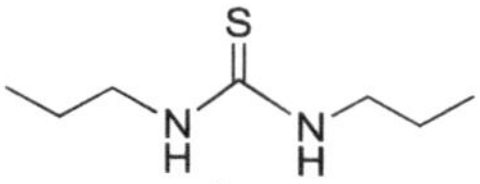

3e

1,3-Dipropylthiourea (**3e**): solid, mp 71–72 °C; yield: 82% (PEG); 27% (water)

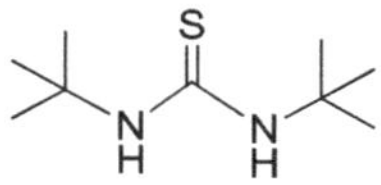

3f

1,3-Di-*tert*-butylthiourea (**3f**): solid, mp 174–175 °C; yield: 72% (PEG); 45% (water)

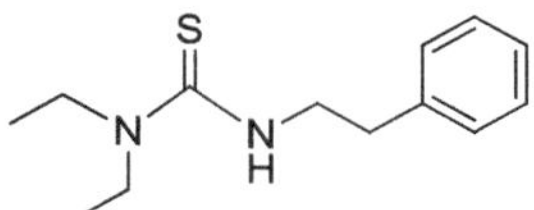

5a

1,1-Diethyl-3-phenethylthiourea (**5a**): solid, mp 67–68 °C; yield: 60% (PEG); 52% (water)

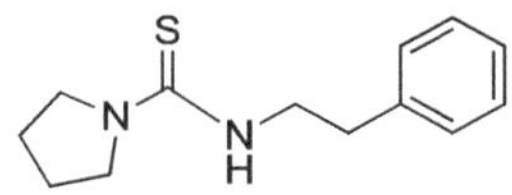

5b

N-Phenethylpyrrolidine-1-carbothioamide (**5b**): oil; yield: 65% (PEG); 60% (water)

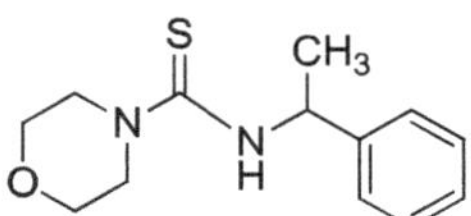

5c

N-(1-Phenylethyl)morpholine-4-carbothioamide (**5c**): oil; yield: 76% (PEG); 80% (water)

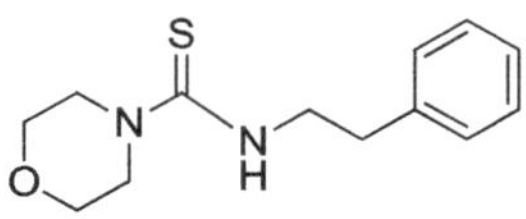

5d

N-Phenethylmorpholine-4-carbothioamide (**5d**): solid, mp 76–78 °C; yield: 68% (PEG); 50% (water)

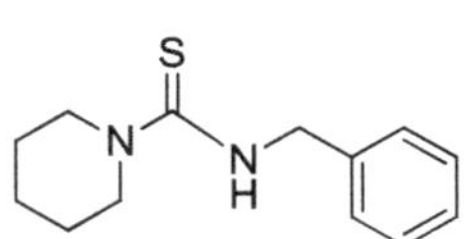

5e

N-Benzylpiperidine-1-carbothioamide (**5e**): solid, mp 88–89 °C; yield: 80% (PEG); 74% (water)

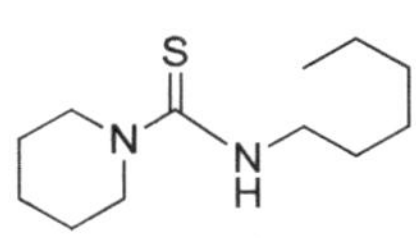

5f

N-Hexylpiperidine-1-carbothioamide (**5f**): solid, mp 87–88 °C; yield: 68% (PEG); 60% (water)

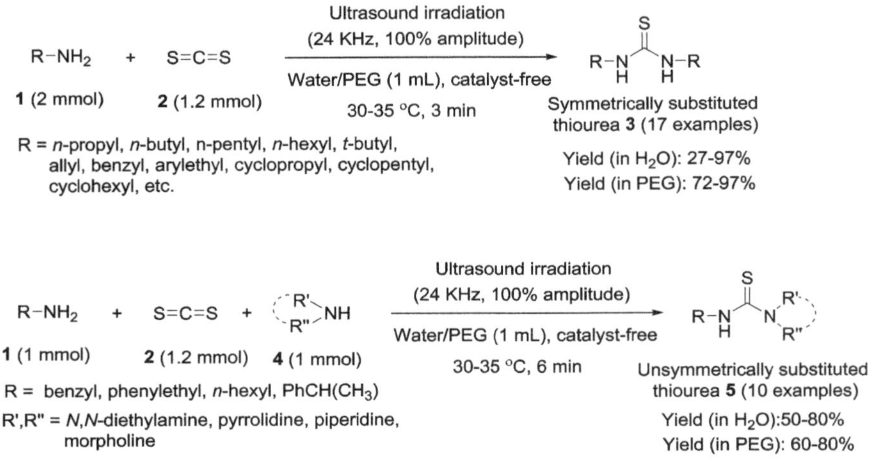

Scheme 5.3 Synthesis of substituted thiourea derivatives (**3**/**5**).

5.2.3.3 Experimental Procedure

To a mixture of carbon disulfide (**2**; 1.2 mmol) and water (1 mL) or PEG (1 mL) was added with amines (**1**/**4**; 2 mmol) placed within a test tube surrounded by water bath for cooling. The mixture was subjected to ultrasound irradiation for 3–6 min at 30–35 °C. On completion of the reaction, the precipitate was filtered off, washed with water, ethanol or diethyl ether and dried in air to afford the corresponding dithiocarbamate derivatives (**3**/**5**) in high purity. No further purification by recrystallization or column chromatography was required in most cases. All compounds were characterized by melting points found to be identical with the respective authentic samples described in the literature.

5.2.3.4 Critical Views

The investigators developed a rapid and straightforward protocol for the catalyst-and additive- free synthesis of both symmetrical and unsymmetrical thiourea derivatives under ultrasound irradiation in water and PEG.[49] Facile and cleaner reaction profiles, higher product yields and reaction rates are the important features of this method. However, this method does not work with aromatic amines. Still the present protocol is more advantageous than the earlier methods[50–70] in terms of greenness, reaction rate and product yield. Thiourea derivatives are important intermediates for the synthesis of natural products and biologically active compounds and they find useful applications particularly in heterocyclic and medicinal chemistry.[71–73]

5.2.4 Entry-4: Synthesis of α-Aminophosphonates

Type of reaction: C–N and C–P bond formation
Reaction conditions: Catalyst-free, solvent-free, room temperature, ultrasound irradiation
Synthetic strategy: One-pot multicomponent reaction
Keywords: Aldehydes, anilines, diethyl phosphonate/triethyl phosphite, solvent- and catalyst-free, room temperature, ultrasonication, α-aminophosphonates

5.2.4.1 *General Reaction Scheme*

In 2007, Xia and Lu[74] reported an ultrasound-promoted solvent- and catalyst-free protocol for a one-pot synthesis of potentially useful α-aminophosphonate derivatives (**4**) from the reaction of aldehydes (**1**), anilines (**2**) and diethyl phosphonate (**3a**) with good to excellent yields (Scheme 5.4). However, Dar *et al.*[75] developed the method at a later date to be more efficient (completed within seconds) by replacing the phosphonating reagent with triethyl phosphite (**3b**) under similar reaction conditions (Scheme 5.5).

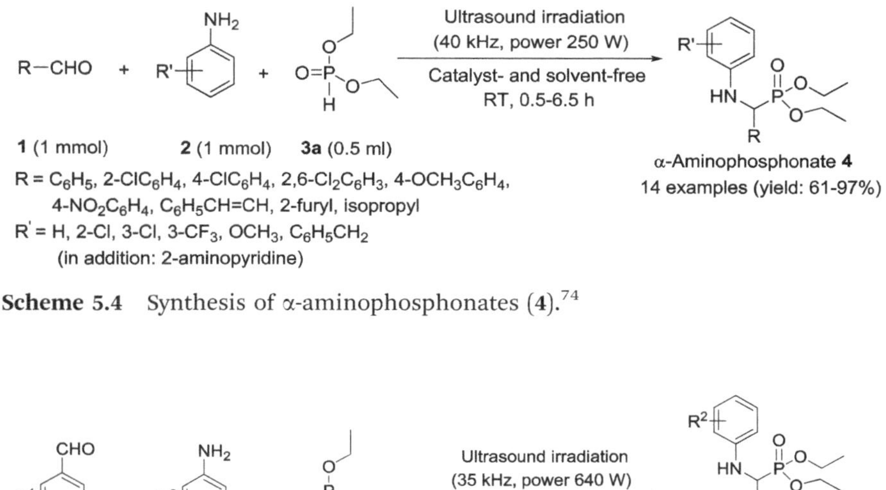

Scheme 5.4 Synthesis of α-aminophosphonates (**4**).[74]

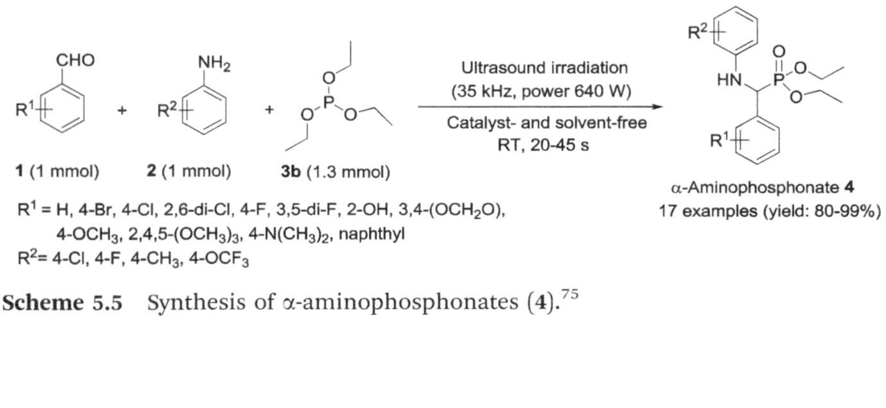

Scheme 5.5 Synthesis of α-aminophosphonates (**4**).[75]

5.2.4.2 Representative Examples

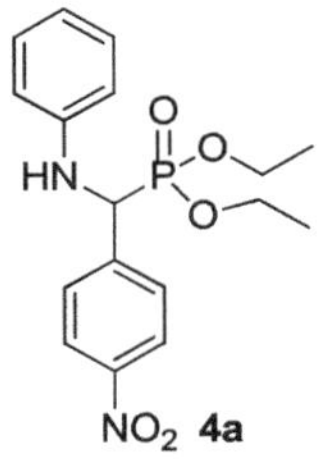

4a

Diethyl ((4-nitrophenyl)(phenylamino)methyl)phosphonate (**4a**): bright yellow solid; time: 2.5 h; yield: 97%

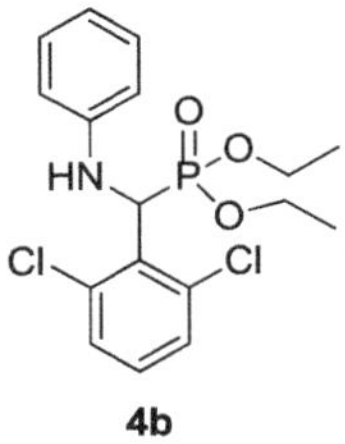

4b

Diethyl ((2,6-dichlorophenyl)(phenylamino)methyl)phosphonate (**4b**): white solid; time: 6.5 h; yield: 61%

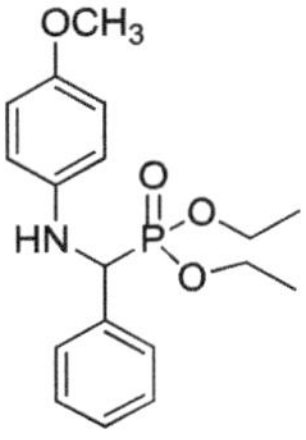

4c

Diethyl (((4-methoxyphenyl)amino)-(phenyl)methyl)phosphonate (**4c**): white solid; time: 1.5 h; yield: 92%

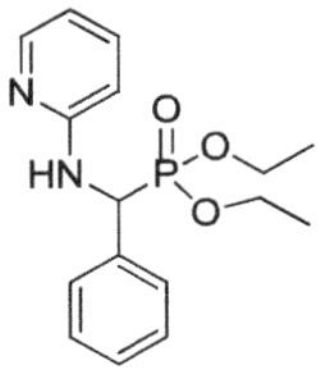

4d

Diethyl (phenyl(pyridin-2-ylamino)methyl)-phosphonate (**4d**): white solid; time: 2.5 h; yield: 84%

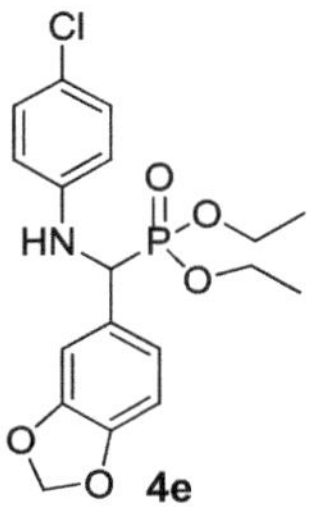

4e

Diethyl (benzo[*d*][1,3]dioxol-5-yl((4-chlorophenyl)amino)methyl)phosphonate (**4e**): solid; time: 20 s; yield: 87%

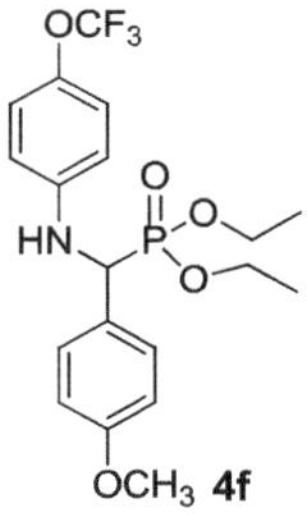

4f

Diethyl ((4-methoxyphenyl)((4-(trifluoromethoxy)phenyl)amino)methyl)phosphonate (**4f**): solid; time: 20 s; yield: 87%

5.2.4.3 Experimental Procedure

A mixture of aldehyde (**1**; 1 mmol), amine (**2**; 1 mmol) and diethyl phosphonate (**3a**; 0.5 mL) or triethyl phosphite (**3b**; 1.3 mmol) was sonicated for an appropriate length of time under ambient conditions. Upon completion of the reaction (as monitored by TLC), the crude mass was extracted with ether or ethyl acetate, concentrated under reduced pressure and then subjected to column chromatographic purification or recrystallization to yield the pure product of α-aminophosphonate (**4**). All the compounds were identified based on their physical and spectral properties.

5.2.4.4 Characterization Data of Two Representative Compounds

Diethyl ((2,6-dichlorophenyl)(phenylamino)methyl)phosphonate (**4b**): white solid; yield: 61%; IR (KBr): 3332, 2983, 2907, 1603, 1500, 1439, 1246, 1030, 970, 777, 749, 694, 580, 564 cm^{-1}; ^1H NMR ($CDCl_3$, 400 MHz): δ 1.12 (3H, t, $J = 7.2$ Hz), 1.33 (3H, t, $J = 7.2$ Hz), 3.84–3.91 (1H, m), 4.03–4.08 (1H, m), 4.20–4.27 (2H, m), 5.33 (1H, s), 5.84 (1H, d, $J = 28.4$ Hz), 6.65 (2H, d, $J = 8.0$ Hz), 6.70 (1H, t, $J = 7.6$ Hz), 7.09–7.14 (3H, m), 7.19 (1H, d, $J = 8.0$ Hz), 7.32 (1H, dd, $J = 8.0$ and 1.0 Hz); CIMS: *m/z* 388 (M^+), 251 (base peak), 205, 134, 112,76, 65; Anal calc. for $C_{17}H_{20}NO_3PCl_2$: C, 52.58; H, 5.15; N, 3.61; found C, 52.66; H, 5.21; N, 3.57.

Diethyl (phenyl(pyridin-2-ylamino)methyl)phosphonate (**4d**): white solid; yield: 84%; IR (KBr): 3307, 3065, 2979, 2935, 1603, 1482, 1420, 1227, 1019, 973, 782, 701, 565 cm^{-1}; ^1H NMR ($CDCl_3$, 400 MHz): δ 1.11 (3H, t, $J = 7.2$ Hz), 1.23 (3H, *t*, $J = 7.2$ Hz), 3.73–3.80 (1H, m), 3.93–3.98 (1H, m), 4.05–4.13 (2H, m), 5.52 (1H, d, $J = 24.4$ Hz), 5.61 (1H, br s), 6.42 (1H, d, $J = 8.0$ Hz), 6.57 (1H, dt, $J = 6.8$ and 1.6 Hz), 7.25–7.28 (1H, m), 7.32–7.35 (3H, m), 7.51 (2H, d, $J = 7.6$ Hz), 8.05 (1H, d, $J = 8.0$ Hz); CIMS: *m/z* 320 (M^+), 183 (base peak), 137, 113, 76, 65, 50; Anal calc. for $C_{16}H_{21}N_2O_3P$: C, 60.00; H, 6.56; N, 8.75; found C, 60.33; H, 6.66; N, 8.81.

5.2.4.5 Critical Views

The ultrasound-assisted synthesis of pharmaceutically promising α-aminophosphonate derivatives[76–84] appeared as a convenient protocol in terms of use of varying catalysts, yields, time, operational simplicity and greenness of the process compared to previously reported methods.[85–101]

5.2.5 Entry-5: Synthesis of 4,4′-(Arylmethylene)bis(3-methyl-1-phenyl-1*H*-pyrazol-5-ol)s

Type of reaction: C–C and C–N bond formation
Reaction conditions: Catalyst-free, aqueous ethanol, ultrasound irradiation, room temperature
Synthetic strategy: One-pot multicomponent reaction
Keywords: Aromatic aldehydes, β-ketoesters, phenyl hydrazines, aqueous ethanol, catalyst-free, ultrasonication, room temperature, 4,4′-(arylmethylene)-bis(3-methyl-1-phenyl-1*H*-pyrazol-5-ol)s

5.2.5.1 General Reaction Scheme

A practical and convenient protocol for the synthesis of a wide range of 4,4′-(arylmethylene)bis(3-methyl-1-phenyl-1*H*-pyrazol-5-ol)s (**4**) *via* a one-pot pseudo-five-component condensation of β-ketoesters (**1**; 2 equiv.) and phenyl hydrazine derivatives (**3**; 2 equiv.) with aromatic aldehydes (**2**; 1 equiv.) was developed by Hasaninejed's group[102] under catalyst-free and ultrasound irradiation conditions at room temperature (Scheme 5.6).

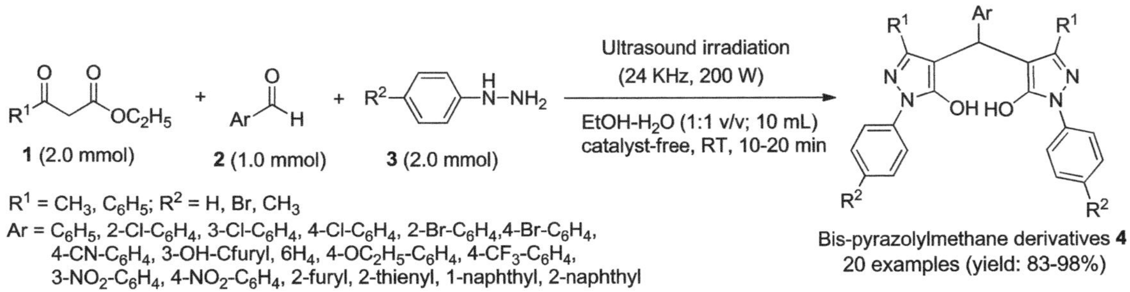

Scheme 5.6 Synthesis of 4,4′-(arylmethylene)bis(3-methyl-1-phenyl-1*H*-pyrazol-5-ol)s (**4**).

5.2.5.2 Plausible Mechanism

The investigators proposed a plausible mechanism for this transformation (Scheme 5.7). Initially, an intermediate (**6**) is formed from the condensation of the phenylhydrazine derivative (**3**) with β-keto ester (**1**), which then tautomerizes to (**7**). This intermediate (**7**) undergoes further condensation with the aromatic aldehyde (**2**) to give the adduct (**8**), which takes part in Michael addition to another molecule of intermediate (**7**) to yield a new adduct (**9**). Finally, after the tautomeric proton shift, the desired product (**4**) is obtained.

5.2.5.3 Representative Examples

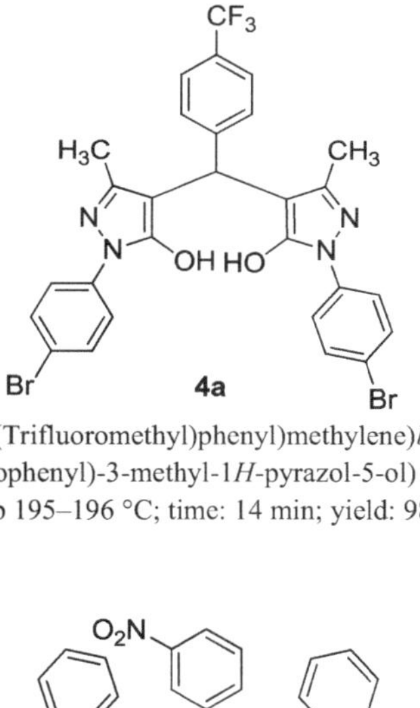

4a

4,4'-((4-(Trifluoromethyl)phenyl)methylene)*bis*(1-(4-bromophenyl)-3-methyl-1*H*-pyrazol-5-ol) (**4a**): solid, mp 195–196 °C; time: 14 min; yield: 98%

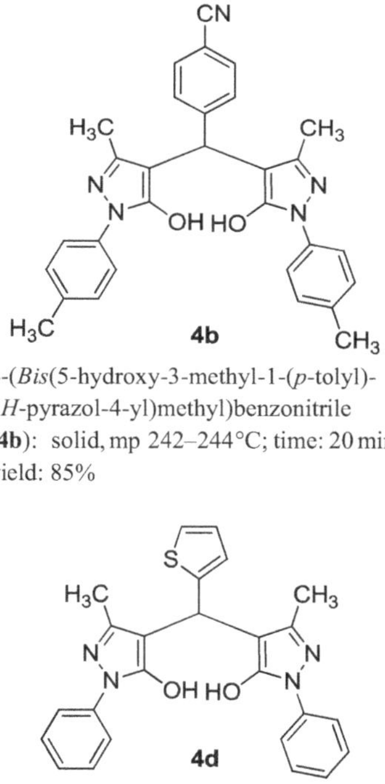

4b

4-(*Bis*(5-hydroxy-3-methyl-1-(*p*-tolyl)-1*H*-pyrazol-4-yl)methyl)benzonitrile (**4b**): solid, mp 242–244°C; time: 20 min; yield: 85%

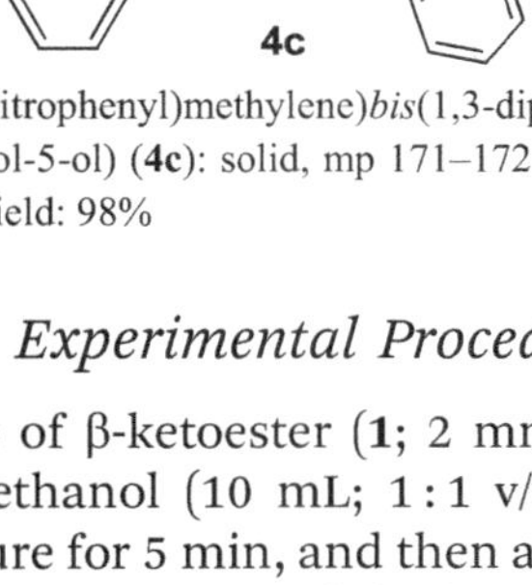

4c

4,4'-((3-Nitrophenyl)methylene)*bis*(1,3-diphenyl-1*H*-pyrazol-5-ol) (**4c**): solid, mp 171–172 °C; time: 10 min; yield: 98%

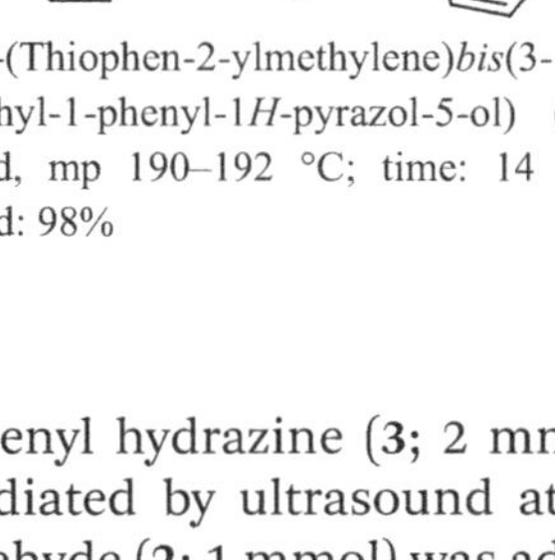

4d

4,4'-(Thiophen-2-ylmethylene)*bis*(3-methyl-1-phenyl-1*H*-pyrazol-5-ol) (**4d**): solid, mp 190–192 °C; time: 14 min; yield: 98%

5.2.5.4 Experimental Procedure

A mixture of β-ketoester (**1**; 2 mmol) and phenyl hydrazine (**3**; 2 mmol) in aqueous ethanol (10 mL; 1 : 1 v/v)) was irradiated by ultrasound at room temperature for 5 min, and then aromatic aldehyde (**2**; 1 mmol) was added to the reaction mixture and the irradiation was continued. The progress of the reaction was monitored by TLC. After the completion of the reaction, the residue was filtered and was recrystallized from hot ethanol to produce

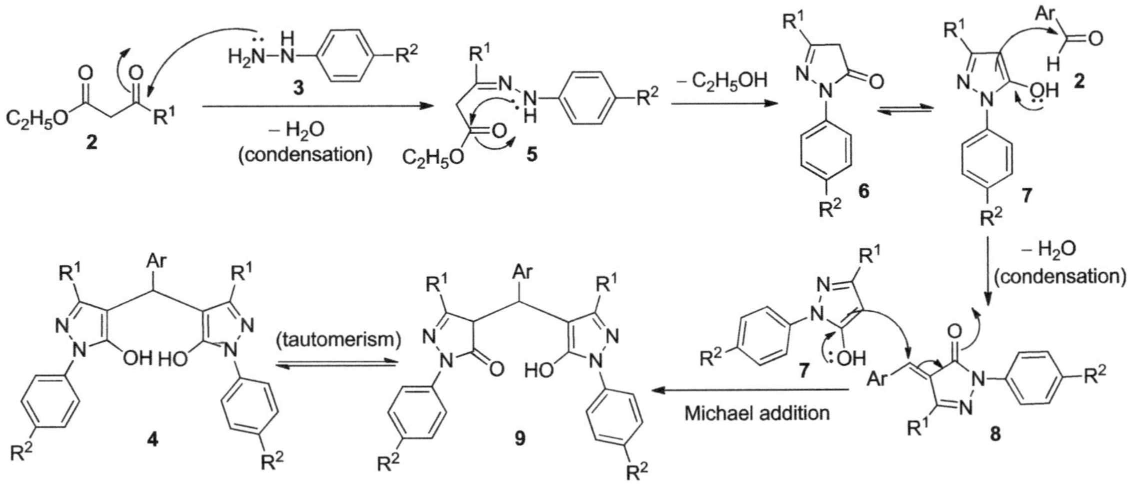

Scheme 5.7 Suggested mechanism for the synthesis of 4,4′-(arylmethylene)bis(3-methyl-1-phenyl-1*H*-pyrazol-5-ol)s (**4**).

the desired pure product of bis-pyrazolylmethane derivatives (**4**) in excellent yields. The structures of the products were confirmed from their physical and spectroscopic data.

5.2.5.5 *Characterization Data of Two Representative Compounds*

4,4′-((4-(Trifluoromethyl)phenyl)methylene)bis(1-(4-bromophenyl)-3-methyl-1*H*-pyrazol-5-ol) (**4a**): solid, mp 195–196 °C; yield: 98%; ^{1}H NMR ($CDCl_3$, 500 MHz): δ 2.3 (6H, s, 2×CH_3), 4.9 (1H, s, CH), 7.34–7.56 (12H, m, Ar–H); MS: *m/z* (%rel) 408 (95), 263 (100), 183 (25),163 (59), 155 (95), 128 (59), 105 (11), 91 (55), 77 (59), 51 (37). Anal. calc. for $C_{28}H_{21}Br_2F_3N_4O_2$: C, 50.78; H, 3.20; N, 8.46%; found: C, 50.59; H, 3.27; N, 8,39%.

4-(Bis(5-hydroxy-3-methyl-1-(*p*-tolyl)-1*H*-pyrazol-4-yl)methyl)benzonitrile (**4b**): solid, mp 242–244 °C; yield: 85%; ^{1}H NMR ($CDCl_3$, 500 MHz): δ 2.3 (6H, s, 2×CH_3), 2.4 (6H, s, 2×CH_3), 4.9 (s, 1H, CH), 7.15 (4H, d, J = 8.2 Hz, Ar–H), 7.35 (2H, d, J = 8.1 Hz, Ar–H), 7.59 – 7.62 (6H, m, Ar–H); ^{13}C NMR ($CDCl_3$, 125 MHz) δ: 10.8, 21.4, 105.6, 122.2, 128.3, 130.2, 131.8, 133.0, 138.6, 144.7, 159.0; MS: *m/z* (%rel) 470 (11), 400 (17), 374 (18), 341 (13), 301 (100), 199 (89), 153 (11), 127 (21), 105 (16), 91 (79), 64 (26). Anal. calc. for $C_{30}H_{27}N_5O_2$: C, 73.60; H, 5.56; N, 14.31%; found: C, 73.81; H, 5.29; N, 14.11%.

5.2.5.6 *Critical Views*

The present method offers a practical and greener alternative way for easy access of such useful organic scaffolds in aqueous ethanol under catalyst-free conditions at room temperature. Compared to the classical methods,[103–110] this protocol is more advantageous in terms of milder reaction conditions, shorter reaction time, and higher yields and use of no other additives.

5.2.6 Entry-6: Synthesis of Substituted Dihydroquinolines

Type of reaction: C–C and C–N bond formation
Reaction conditions: Catalyst-free, water, ultrasound irradiation, heating
Synthetic strategy: One-pot multicomponent reaction
Keywords: Aromatic aldehydes, malononitrile, 2-naphthol/resorcinol, catalyst-free, water, ultrasonication, dihydroquinolines

5.2.6.1 *General Reaction Scheme*

An ultrasound-assisted one-pot synthesis of substituted dihydroquinolines (**5**/**5′**) from catalyst-free multicomponent reaction between aromatic aldehydes (**1**), malononitrile (**2**), ammonium acetate (**3**), and 2-naphthol (**4**) or resorcinol (**4′**), in aqueous medium at 60 °C was developed by

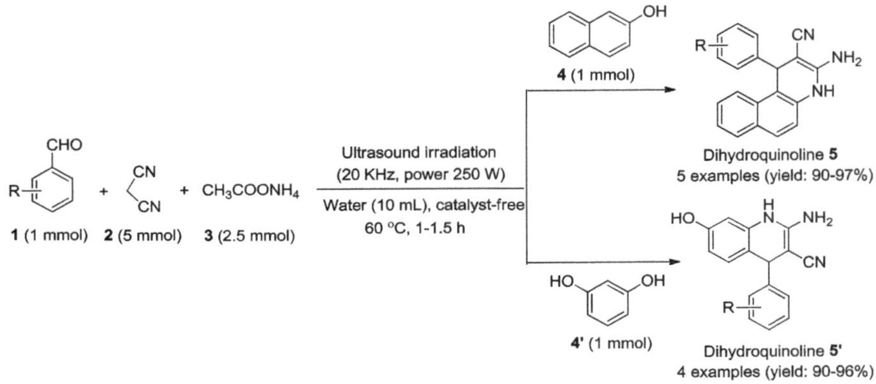

Scheme 5.8 Synthesis of substituted dihydroquinolines (5/5′).

Pagadala *et al.*[111] (Scheme 5.8). The dihydroquinolines were obtained in good yields within 1–1.5 h.

5.2.6.2 Representative Examples

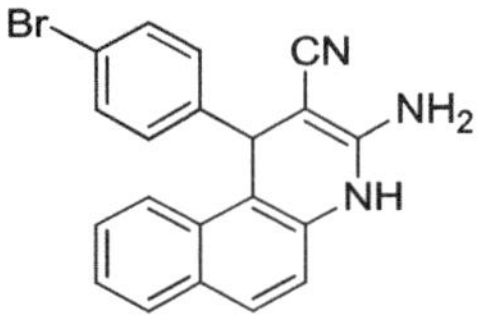

5a

3-Amino-1-(4-bromophenyl)-1,4-dihydrobenzo[*f*]quinoline-2-carbonitrile (**5a**): white solid, mp 195 °C; time: 1 h; yield: 95%

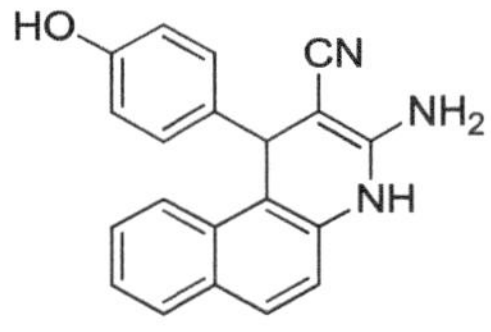

5b

3-Amino-1-(4-hydroxyphenyl)-1,4-dihydrobenzo[*f*]quinoline-2-carbonitrile (**5b**): white solid, mp 217 °C; time: 1.5 h; yield: 90%

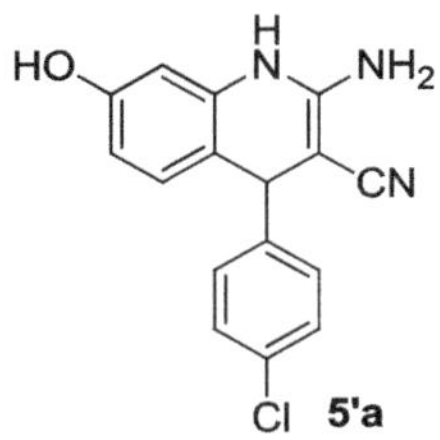

5′a

2-Amino-4-(4-chlorophenyl)-7-hydroxy-1,4-dihydroquinoline-3-carbonitrile (**5′a**): white solid, mp 257 °C; time: 1 h; yield: 92%

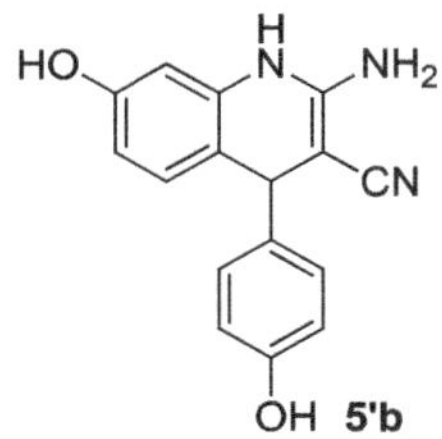

5′b

2-Amino-7-hydroxy-4-(4-hydroxyphenyl)-1,4-dihydroquinoline-3-carbonitrile (**5′b**): off-white solid, mp 289 °C; time: 1 h; yield: 90%

5.2.6.3 Experimental Procedure

Aromatic aldehyde (**1**; 1.0 mmol), malononitrile (**2**; 1.0 mmol), 2-naphthol/resorcinol (**4**/**4′**; 1.0 mmol), water (10 mL) and ammonium acetate (**3**;

2.5 mmol) were added within a 100 mL conical flask in a sequential manner. The reaction flask was then kept in an ultrasonic bath at 60 °C, and irradiated for the period of 1–1.5 h. On completion of the reaction, a solid deposit was collected by the filtration, washed with water and dried. The crude product was recrystallized from ethanol to offer pure product of dihydroquinoline **5**/**5′**. The products were characterized based on their spectral studies.

5.2.6.4 Characterization Data of Two Representative Compounds

3-Amino-1-(4-hydroxyphenyl)-1,4-dihydrobenzo[*f*]quinoline-2-carbonitrile (**5b**): white solid, mp 217 °C; time: 1.5 h; yield: 90%; IR (KBr): 3442, 3321, 3290, 2188 cm^{-1}; ^1H NMR (DMSO-d_6, 400 MHz): δ 5.30 (1H, s), 6.96 (2H, s, $-NH_2$), 7.14–7.29 (4H, m), 7.38–7.49 (3H, m), 7.52 (1H, s, –NH), 7.81–7.95 (3H, m), 9.65 (1H, s, –OH); ^{13}C NMR (DMSO-d_6, 100 MHz): δ 37.2, 57.1, 115.0, 116.7, 120.4, 123.1, 124.2, 126.2, 127.4, 127.7, 128.6, 128.8, 129.4, 130.4, 130.8, 145.3, 146.7, 159.6; EIMS: $m/z = 336$ (M + Na, 100%); Anal. calc. for $C_{20}H_{15}N_3O$: C,76.66; H, 4.82; N, 13.41%; found: C, 76.61; H, 4.85; N, 13.36%.

2-Amino-7-hydroxy-4-(4-hydroxyphenyl)-1,4-dihydroquinoline-3-carbonitrile (**5′b**): Off-white solid, mp 289 °C; time: 1 h; yield: 90%; IR (KBr): 3478, 3348, 3193, 2191 cm^{-1}; ^1H NMR (DMSO-d_6, 400 MHz): δ 4.48 (1H, s), 6.37 (1H, s), 6.45 (1H, d, $J = 8.4$ Hz), 6.65–6.77 (5H, m, Ar–H & $-NH_2$), 6.93 (2H, d, $J = 8.3$ Hz), 7.56 (1H, s, –NH), 9.27 (1H, s, –OH), 9.65 (1H, s, –OH); ^{13}C NMR (DMSO-d_6, 100 MHz): δ 38.0, 56.7, 102.0, 112.2, 114.2, 115.1, 120.7, 128.3, 129.8, 136.7, 148.7, 155.9, 156.8, 160.0; EIMS: $m/z = 280$ (M + 1, 100%); Anal. calc. for $C_{16}H_{13}N_3O_2$: C, 68.81; H, 4.69; N, 15.05%; found: C, 68.86; H, 4.61; N, 15.12%.

5.2.6.5 Critical Views

The present protocol is advantageous over the earlier conventional methods[112–118] in regard to its operational simplicity, short reaction time, use of water as the solvent, and avoidance of a catalyst. This high-yielding process for the biologically important quinoline derivatives may be explored to synthesize a greater variety of such compounds of interest.

5.2.7 Entry-7: Synthesis of Substituted 1,4-Dihydropyridines

Type of reaction: C–C and C–N bond formation
Reaction conditions: Catalyst-free, solvent-free, ultrasound irradiation, room temperature
Synthetic strategy: One-pot multicomponent reaction
Keywords: Aromatic aldehydes, ethyl acetoacetate, ammonium acetate, solvent-free, catalyst-free, ultrasonication, room temperature, 1,4-dihydropyridines (1,4-DHPs)

5.2.7.1 General Reaction Scheme

Wang *et al.*[119] accomplished an ultrasound-promoted solvent- and catalyst-free protocol for the one-pot synthesis of a series of 1,4-dihydropyridines (**4**) from the condensation of aromatic aldehydes (**1**), ethyl acetoacetate (**2**) and ammonium acetate (**3**) in good yields at room temperature (Scheme 5.9).

5.2.7.2 Representative Examples

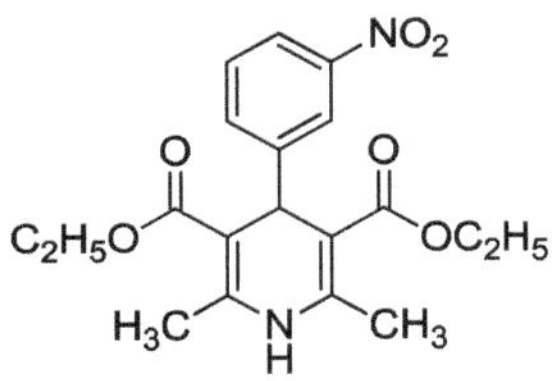

4a

Diethyl 2,6-dimethyl-4-(3-nitrophenyl)-1,4-dihydropyridine-3,5-dicarboxylate (**4a**): yellow solid, mp 161–162 °C; time: 25 min; yield: 98%

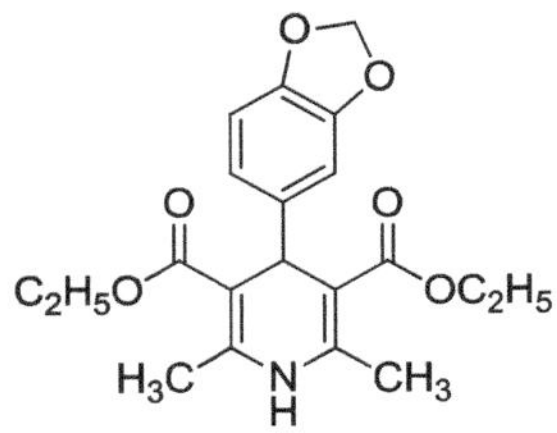

4b

Diethyl 4-(benzo[*d*][1,3]dioxol-5-yl)-2,6-dimethyl-1,4-dihydropyridine-3,5-dicarboxylate (**4b**): yellow solid, mp 125–126 °C; time: 70 min; yield: 90%

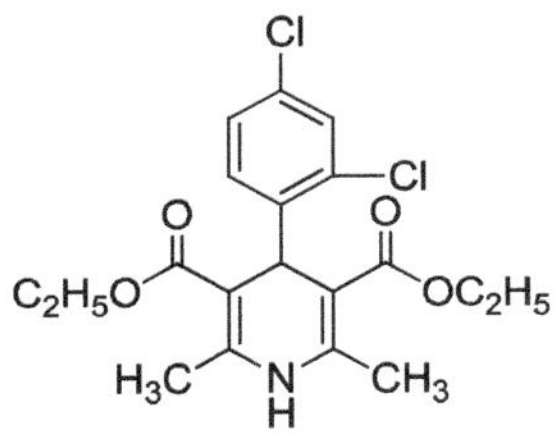

4c

Diethyl 4-(2,4-dichlorophenyl)-2,6-dimethyl-1,4-dihydropyridine-3,5-dicarboxylate (**4c**): yellow solid, mp 109–110 °C; time: 70 min; yield: 82%

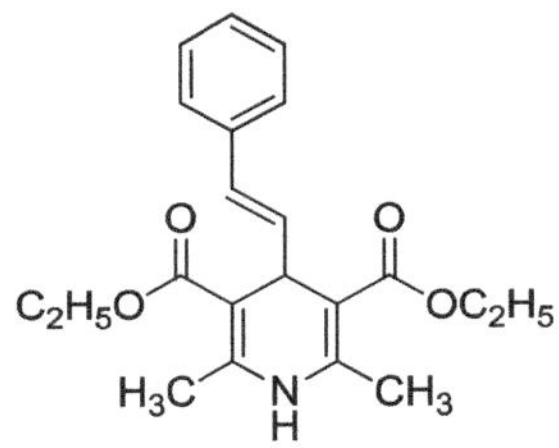

4d

(*E*)-Diethyl 2,6-dimethyl-4-styryl-1,4-dihydropyridine-3,5-dicarboxylate (**4d**): yellow solid, mp 141–143 °C; time: 30 min; yield: 99%

5.2.7.3 Experimental Procedure

A mixture of aldehyde (**1**, 1 mmol), ethyl acetoacetate (**2**, 2.5 mmol) and ammonium acetate (**3**; 1.2 mmol), placed in a 50 mL Pyrex flask, was irradiated in the water bath of the ultrasonic cleaner at 26–35 °C for a stipulated period (25–70 min), and the progress of the reaction was monitored by TLC. After completion of the reaction, the mixture was cooled to room temperature; ice water was added and the product was extracted with ethyl acetate. The organic layer was dried over anhydrous sodium sulfate and concentrated *in vacuo*. The crude products were purified by column chromatography using silica gel (200–300 mesh) and a mixture of petroleum ether and ethyl acetate (3 : 1 v/v) as the eluent to afford pure products of 1,4-dihydropyridines (**4**), characterized based on their physical, elemental and spectral studies.

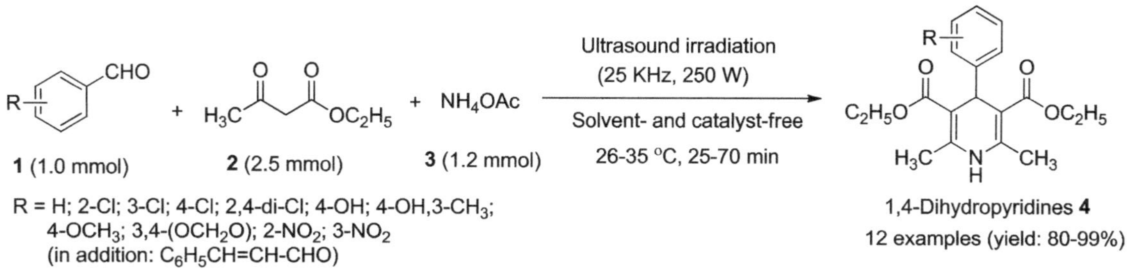

Scheme 5.9 Synthesis of substituted 1,4-dihydropyridines (**4**).

5.2.7.4 *Characterization Data of Two Representative Compounds*

Diethyl 4-(2,4-dichlorophenyl)-2,6-dimethyl-1,4-dihydropyridine-3,5-dicarboxylate (**4c**): yellow solid, mp 109–110 °C; yield: 82%; IR (KBr): 3291, 2979, 1696, 1669, 819 cm^{-1}; ^{1}H NMR ($CDCl_3$, 400 MHz): δ 1.24 (6H, t, 2×CH_2CH_3), 2.13 (3H, s, pyr–CH_3), 2.49 (3H, s, pyr–CH_3), 4.13 (4H, q, 2×CH_2CH_3), 5.58 (1H, s, NH), 5.94 (1H, d, pyr–H), 7.22–7.39 (3H, m, Ph–H); ^{13}C NMR ($CDCl_3$, 100 MHz): δ 14.6, 14.8, 19.8, 21.6, 51.6, 60.2, 60.4, 104.5, 108.2, 127.9, 129.9, 130.7, 133.2, 134.7, 137.1, 149.2, 154.3, 166.5, 167.9. Anal. calc. for $C_{19}H_{21}NO_4Cl_2$: C, 57.53; H, 5.16; N, 3.38; found: C, 57.30; H, 5.31; N, 3.52.

(*E*)-Diethyl 2,6-dimethyl-4-styryl-1,4-dihydropyridine-3,5-dicarboxylate (**4d**): yellow solid, mp 141–143 °C; yield: 99%; IR (KBr): 3335, 2981, 1688, 1645, 759 cm^{-1}; ^{1}H NMR ($CDCl_3$, 400 MHz): δ 1.29 (6H, t, 2×CH_2CH_3), 2.32 (6H, s, pyr–CH_3), 4.19 (4H, q, 2×CH_2CH_3), 4.62 (1H, d, pyr–H), 5.74 (1H, s, NH), 6.17 (1H, q, pyr–CH=CH–), 6.23 (1H, d, Ph–CH=CH–), 7.13–7.31 (5H, m, Ph–H); ^{13}C NMR ($CDCl_3$, 100 MHz): δ 14.9, 19.9, 36.9, 60.2, 101.9, 126.6, 127.3, 128.5, 128.8, 132.2, 138.2, 145.2, 168.0. Anal. calc. for $C_{21}H_{25}NO_4$: C, 71.15; H, 8.93; N, 3.78; found: C, 71.96; H, 7.09; N, 3.94.

5.2.7.5 *Critical Views*

1,4-Dihydropyridines (1,4-DHPs) find useful applications for their wide range of pharmaceutical and biological properties.[120] The present protocol for the synthesis of such organic scaffolds, using ultrasound irradiation without a solvent or catalyst at room temperature, is an improved version of the Hantzsch reaction offering certain advantages over the conventional methods,[121–138] particularly in terms of a cleaner reaction profile, milder conditions, shorter reaction time, catalyst-use, and higher yields.

5.2.8 Entry-8: Synthesis of Substituted Quinoxalines

Type of reaction: C–C and C–N bond formation
Reaction conditions: Catalyst-free, absolute ethanol/absolute ethanol–acetic acid, ultrasound irradiation, room temperature
Synthetic strategy: Condensation reaction
Keywords: 1,2-Diketones, 1,2-diamines, aromatic amines, absolute ethanol/absolute ethanol–acetic acid, catalyst-free, ultrasonication, room temperature, quinoxalines

5.2.8.1 *General Reaction Scheme*

Guo *et al.*[139] reported an ultrasound-promoted highly efficient and facile method for the synthesis of quinoxaline derivatives in good to excellent yields by a condensation reaction of various 1,2-diketones with 1,2-diamines, in absolute ethanol or ethanol–acetic acid medium without the aid of a catalyst, at room temperature (Scheme 5.10).

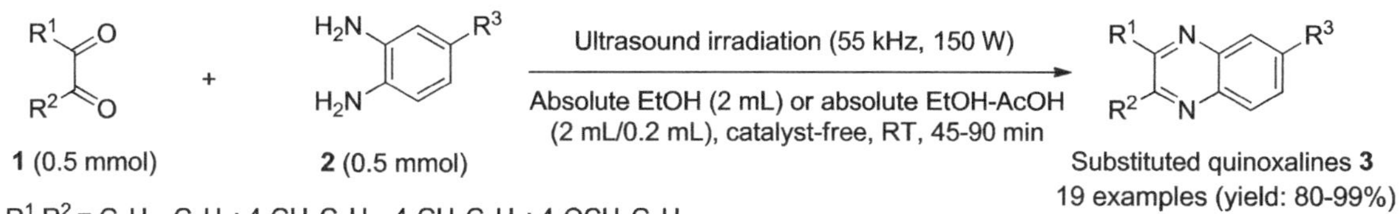

R^1,R^2 = C_6H_5, C_6H_5; 4-$CH_3C_6H_4$, 4-$CH_3C_6H_4$; 4-$OCH_3C_6H_4$, 4-$OCH_3C_6H_4$; C_6H_5, 4-ClC_6H_4; furyl, furyl; C_2H_5, C_2H_5

R^3 = H, Br, CH_3, NO_2, C_6H_5CO (in addition: naphthalene-1,2-diamine)

Scheme 5.10 Synthesis of substituted quinoxalines (**3**).

5.2.8.2 Representative Examples

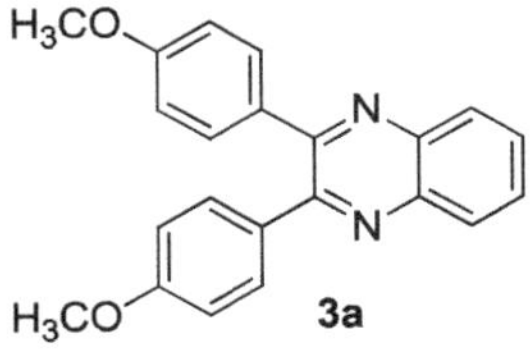

3a

2,3-*Bis*(4-methoxyphenyl)quinoxaline (**3a**): white solid, mp 146–148 °C; time: 60 min; yield: 90%

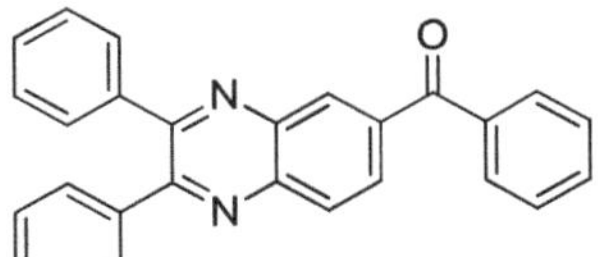

3b

(2,3-Diphenylquinoxalin-6-yl)(phenyl)methanone (**3b**): white solid, mp 139–140 °C; time: 60 min; yield: 95%

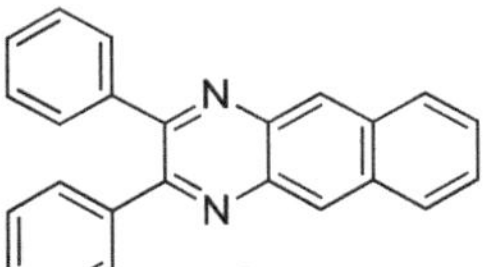

3c

2,3-Diphenylbenzo[*g*]quinoxaline (**3c**): yellow solid, mp 188–189 °C; time: 90 min; yield: 93%

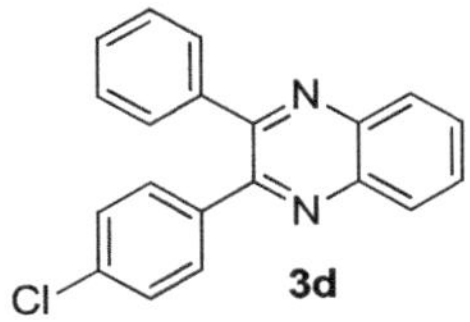

3d

2-(4-Chlorophenyl)-3-phenylquinoxaline (**3d**): white solid, mp 140–142 °C; time: 45 min; yield: 94%

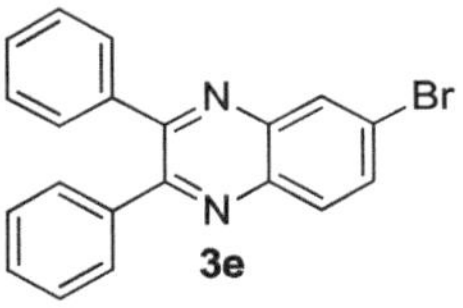

3e

6-Bromo-2,3-diphenylquinoxaline (**3e**): pale yellow solid, mp 154–155 °C; time: 60 min; yield: 97%

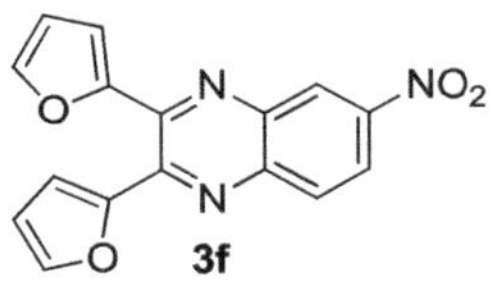

3f

2,3-Di(furan-2-yl)-6-nitroquinoxaline (**3f**): orange solid, mp 164–166 °C; time: 60 min; yield: 96%

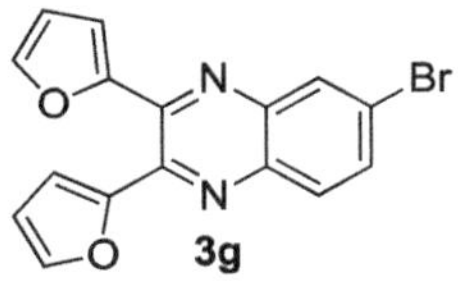

3g

6-Bromo-2,3-di(furan-2-yl)quinoxaline (**3g**): dark brown solid, mp 134–136 °C; time: 60 min; yield: 92%

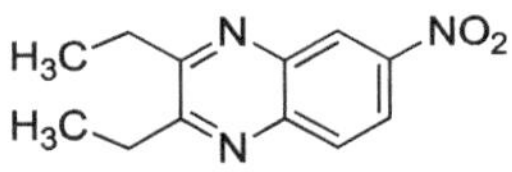

3h

2,3-Diethyl-6-nitroquinoxaline (**3e**): pale yellow solid, mp 98–100 °C; time: 60 min; yield: 90%

5.2.8.3 Experimental Procedure

A mixture of 1,2-diketone (**1**; 0.5 mmol), 1,2-diamine (**2**; 0.5 mmol), and absolute ethanol (2 mL) or absolute ethanol–acetic acid (2 mL:0.2 mL), was irradiated with ultrasound in an open Schlenk tube at room temperature (22–25 °C) until completion of the reaction (45–90 min, as monitored by TLC). After the reaction was completed, the reaction mixture was concentrated *in vacuo* and the residue was purified by flash column chromatography over silica gel (300–400 mesh) using a petroleum ether–ethyl acetate mixture as the eluent to obtain the pure product of quinoxalines (**3**), characterized based on physical and spectral studies.

5.2.8.4 Characterization Data of Two Representative Compounds

(2,3-Diphenylquinoxalin-6-yl)(phenyl)methanone (**3b**): white solid, mp 139–140 °C; yield: 95%; IR (KBr): 3432, 3052, 1661, 1438, 1396, 1344, 1270, 698 cm^{-1}; ^{1}H NMR ($CDCl_3$, 300 MHz): δ 8.54 (1H, s), 8.28 (2H, s), 7.91 (2H, d, $J = 7.4$ Hz), 7.63 (1H, d, $J = 7.3$ Hz), 7.51–7.57 (6H, m), 7.34–7.40 (6H, m); ^{13}C NMR ($CDCl_3$, 75 MHz): δ 195.74, 155.12, 154.58, 142.96, 140.18, 138.63, 138.30, 137.20, 132.41, 130.09, 129.80, 129.74, 129.25, 129.11, 128.48, 128.33.

2,3-Di(furan-2-yl)-6-nitroquinoxaline (**3f**): orange solid, mp 164–166 °C; yield: 96%; IR (KBr): 3388, 1574, 1522, 1477, 1337, 749 cm^{-1}; ^{1}H NMR ($CDCl_3$, 300 MHz): δ 8.98 (1H, d, $J = 2.4$ Hz), 8.47 (1H, dd, $J = 9.2$ and 2.5 Hz,), 8.20 (1H, d, $J = 9.2$ Hz), 7.65–7.68 (2H, m), 6.85 (2H, dd, $J = 16.4$ and 3.5 Hz,), 6.60–6.63 (2H, m); ^{13}C NMR ($CDCl_3$, 75 MHz): δ 149.81, 149.75, 147.57, 145.14, 144.66, 144.34, 143.84, 142.63, 138.87, 130.10, 124.99, 123.99, 123.28, 115.04, 114.17, 112.10, 111.96.

5.2.8.5 Critical Views

Quinoxalines are important heterocyclic compounds finding immense application in organic and medicinal chemistry. Compared to previously reported methodologies,[140–162] the present protocol offers a handful of benefits including operational simplicity, milder reaction conditions at ambient temperature, higher reaction rates and excellent yields, and eco-friendliness.

5.2.9 Entry-9: Synthesis of Substituted Dispiropyrrolizidines

Type of reaction: C–C and C–N bond formation
Reaction conditions: Catalyst-free, methanol, ultrasound irradiation, room temperature
Synthetic strategy: 1,3-Dipolar cycloaddition reaction
Keywords: Substituted (*E*)-3-benzylidene-3,4-dihydroquinolin-2(1*H*)-ones (dipolarophiles), isatin, L-proline, methanol, catalyst-free, ultrasonication, room temperature, 1,3-dipolar cycloaddition, dispiropyrrolizidines, regioselectivity, stereoselectivity

5.2.9.1 General Reaction Scheme

An ultrasound-assisted catalyst-free synthesis of a novel tetracyclic framework of dispiropyrrolizidines (**4**), in moderate to good yields, was developed by Ge *et al.*[163] This was achieved *via* a 1,3-dipolar cycloaddition reaction of azomethine ylide (derived from the reaction between isatin, **2**, and L-proline, **3**) with substituted (*E*)-3-benzylidene-3,4-dihydroquinolin-2(1*H*)-ones acting as dipolarophiles (**1**) in methanol at room temperature (Scheme 5.11).

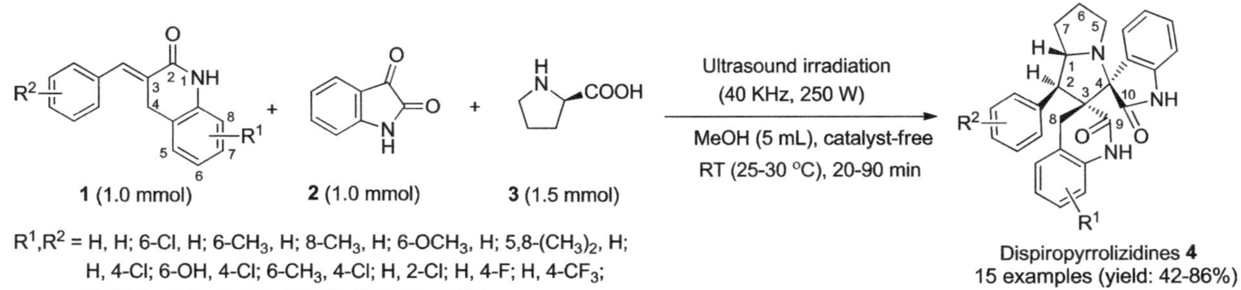

R^1,R^2 = H, H; 6-Cl, H; 6-CH_3, H; 8-CH_3, H; 6-OCH_3, H; 5,8-$(CH_3)_2$, H; H, 4-Cl; 6-OH, 4-Cl; 6-CH_3, 4-Cl; H, 2-Cl; H, 4-F; H, 4-CF_3; 6-OCH_3, 4-Br; 6-CH_3, 2,4-Cl_2; 6-OCH_3, 4-CH_3

Scheme 5.11 Synthesis of substituted dispiropyrrolizidines (**4**).

5.2.9.2 Representative Examples

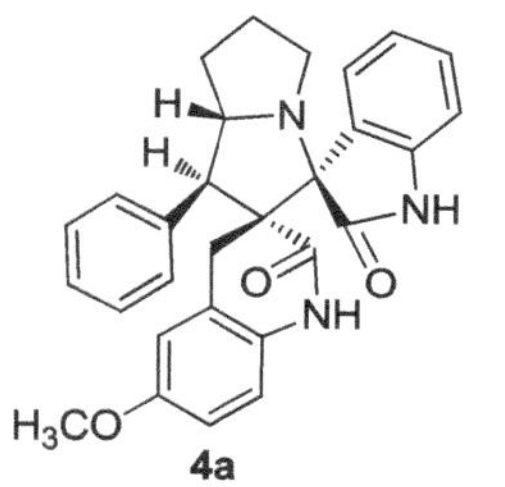

4a

Dispiropyrrolizidine **4a**: yellowish solid, mp 134–135 °C; time: 40 min; yield: 80%

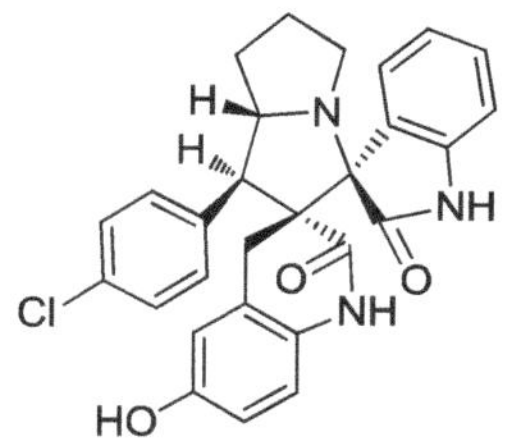

4b

Dispiropyrrolizidine **4b**: yellow solid, mp 130–132 °C; time: 30 min; yield: 84%

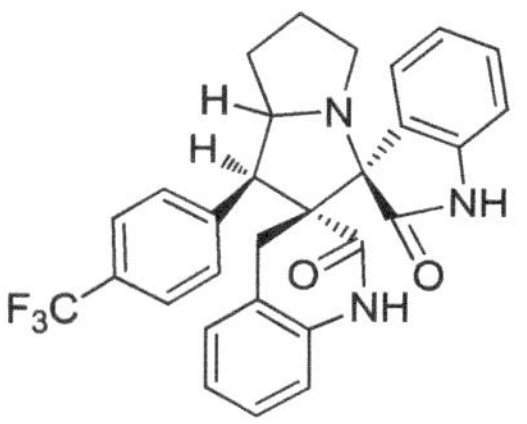

4c

Dispiropyrrolizidine **4c**: yellow solid, mp 136–137 °C; time: 20 min; yield: 86%

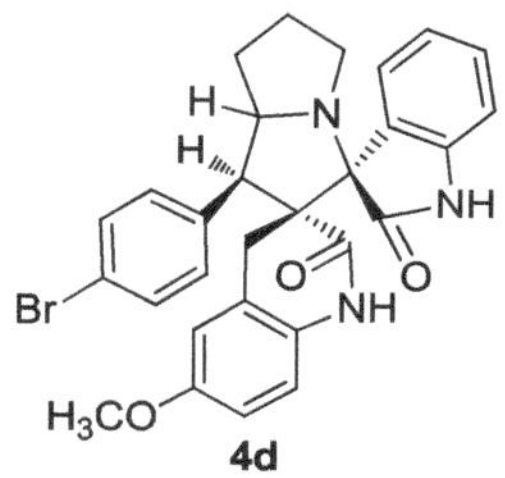

4d

Dispiropyrrolizidine **4d**: yellow solid, mp 129–130 °C; time: 35 min; yield: 75%

5.2.9.3 Experimental Procedure

Substituted (*E*)-3-benzylidene-3,4-dihydroquinolin-2(1*H*)-one (**1**; the dipolarophile derived from aza-Claisen rearrangement of Baylis–Hillman amines; 1.0 mmol) was added to the solution of isatin (**2**; 1.0 mmol) and L-proline (**3**; 1.5 mmol) in methanol (10 mL) at room temperature. The resulting mixture was then irradiated with ultrasound for an appropriate time (indicated by TLC) at 25–30 °C. Upon completion of the reaction, water was added to the solution and ethyl acetate (2×10 mL) was used to extract the mixture. The combined organic layers were dried over anhydrous sodium sulfate and the solvent was removed on a rotary evaporator. The residue was purified using silica column chromatography with ether–dichloromethane (1 : 5–1 : 2) as the eluent to afford the product of dispiropyrrolizidine (**4**) in moderate to good yields. All the synthesized compounds were characterized with the help of detailed spectral studies.

5.2.9.4 Characterization Data of Two Representative Compounds

Dispiropyrrolizidine (**4b**): yellow solid, mp 130–132 °C; yield: 84%; IR (KBr): 3383 (NH), 3257 (NH), 1706 (C=O),1667 (C=O) cm^{-1}; ^{1}H NMR (DMSO-d_6, 400 MHz): δ 1.56 (1H, m, H-7a), 1.69 (1H, m, H-6a), 1.79 (1H, d, J = 15.6 Hz, H-8a), 1.94 (2H, m, H-6b,7b), 2.33 (1H, t, J = 7.2 Hz, H-5a), 3.05 (1H, m, H-5b), 3,09 (1H, d, J = 16.0 Hz, H-8b), 4.48 (1H, d, J = 10.8 Hz, H-2), 4.54 (1H, m, H-1), 5.94 (1H, d, J = 7.6 Hz, Ar–H), 6.19 (1H, q, J = 2.0 Hz,

Ar–H), 6.28 (1H, s, Ar–H), 6.37 (1H, d, $J=7.6$ Hz, Ar–H), 6.84 (1H, t, $J=7.6$ Hz, Ar–H), 7.02 (1H, d, $J=7.6$ Hz, Ar–H), 7.23 (1H, d, $J=7.6$ Hz, Ar–H), 7.42 (4H, dd, $J=8.8$ and 16.8 Hz, Ar–H), 9.87 (1H, s, NH), 9.99 (1H, s, NH); ^{13}C NMR (DMSO-d_6, 100 MHz): δ 24.58 (C-6), 27.80 (C-7), 32.65 (C-8), 47.31(C-2), 52,41 (C-5), 59.65 (C-3), 65.22 (C-1), 73.33 (C-4), 108.48 (C–Ar), 113.50 (C–Ar), 114.17 (C–Ar), 114.77 (C–Ar), 120.38 (C–Ar), 122.88 (C–Ar), 126.15 (C–Ar), 126.42 (C–Ar), 128.40 (C–Ar), 128.63 (C–Ar), 129.03 (C–Ar), 131.07 (C–Ar), 131.41(C–Ar), 137.38 (C–Ar), 142.25 (C–Ar), 152.26 (C–Ar), 171.06 (C-10), 176.27 (C-9); EIMS: *m/z* (%rel): 485(M^+, 1), 284 (M–1–$C_{12}H_{12}N_2O$, 98), 268 (36), 250 (50), 232 (41), 220 (15), 200 (M–$C_{16}H_{12}NO_2Cl$, 100), 172 (10), 144 (21), 132 (18), 125 (34), 89 (10), 77(9).

Dispiropyrrolizidine **4c**: yellow solid, mp 136–137 °C; yield: 86%; IR (KBr): 3219 (NH), 1705 (C=O), 1669 (C=O) cm^{-1}; ^{1}H NMR (DMSO-d_6, 400 MHz): δ 1.58 (1H, m, H-7a), 1.71 (1H, m, H-6a), 1.83 (1H, d, $J=15.6$ Hz, H-8a), 1.96 (2H, m, H-6b,7b), 2.36 (1H, t, $J=8.0$ Hz, H-5a), 3.12 (1H, dd, $J=8.0$ and 8.0 Hz, H-5b), 3.21 (1H, d, $J=15.2$ Hz, H-8b), 4.58 (1H, d, $J=8.4$ Hz, H-2), 4.66 (1H, m, H-1), 6.13 (1H, d, $J=7.2$ Hz, Ar–H), 6.30 (1H, d, $J=7.6$ Hz, Ar–H), 6.75 (2H, m, Ar–H), 6.87 (2H, t, $J=7.2$ Hz, Ar–H), 7.01 (1H, t, $J=7.6$ Hz, Ar–H), 7.23 (1H, d, $J=7.6$ Hz, Ar–H), 7.63 (2H, d, $J=8.0$ Hz, Ar–H), 7.76 (2H, d, $J=8.0$ Hz, Ar–H), 9.98 (s, 1H, NH), 10.17 (1H, s, NH); ^{13}C NMR (DMSO-d_6, 100 MHz): δ 24.55 (C-6), 27.83 (C-7), 32.61 (C-8), 47.33 (C-2), 53.05 (C-5), 59.81 (C-3), 65.50 (C-1), 73.55 (C-4), 108.54 (C–Ar), 113.20 (C–Ar), 120.53 (C–Ar), 121.51 (C–Ar), 121.83 (C–Ar), 125.31 (C–Ar), 125.87 (C–Ar), 126.26 (C–Ar), 127.00 (C–Ar), 127.34 (C–Ar), 128.08 (C–Ar), 128.78 (C–Ar), 130.16 (C–Ar), 136.93 (C–Ar), 142.16 (C–Ar), 143.31 (C–Ar), 171.94 (C-10), 176.25 (C-9); EIMS: *m/z* (%rel): 503 (M^+, 2), 400 (5), 303(M–$C_{12}H_{12}N_2O$, 42), 284 (9), 200 (M–$C_{17}H_{12}NOF_3$, 100), 171 (8), 144 (9), 133 (8), 117 (7), 106 (7), 91 (6), 77 (7).

5.2.9.5 Critical Views

Pyrrolizidines, a class of important alkaloids, are widely distributed in natural products and exhibit a variety of remarkable and attractive bioactivities. In the present method, the investigators were successful in the effective catalyst-free synthesis of a series of novel tetracyclic frameworks of biologically relevant dispiropyrrolizidines, *via* the 1,3-dipolar cycloaddition of azomethine ylide with substituted dipolarophiles, using ultrasound irradiation and room temperature. Mild reaction conditions, a clean reaction profile, considerable reaction rates and good yields, as well as high regio- and stereoselectivity are the notable advantages of this present protocol.[163]

5.2.10 Entry-10: Synthesis of 7-Methyl-substituted Pyrido[4,3-*d*]pyrimidine Derivatives

Type of reaction: C–C and C–N bond formation
Reaction conditions: Catalyst-free, ethanol, ultrasound irradiation, room temperature

Synthetic strategy: Three-component aza-Diels–Alder reaction
Keywords: 6-[2-(Dimethylamino)prop-1-enyl]-1,3-dimethyluracil, aromatic aldehydes, ammonium acetate, ethanol, catalyst-free, ultrasonication, room temperature, 7-methyl-substituted pyrido[4,3-*d*]pyrimidine derivatives

5.2.10.1 General Reaction Scheme

Ultrasound-assisted synthesis of a series of substituted pyrido[4,3-*d*]pyrimidine derivatives (**4**) was accomplished by Sarmah and Prajapati.[164] *via* the three-component 1,3-dipolar cycloaddition reaction of isatin (**1**; 1 mmol), sarcosine (**2**; 1 mmol) and 5-arylidene-1,3-thiazolidine-2,4-dione/5-arylidene-4-thioxo-1,3-thiazolidine-2-one (**3**; 1 mmol) in ethanol at ambient conditions without the aid of a catalyst (Scheme 5.12).

5.2.10.2 Representative Examples

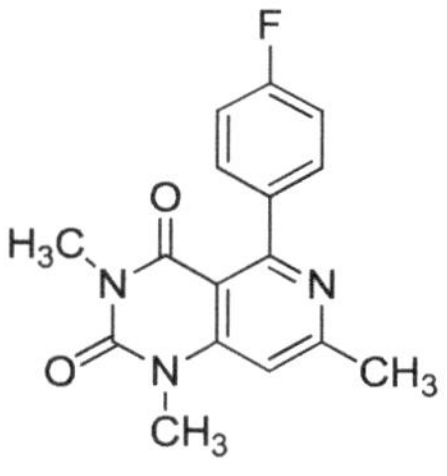

4a

5-(4-Fluorophenyl)-1,3,7-trimethylpyrido[4,3-*d*]pyrimidine-2,4(1*H*,3*H*)-dione (**4a**): off-white solid, mp 217–219 °C; yield: 85%

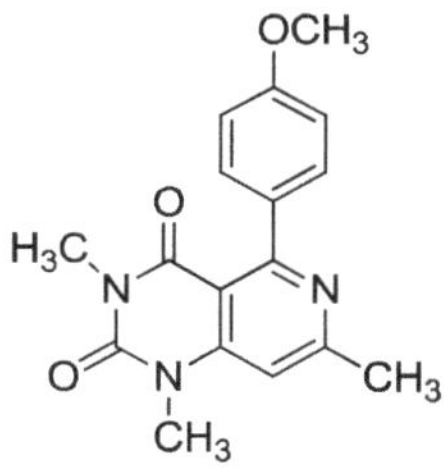

4b

5-(4-Methoxyphenyl)-1,3,7-trimethylpyrido[4,3-*d*]pyrimidine-2,4(1*H*,3*H*)-dione (**4b**): yellow solid, mp 155–156 °C; yield 83%

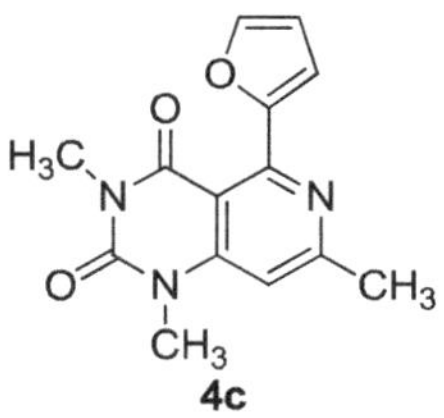

4c

5-(Furan-2-yl)-1,3,7-trimethylpyrido[4,3-*d*]pyrimidine-2,4(1*H*,3*H*)-dione (**4c**): dark brown solid, mp 151–153 °C; yield: 79%

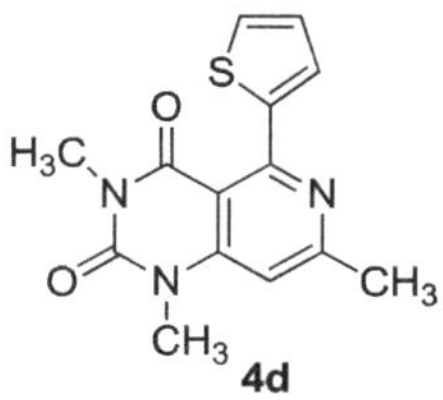

4d

1,3,7-Trimethyl-5-(thiophen-2-yl)pyrido[4,3-*d*]pyrimidine-2,4(1*H*,3*H*)-dione (**4d**): dark brown solid, mp 179–180 °C; yield: 80%

5.2.10.3 Experimental Procedure

A mixture of 6-[2-(dimethylamino)prop-1-enyl]-1,3-dimethyluracil (**1**; 1 mmol), aldehydes (**2**; 1 mmol), and ammonium acetate (**3**; 2 mmol) in ethanol (5 mL) was ultrasonicated for 30 min to complete the reaction. Upon

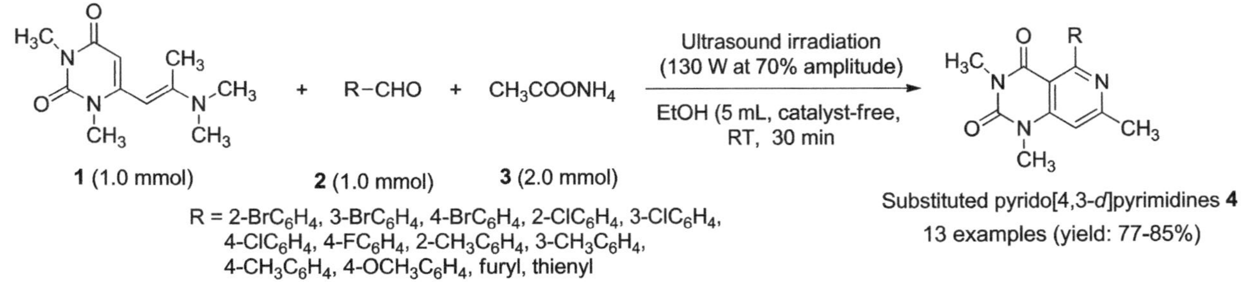

Scheme 5.12 Synthesis of 7-methyl-substituted pyrido[4,3-*d*]pyrimidine derivatives (**4**).

completion, the solvent was removed under reduced pressure, and the crude products were purified by column chromatography eluting with ethyl acetate–hexane (2 : 1) to afford pure products of 7-methyl-substituted pyrido[4,3-*d*]pyrimidines (**4**). All the products were characterized from their analytical and spectral measurements.

5.2.10.4 Characterization Data of Two Representative Compounds

5-(4-Fluorophenyl)-1,3,7-trimethylpyrido[4,3-*d*]pyrimidine-2,4(1*H*,3*H*)-dione (**4a**): off-white solid, mp 217–219 °C; yield: 85%; ^{1}H NMR ($CDCl_3$, 300 MHz): δ 7.37–7.04 (4 H, m, Ar–H), 6.92 (1 H, s, CH=CCH_3), 3.57 (3 H, s, NCH_3), 3.30 (3 H, s, NCH_3), 2.66 (3 H, s, CH_3); ^{13}C NMR ($CDCl_3$,75 MHz): δ 164.6, 162.2, 161.6, 160.1, 150.9, 148.0, 136.4, 130.4 (2C), 114.9 (2C), 106.9, 106.2, 31.1, 28.6, 25.2; MS (GCMS): *m/z* 299 $[M]^+$. Anal. calc. for $C_{16}H_{14}FN_3O_2$: C, 64.21; H, 4.71; N, 14.04; O, 10.69; found: C, 64.20; H, 4.68; N, 14.00; O, 10.61.

1,3,7-Trimethyl-5-(thiophen-2-yl)pyrido[4,3-*d*]pyrimidine-2,4(1*H*,3*H*)-dione (**4d**): dark brown solid, mp 179–180 °C; yield: 80%; ^{1}H NMR ($CDCl_3$, 300 MHz): δ 7.93–7.13 (3 H, m, Ar–H), 6.89 (1 H, s, CH=CCH_3), 3.43 (3H, s, NCH_3), 3.17 (3H, s, NCH_3), 2.47 (3H, s, CH_3); ^{13}C NMR ($CDCl_3$,75 MHz): δ 162.5, 152.6, 149.7, 136.4, 135.8, 134.0, 129.2, 128.3, 126.9, 107.3, 103.4, 32.4, 28.1, 26.2; MS (GCMS): *m/z* 287 $[M]^+$. Anal. calc. for $C_{14}H_{13}N_3O_2S$: C, 58.52; H, 4.56; N, 14.62; O, 11.14; found: C, 58.51; H, 4.54; N, 14.62; O, 11.13.

5.2.10.5 Critical Views

The investigators developed a convenient multicomponent strategy for the preparation of a series of 7-methyl-substituted pyrido[4,3-*d*]pyrimidine derivatives at ambient conditions in the absence of any catalyst under ultrasonication. They also performed the same transformation under refluxing (for 4 h) the mixture of the corresponding substrates in ethanol, but the ultrasound-assisted protocol was found to be more efficient in regard to the respective yield and reaction rate.[164]

5.2.11 Entry-11: Synthesis of 6*H*-1-Benzopyrano[4,3-*b*]quinolin-6-ones

Type of reaction: C–C and C–N bond formation
Reaction conditions: Catalyst-free, methanol, ultrasound irradiation, room temperature
Synthetic strategy: Condensation reaction
Keywords: 4-Chloro-2-oxo-2*H*-chromene-3-carbaldehyde, aromatic amines, methanol, catalyst-free, ultrasonication, room temperature, 6*H*-1-benzopyrano[4,3-*b*]quinolin-6-ones

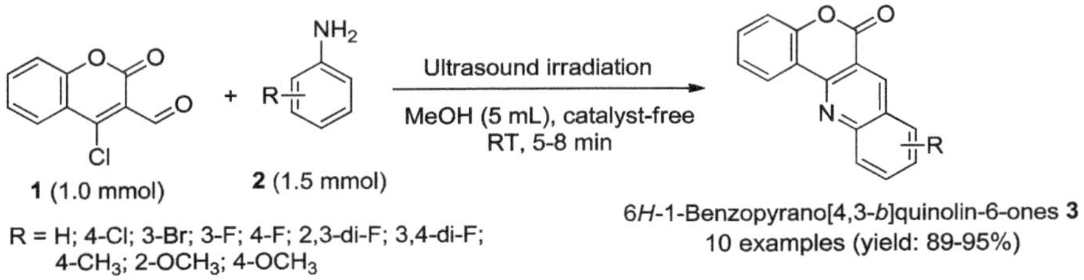

Scheme 5.13 Synthesis of 6*H*-1-benzopyrano[4,3-*b*]quinolin-6-ones (**3**).

5.2.11.1 General Reaction Scheme

An ultrasound-assisted, simple and catalyst-free synthesis of a novel series of 6*H*-1-benzopyrano[4,3-*b*]quinolin-6-ones (**3**) was accomplished with excellent yields by Mulakayala *et al.*[165] from the condensation of 4-chloro-2-oxo-2*H*-chromene-3-carbaldehyde (**1**; 1 mmol) with various aromatic amines (**2**; 1.5 mmol) in methanol at room temperature (Scheme 5.13).

5.2.11.2 Representative Examples

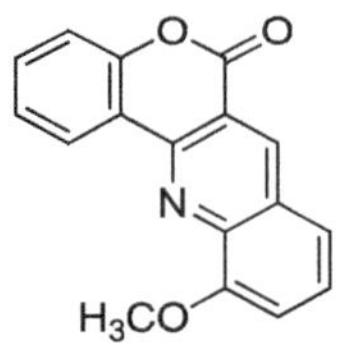

3a

11-Methoxy-6*H*-chromeno[4,3-*b*]quinolin-6-one (**3a**): white solid, mp 236–238 °C; time: 7 min; yield: 92%

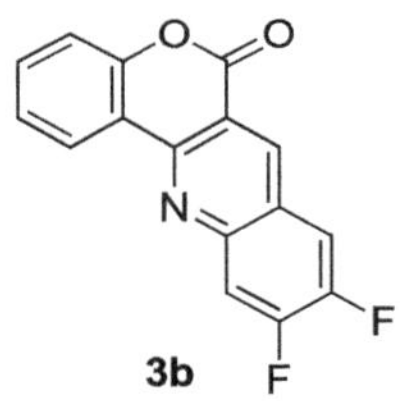

3b

9,10-Difluoro-6*H*-chromeno[4,3-*b*]quinolin-6-one (**3b**): white solid, mp 252–254 °C; time: 8 min; yield: 90%

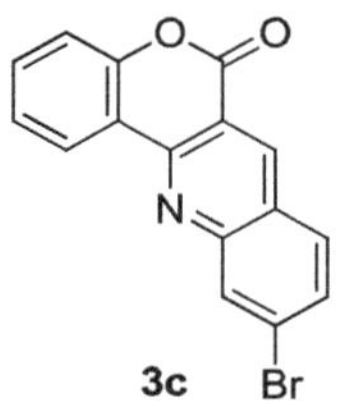

3c

10-Bromo-6*H*-chromeno[4,3-*b*]quinolin-6-one (**3c**): white solid, mp 246–248 °C; time: 7 min; yield: 89%

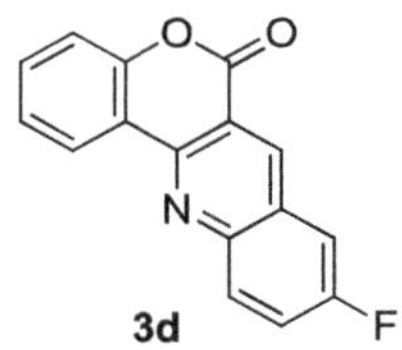

3d

9-Fluoro-6*H*-chromeno[4,3-*b*]quinolin-6-one (**3d**): white solid, mp 225–227 °C; time: 6 min; yield: 93%

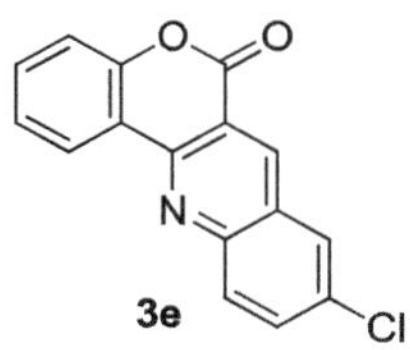

3e

9-Chloro-6*H*-chromeno[4,3-*b*]quinolin-6-one (**3e**): white solid, mp 243–245 °C; time: 6 min; yield: 93%

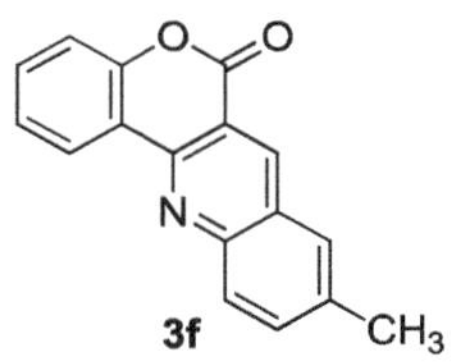

3f

9-Methyl-6*H*-chromeno[4,3-*b*]quinolin-6-one (**3f**): white solid, mp 233–235 °C; time: 5 min; yield: 89%

5.2.11.3 Experimental Procedure

A solution of 4-chloro-2-oxo-2*H*-chromene-3-carbaldehyde (**1**; 1 mmol) and aromatic amine (**2**; 1.5 mmol) in methanol (5 mL) placed within a capped reaction vial (10 mL) was irradiated with ultrasound for 5–8 min. After completion of the reaction (confirmed by TLC), the reaction mixture was concentrated under vacuum, and the crude residue was purified by flash chromatography (*n*-hexane–ethylacetate 3 : 1) to afford the desired product of 6*H*-1-benzopyrano[4,3-*b*]quinolin-6-one (**3**). The products were identified from their detailed spectral studies.

5.2.11.4 Characterization Data of Two Representative Compounds

11-Methoxy-6*H*-chromeno[4,3-*b*]quinolin-6-one (**3a**): white solid, mp 236–238 °C; yield: 92%; IR (KBr): 3063, 2836, 1735, 1604, 1380, 1177, 751 cm^{-1}; ^{1}H NMR (DMSO-d_6, 400 MHz): δ 9.2 (1H, s), 8.86 (1H, dd, $J = 7.6$ and 1.2 Hz), 7.57–7.63 (3H, m), 7.39–7.46 (2H, m), 7.26 (1H, d, $J = 5.6$ Hz,), 4.17 (3H, s); ^{13}C NMR ($CDCl_3$, 100 MHz): δ 160.7, 153.3, 152.4, 150.1, 136.0, 129.8, 129.7, 127.8, 127.0, 126.6, 125.9, 124.4, 122.1, 110.0, 114.9, 112.3, 56.2; ESMS: *m/z* 278 (M + 1, 10%); HRMS: calc. for $C_{17}H_{12}NO_3$: 278.0817; found 278.0819.

9,10-Difluoro-6*H*-chromeno[4,3-*b*]quinolin-6-one (**3b**): white solid, mp 252–254 °C; yield: 90%; IR (KBr): 3070, 2930, 1740, 1610, 1470, 1185 cm^{-1}; ^{1}H NMR ($CDCl_3$, 400 MHz): δ 9.24 (1H, s), 8.85 (1H, dd, $J = 8.0$ and 1.6 Hz), 7.82–7.86 (1H, m), 7.62–7.66 (1H, m), 7.40–7.57 (3H, m); ^{13}C NMR ($CDCl_3$, 100 MHz): δ 160.8, 152.7, 150.0, 140.1, 140.0, 132.7, 125.2, 125.1, 119.1, 117.4, 115.8, 115.6, 114.4, 114.3, 114.2, 114.1; HRMS: calc. for $C_{16}H_8F_2NO_2$: 284.0467; found: 284.0465.

5.2.11.5 Critical Views

The investigators synthesized a novel series of 6*H*-1-benzopyrano[4,3-*b*]quinolin-6-ones by reacting 4-chloro-2-oxo-2*H*-chromene-3-carbaldehyde with various aromatic amines *via* a catalyst-free method in the presence of ultrasound under ambient conditions. The reaction proceeds very fast (within 5–8 min) and offers excellent yields of products. Some of the products showed potent anticancer activity *in vitro*.

5.2.12 Entry-12: Synthesis of Rhodanines

Type of reaction: C–N and C–S bond formation
Reaction conditions: Catalyst-free, water, room temperature, ultrasound irradiation
Synthetic strategy: One-pot multicomponent reaction
Keywords: Primary amines, carbon disulfide, dimethyl/diethyl acetylenedicarboxylate, catalyst-free, water, room temperature, ultrasonication, rhodanines

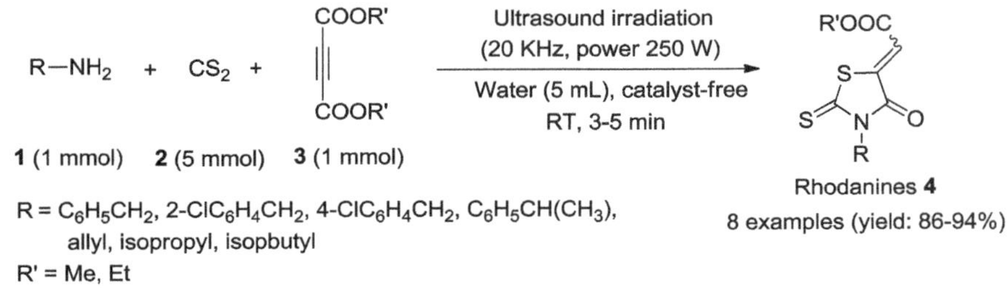

Scheme 5.14 Synthesis of rhodanines (**4**).

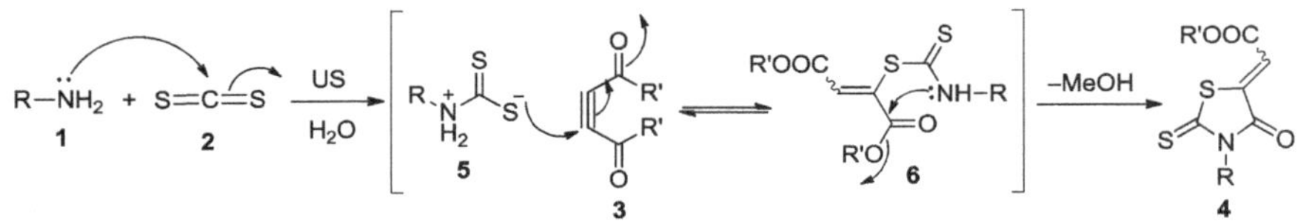

Scheme 5.15 Suggested mechanism for the synthesis of rhodanines (**4**).

5.2.12.1 General Reaction Scheme

Rostamnia and Lamei[166] developed a rapid and high-yielding ultrasound-assisted green protocol for the one-pot synthesis of biologically potent rhodanines (**4**). This was achieved with a three-component reaction of primary amines (**1**), carbon disulfide (**2**) and dimethyl/diethyl acetylenedicarboxylate (**3**) in water under catalyst-free ambient conditions (Scheme 5.14). This straightforward procedure affords rhodanine products just by simple filtration.

5.2.12.2 Plausible Mechanism

The investigators proposed a plausible mechanism for the sonochemical synthesis of rhodanines (**4**) and is shown in Scheme 5.15. Initially, a dithiocarbamate species (**5**) is formed from the reaction between the amine (**1**) and carbon disulfide (**2**), which then undergoes addition of the negatively charged sulfur to the alkyne carbone of the acetylenedicarboxylate (**3**) to afford intermediate vinyl dithiocarbamate (**6**). Finally, intramolecular cyclization of **6** yields the rhodanine (**4**).

5.2.12.3 Representative Examples

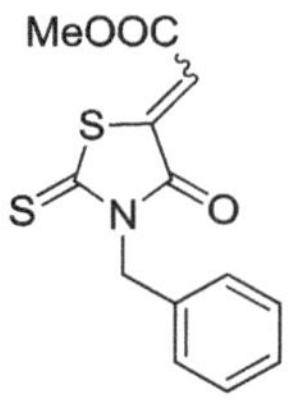

4a

Methyl 2-(3-benzyl-4-oxo-2-thioxothiazolidin-5-ylidene)acetate (**4a**): orange powder, mp 128–130 °C; time: 3 min; yield: 94%

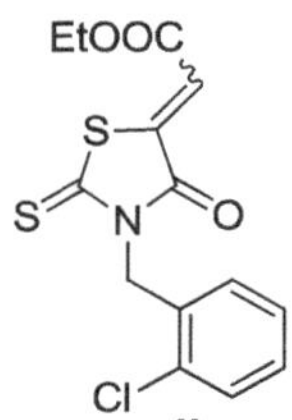

4b

Ethyl 2-(3-(2-chlorobenzyl)-4-oxo-2-thioxothiazolidin-5-ylidene)acetate (**4b**): yellow powder, mp 104–106 °C; time: 4 min; yield: 92%

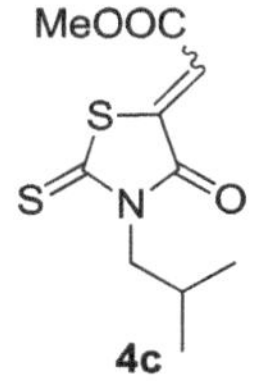

4c

Methyl 2-(3-isobutyl-4-oxo-2-thioxothiazolidin-5-ylidene)acetate (**4c**): yellow powder, mp 88–90 °C; time: 5 min; yield: 88%

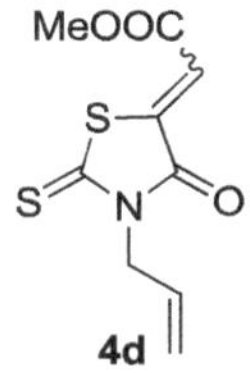

4d

Methyl 2-(3-allyl-4-oxo-2-thioxothiazolidin-5-ylidene)acetate (**4d**): orange powder, mp 140–142 °C; time: 5 min; yield: 86%

5.2.12.4 Experimental Procedure

To a mixture of a primary amine (**1**; 1 mmol) and carbon disulfide (CS_2; **2**; 5 mmol) in water (5 mL) was added dialkyl acetylenedicarboxylate (**3**; 1 mmol), and the resulting mixture was subjected to ultrasonic irradiation for 3–5 min using a high-density ultrasonic probe. Upon completion of the reaction as indicated by TLC, the mixture was filtered and the crude residue was washed with water, followed by drying to afford pure rhodanine (**4**) in excellent yields. All the products were characterized by comparison of their spectral and physical data with those of authentic samples previously reported by the same group of investigators.[167]

5.2.12.5 Characterization Data of Two Representative Compounds

Ethyl 2-(3-(2-chlorobenzyl)-4-oxo-2-thioxothiazolidin-5-ylidene)acetate (**4b**): yellow powder, mp 104–106 °C; yield: 92%; IR (KBr): 1711 (C=O), 1683 (C=C), 1335 and 1192 (C=S) cm^{-1}; ^1H NMR ($CDCl_3$, 500.13 MHz): δ 1.37 (3H, t, $J = 7.2$ Hz, OCH_2CH_3), 4.34 (2H, q, $J = 7.2$ Hz, OCH_2CH_3), 5.41 (2H, s, $ArCH_2$), 6.88 (1H, s, C=CH), 6.90 (1H, d, $J = 7.6$ Hz, Ar–H), 7.18 (1H, t, $J = 7.4$ Hz, Ar–H), 7.23 (1H, t, $J = 7.7$ Hz, Ar–H), 7.4 (1H, d, $J = 7.8$ Hz, Ar–H); ^{13}C NMR ($CDCl_3$, 125.7 MHz): δ 14.15 (OCH_2CH_3), 45.10 (OCH_2CH_3), 62.23 ($ArCH_2$), 117.90 (C=CH), 126.97, 127.11, 128.98 and 129.86 (4 CH of Ar), 131.50 (C_{ipso}–Cl), 132.98 (C_{ipso}–CH_2), 141.50 (C=CH), 165.04 (CON), 166.35 (CO_2Et), 195.32 (C=S); EIMS: *m/z* (% rel.) 342 ($[M+1]^+$, 2), 341 (M^+, 2), 306 (100), 278 (26), 182 (6), 148 (7), 125 (26), 89 (14), 63 (4), 53 (4). Anal. calc. for $C_{14}H_{12}ClNO_3S_2$ (341.83): C, 49.19; H, 3.54; N, 4.10; found: C, 49.50; H, 3.50; N, 4.00.

Methyl 2-(3-allyl-4-oxo-2-thioxothiazolidin-5-ylidene)acetate (**4d**): orange powder, mp 140–142 °C; yield: 86%; IR (KBr): 1716 (C=O), 1683 (C=C), 1331 and 1199 (C=S) cm^{-1}; ^1H NMR ($CDCl_3$, 500.13 MHz): δ 3.88 (3H, s, OCH_3), 4.71–4.72 (2H, m, $CH_2CH{=}CH_2$), 5.25–5.30 (2H, m, $CH_2CH{=}CH_2$), 5.75–5.90 (1H, m, $CH_2CH{=}CH_2$), 6.84 (1H, s, C=CH); ^{13}C NMR ($CDCl_3$, 125.7 MHz): δ 46.19 (OCH_3), 52.83 ($CH_2CH{=}CH_2$), 116.84 ($C{=}CHCO_2CH_3$), 119.78

($CH_2CH{=}CH_2$), 129.2 ($CH_2CH{=}CH_2$), 142.08 ($C{=}CHCO_2CH_3$), 165.52 (CON), 166.14 (CO_2CH_3), 195.15 (C=S); EIMS: *m/z* (%rel.) 243 (M^+, 47), 228 (100), 116 (28), 98 (25), 85 (79), 72 (32), 59 (26), 43 (61), 41 (79). Anal. calc. for $C_9H_9NO_3S_2$ (243.29): C, 44.43; H, 3.73; N, 5.76; found: C, 44.50; H, 3.60; N, 5.60.

5.2.12.6 Critical Views

The present ultrasound-promoted method offers a green and straightforward way to synthesize rhodanine derivatives in water without the aid of any catalyst in a one-pot. The reaction is not only very fast (completes within few minutes), but it avoids any further purification and thus a notable alternative to multistep classical methods.[168–173] Rhodanines are very useful molecules found to inhibit numerous targets such as HCV NS3 protease,[174] β-lactamase,[175] PMT1 mannosyl transferase,[176] and PRL-3 and JSP-1 phosphatases.[177,178]

5.2.13 Entry-13: Synthesis of Formamidines

Type of reaction: C–N bond formation
Reaction conditions: Catalyst-free, acetonitrile, ultrasound irradiation
Synthetic strategy: Addition reaction
Keywords: 2-Mercaptobenzothiazole/2-mercaptobenzoxazole, isocyanide, acetonitrile, catalyst-free, ultrasonication, formamidines

5.2.13.1 General Reaction Scheme

The catalyst-free synthesis of a series of formamidine scaffolds (**3**) from the addition reaction of isocyanides (**2**) with 2-mercaptobenzothiazole/2-mercaptobenzoxazole (**1**) in moderate to high yields under ultrasound irradiation at reflux temperature (50 °C) was reported by Zhu *et al.*[179] (Scheme 5.16).

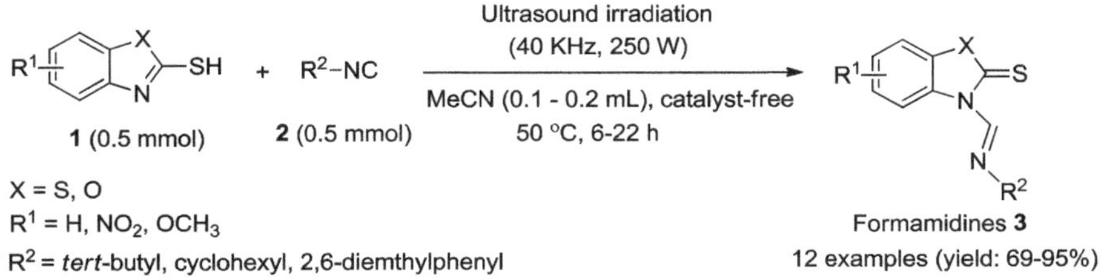

Scheme 5.16 Synthesis of formamidines (**3**).

5.2.13.2 Representative Examples

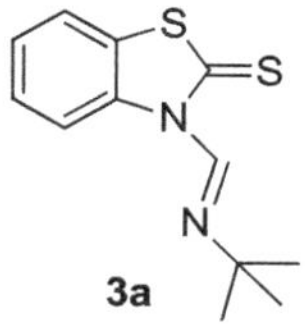

3a

(*E*)-3-((*tert*-Butylimino)methyl)benzo[*d*]thiazole-2(3*H*)-thione (**3a**): white solid, mp 105–106 °C; yield: 83%

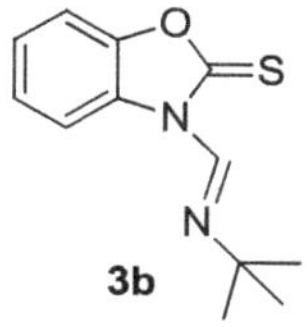

3b

(*E*)-3-((*tert*-Butylimino)methyl)benzo[*d*]oxazole-2(3*H*)-thione (**3b**): white solid, mp 144–146 °C; yield: 73%

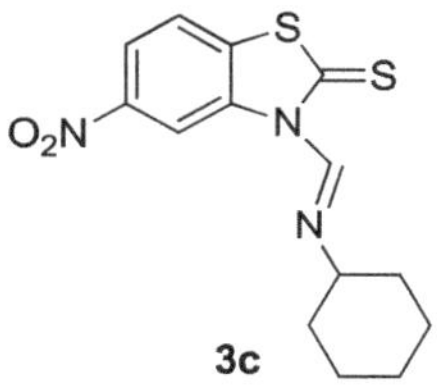

3c

(*E*)-3-((Cyclohexylimino)methyl)-5-nitrobenzo[*d*]thiazole-2(3*H*)-thione (**3c**): yellow solid, mp 242–244 °C; yield: 95%

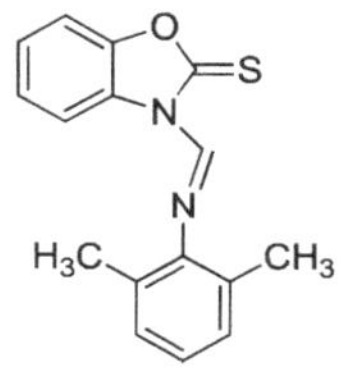

3d

(*E*)-3-(((2,6-Dimethylphenyl)imino)methyl)benzo[*d*]oxazole-2(3*H*)-thione (**3d**): white solid, mp 166–167 °C; yield: 69%

5.2.13.3 Experimental Procedure

To a mixture of 2-mercaptobenzothiazole/2-mercaptobenzoxazole (**1**; 0.5 mmol) and isocyanide (**2**; 0.5 mmol) was added acetonitrile (0.1–0.2 mL) within a reaction vessel, which was then irradiated by ultrasound in an appropriate time (6–22 h) at 50 °C until the thio-compound was completely consumed (as monitored by TLC). Upon completion of the reaction, the solvent was evaporated under the reduced pressure, and the residue was purified by flash column chromatography with ethyl acetate and petroleum ether as eluent to afford pure product of formamidines (**3**). The synthesized products were characterized based on spectral studies.

5.2.13.4 Characterization Data of Two Representative Compounds

(*E*)-3-((*tert*-Butylimino)methyl)benzo[*d*]thiazole-2(3H)-thione (**3a**): white solid, mp 105–106 °C; yield: 83%; ^{1}H NMR ($CDCl_3$, 400 MHz): δ 9.23 (1H, s, –N=CH–N–), 8.88 (1H, d, $J = 8.5$ Hz, Ar–H), 7.46–7.37 (2H, m, Ar–H), 7.34 (1 H, d, $J = 7.6$ Hz, Ar–H), 1.40 (9H, s, $3\times CH_3$); ^{13}C NMR ($CDCl_3$, 75 MHz): δ 193.21, 144.47, 140.21, 127.33, 126.48, 125.63, 120.69, 116.97, 57.76, 30.26; HRMS: *m*/*z* calc. for $C_{12}H_{14}N_2S_2$ (M^+): 250.0598; found: 250.0606.

(*E*)-3-(((2,6-Dimethylphenyl)imino)methyl)benzo[*d*]oxazole-2(3*H*)-thione (**3d**): white solid, mp 166–167 °C; yield: 69%; ^{1}H NMR ($CDCl_3$, 400 MHz): δ 8.92 (1H, s, –N=CH–N–), 8.45–8.42 (1H, m, Ar–H), 7.43–7.34 (3H, m, Ar–H), 7.12 (2H, d, J = 7.6 Hz, Ar–H), 7.05–6.99 (1H, m, Ar–H), 2.23 (6 H, s, 2×CH_3); ^{13}C NMR ($CDCl_3$, 75 MHz): δ 176.59, 142.37, 142.28, 141.92, 124.36, 123.89, 123.24, 121.28, 121.15, 120.17, 110.41, 105.55, 14.05; HRMS: *m/z* calc. for $C_{16}H_{14}N_2OS$ (M^+): 282.0827; found: 282.0824.

5.2.13.5 Critical Views

The formamidine scaffold is an important building block for numerous biologically potent compounds.[180–195] The present protocol[179] offers a practical way to access such an interesting framework by direct reactions of isocyanides with compounds having active N–H bonds under mild and catalyst-free conditions without the need for an inert atmosphere. Ultrasound irradiation was found to promote the reaction effectively. However, the reaction time is too long in some cases (22 h).

5.2.14 Entry-14: Synthesis of Thiazolo[3,2-*a*]pyrimidines

Type of reaction: C–C bond formation
Reaction conditions: Catalyst-free, methanol, ultrasound irradiation, room temperature
Synthetic strategy: Condensation followed by cyclization
Keywords: Dihydropyrimidinone, diethyl and dimethyl acetylene dicarboxylates, methanol, catalyst-free, ultrasonication, room temperature, ethyl-5-(aryl)-2-(2-alkoxy-2-oxoethylidene)-7-methyl-3-oxo-3,5-dihydro-2*H*-thiazolo-[3,2-*a*]pyrimidine-6-carboxylate derivatives

5.2.14.1 General Reaction Scheme

Darehkordi and Ghazi[196] have achieved an efficient ultrasonic-assisted synthesis of biologically interesting thiazolo-pyrimidine derivatives (**3**) from the reaction between pharmaceutically potent dihydropyrimidinone (**1**) and diethyl/dimethyl acetylene carboxylates (**2**) in methanol without the aid of any catalyst at room temperature (Scheme 5.17).

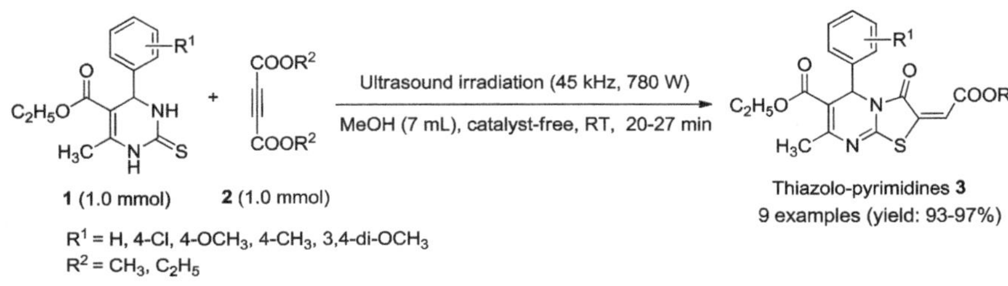

Scheme 5.17 Synthesis of thiazolo[3,2-*a*]pyrimidines (**3**).

5.2.14.2 Representative Examples

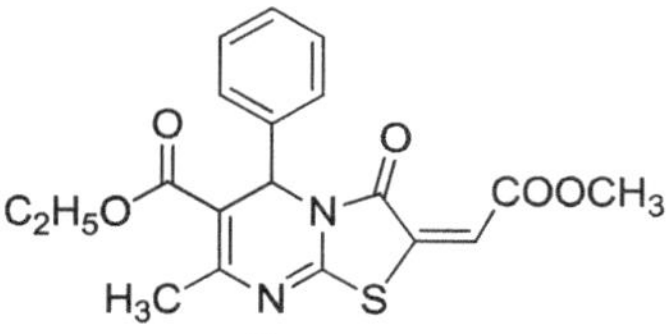

3a

(*E*)-Ethyl 2-(2-methoxy-2-oxoethylidene)-7-methyl-3-oxo-5-phenyl-3,5-dihydro-2*H*-thiazolo[3,2-*a*]pyrimidine-6-carboxylate (**3a**): yellow powder, mp 168–170 °C; time: 26 min; yield: 95%

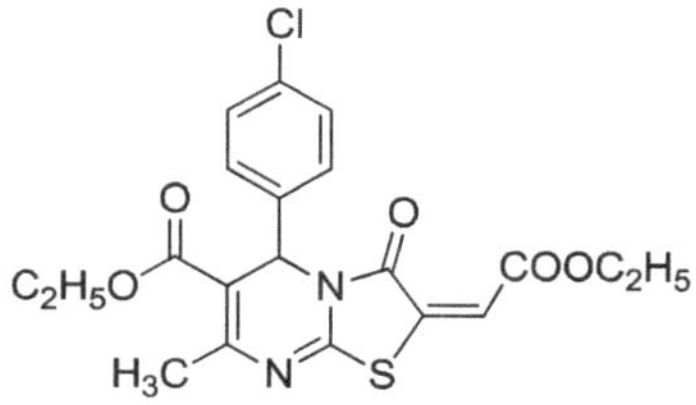

3b

(*E*)-Ethyl 5-(4-chlorophenyl)-2-(2-ethoxy-2-oxoethylidene)-7-methyl-3-oxo-3,5-dihydro-2*H*-thiazolo[3,2-*a*]pyrimidine-6-carboxylate (**3b**): yellow powder, mp 167–169 °C; time: 23 min; yield: 97%

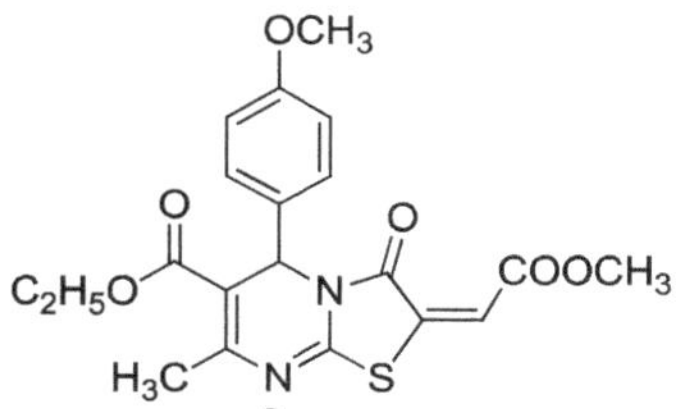

3c

(*E*)-Ethyl 2-(2-methoxy-2-oxoethylidene)-5-(4-methoxyphenyl)-7-methyl-3-oxo-3,5-dihydro-2*H*-thiazolo[3,2-*a*]pyrimidine-6-carboxylate (**3c**): yellow powder, mp 131–132 °C; time: 24 min; yield: 96%

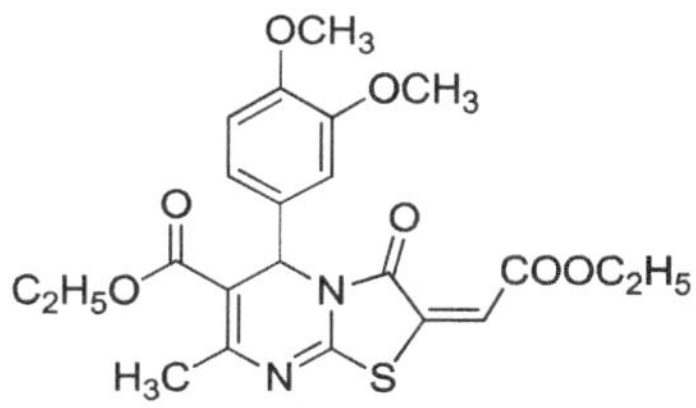

3d

(*E*)-Ethyl 5-(3,4-dimethoxyphenyl)-2-(2-ethoxy-2-oxoethylidene)-7-methyl-3-oxo-3,5-dihydro-2*H*-thiazolo[3,2-*a*]pyrimidine-6-carboxylate (**3d**): yellow powder, mp 168–169 °C; time: 27 min; yield: 93%

5.2.14.3 Experimental Procedure

A solution of diethyl/dimethyl acetylene carboxylate (**2**; 1.0 mmol) in methanol (3 mL) was added to a solution of dihydropyrimidinone derivatives (**1**; 1.0 mmol) in methanol (4 mL) in small portions at room temperature. This reaction mixture was then irradiated with ultrasound for 20–27 min to complete the transformation. After the completion of reaction, the resulting yellow precipitate was filtered and washed with cold methanol or ethanol to afford pure products of thiazo-pyrimidines (**3**) as yellow powder in excellent yields. All the products were characterized from their analytical and spectral studies.

5.2.14.4 Characterization Data of Two Representative Compounds

(*E*)-Ethyl 5-(4-chlorophenyl)-2-(2-ethoxy-2-oxoethylidene)-7-methyl-3-oxo-3,5-dihydro-2*H*-thiazolo[3,2-*a*]pyrimidine-6-carboxylate (**3b**): yellow powder, mp

167–169 °C; yield: 97%; IR (KBr): 3070, 2908, 2986, 1729, 1708, 1628, 1487 cm^{-1}; ^{1}H NMR (DMSO-d_6, 500 MHz): δ 1.1 (3H, t, J = 7.0 Hz, CH_3), 1.25 (3H, t, J = 7.0 Hz, CH_3), 2.39 (3H, s, CH_3), 4.05 (2H, q, J = 7.0 Hz, OCH_2), 4.23 (2H, q, J = 7.0 Hz, OCH_2), 5.99 (1H, s, CH), 6.83 (1H, s, CH), 7.32 (2H, d, J = 7.0 Hz, Ar–H) , 7.42 (2H, d, J = 7.0 Hz, Ar–H); ^{13}C NMR (DMSO-d_6, 125 MHz): δ 14.69 (CH_3), 14.80 (CH_3), 23.19 (CH_3), 55.47 (OCH_2), 61.27 (OCH_2), 62.71 (OCH_2), 118.72, 129.57, 130.58, 134.13, 139.29, 140.03, 151.312, 156.46, 163.69, 165.41, 166.12. Anal. calc. for $C_{20}H_{19}N_2O_5SCl$: C, 55.24; H, 4.40; N, 6.30%; found; C, 55.36; H, 4.66; N, 6.42%.

(*E*)-Ethyl 5-(3,4-dimethoxyphenyl)-2-(2-ethoxy-2-oxoethylidene)-7-methyl-3-oxo-3,5-dihydro-2*H*-thiazolo[3,2-*a*]pyrimidine-6-carboxylate (**3d**): yellow powder, mp 168–169 °C; yield: 93%; IR (KBr): 3058, 2984, 2936 1717, 1711, 1622, 1464 cm^{-1}; ^{1}H NMR (DMSO-d_6, 500 MHz): δ 1.14 (3H, t, J = 7.1 Hz, CH_3), 1.25 (3H, t, J = 7.0 Hz, CH_3), 2.39 (3H, s, OCH_3), 3.72 (6H, s, OCH_3), 4.06 (2H, q, J = 7.1 Hz, OCH_2), 4.20 (2H, q, J = 7.0 Hz, OCH_2), 5.95 (1H, s, HC–N), 6.70 (1H, s, CH), 6.83 (2H, s, Ar–H), 6.90 (1H, Ar–H). Anal. calc. for $C_{22}H_{24}N_2O_7S$: C, 57.38; H, 5.24; N, 6.08; found; C, 57.64; H, 5.36; N, 6.12%.

5.2.14.5 *Critical Views*

The present protocol offers a rapid, eco-friendly and practical route in accessing some novel and biologically interesting thiazolo-pyrimidine derivatives in excellent yields under ultrasound irradiation at room temperature in methanol without the aid of any other additives and column chromatographic purification.[196]

5.2.15 Entry-15: Synthesis of Thiazolo[3,2-*a*]pyrimidines

Type of reaction: C–C and C–N bond formation
Reaction conditions: Catalyst-free, ethanol, ultrasound irradiation, room temperature
Synthetic strategy: Three-component 1,3-dipolar cycloaddition reaction
Keywords: Isatin, 5-arylidene-1,3-thiazolidine-2,4-dione/5-arylidene-4-thioxo-1,3-thiazolidine-2-one, sarcosine, ethanol, catalyst-free, ultrasonication, room temperature, thiazolo[3,2-*a*]pyrimidine derivatives

5.2.15.1 *General Reaction Scheme*

Ultrasound-assisted synthesis of a series of thiazolo[3,2-*a*]pyrimidine derivatives (**4**) was accomplished by Liu *et al.*[197] *via* the three-component 1,3-dipolar cycloaddition reaction of isatin (**1**; 1 mmol), sarcosine (**2**; 1 mmol) and 5-arylidene-1,3-thiazolidine-2,4-dione/5-arylidene-4-thioxo-1,3-thiazolidine-2-one (**3**; 1 mmol) in ethanol at ambient conditions without the aid of any catalyst (Scheme 5.18).

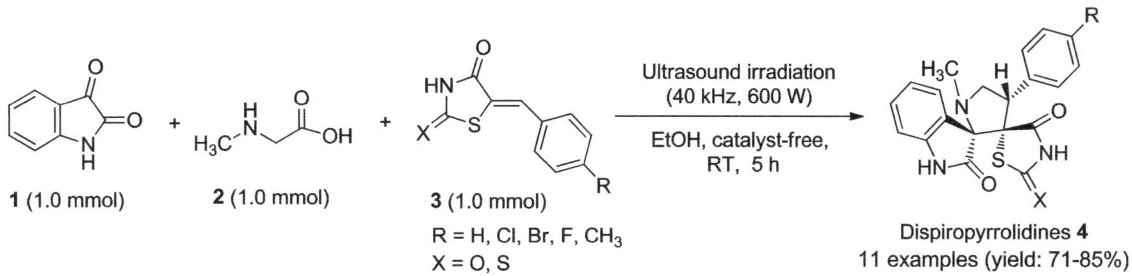

Scheme 5.18 Synthesis of thiazolo[3,2-*a*]pyrimidines (**4**).

5.2.15.2 Representative Examples

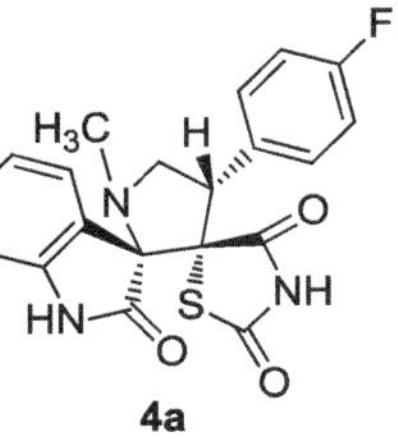

4a

1-*N*-Methyl-spiro[2,3′]oxindole-spiro[3,5″]thiazolidine-2″,4″-dione-4-(4-fluorophenyl)-pyrrolidine (**4a**): light yellow powder, mp 154–156 °C; yield: 71%

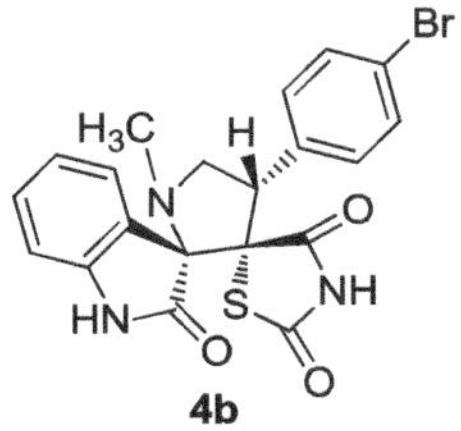

4b

1-*N*-Methyl-spiro[2,3′]oxindole-spiro[3,5″]thiazolidine-2″,4″-dione-4-(4-bromophenyl)-pyrrolidine (**4b**): light yellow powder, mp 160–161 °C; yield 78%

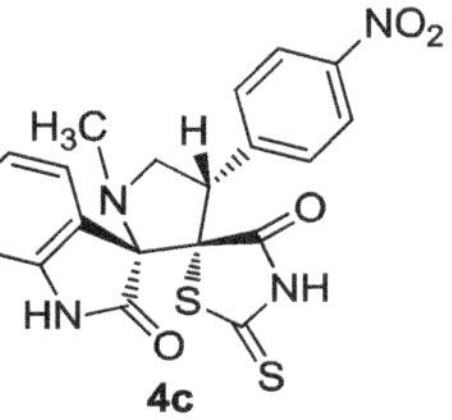

4c

1-*N*-Methyl-spiro[2,3′]oxindole-spiro[3,5″]-4″-thioxo-thiazolidine-2″-one-4-(4-nitrophenyl)-pyrrolidine (**4c**): yellow powder, mp 158–159 °C; yield: 85%

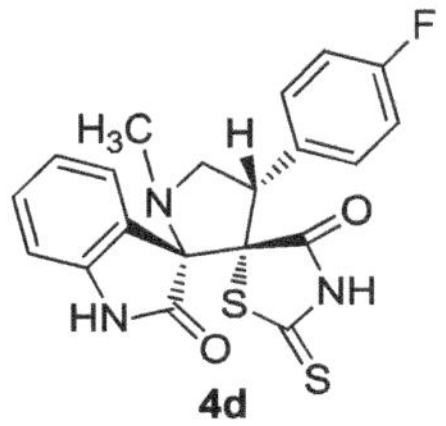

4d

1-*N*-Methyl-spiro[2,3′]oxindole-spiro[3,5″]-4″-thioxo-thiazolidine-2″-one-4-(4-fluorophenyl)-pyrrolidine (**4d**): light yellow powder, mp 142–143 °C; yield: 81%

5.2.15.3 Experimental Procedure

A mixture of isatin (**1**; 1 mmol), sarcosine (**2**; 1 mmol), and 5-arylidene-1,3-thiazolidine-2,4-dione or 5-arylidene-4-thioxo-1,3-thiazolidine-2-one (**3**; 1 mmol) in ethanol, was placed in a 100 mL flask and sonicated in the water bath of an ultrasonic cleaner under ambient conditions for 5 h. After completion of the reaction, as monitored by TLC, the reaction mixture was filtered and the precipitate washed with water (10 mL) and then recrystallized from ethanol to afford the pure product of dispiropyrrolidine (**4**). All the products were characterized from their spectral measurements.

5.2.15.4 Characterization Data of Two Representative Compounds

1-*N*-Methyl-spiro[2,3′]oxindole-spiro[3,5″]thiazolidine-2″,4″-dione-4-(4-bromophenyl)-pyrrolidine (**4b**): light yellow powder, mp 160–161 °C; yield 78%; IR

(KBr): 3195, 3077, 2954, 2876, 1756, 1703, 1615, 1468, 1317, 1159, 823, 762 cm^{-1}; ^1H NMR (DMSO-d_6, 300 MHz): δ 2.05 (3H, s, CH_3), 3.46 (1H, t, $J=8.1$ Hz, CH–H), 3.80 (1H, t, $J=9.3$ Hz, CH–H), 4.42 (1H, t, $J=8.7$ Hz, CH), 6.85 (1H, d, $J=7.8$ Hz, Ar–H), 7.01 (1H, t, $J=7.5$ Hz, Ar–H), 7.18 (1H, d, $J=7.5$ Hz, Ar–H), 7.30 (1H, t, $J=7.8$ Hz, Ar–H), 7.38 (2H, d, $J=8.4$ Hz, Ar–H), 7.55 (2H, d, $J=8.4$ Hz, Ar–H), 10.78 (1H, s, NH), 12.03 (1H, s, NH); HRMS: calc. for $C_{20}H_{17}BrN_3O_3S$: [M + H] 458.0169; found 458.0163.

1-*N*-Methyl-spiro[2,3′]oxindole-spiro[3,5″]-4″-thioxo-thiazolidine-2″-one-4-(4-fluorophenyl)-pyrrolidine (**4d**): light yellow powder, mp 142–143 °C; yield: 81%; IR (KBr): 3372, 3219, 3044, 2950, 2858, 1719, 1612, 1511, 1426, 1322, 1208, 835, 757 cm^{-1}; ^1H NMR (DMSO-d_6, 300 MHz): δ 2.04 (3H, s, CH_3), 3.46 (1H, t, $J=8.4$ Hz, CH–H), 3.78 (1H, t, $J=6.3$ Hz, CH–H), 4.39–4.44 (1H, m, CH), 6.84 (1H, d, $J=7.8$ Hz, Ar–H), 7.02 (1H, t, $J=7.5$ Hz, Ar–H), 7.16–7.22 (3H, m, Ar–H), 7.30 (1H, t, $J=7.2$ Hz, Ar–H), 7.45–7.49 (2H, m, Ar–H), 10.84 (1H, s, NH), 13.13 (1H, s, NH); HRMS: calc. for $C_{20}H_{17}FN_3O_2S_2$: [M + H] 414.0741; found 414.0743.

5.2.15.5 *Critical Views*

The present protocol offers a facile route for the synthesis of a series of dispiropyrrolidines under mild reaction conditions with higher yields, and shorter reaction time. Further, no column chromatographic purification was needed.

5.2.16 Entry-16: Synthesis of Spiro[indoline-3,4′-pyrazolo[3,4-*e*][1,4]thiazepine]diones

Type of reaction: C–C, C–N and C–S bond formation
Reaction conditions: Catalyst-free, water, ultrasound irradiation, room temperature
Synthetic strategy: One-pot multicomponent domino reaction
Keywords: Isatins, 5-amino-3-methylpyrazole, alpha-mercaptocarboxylic acids, water, catalyst-free, ultrasonication, room temperature, spiro[indoline-3,4′-pyrazolo[3,4-*e*][1,4]thiazepine]diones

5.2.16.1 *General Reaction Scheme*

Dandia *et al.*[198] developed a practical and efficient ultrasound-assisted protocol for the synthesis of a series of biologically promising spiro[indoline-3,4′-pyrazolo[3,4-*e*][1,4]thiazepines] (**4**) out of catalyst-free aqueous mediated multicomponent domino reaction of isatins (**1**), 5-amino-3-methylpyrazole (**2**) and alpha-mercaptocarboxylic acids (**3**) (Scheme 5.19).

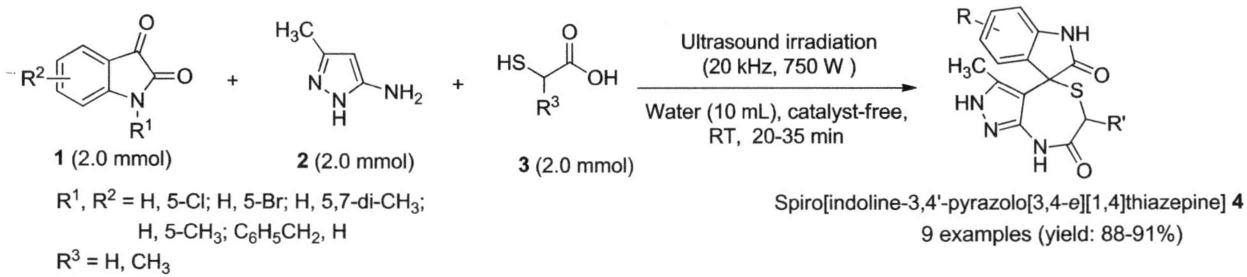

Scheme 5.19 Synthesis of spiro[indoline-3,4′-pyrazolo[3,4-*e*][1,4]thiazepine]diones (**4**).

5.2.16.2 Representative Examples

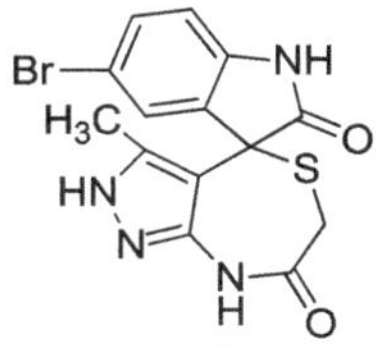

4a

5-Bromo-3'-methyl-6',8'-dihydrospiro[indoline-3,4'-pyrazolo[3,4-*e*][1,4]thiazepine]-2,7'(2'*H*)-dione (**4a**): solid, mp 329–331 °C; time: 35 min; yield: 88%

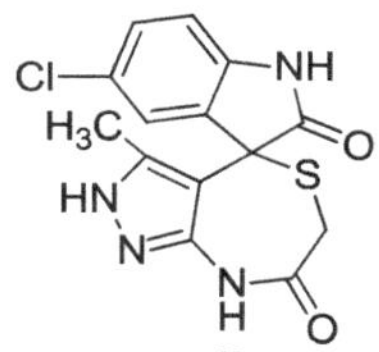

4b

5-Chloro-3'-methyl-6',8'-dihydrospiro[indoline-3,4'-pyrazolo[3,4-*e*][1,4]thiazepine]-2,7'(2'*H*)-dione (**4b**): solid, mp 194–196 °C; time: 25 min; yield: 90%

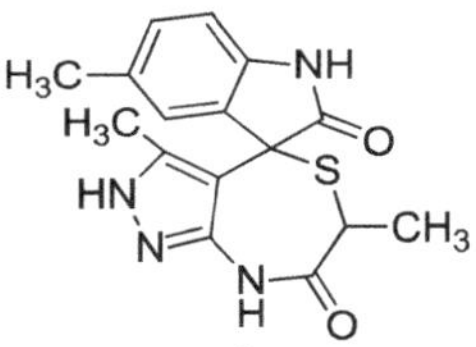

4c

3',5,6'-Trimethyl-6',8'-dihydrospiro[indoline-3,4'-pyrazolo[3,4-*e*][1,4]thiazepine]-2,7'(2'*H*)-dione (**4c**): solid, mp 317–319 °C; time: 30 min; yield: 89%

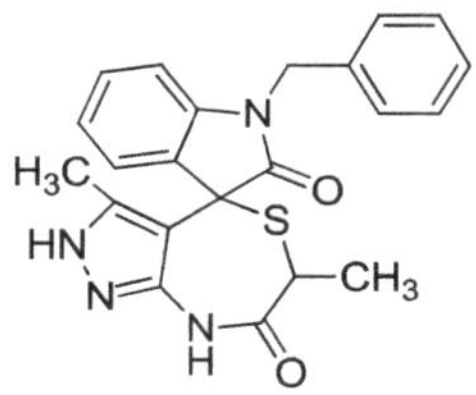

4d

1-Benzyl-3',6'-dimethyl-6',8'-dihydrospiro[indoline-3,4'-pyrazolo[3,4-*e*][1,4]thiazepine]-2,7'(2'*H*)-dione (**4d**): solid, mp 255–257 °C; time: 30 min; yield: 90%

5.2.16.3 Experimental Procedure

A reaction flask was charged with isatins (**1**; 2 mmol), 5-amino-3-methylpyrazole (**2**; 2 mmol), α-mercaptocarboxylic acids (**3**; 2 mmol) and water (10 mL), and the resulting mixture was sonicated for 20–35 min with a probe sonicator at 50% power of the processor at 4 s pulse mode. After completion of the reaction as monitored by TLC, the flask was detached from the probe and the contents were transferred into a beaker. The crude products formed were filtered and washed well with water to afford the pure crystalline products of spiro[indoline-3,4′-pyrazolo[3,4-*e*][1,4]thiazepines] (**4**), identified by means of their analytical and spectral measurements.

5.2.16.4 Characterization Data of Two Representative Compounds

5-Bromo-3′-methyl-6′,8′-dihydrospiro[indoline-3,4′-pyrazolo[3,4-*e*][1,4]thiazepine]-2,7′(2′H)-dione (**4a**): solid, mp 329–331 °C; yield: 88%; IR (KBr): 3336, 3145, 2964, 1709, 1670, 1438, 1415, 834 cm^{-1}; ^1H NMR (DMSO-d_6, 400 MHz): δ 1.52 (3H, s, CH_3), 3.39 (1H, d, $J = 15.2$ Hz, CH), 4.61 (1H, d, $J = 15.2$ Hz, CH), 6.83–7.20

(3H, m, Ar–H), 10.04 (1H, s, NH), 10.50 (1H, s, NH), 12.05 (1H, s, NH); ^{13}C NMR (DMSO-d_6, 100 MHz): δ 9.9, 23.9, 44.5, 105.6, 110.5, 123.1, 124.4, 126.3, 131.4, 145.1, 150.2, 154.4, 171.4, 176.3; MS (ESI): *m/z* 380 $[M+H]^+$. Anal. calc. for $C_{14}H_{11}BrN_4O_2S$: C, 44.34; H, 2.92; N, 14.77%; found: C, 44.40; H, 2.90; N, 14.72%.

1-Benzyl-3′,6′-dimethyl-6′,8′-dihydrospiro[indoline-3,4′-pyrazolo[3,4-*e*][1,4]-thiazepine]-2,7′(2′H)-dione (**4d**): solid, mp 255–257 °C; yield: 90%; IR (KBr): 3326, 2982, 2914, 1710, 1680, 1473, 1421, 785 cm^{-1}; ^{1}H NMR (DMSO-d_6, 400 MHz): δ 1.01 (3H, s, CH_3), 1.42 (3H, d, $J=7.6$ Hz, CH_3), 3.98 (1H, q, $J=7.2$ Hz, CH), 4.79 (1H, d, $J=15.2$ Hz, CH), 5.10 (1H, d, $J=15.6$ Hz, CH), 6.42–6.75 (9H, m, Ar–H), 9.91 (1H, s, NH), 11.96 (1H, s, NH); ^{13}C NMR (DMSO-d_6, 100 MHz): δ 9.5, 15.1, 34.0, 48.8, 49.0, 109.7, 122.0, 124.2, 128.2, 128.9, 140.7, 173.5, 178.1; MS (ESI) *m/z*: 405 $[M+H]^+$. Anal. calc. for $C_{22}H_{20}N_4O_2S$: C, 65.33; H, 4.98; N, 13.85%; found: C, 65.28; H, 4.93; N, 13.80%.

5.2.16.5 Critical Views

This ultrasound-promoted protocol offers a practical and convenient route for the synthesis of such biologically interesting spiro[indoline-3,4′-pyrazolo[3,4-*e*][1,4]thiazepines] in water under ambient conditions without using any additional catalyst. Compared with traditional methods,[199–201] this procedure has several advantages including excellent yields, shorter reaction time, clean reactions and eco-friendlier.

5.3 Concluding Remarks

The sonochemical technique is considered within the green arena when irradiation of ultrasound is associated with both enhanced reactivity and acceleration of a chemical process, thereby causing energy savings and cleaner products with little or no byproducts. The benefits of this green tool have now been explored for the design of catalyst-free organic synthesis, as well as by the synthetic chemists as part of their ongoing endeavors in this direction.

References

1. K. S. Suslick, *J. Am. Chem. Soc.*, 1989, **111**, 2342.
2. K. S. Suslick, *Science*, 1990, **247**, 1439.
3. K. S. Suslick, *New Sci.*, 1990, **125**, 50.
4. K. S. Suslick, *Ultrasound: Its Chemical, Physical and Biological Effects*, VCH, New York, 1988.
5. S. V. Ley and C. M. R. Low, *Ultrasound in Synthesis*, Springer-Verlag, Berlin, 1989.
6. T. J. Mason, *Ultrasonics*, 1986, **24**, 245.
7. T. J. Mason, *Chemistry with Ultrasound*, Elsevier Applied Science, New York, 1990.

8. T. J. Mason, *Sonochemistry: The Uses of Ultrasound in Chemistry*, Royal Society of Chemistry, Cambridge, UK, 1990.
9. T. J. Mason, *Practical Sonochemistry: User's Guide to Applications in Chemistry and Chemical Engineering*, Ellis Horwood Limited, Chichester, England, 1991.
10. T. J. Mason, Sonochemistry: Current Trends and Future Prospects, in *Current Trends in Sonochemistry*, ed. G. J. Price, Royal Society of Chemistry, Cambridge, UK, 1992.
11. T. J. Mason, *Chem. Soc. Rev.*, 1997, **26**, 443.
12. T. J. Mason and J. P. Lorimer, *Endeavour*, 1989, **13**, 123.
13. L. H. Thompson and L. K. Doraiswamy, *Ind. Eng. Chem. Res.*, 1999, **38**, 1215.
14. J.-L. Luche, *Synthetic Organic Sonochemistry*, Plenum Press, New York, 1998.
15. T. J. Mason and J. P. Lorimer, *Applied Sonochemistry*, Wiley-VCH Verlag GmbH, Weinheim, 2002.
16. G. Cravotto and P. Cintas, *Chem. Soc. Rev.*, 2006, **35**, 180.
17. S. K. Sharma, A. Mudhoo and W. Zhang, Green Chemistry and Engineering. A Versatile Research, in *Green Chemistry for Environmental Sustainability*, ed. S. K. Sharma and A. Mudhoo, CRC Press Taylor & Francis, 2010, pp. 1–52.
18. R. B. N. Baig and R. S. Varma, *Chem. Soc. Rev.*, 2012, **41**, 1559.
19. L. Pizzuti, M. S. F. Franco, A. F. C. Flores, F. H. Quina and C. M. P. Pereira, Recent Advances in the Ultrasound-assisted Synthesis of Azoles, in *Green Chemistry – Environmentally Benign Approaches*, ed. M. Kidwai, InTech, Rijeka, 2012.
20. S. Puri, B. Kaur, A. Parmar and H. Kumar, *Curr. Org. Chem.*, 2013, **17**, 1790.
21. M. A. Schiel, A. B. Chopa, G. F. Silbestri, M. B. Alvarez, A. G. Lista and C. E. Domini, Use of Ultrasound in the Synthesis of Heterocycles of Medicinal Interest, in *Green Synthetic Approaches for Biologically Relevant Heterocycles*, ed. G. Brahmachari, Elsevier, Amsterdam, 2015.
22. G. Chatel, *Sonochemistry: New Opportunities for Green Chemistry*, World Scientific Publishing Co., Singapore, 2015.
23. D. Habibi and M. Nasrollahzadeh, *Comptes Rendus Chimie*, 2013, **16**, 1008.
24. F. F. Blicke and C. J. Lu, *J. Am. Chem. Soc.*, 1952, **74**, 3933.
25. H. L. Yale, *J. Org. Chem.*, 1971, **36**, 3238.
26. J. Waki and J. Meienhofer, *J. Org. Chem.*, 1977, **42**, 2019; F. M. F. Chen and N. L. Benoiton, *Synthesis*, 1979, **9**, 709.
27. P. Strazzolini, A. G. Giumanini and S. Cauci, *Tetrahedron*, 1990, **46**, 1081.
28. M. Neveux, C. Bruneau and P. H. Dixneuf, *J. Chem. Soc., Perkin Trans. I*, 1991, 1197.
29. P. G. Reddy, G. D. K. Kumar and S. Baskaran, *Tetrahedron Lett.*, 2000, **41**, 9149.

30. D. R. Hill, C.-N. Hasiao, R. Kurukulasuriya and S. J. Wittenberger, *Org. Lett.*, 2002, **4**, 111.
31. M. Hosseni-Sarvari and H. Sharghi, *J. Org. Chem.*, 2006, **71**, 6652.
32. D. Biswanath, K. Meddeboina, P. Balasubramanayam, V. D. Boyapati and K. D. Nandan, *Tetrahedron Lett.*, 2008, **49**, 2225.
33. G. Brahmachari and S. Laskar, *Tetrahedron Lett.*, 2010, **51**, 2319.
34. M. M. Mojtahedi, M. Saeed Abaee, V. Hamidi and A. Zolfaghari, *Ultrasonics Sonochem.*, 2007, **14**, 596.
35. J. Cassy and P. Pale, *Tetrahedron Lett.*, 1987, **28**, 6039.
36. T. T. Upadhya, T. Daniel, A. Sudalai, T. Ravindranathan and K. R. Sabu, *Synth. Commun.*, 1996, **26**, 4539.
37. Z. H. Zhang, T. S. Li, F. Yang and C. G. Fu, *Synth. Commun.*, 1998, **28**, 3105.
38. W. Mormann and J. Demeter, *Macromolecules*, 1999, **32**, 1706; M. Curini, F. Epifano, M. C. Marcotullio, O. Rosati and U. Costantino, *Synth. Commun.*, 1999, **29**, 541.
39. B. E. Blass, C. L. Harris and D. E. Portlock, *Tetrahedron Lett.*, 2001, **42**, 1611.
40. M. M. Mojtahedi, M. R. Saidi, M. Bolourtchian and M. M. Heravi, *Phosphorus Sulfur Silicon*, 2002, **177**, 289.
41. H. Firouzabadi, N. Iranpoor, K. Amani and F. Nowrouzi, *J. Chem. Soc. Perkin Trans.*, 2002, **1**, 2601.
42. T. S. Jin, Y. W. Li, G. Sun and T. S. Li, *J. Chem. Res.*, 2002, 456.
43. H. Firouzabadi, N. Iranpoor, S. Sobhani, S. Ghassamipour and Z. Amoozgar, *Tetrahedron Lett.*, 2003, **44**, 891.
44. H. Firouzabadi, N. Iranpoor, S. Sobhani and S. Ghassamipour, *J. Organomet. Chem.*, 2004, **689**, 3197.
45. N. Azizi and M. R. Saidi, *Organometallics*, 2004, **23**, 1457.
46. B. Akhlaghinia, M. Asadi, E. Safaee and M. Heydarpoor, *Phosphorus Sulfur Silicon*, 2004, **179**, 2099.
47. H. Firouzabadi, N. Iranpoor, S. Sobhani and S. Ghassamipour, *Synthesis*, 2005, 595.
48. B. Akhlaghinia and S. Tavakoli, *Synthesis*, 2005, 1775.
49. N. Azizi, A. Rahimzadeh-Oskooee, Z. Yadollahy and A. G. Ourimi, *Monatsh Chem.*, 2014, **145**, 1675.
50. N. Yamazaki, T. Tomioka and F. Higashi, *Synthesis*, 1975, 384.
51. S. Sharma, *Synthesis*, 1978, 803.
52. C. Levallet, J. Lerpiniere and S. Y. Ko, *Tetrahedron*, 1997, **53**, 5291.
53. M. Ballini, R. Bigi, F. Maggi, R. Parrini, G. Predieri and G. Sartori, *J. Org. Chem.*, 1999, **64**, 1029.
54. P. K. Mohanta, S. Dhar, S. Samal, H. Ila and H. Junjappa, *Tetrahedron*, 2000, **56**, 629.
55. M. Kidwai, R. Venkataramanan and B. Dave, *Green Chem.*, 2001, **3**, 278.
56. B. C. Ranu, S. S. Dey and S. Bag, *Arkivoc*, 2003, **ix**, 14.
57. P. R. Schreiner, *Chem. Soc. Rev.*, 2003, **32**, 289.

58. A. R. Katritzky, S. Ledoux, R. M. Witek and S. K. Nair, *J. Org. Chem.*, 2004, **69**, 2976.
59. M. Kodomari, M. Suzuki, K. Tanigawaa and T. Aoyama, *Tetrahedron Lett.*, 2005, **46**, 5841.
60. S. J. Connon, *Chem. Eur. J.*, 2006, **12**, 5418.
61. D. Chaturvedi and S. Ray, *Tetrahedron Lett.*, 2006, **47**, 1307.
62. M. Maddani and K. R. Prabhu, *Tetrahedron Lett.*, 2007, **48**, 7151.
63. D. Chaturvedi, N. Mishra and V. Mishra, *Synthesis*, 2008, 355.
64. D. Chaturvedi and S. Ray, *Tetrahedron Lett.*, 2008, **48**, 149.
65. S. F. Gan, J. P. Wan, Y. J. Pan, C. R. Sun, F. Liang, J. Tan, C. Piao and Q. Liu, *Synthesis*, 2008, 3579; A. Ziyaei Halimehjani, Y. Pourshojaei and M. R. Saidi, *Tetrahedron Lett.*, 2009, **50**, 32.
66. M. Maddani and K. R. Prabhu, *J. Org. Chem.*, 2010, **75**, 2327.
67. S. Sahu, P. Rani Sahoo, S. Patel and B. K. Mishra, *J. Sulfur Chem.*, 2011, **32**, 171.
68. S. F. Gan, J. P. Wan, Y. J. Pan and C. R. Sun, *Mol. Divers.*, 2011, **15**, 809.
69. N. Azizi, A. K. Amiri, H. Ghafuri and M. Bolourtchian, *Mol. Divers.*, 2011, **15**, 157.
70. P. P. Kumavat, A. D. Jangale, D. R. Patil, K. S. Dalal, J. S. Meshram and D. S. Dalal, *Environ. Chem. Lett.*, 2013, **11**, 177.
71. W. Du and D. P. Curran, *Org. Lett.*, 2003, **5**, 1765.
72. I. Hisaki, S. I. Sasaki, K. Hirose and T. Yoshito, *Eur. J. Org. Chem.*, 2007, 607.
73. S. Ghosh, *Synlett*, 2009, 2882.
74. M. Xia and Y.-D. Lu, *Ultrasonics Sonochemistry*, 2007, **14**, 235.
75. B. Dar, A. Singh, A. Sahu, P. Patidar, A. Chakraborty, M. Sharma and B. Singh, *Tetrahedron Lett.*, 2012, **53**, 5497.
76. J. C. Craig and P. E. Person, *J. Med. Chem.*, 1971, **14**, 1221.
77. F. R. Atherton, C. H. Hassal and R. W. Lambert, *J. Med. Chem.*, 1986, **29**, 29.
78. I. A. Natchev, *Liebigs Ann. Chem.*, 1988, 861; N. M. Sukhova, M. Lidak, A. Zidermane, I. S. Pelevina and S. S. Voronia, *Khim.-Farm. Zh.*, 1989, **23**, 1226.
79. M. C. Allen, W. Fuhrer, B. Tuck, R. Wade and J. M. Wood, *J. Med. Chem.*, 1989, **32**, 1652.
80. H. V. Patel, K. V. Vyas and P. S. Fernandes, *Indian J. Chem.*, 1990, **29B**, 836.
81. L. Maier, *Phosphorus Sulfur Silicon Rel. El*m., 1990, **53**, 43.
82. P. Kafarski and B. Lejczak, *Phosphorus Sulfur Silicon Rel. Elm.*, 1991, **63**, 193.
83. R. Y. Chen and L. J. Mao, *Phosphorus Sulfur Silicon Rel. Elm.*, 1994, **89**, 97.
84. K. Moonen, I. Laureyn and C. V. Stevens, *Chem. Rev.*, 2004, **104**, 6177.
85. H. J. Ha and G. S. Nam, *Synth. Commun.*, 1992, **22**, 1143.
86. T. Yokomatsu, Y. Yoshida and S. Shibuya, *J. Org. Chem.*, 1994, **59**, 7930.
87. T. Yokomatsu, Y. Yoshida and S. Shibuya, *J. Org. Chem.*, 1994, **59**, 7930.

88. J. S. Yadav, B. V. S. Reddy, K. S. Raj, K. B. Reddy and A. R. Prasad, *Synthesis*, 2001, 2277.
89. S. Chandrasekhar, S. J. Prakash, V. Jagadeshwar and C. Narsihmulu, *Tetrahedron Lett.*, 2001, **42**, 5561.
90. A. Vinu, P. Kalita, V. V. Balasubramanian, H. Oveisi, T. Selvan, A. Mano, M. A. Chari and B. V. S. Reddy, *Tetrahedron Lett.*, 2009, **50**, 7132.
91. J. J. Yang, N. Dang and Y. W. Chang, *Lett. Org. Chem.*, 2009, **6**, 470.
92. S. Sobhani and Z. Tashrifi, *Heteroat. Chem.*, 2009, **20**, 109.
93. M. T. Maghsoodlou, S. M. Habibi-Khorassani, R. Heydari, N. Hazeri, S. S. Sajadikhah and M. Rostamizadeh, *Chin. J. Chem.*, 2010, **28**, 285.
94. A. A. Jafari, M. Nazarpour and M. Abdollahi-Alibeik, *Heteroat. Chem.*, 2010, **21**, 6.
95. K. U. M. Rao, S. H. Jayaprakash, S. K. Nayak and C. S. Reddy, *Catal. Sci. Technol.*, 2011, **1**, 1665.
96. J. T. Hou, J. W. Gao and Z. H. Zhang, *Appl. Organometal. Chem.*, 2011, **25**, 47.
97. R. G. de-Noronha, C. C. Romão and A. C. Fernandes, *Catal. Commun.*, 2011, **12**, 337.
98. S. D. Dindulkar, M. V. Reddy and Y. T. Jeong, *Catal. Commun.*, 2012, **17**, 114.
99. B. V. S. Reddy, A. S. Krishna, A. V. Ganesh and G. G. K. S. N. Kumar, *Tetrahedron Lett.*, 2011, **52**, 1359.
100. M. V. Reddy, S. D. Dindulkar and Y. T. Jeong, *Tetrahedron Lett.*, 2011, **52**, 4764.
101. S. T. Disale, R. K. Sandip, S. S. Kahandal, T. G. Srinivasan and R. V. Jayaram, *Tetrahedron Lett.*, 2012, **53**, 2277.
102. A. Hasaninejed, M. R. Kazerooni and A. Zare, *ACS Sustainable Chem. Eng.*, 2013, **1**, 679.
103. D. Singh and D. Singh, *J. Chem. Eng. Data*, 1984, **29**, 355.
104. W. Wang, S. X. Wang, X. Y. Qin and J. T. Li, *Synth. Commun.*, 2005, **35**, 1263.
105. D. Q. Shi, J. Chen, N. Wu, Q. Y. Zhuang and X. S. Wang, *J. Org. Chem.*, 2005, **25**, 405.
106. M. N. Elinson, A. S. Dorofeev, R. F. Nasybullin and G. I. Nikishin, *Synthesis*, 2008, 1933.
107. K. Niknam, D. Saberi, M. Sadegheyan and A. Deris, *Tetrahedron Lett.*, 2010, **51**, 692.
108. A. Hasaninejad, M. Shekouhy, A. Zare, S. M. S. Hoseini Ghattali and N. Golzar, *J. Iran. Chem. Soc.*, 2011, **8**, 411.
109. A. Hasaninejad, A. Zare, M. Shekouhy and N. Golzar, *Org. Prep. Proced. Int.*, 2011, **43**, 131.
110. K. Sujatha, G. Shanthi, N. P. Selvam, S. Manoharan, P. T. Perumal and M. Rajendran, *Bioorg. Med. Chem. Lett.*, 2009, **19**, 4501.
111. R. Pagadala, S. Maddila and S. B. Jonnalagadda, *Green Chem. Lett. Rev.*, 2014, **7**, 131.

112. I. L. Baraznenok, V. G. Nenajdenko and E. S. Balenkova, *Eur. J. Org. Chem.*, 1999, 937.
113. B. Crousse, J.-P. Begue and D. Bonnet-Delpon, *J. Org. Chem.*, 2000, **65**, 5009.
114. M. M. Ali, K. C. Tasneem, P. K. Rajanna and S. Prakash, *Synlett.*, 2001, 251.
115. C. S. Cho, B. T. Kim, T. J. Kim and S. C. Shim, *Chem. Commun.*, 2001, 2576.
116. M. Demeunynck, C. Moucheron and A. K.-D. Mesmaeker, *Tetrahedron Lett.*, 2002, **43**, 261.
117. X. F. Lin, S. L. Cui and Y. G. Wang, *Tetrahedron Lett.*, 2006, **47**, 3127.
118. X. F. Lin, S. L. Cui and Y. G. Wang, *Tetrahedron Lett.*, 2006, **47**, 4509.
119. S.-X. Wang, Z.-Y. Li, J.-C. Zhang and J.-T. Li, *Ultrasonics Sonochem.*, 2008, **15**, 677.
120. C. O. Kappe, *Tetrahedron*, 1993, **49**, 6937.
121. A. P. Phillips, *J. Am. Chem. Soc.*, 1949, **71**, 4003.
122. G. J. R. Anderson and G. Berkelhammer, *J. Am. Chem. Soc.*, 1958, **80**, 992.
123. H. Singh, D. S. S. Chimni and S. Kumar, *Tetrahedron*, 1995, **5**, 12775.
124. B. M. Khadikar, V. G. Gaikar and A. A. Chitnavis, *Tetrahedron Lett.*, 1995, **36**, 8083.
125. M. F. Gordeev, D. V. Patel and E. M. Gordon, *J. Org. Chem.*, 1996, **61**, 924.
126. J. G. Breitenbucher and G. Figliozzi, *Tetrahedron Lett.*, 2000, **41**, 4311.
127. L. Ohberg and J. Westman, *Synlett*, 2001, 1296.
128. J.-C. Liang, J.-L. Yeh, C.-S. Wang, S.-F. Liou, C.-H. Tasi and I.-J. Chen, *Bioorg. Med. Chem.*, 2002, **10**, 719.
129. R. Miri, H. Niknahad, Gh. Vesal and A. Shafiee, *IL Farmaco*, 2002, **57**, 123.
130. A. Dondoni, A. Massi, E. Minghini, S. Sabbatini and V. Bertoasi, *J. Org. Chem.*, 2003, **68**, 6172.
131. G. Sabitha, G. S. K. K. Reddy, Ch. S. Reddy and J. S. Yadav, *Tetrahedron Lett.*, 2003, **44**, 4129.
132. S. J. Ji, Z. Q. Jiang, J. Lu and T. P. Loh, *Synlett*, 2004, 831.
133. A. Dondoni, A. Massi, E. Minghini and V. Bertoasi, *Tetrahedron*, 2004, **60**, 2311.
134. N. Tewari, N. Dwivedi and R. P. Tripathi, *Tetrahedron Lett.*, 2004, **45**, 9011.
135. A. Agarwal and P. M. S. Chauhan, *Tetrahedron Lett.*, 2005, **46**, 1345.
136. J. D. Moseley, *Tetrahedron Lett.*, 2005, **46**, 3179.
137. L. M. Wang, J. Sheng, L. Zhang, J. W. Han, Z. Fan, H. Tian and C. T. Qian, *Tetrahedron*, 2005, **61**, 1539.
138. R. Sridhar and P. T. Perumal, *Tetrahedron*, 2005, **61**, 2465.
139. W.-X. Guo, H.-L. Jin, J.-X. Chen, F. Chen, J.-C. Dinga and H.-Y. Wu, *J. Braz. Chem. Soc.*, 2009, **20**, 1674.

140. A. E. A. Porter, *Comprehensive Heterocyclic Chemistry*, ed. A. R. Katritsky and C. W. Rees, Pergamon: Oxford, 1984, 157.
141. G. H. C. Woo, J. K. Snyder and Z. K. Wan, *Prog. Heterocycl. Chem.*, 2002, **14**, 279.
142. Z. Wu and N. J. Ede, *Tetrahedron Lett.*, 2001, **42**, 8115.
143. S. Antoniotti and E. Donach, *Tetrahedron Lett.*, 2002, **43**, 3971.
144. A. R. Steven, D. W. Cecilia and J. K. T. Richard, *Chem. Commun.*, 2003, 2286.
145. Z. Zhao, D. D. Wisnoski, S. E. Wolkenberg, W. H. Leister, Y. Wang and C. W. Lindsley, *Tetrahedron Lett.*, 2004, **45**, 4873.
146. S. A. Raw, C. D. Wilfred and R. J. K. Taylor, *Org. Biomol. Chem.*, 2004, **2**, 788.
147. R. S. Robinson and R. J. K. Taylor, *Synlett*, 2005, 1003.
148. C. Venkatesh, B. Singh, P. K. Mahata, H. Ila and H. Junjappa, *Org. Lett.*, 2005, **7**, 2169.
149. R. S. Bhosale, S. R. Sarda, S. S. Ardhapure, W. N. Jadhav, S. R. Bhusare and R. P. Pawar, *Tetrahedron Lett.*, 2005, **46**, 7183.
150. R. S. Bhosale, S. R. Sarda, S. S. Ardhapure, W. N. Jadhav, S. R. Bhusare and R. P. Pawar, *Tetrahedron Lett.*, 2005, **46**, 7183.
151. S. V. More, M. N. V. Sastry, C. C. Wang and C. F. Yao, *Tetrahedron Lett.*, 2005, **46**, 6345.
152. S. V. More, M. N. V. Sastry and C. F. Yao, *Green Chem.*, 2006, **8**, 91.
153. A. Shaabani and A. Maleki, *Chin. J. Chem.*, 2007, **25**, 818.
154. F. Mohsenzadeh, K. Aghapoor and H. R. Darabi, *J. Braz. Chem. Soc.*, 2007, **18**, 297.
155. M. M. Heravi, S. Taheri, K. Bakhtiari and H. A. Oskooie, *Catal. Commun.*, 2007, **8**, 211.
156. H. R. Darabi, S. Mohandessi, K. Aghapoor and F. Mohsenzadeh, *Catal. Commun.*, 2007, **8**, 389.
157. J. Gris, R. Glisoni, L. Fabian, B. Fernández and A. G. Moglioni, *Tetrahedron Lett.*, 2008, **49**, 1053.
158. D. Fang, K. Gong, Z. Fei, X. Zhou and Z. Liu, *Catal. Commun.*, 2008, **9**, 317.
159. A. Kumar, S. Kumar, A. Saxena, A. De and S. Mozumdar, *Catal. Commun.*, 2008, **9**, 778.
160. T. K. Huang, R. Wang, L. Shi and X. X. Lu, *Catal. Commun.*, 2008, **9**, 1143.
161. T. M. Potewar, S. A. Ingale and K. V. Srinivasan, *Synth. Commun.*, 2008, **38**, 3601.
162. H. R. Darabi, F. Tahoori, K. Aghapoor, F. Taala and F. Mohsenzadeh, *J. Braz. Chem. Soc.*, 2008, **19**, 1646.
163. S.-Q. Ge, Y.-Y. Hua and M. Xia, *Ultrasonics Sonochem.*, 2009, **16**, 232.
164. M. M. Sarmah and D. Prajapati, *Synlett.*, 2015, **26**, 91.
165. N. Mulakayala, D. Rambabu, M. Rao Raja, M. Chaitanya, C. Suresh Kumar, A. M. Kalle, G. Rama Krishna, C. Malla Reddy, M. V. Basaveswara Rao and M. Pal, *Bioorg. Med. Chem.*, 2012, **20**, 759.

166. S. Rostamnia and K. Lamei, *Synthesis*, 2011, 3080.
167. A. Alizadeh, S. Rostamnia, N. Zohreh and R. Hosseinpour, *Tetrahedron Lett.*, 2009, **50**, 1533.
168. F. C. Brown, *Chem. Rev.*, 1961, **61**, 463.
169. W.-D. Rudorf and R. Schwarz, *Heterocycles*, 1986, **24**, 3459.
170. S. P. Singh, S. S. Parmar, K. Raman and V. I. Stenberg, *Chem. Rev.*, 1981, **81**, 175.
171. B. B. Lohray, V. Bhushan, P. B. Rao, G. R. Madhavan, N. Murali, K. N. Rao, K. A. Reddy, B. M. Rajesh, P. G. Reddy, R. Chakrabarti and R. Rajagopalan, *Bioorg. Med. Chem. Lett.*, 1997, **7**, 785.
172. C. L. Lee and M. M. Sim, *Tetrahedron Lett.*, 2000, **41**, 5729.
173. A. I. Khodair and J. Nielsen, *Heterocycles*, 2002, **57**, 1017.
174. W. T. Sing, C. L. Lee, S. L. Yeo, S. P. Lim and M. M. Sim, *Bioorg. Med. Chem. Lett.*, 2001, **11**, 91.
175. E. B. Grant, D. Guiadeen, E. Z. Baum, B. D. Foleno, H. Jin, D. A. Montenegro, E. A. Nelson, K. Bush and D. J. Hlasta, *Bioorg. Med. Chem. Lett.*, 2000, **10**, 2179.
176. M. G. Orchard, J. C. Neuss, C. M. S. Galley, A. Carr, D. W. Porter, P. Smith, D. I. C. Scopes, D. Haydon, K. Vousden, C. R. Stubberfield, K. Young and M. Page, *Bioorg. Med. Chem. Lett.*, 2004, **14**, 3975.
177. N. S. Cutshall, C. O'Day and M. Prezhdo, *Bioorg. Med. Chem. Lett.*, 2005, **15**, 3374.
178. J. H. Ahn, S. J. Kim, W. S. Park, S. Y. Cho, J. D. Ha, S. S. Kim, S. K. Kang, D. G. Jeong, S.-K. Jung, S.-H. Lee, H. M. Kim, S. K. Park, K. H. Lee, C. W. Lee, S. E. Ryu and J.-K. Choi, *Bioorg. Med. Chem. Lett.*, 2006, **16**, 2996.
179. T.-H. Zhu, X. Zhu, X.-P. Xu, T. Chen and S.-J. Ji, *Tetrahedron Lett.*, 2011, **52**, 2771.
180. A. I. Meyers and W. T. Hoeve, *J. Am. Chem. Soc.*, 1980, **102**, 7125.
181. V. E. Shashoua, *Biochemistry*, 1964, **3**, 1719.
182. H. Geissbuhler, K. Kossmann and I. Baunok, *J. Agric. Food Chem.*, 1971, **19**, 365.
183. I. R. Harrison, A. Kozlik and J. F. McCarthy, *Pestic. Sci.*, 1973, **4**, 901.
184. K. M. Chang and C. O. Knowles, *J. Agric. Food Chem.*, 1977, **25**, 493.
185. R. E. Hornish, *J. Agric. Food Chem.*, 1984, **32**, 114.
186. B. K. Eya and T. R. Fukuto, *J. Agric. Food Chem.*, 1986, **34**, 947.
187. A. I. Meyers and G. Milot, *J. Am. Chem. Soc.*, 1993, **115**, 6652.
188. M. A. Matulenko and A. I. Meyers, *J. Org. Chem.*, 1996, **61**, 573.
189. M. E. Kopach and A. I. Meyers, *J. Org. Chem.*, 1996, **61**, 6764.
190. M. E. Kopach, A. H. Fray and A. I. Meyers, *J. Am. Chem. Soc.*, 1996, **118**, 9876.
191. F. A. Cotton, J. H. Matonic and C. A. Murillo, *Inorg. Chem.*, 1996, **35**, 498.
192. N. Tsukada, O. Tamura and Y. Inoue, *Organometallics*, 2002, **21**, 2521.
193. J. M. Casas, B. E. Diosdado, L. R. Falvello, J. Fornies and A. Martin, *Inorg. Chem.*, 2005, **44**, 9444.

194. C. B. Divito, S. Davies, S. Masoudi and C. N. Muhoro, *J. Agric. Food Chem.*, 2007, **55**, 5377.
195. P. E. Zhichkin, L. H. Peterson, C. M. Beer and W. M. Rennells, *J. Org. Chem.*, 2008, **73**, 8954.
196. A. Darehkordi and S. Ghazi, *Arabian J. Chem.*, 2015, DOI: 10.1016/j.arabjc.2015.01.010.
197. H. Liu, Y. Zou, Y. Hu and D.-Q. Shia, *J. Heterocyclic Chem.*, 2011, **48**, 877.
198. A. Dandia, R. Singh, J. Joshi, S. Maheshwari and P. Soni, *RSC Adv.*, 2013, **3**, 18992.
199. H. Chen and D. Shi, *Tetrahedron*, 2011, **67**, 5686.
200. K. Karnakar, S. N. Murthy, K. Ramesh, K. H. V. Reddy, Y. V. Nageswar, U. Chandrakala, B. L. A. Prabhavathi Devi and R. B. N. Prasad, *Tetrahedron Lett.*, 2012, **53**, 3497.
201. A. Dandia, M. Sati, K. Arya, P. Sarawgi and A. Loupy, *ARKIVOC*, 2005, **i**, 105.

CHAPTER 6

Catalyst-free Organic Reactions with Ball Milling

6.1 Introduction

The term, *mechanochemistry*, is directly associated with chemical transformations induced by mechanical energy arising out of friction, impact, and collision. In connection with this, we can recall the use of the simple '*mortar* and *pestle*' for generating mechanical energy (this technique is furthermore defined as *'grinding'*) for reactants to effect a chemical transformation in its simplest possible way, which can be traced to the remote past during the early age of the evaluation of chemistry! Because of variable and relatively low grinding strength and speed, this traditional technique is not very effective, and as a result the new generation of mechanical grinder – mixer/shaker mill or planetary mill – has been developed to facilitate the process. Mixing/grinding in such a mill is usually referred to as 'milling' and the mechanical grinding is aided by metal balls, thus, the whole process of mechanical grinding in a mill is known as 'ball milling'. This area is a relatively new branch of research in modern synthetic protocols and is attaining much interest among the researchers at large. An appreciable amount of literature including reviews, comments, reports, books and book chapters are now available in this particular area, which delineates various aspects of ball-milling techniques. It is suggested that readers follow this literature[1–50] to broaden their knowledge of the idea and concept. It has been observed that ball milling can also be effective in accomplishing certain organic transformations in the absence of any catalyst. With this view, the present chapter offers a detailed description of a few comprehensively screened catalyst-free ball-milling-assisted organic synthetic protocols.

Green Chemistry Series No. 51
Catalyst-free Organic Synthesis
By Goutam Brahmachari

Published by the Royal Society of Chemistry, www.rsc.org

6.2 Catalyst-free Organic Transformations with Ball Milling

A few catalyst-free organic reactions leading to the synthesis of useful organic compounds under ball milling was reported so far. This section presents such organic transformations as entry-based format highlighting the key aspects for each of them.

6.2.1 Entry-1: Synthesis of Pyrroles

Type of reaction: C–C and C–N bond formation
Reaction conditions: Catalyst-free, ball milling
Synthetic strategy: One-pot cascade reactions
Keywords: Enamine esters or enamine ketone, *trans*-1,2-dibenzoylethene, solvent- and catalyst-free, one-pot cascade reactions, solid-state reaction, ball milling, pyrrole derivatives

6.2.1.1 General Reaction Schemes

Kaupp *et al.*[51] were the first group of investigators who used the ball-milling technique for the solid-state cascade reactions between enamine esters (**1**) or enamine ketone (**4**) and *trans*-1,2-dibenzoylethene (**2**) to prepare pyrrole derivatives (**3**/**5**) in quantitative yields without the aid of any catalyst (Scheme 6.1).

6.2.1.2 Possible Mechanism

The authors proposed cascade reactions involving four sequential reaction steps in generating the product as shown in Scheme 6.2. Initially, a vinylogous Michael addition between enamine ester (**1**) and *trans*-1,2-dibenzoylethene (**2**) takes place to give the Michael adduct (**6**), which in the next step undergoes imine-enamine rearrangement to form an intermediate (**7**). The amino group, thus formed, adds to the more favorably situated carbonyl

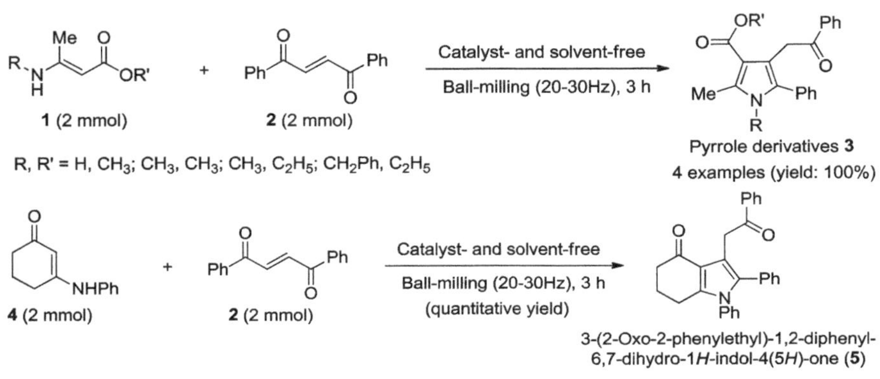

Scheme 6.1 Synthesis of pyrroles (3/5).

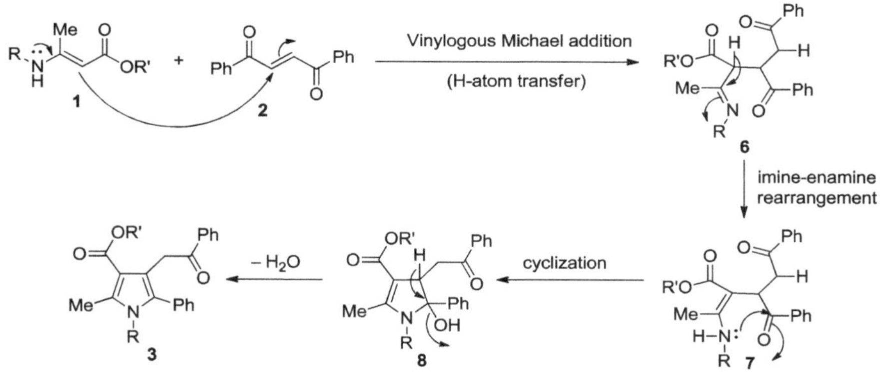

Scheme 6.2 Suggested mechanism for the synthesis of pyrroles (**3**).

function, which leads to the formation of a five-membered ring to give **8**; on eliminating a molecule of water, the pyrrole derivative (**3**) is formed.

6.2.1.3 *Representative Examples*

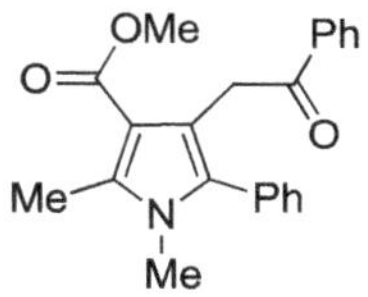

3a

Methyl 1,2-dimethyl-4-(2-oxo-2-phenylethyl)-5-phenyl-1*H*-pyrrole-3-carboxylate (**3a**): solid, mp 152 °C; yield: 100%

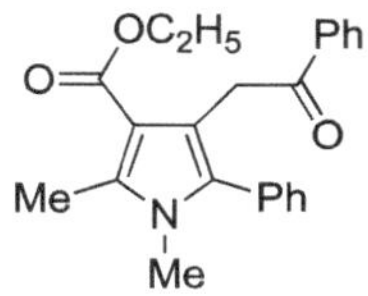

3b

Ethyl 1,2-dimethyl-4-(2-oxo-2-phenylethyl)-5-phenyl-1*H*-pyrrole-3-carboxylate (**3b**): solid, mp 122–124 °C; yield: 100%

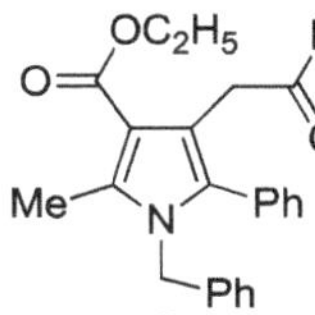

3c

Ethyl 1-benzyl-2-methyl-4-(2-oxo-2-phenylethyl)-5-phenyl-1*H*-pyrrole-3-carboxylate (**4c**): solid, mp 130–131 °C; yield: 100%

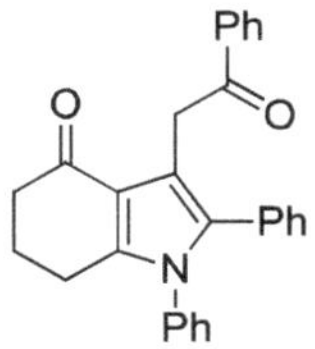

5

3-(2-Oxo-2-phenylethyl)-1,2-diphenyl-6,7-dihydro-1*H*-indol-4(5*H*)-one (**5**) (**4d**):solid, mp 240–241 °C; yield: 100%

6.2.1.4 *Experimental Procedure*

A mixture of enamine esters (**1**; 2 mmol) or enamine ketone (**4**; 2 mmol) and *trans*-1,2-dibenzoylethene (**2**; 2 mmol) was ground in the solid phase with the help of a heatable/coolable ball mill (Retsch MM2000) with 10 mL mill beakers made from stainless steel and two stainless steel balls (6.5 g) at 20–30 Hz for 3 h in order to achieve quantitative conversions. Milling of the starting materials gave dust-dry powders that were heated to 80 °C for removal of the water of reaction, thereby yielding the pyrrole derivative (**3**/**5**). The compounds obtained were characterized by means of spectral studies.

6.2.1.5 Characterization Data of Two Representative Compounds

Methyl 1,2-dimethyl-4-(2-oxo-2-phenylethyl)-5-phenyl-1*H*-pyrrole-3-carboxylate (**3a**): solid, mp 152 °C; yield: 100%; IR (KBr): 1691 (C=O) cm^{-1}; ^{1}H NMR ($CDCl_3$, 300 MHz): δ 8.00 (2H, d), 7.35 (8H, m), 4.22 (2H, s), 3.54 (3H, s), 3.41 (3H, s), 2.61 (3H, s); ^{13}C NMR ($CDCl_3$, 75 MHz): δ 199.12, 165.93, 137.43, 136.91, 133.13, 132.53, 131.33, 130.66 (2C), 128.43 (2C), 128.36 (2C), 128.01 (2C), 127.87, 115.10, 110.13, 50.24, 36.53, 31.77, 12.01; HRMS: calc. for $C_{22}H_{21}NO_3$: 347.1521; found: 347.1521.

3-(2-Oxo-2-phenylethyl)-1,2-diphenyl-6,7-dihydro-1*H*-indol-4(5*H*)-one (**5**) (**4d**): solid, mp 240–241 °C; yield: 100%; IR (KBr): 1695 (C=O), 1652 (C=O) cm^{-1}; ^{1}H NMR ($CDCl_3$, 300 MHz): δ 8.03 (2H, d), 7.51 (3H, m), 7.30 (3H, m), 7.11 (7H, m), 4.42 (2H, s), 2.69 (2H, t, $J=6.05$ Hz), 2.48 (2H, t, $J=6.6$ Hz), 2.14 (2H, quint, $J=6.05$ Hz); ^{13}C NMR ($CDCl_3$, 75 MHz): δ 198.69, 195.12, 144.55, 137.67, 137.51, 132.71, 130.71, 130.31 (2C), 129.07 (3C), 128.48 (2C), 128.39 (2C), 128.19 (2C), 127.92, 127.80 (2C), 127.43, 119.16, 114.19, 38.33, 35.95, 23.70, 23.26; HRMS: calc. for $C_{28}H_{23}NO_2$: 405.1667; found: 405.1698.

6.2.1.6 Critical Views

This is the first instance of using the ball-milling technique to perform a cascade reaction in the solid phase without the aid of any catalyst. It was found to be excellent in achieving quantitative yields of N-heterocycles.

6.2.2 Entry-2: Synthesis of Substituted Benzimidazolidine-2-thiones

Type of reaction: C–N bond formation
Reaction conditions: Catalyst-free, ball milling
Synthetic strategy: One-pot condensation–cyclization reaction
Keywords: *ortho*-Phenylene diisothiocyanate, anilines, solvent- and catalyst-free, one-pot reaction, condensation–cyclization, solid-state reaction, ball milling, substituted benzimidazolidine-2-thiones

6.2.2.1 General Reaction Scheme

Kaupp and coworkers[52] demonstrated that the solid-phase reaction of *ortho*-phenylene diisothiocyanate (**1**) with anilines (**2**) in a ball mill, at room

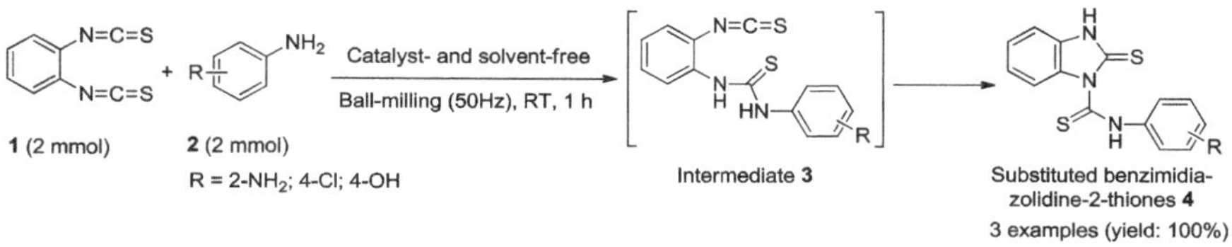

Scheme 6.3 Synthesis of substituted benzimidazolidine-2-thiones (**4**).

temperature, produced substituted (anilino-thiocarbonyl)-benzimidazolidine-2-thiones (**4**) *via* the intermediate (**3**) after 1 h of milling. A yield of 100% was achieved in the absence of any catalyst (Scheme 6.3).

6.2.2.2 Representative Examples

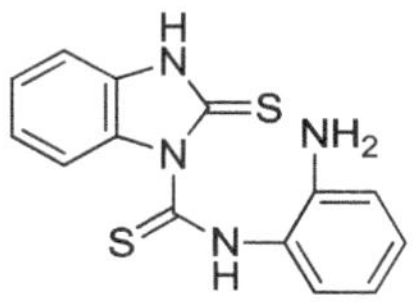

4a

N-(2-Aminophenyl)-2-thioxo-2,3-dihydro-1*H*-benzo[*d*]imidazole-1-carbothioamide (**4a**): solid, mp 109 °C; yield: 100%

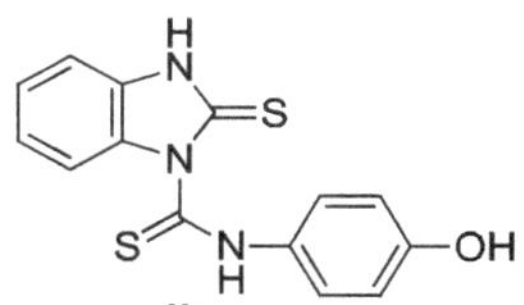

4b

N-(4-Hydroxyphenyl)-2-thioxo-2,3-dihydro-1*H*-benzo[*d*] imidazole-1-carbothioamide (**4b**): solid, mp 177 °C; yield: 100%

6.2.2.3 Experimental Procedure

A mixture of *ortho*-phenylene diisothiocyanate (**1**; 2 mmol) and aniline derivatives (**2**; 2 mmol) was ground in the solid phase with the help of a ball mill (Retsch MM2000), with 10 mL stainless steel mill beakers and stainless steel balls, at 50 Hz for 1 h at room temperature, in order to achieve quantitative yields of substituted (anilino-thiocarbonyl)-benzimidazolidine-2-thiones (**4**) just on washing. The compounds obtained were identified by comparing their physical and spectral properties with those reported in the literature.[53]

6.2.2.4 Critical Views

The present mechanochemical method offers an alternative approach to synthesize such an important series of heterocycles under catalyst- and solvent-free conditions with quantitative yields; similar reactions in solution were reported to give yields ranging from 81 to 95% and also required a workup process.[53]

6.2.3 Entry-3: Synthesis of Quinoxaline Derivatives

Type of reaction: C–N bond formation
Reaction conditions: Catalyst-free, ball milling

Synthetic strategy: One-pot cascade reactions
Keywords: *o*-Phenylenediamines, 1,2-dicarbonyl compounds (benzils, 2-hydroxy-1,4-naphthoquinone, 2-oxoglutaric acid and alloxane hydrate), solvent- and catalyst-free, one-pot cascade reactions (condensation, addition and elimination), solid-state reaction, ball milling, quinoxaline derivatives

6.2.3.1 General Reaction Schemes

A series of solid-phase condensation cascades of *o*-phenylenediamines (**1**) with varying 1,2-dicarbonyl compounds, such as benzyls (**2**), 2-hydroxy-1,4-naphthoquinone (**4**), 2-oxoglutaric acid (**6**) and alloxane hydrate (**8**), in a stoichiometric ratio were demonstrated by Kaupp and his group[54] on a ball mill to prepare quinoxaline derivatives (**3**/**5**/**7**/**11**) under catalyst-free conditions in quantitative yields in most of the cases (Scheme 6.4).

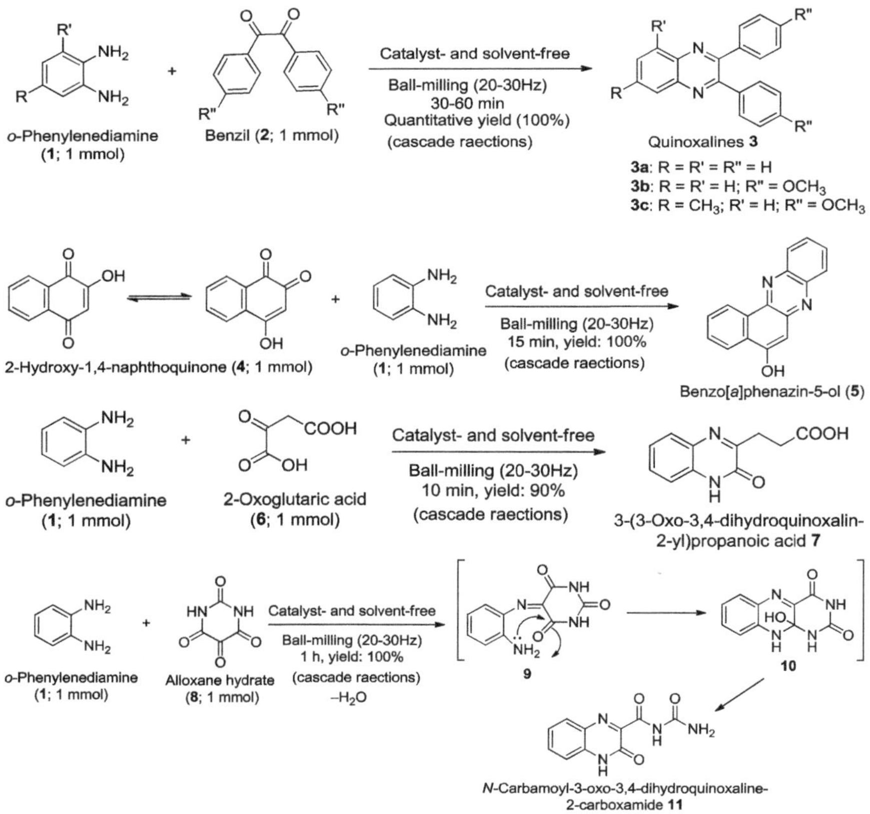

Scheme 6.4 Synthesis of quinoxaline derivatives (**3**/**5**/**7**/**11**).

6.2.3.2 Representative Examples

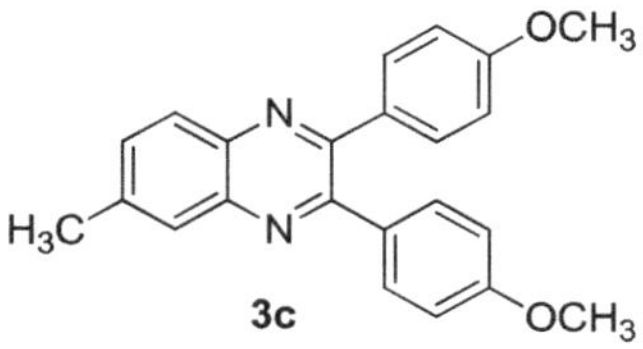

3c

2,3-*Bis*(4-methoxyphenyl)-6-methylquinoxaline (**3c**): solid, mp 123–125 °C; yield: 100%

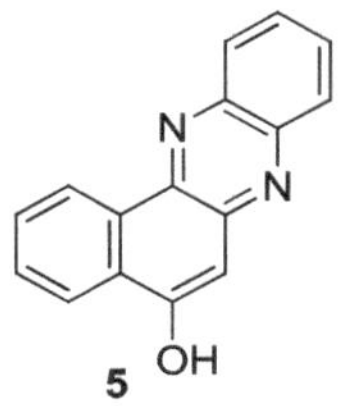

5

Benzo[*a*]phenazin-5-ol (**5**): solid, mp >300 °C; yield: 100%

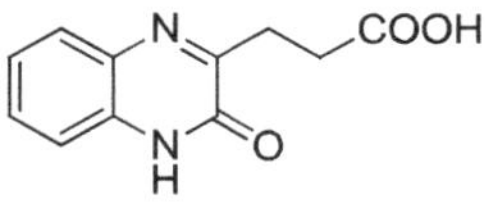

7

3-(3-Oxo-3,4-dihydroquinoxalin-2-yl)propanoic acid (**7**): solid, mp 252–254 °C; yield: 90%

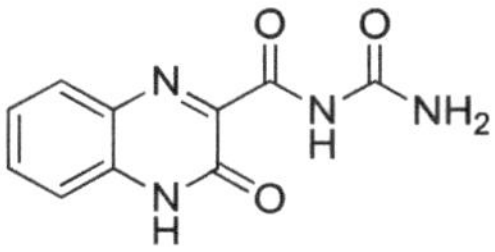

11

N-Carbamoyl-3-oxo-3,4-dihydroquinoxaline-2-carboxamide (**11**): solid, mp 254–256 °C; yield: 100%

6.2.3.3 Experimental Procedure

A mixture of *o*-phenylenediamines (**1**; 2 mmol) and 1,2-dicarbonyl compounds (**2**/**4**/**6**/**8**; 2 mmol), in a stoichiometric ratio, was ball milled for a stipulated period of time (10–60 min) (Retsch MM2000 fitted with 10 mL stainless steel mill beakers and two stainless steel balls of 12 mm diameter) at 20–30 Hz in order to achieve quantitative conversions to the desired quinoxaline derivatives in almost pure form. The compounds were identified based on their physical and spectral properties.

6.2.3.4 Characterization Data of Representative Compounds

2,3-Bis(4-methoxyphenyl)-6-methylquinoxaline (**3c**): solid, mp 123–125 °C; yield: 100%; ^{1}H NMR ($CDCl_3$, 300 MHz): δ 2.59 (3H, s), 3.82 (6H, s), 6.86 (4H, AA′BB′), 7.47 (4H, BB′AA′), 7.54 (1H, m), 7.89 (1H, m), 8.00 (1H, m); ^{13}C NMR ($CDCl_3$, 75 MHz): δ 21.8, 55.2 (2C), 113.7 (4C), 127.8, 128.5, 131.2 (4C), 131.8, 131.9, 139.5, 139.9 (2C), 141.1, 152.1, 152.8, 160.0 (2C); HRMS (EI): calc. for $C_{23}H_{20}N_2O_2$ 356.1525; found 356.1509.

Benzo[*a*]phenazin-5-ol (**5**): solid, mp >300 °C; yield: 100%; ^{1}H NMR ($CDCl_3/CF_3COOH$, 300 MHz): δ 7.35 (1H, s), 7.85–8.18 (5H, m), 8.43 (1H, m), 8.50 (1H, m), 9.26 (1H, m); ^{13}C NMR ($CDCl_3/CF_3COOH$, 75 MHz): δ 96.0, 118.6, 125.2, 126.0, 127.8, 129.1, 130.8, 131.1, 131.4, 132.8, 132.9, 136.3, 137.6, 140.7, 144.0, 168.7.

3-(3-Oxo-3,4-dihydroquinoxalin-2-yl)propanoic acid (**7**): solid, mp 252–254 °C; yield: 90%; ^{1}H NMR (DMSO-d_6, 300 MHz): δ 2.70 (2H, t, $J = 7.0$ Hz), 3.00 (2H, t,

$J = 7$ Hz), 7.22–7.28 (2H, m), 7.45 (1H, m), 7.67 (1H, m), 12.26 (2H, br peak, 2NH); ^{13}C NMR (DMSO-d_6, 75 MHz): δ 27.6, 29.6, 115.2, 123.0, 128.0, 129.4, 131.4, 131.6, 154.5, 160.2, 173.8.

N-Carbamoyl-3-oxo-3,4-dihydroquinoxaline-2-carboxamide (**11**): solid, mp 254–256 °C; yield: 100%; ^{1}H NMR ($CDCl_3/CF_3COOH$, 300 MHz): δ 7.30 (1H, br. s, NH), 7.50 (1H, pseudo-d), 7.68 (1H, m), 7.90 (1H, m), 8.20 (1H, pseudo-d), 8.85 (1H, br s, NH); ^{13}C NMR ($CDCl_3/CF_3COOH$, 75 MHz): δ 116.6, 128.1, 131.3, 132.0, 133.2, 137.0, 140.0, 156.3, 157.1, 163.0.

6.2.3.5 *Critical Views*

The present ball-mill assisted protocol reported on various condensation cascades between *o*-phenylenediamines and 1,2-dicarbonyl compounds in stoichiometric ratios in the solid-state that obtained 100% yields in most of the cases without producing any waste. Cascade reactions, with their high atom economy, are particularly benign if they proceed quantitatively from stoichiometric mixtures of the reagents. It is very much noteworthy that despite numerous reaction steps, most of these uncatalyzed stoichiometric reactions proceeded quantitatively in the solid state to furnish only the desired product (with water as a byproduct), with an unsurpassed atom economy.[54]

6.2.4 Entry-4: Synthesis of 2-Oxo/thioxo-1,2,3,4-tetrahydropyrimidine-5-carbonitriles

Type of reaction: C–N and C–C bond formation
Reaction conditions: Catalyst-free, solvent-free, ball milling
Synthetic strategy: One-pot multicomponent reaction
Keywords: Aromatic aldehydes, malononitrile, urea/thiourea, solvent- and catalyst-free, one-pot multicomponent reaction, solid-state reaction, ball milling, 2-oxo and 2-thioxo-1,2,3,4-tetrahydropyrimidine-5-carbonitriles

6.2.4.1 *General Reaction Scheme*

M'hamed and Alduaij[55] synthesized a series of 2-oxo and 2-thioxo-1,2,3,4-tetrahydropyrimidine-5-carbonitriles (**4**) from a one-pot three-component reaction of an equimolar amount of aromatic aldehyde (**1**), malononitrile (**2**), and thiourea/urea (**3**) on a ball mill under solvent-free and catalyst-free conditions with excellent yield and purity (Scheme 6.5).

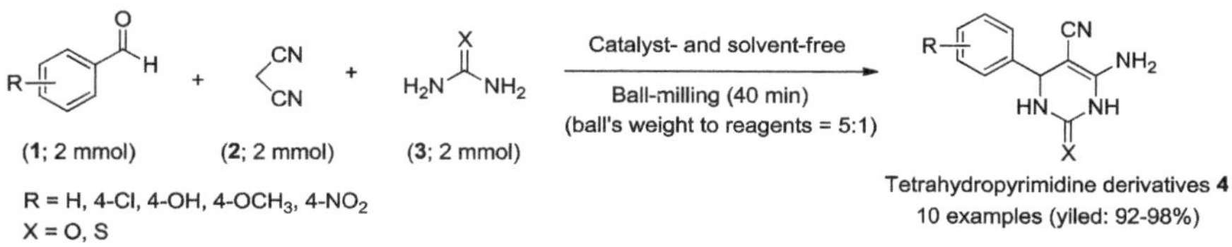

Scheme 6.5 Synthesis of 2-oxo/thioxo-1,2,3,4-tetrahydropyrimidine-5-carbonitriles (**4**).

6.2.4.2 Representative Examples

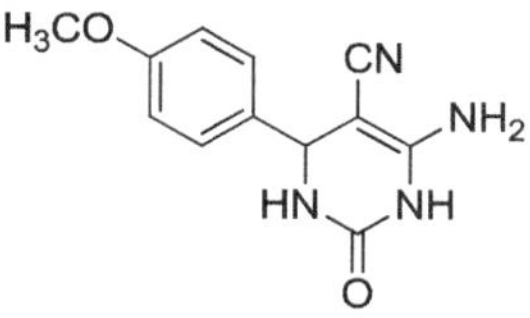

4a

6-Amino-4-(4-methoxyphenyl)-2-oxo-1,2,3,4-tetrahydropyrimidine-5-carbonitrile (**4a**): solid, mp 142–143 °C; yield: 96%

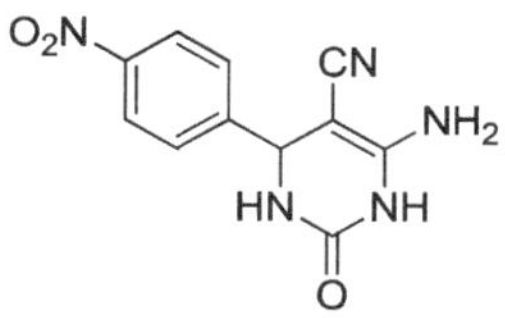

4b

6-Amino-4-(4-nitrophenyl)-2-oxo-1,2,3,4-tetrahydropyrimidine-5-carbonitrile (**4b**): solid, mp 198–199 °C; yield: 93%

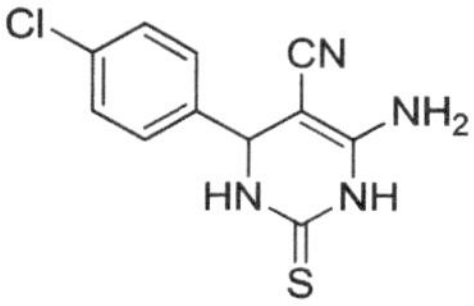

4c

6-Amino-4-(4-chlorophenyl)-2-thioxo-1,2,3,4-tetrahydropyrimidine-5-carbonitrile (**4c**): solid, mp 202–203 °C; yield: 95%

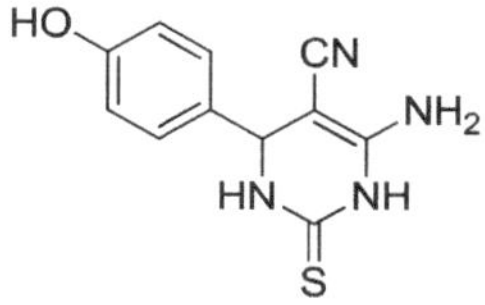

4d

6-Amino-4-(4-hydroxyphenyl)-2-thioxo-1,2,3,4-tetrahydropyrimidine-5-carbonitrile (**4d**): solid, mp 220–221 °C; yield: 96%

6.2.4.3 Experimental Procedure

A mixture of equimolar amounts (2 mmol) of aldehyde (**1**), malononitrile (**2**) and urea or thiourea (**3**) was placed into stainless steel vials with stainless steel balls (12 mm in diameter and in a weight ratio of five times to the amount of reagents) in a SPEX 8000 mixer. On grinding for 40 min, the desired product (**4**) was obtained in pure. Each of the products was identified based on their elemental analyses and spectral studies.

6.2.4.4 Characterization Data of Representative Compounds

6-Amino-4-(4-methoxyphenyl)-2-oxo-1,2,3,4-tetrahydropyrimidine-5-carbonitrile (**4a**): solid, mp 142–143 °C; yield: 96%; IR (KBr): 3415, 3340, 3230, 2222, 1678 cm^{-1}; ^{1}H NMR (DMSO-d_6, 400 MHz): δ 7.89 (1H, s, NH), 7.24–7.04 (4H, m, Ar–H), 6.35 (2H, br s, NH_2), 6.12 (1H, s, NH), 5.18 (1H, s, CH), 3.84 (3H, s, OCH_3); ^{13}C NMR (100 MHz. DMSO-d_6): δ 162.65, 160.17, 152.63, 132.17, 127.05, 116.44, 115.43, 72.91, 56.19, 48.29. Anal. calc. for $C_{12}H_{12}N_4O_2$: C, 59.01; H, 4.95; N, 22.94; O, 13.10; found: C, 58.92; H, 4.91; N, 22.88; O, 13.26.

6-Amino-4-(4-hydroxyphenyl)-2-thioxo-1,2,3,4-tetrahydropyrimidine-5-carbonitrile (**4d**): solid, mp 220–221 °C; yield: 96%; IR (KBr): 4510, 3430, 3360, 3220, 2221, 1260 cm^{-1}; ^{1}H NMR (DMSO-d_6, 400 MHz): δ 7.26–7.24 (2H, m, Ar–H), 6.80–6.78 (2H, m, Ar–H), 6.60 (5H, br s, 2NH, NH_2 and OH), 5.51 (1H, s, CH); ^{13}C NMR (100 MHz, DMSO-d_6): δ 167.65, 163.27, 159.39, 136.83,

127.68, 118.59, 117.89, 81.76, 52.16. Anal. calc. for $C_{11}H_{10}N_4OS$: C, 53.64; H, 4.09%; N, 22.75; O, 6.50; S, 13.02; found: C, 53.24; H, 4.11; N, 22.80; O, 6.13; S, 12.96.

6.2.4.5 *Critical Views*

This ball-milling protocol offers a solvent- and catalyst-free one-pot synthesis of biologically relevant functionalized pyrimidine derivatives with excellent yields and purity, with a short reaction time. No further purification of the product is required. However, the scope of the present protocol needs to be explored more with varied starting compounds.

6.2.5 Entry-5: Synthesis of Phenyl Boronate Esters

Type of reaction: N–B and O–B bond formation
Reaction conditions: Catalyst-free, solvent-free, ball milling
Synthetic strategy: Solid-state protection of diamino/diol groups with phenyl boronic acid
Keywords: Diamines, diols, anthranilic acid, phenyl boronic acid, solvent- and catalyst-free, solid-state protection, ball milling, phenyl boronate esters

6.2.5.1 *General Reaction Scheme*

Kaupp and his group[56] carried out the solid-state protection of diamine (**1**), anthranilic acid (**4**) and diols (**6, 8, 10**) with phenyl boronic acid (**2**), from the reactions of the corresponding starting materials, on a ball mill to produce the respective boronic esters (**3, 5, 7, 9, 11**) in quantitative yields under catalyst- and waste-free conditions (Scheme 6.6).

6.2.5.2 *Experimental Procedure*

A mixture of equimolar amounts (1 mmol) of diamine (**1**), anthranilic acid (**4**) or diols (**6, 8, 10**) and phenyl boronic acid (**2**) was ball milled at prescribed temperature (schemes) for 1 h to afford the desired product (**3/5/7/9/11**) with almost quantitative yields, identified on the basis of spectral measurements.

6.2.5.3 *Characterization Data of Two Representative Phenyl Boronic Esters*

2-Phenyl-2,3-dihydro-1*H*-2-boraperimidine or 2-phenyl-2,3-dihydro-1*H*-naphtho[1,8-de][1,3,2]diazaborinine (**3**): solid, mp 90.5–91.5 °C; yield: 100%; IR (KBr): 2433, 1628, 1603, 1515, 1485, 1414 cm^{-1}; ^1H NMR ($CDCl_3$, 300 MHz): δ 7.70–7.60 (2H, m), 7.50–7.35 (3H, m), 7.15–7.08 (2H, m), 7.05–7.00 (2H, m), 6.42 (2H, d), 6.15 (2H, br s, 2NH); ^{13}C NMR ($CDCl_3$, 75 MHz): δ 141.1 (2C), 136.3, 134.0, 131.4 (2C), 130.1, 128.1 (2C), 127.5 (2C), 119.8, 117.6

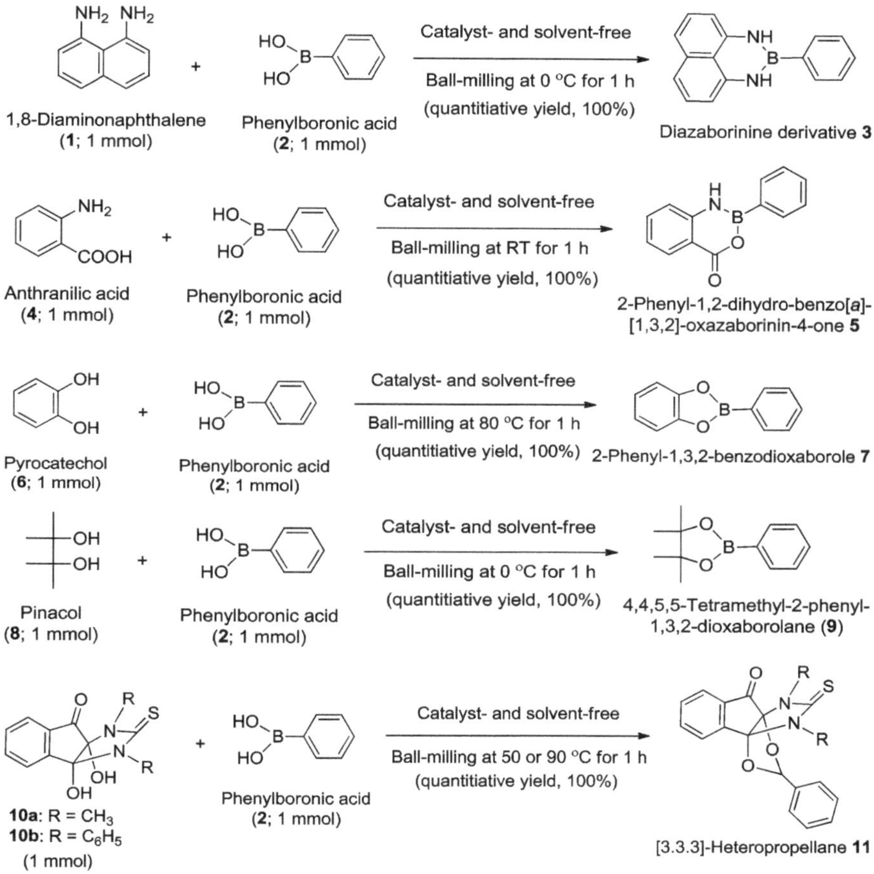

Scheme 6.6 Synthesis of phenyl boronic esters (**3**, **5**, **7**, **9**, **11**).

(2C), 105.9 ppm (2C); HRMS (70 eV): calc. for $C_{16}H_{13}BN_2$: 244.1172; found 244.1170.

(3a*R*,8a*R*)-1,3-Dimethyl-10-phenyl-2-thioxo-2,3-dihydro-3a,8a-(epoxymethanooxy)-indeno[1,2-d]imidazol-8(1*H*)-one (**11a**): solid, mp 222–224 °C (decomp.); yield: 100%; IR (KBr): 1733, 1603, 1333, 1076, 1033, 750, 698, 661, 619, 510 cm^{-1}; ^{1}H NMR ($CDCl_3$, 300 MHz): δ 8.00–7.78 (5H, m), 7.79 (1H, m), 7.50 (1H, m), 7.38 (2H, m), 3.52 (3H, s), 3.49 (3H, s); ^{13}C NMR ($CDCl_3$, 75 MHz): δ 198.0, 189.9, 144.6, 137.3, 135.3 (2C), 134.8, 132.8, 131.8, 128.0 (2C), 126.2, 124.9, 100.6, 97.8, 30.1, 29.8 ppm; HRMS (70 eV): calc. for $C_{18}H_{15}BN_2O_3S$: 350.0896; found 350.0896.

6.2.5.4 *Critical Views*

The investigators were successful in demonstrating waste- and catalyst-free ball-mill assisted solid-state protection of varying amines and diols with phenyl boronic acid. They were also able to isolate the protected products without the necessity of any further purification during the workup. Easy and facile access to such diverse phenyl boronate esters is of potential synthetic significance.

6.2.6 Entry-6: Synthesis of Boronic Acid Esters

Type of reaction: B–O bond formation
Reaction conditions: Catalyst-free, solvent-free, ball milling
Synthetic strategy: Mechanochemical esterification of boronic acids with diols
Keywords: Boronic acids, diols, pinacol, 2,2-dimethylpropane-1,2-diol, solvent- and catalyst-free, ball milling, boronic acid esters

6.2.6.1 General Reaction Scheme

A facile and eco-friendly protocol for the synthesis of a diverse range of boronic acid esters (**4**/**5**), from the reaction of the corresponding boronic acids (**1**) with diols (pinacol, **2**; 2,2-dimethylpropane-1,2-diol, **3**) on a ball mill under neat conditions, was demonstrated by Schnürch and coworkers[57] (Scheme 6.7).

6.2.6.2 Representative Examples

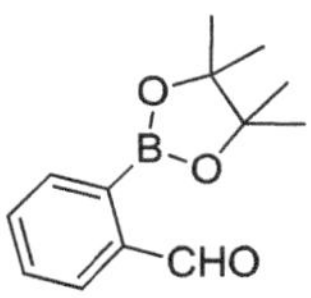

4a

2-(4,4,5,5-Tetramethyl-1,3,2-dioxaborolan-2-yl)benzaldehyde (**4a**): beige liquid; yield: 88%

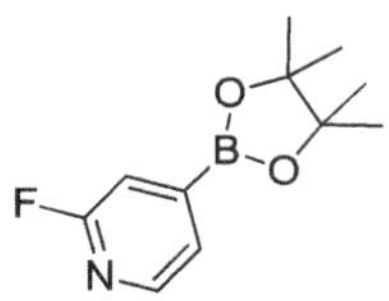

4b

2-Fluoro-4-(4,4,5,5-tetramethyl-1,3,2-dioxaborolan-2-yl)pyridine (**4b**): solid, mp 37–45 °C; yield: 89%

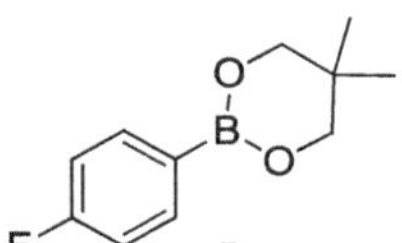

5a

2-(4-Fluorophenyl)-5,5-dimethyl-1,3,2-dioxaborinane (**5a**): colorless solid, mp 61–55 °C; yield: 85%

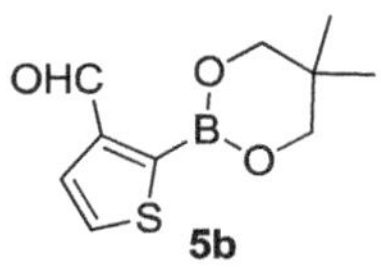

5b

2-(5,5-Dimethyl-1,3,2-dioxaborinan-2-yl)thiophene-3-carbaldehyde (**4d**): yellow liquid; yield: 84%

6.2.6.3 Experimental Procedure

An equimolar mixture of boronic acid (**1**) and pinacol (**2**) or 2,2-dimethyl-propane-1,3-diol (**3**) was ball milled for 1 hour without the aid of any solvent or catalyst; the resulting crude product was then dissolved in diethyl ether

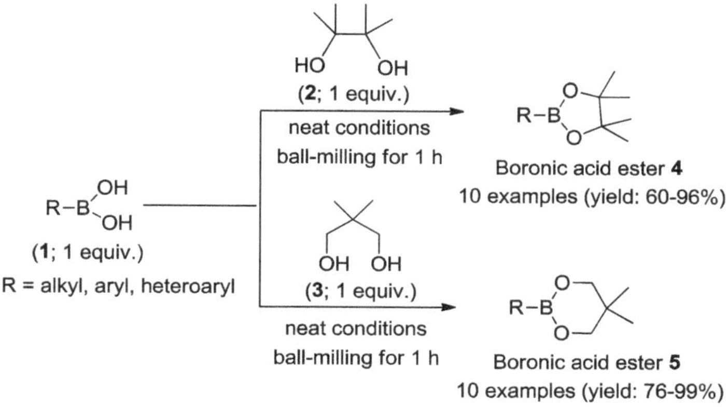

Scheme 6.7 Synthesis of boronic acid esters (**4**/**5**).

or, in the case of a solid product, in ethanol. The solution was dried over sodium sulfate, filtered, and evaporated under reduced pressure to obtain almost pure products (**4**/**5**). All the boronic acid esters were characterized based on spectral properties.

6.2.6.4 Characterization Data of a Representative Compound

2-(4-Fluorophenyl)-5,5-dimethyl-1,3,2-dioxaborinane (**5a**): colorless solid, mp 61–55 °C; yield: 85%; ^{1}H NMR ($CDCl_3$, 200 MHz): δ 1.02 (6H,s, 2×CH_3), 3.76 (4H,s, 2×OCH_2), 7.04 (2H, t, J = 8.9 Hz, 2×CH), 7.79 (2H, t, J = 7.4 Hz, 2×CH); ^{13}C NMR ($CDCl_3$, 50 MHz): δ 21.9 (2×q), 31.84 (s), 72.3 (2×t), 114.5 (dd, J = 20.1 Hz), 136.0 (dd, J = 7.8 Hz), 164.8 (ds, J = 164.8 Hz).

6.2.6.5 Critical Views

Boronic acids and esters are important building blocks in synthetic organic chemistry now due to their diverse applications, including in Suzuki–Miyaura cross-coupling reactions.[58–66] The present mechanochemical practically waste-free method appears to be the most general, facile and eco-friendly compared to the existing classical methods to form boronic acid esters from boronic acids, which require catalysts (Pd-catalyzed reactions frequently suffer from homo-coupling byproducts) and greater processing time. The key advantages of this method include the use of neat samples, no catalyst, no tedious purification of products, water as byproduct, high atom-efficiency and excellent yields within a short reaction time.[57]

6.3 Concluding Remarks

The broad field of ball-milling procedures applicable to synthetic organic chemistry is gaining potential interest in recent times as the ball-milling technique offers several key advantages over conventional processes, thus, making it an interesting tool in organic synthesis. Although catalyst-free organic synthesis under this technique is still not flourished, it is well anticipated that in the very near future this area would be explored and enriched with the overall growth of ball-milling research.

References

1. *Ball Milling Towards Green Synthesis: Applications, Projects, Challenges*, ed. B. C. Ranu and A. Stolle, Royal Society of Chemistry, Cambridge, UK, 2014.
2. *Mechanochemical Organic Synthesis*, ed. D. Margetic and V. Štrukil, Elsevier, Amsterdam, 2016.
3. A. J. Lynch and C. A. Rowland, *The History of Grinding, Society of Mining, Metallurgy and Exploration, Inc.*, Littleton, CO, 2005.
4. Z. V. Todres, *Organic Mechanochemistry and its Practical Applications*, Taylor & Francis, Boca Raton, 2006.
5. E. G. Avvakumov, *Mechanical Methods for the Activation of Chemical Processes*, Izd. Hauka, Novosibirsk, 1986 (in Russian).
6. V. V. Boldyrev, *Russ. Chem. Rev.*, 2006, **75**, 177.
7. G. Kaupp, *CrystEngComm*, 2009, **11**, 388.
8. K. Tanaka and F. Toda, *Chem. Rev.*, 2000, **100**, 1025.
9. K. Komatsu, *Top. Curr. Chem.*, 2005, **254**, 185.
10. S. L. James, C. J. Adams, C. Bolm, D. Braga, P. Collier, T. Friščić, F. Grepioni, K. D. M. Harris, G. Hyett, W. Jones, A. Krebs, J. Mack, L. Maini, A. G. Orpen, I. P. Parkin, W. C. Shearouse, J. W. Steed and D. C. Waddell, *Chem. Soc. Rev.*, 2012, **41**, 413.
11. L. Takacs, *Chem. Soc. Rev.*, 2013, **42**, 7649.
12. A. Stolle, T. Szuppa, S. E. S. Leonhardt and B. Ondruschka, *Chem. Soc. Rev.*, 2013, **42**, 2317.
13. P. Baláž, M. Achimovičová, M. Baláž, P. Billik, Z. Cherkezova-Zheleva, J. M. Criado, F. Delogu, E. Dutková, E. Gaffet, F. J. Gotor, R. Kumar, I. Mitov, T. Rojac, M. Senna, A. Streletskiik and K. Wieczorek-Ciurowa, *Chem. Soc. Rev.*, 2013, **42**, 7571.
14. E. Boldyreva, *Chem. Soc. Rev.*, 2013, **42**, 7719.
15. P. G. Fox, *J. Mater. Sci.*, 1975, **10**, 340.
16. V. V. Boldyrev and K. Tkáčova, *J. Mater. Synth. Process*, 2000, **8**, 121.
17. M. K. Beyer and H. Clausen-Schaumann, *Chem. Rev.*, 2005, **105**, 2921.
18. A. Beillard, E. Golliard, V. Gillet, X. Bantreil, T.-X. Métro, J. Martinez and F. Lamaty, *Chem. – Eur. J.*, 2015, **15**, 17614.
19. C. Kajdas, *Mat. Sci. Appl.*, 2015, **6**, 60.
20. Y. Wang, E. L. Botvinick, Y. Zhao, M. W. Berns, S. Usami, R. Y. Tsien and S. Chien, *Nature*, 2005, **434**, 1040.
21. L. Takacs, *Bull. Hist. Chem.*, 2003, **28**, 26.
22. L. Takacs, *J. Mater. Sci.*, 2003, **39**, 4987.
23. L. Takacs, *J. Metals*, 2000, **52**, 12.

24. A. Bruckmann, A. Krebs and C. Bolm, *Green Chem.*, 2008, **10**, 1131.
25. A. Lazuen-Garay, A. Pichon and S. L. James, *Chem. Soc. Rev.*, 2007, **36**, 846.
26. F. Schneider, A. Stolle, B. Ondruschka and H. Hopf, *Org. Process Res. Dev.*, 2009, **13**, 44.
27. W. C. Shearouse, C. M. Korte and J. Mack, *Green Chem.*, 2011, **13**, 598.
28. V. Šepelák, S. Bégin-Colin and G. Le Caër, *Dalton Trans.*, 2012, **41**, 11927.
29. S. L. James and T. Friščić, *Chem. Commun.*, 2013, **49**, 5349.
30. D. Braga, F. Grepioni, L. Maini, D. Capucci, S. Nanna, J. Wouters, L. Aerts and L. Quéré, *Chem. Commun.*, 2012, **48**, 8219.
31. M. D. Eddleston, M. Arhangelskis, T. Friščić and W. Jones, *Chem. Commun.*, 2012, **48**, 11340.
32. K. Crossey, C. Hardacre and M. E. Migaud, *Chem. Commun.*, 2012, **48**, 11969.
33. V. Štrukil, D. Margetić, M. D. Igrc, M. Eckert-Maksić and T. Friščić, *Chem. Commun.*, 2012, **48**, 9705.
34. M. Pascu, A. Ruggi, R. Scopelliti and K. Severin, *Chem. Commun.*, 2013, **49**, 45.
35. L. Loots, B. H. Wahl, L. van der Westhuizen, B. D. A. Haynes and T. le Roex, *Chem. Commun.*, 2012, **48**, 11507.
36. V. Declerck, E. Colacino, X. Bentreil, J. Martinez and F. Lamaty, *Chem. Commun.*, 2012, **48**, 11778.
37. J. Stojaković, B. S. Farris and L. R. MacGillivray, *Chem. Commun.*, 2012, **48**, 7958.
38. T. Friščić, I. Halasz, P. J. Beldon, A. M. Belenguer, F. Adams, S. A. J. Kimber, V. Honkimäki and R. E. Dinnebier, *Nat. Chem.*, 2012, **5**, 66.
39. I. A. Tumanov, A. F. Achkasov, E. V. Boldyreva and V. V. Boldyrev, *CrystEngComm*, 2011, **13**, 2213.
40. J. Ribas-Arino and D. Marx, *Chem. Rev.*, 2012, **112**, 5412.
41. D. Braga and F. Grepioni, *Angew. Chem., Int. Ed.*, 2004, **43**, 4002.
42. S. Kipp, V. Šepelák and K. D. Becker, *Chem. Unserer Zeit*, 2005, **39**, 384.
43. V. P. Balema, J. W. Wiench, M. Pruski and V. K. Pecharsky, *Chem. Commun.*, 2002, 724.
44. C. F. Burmeister and A. Kwade, *Chem. Soc. Rev.*, 2013, **42**, 7660.
45. P. R. Patil and K. P. R. Kartha, *Green Chem.*, 2009, **11**, 953.
46. A. Pichon, A. Lazuen-Garay and S. L. James, *CrystEngComm*, 2006, **8**, 211.
47. E. L. Fokina, N. I. Budim, V. G. Kochnev and G. G. Chermik, *J. Mater. Sci.*, 2004, **39**, 5217.
48. P. P. Chattopadhyay, I. Manna, S. Talapatra and S. K. Pabi, *Mater. Chem. Phys.*, 2000, **68**, 85.
49. F. Schneider, T. Szuppa, A. Stolle, B. Ondruschka and H. Hopf, *Green Chem.*, 2009, **11**, 1894.
50. T. Szuppa, A. Stolle, B. Ondruschka and W. Hopfe, *Green Chem.*, 2010, **12**, 1288.
51. G. Kaupp, J. Schmeyers, A. Kuse and A. Atfeh, *Angew. Chem., Int. Ed.*, 1999, **38**, 2896.

52. G. Kaupp, J. Schmeyers and J. Boy, *Tetrahedron*, 2000, **56**, 6899.
53. D. Griffiths, R. Hull and T. P. Seden, *J. Chem. Soc., Perkin Trans. 1*, 1980, 1240.
54. G. Kaupp and M. Reza Naimi-Jamal, *Eur. J. Org. Chem.*, 2002, 1368.
55. M. O. M'hamad and O. K. Alduaij, *Phosphorus, Sulphur, Silicon*, 2014, **189**, 235.
56. G. Kaupp, M. Reza Naimi-Jamal and V. Stepanenko, *Chem. - Eur. J.*, 2003, **9**, 4156.
57. M. Schnürch, M. Holzweber, M. D. Mihovilovic and P. Stanetty, *Green Chem.*, 2007, **9**, 139.
58. N. Miyaura and A. Suzuki, *J. Chem. Soc., Chem. Commun.*, 1979, 866.
59. N. Miyaura, K. Yamada and A. Suzuki, *Tetrahedron Lett.*, 1979, **20**, 3437.
60. N. Miyaura and A. Suzuki, *Chem. Rev.*, 1995, **95**, 2457.
61. A. Suzuki, *J. Organomet. Chem.*, 1999, **576**, 147.
62. S. Kotha, K. Lahiri and D. Kashinath, *Tetrahedron*, 2002, **58**, 9633.
63. A. R. Martin and Y. Yang, *Acta Chem. Scand.*, 1993, **47**, 221.
64. A. Suzuki, *Chem. Commun.*, 2005, **38**, 4759.
65. D. M. T. Chan, K. L. Monaco, R.-P. Wang and M. P. Winters, *Tetrahedron Lett.*, 1998, **39**, 2933.
66. D. M. T. Chan, K. L. Monaco, R. Li, D. Bonne, C. G. Clark and P. Y. S. Lam, *Tetrahedron Lett.*, 2003, **44**, 3863.

Subject Index